Distributed G
Implications for th

T0229075

Distributed Generation and its Implications for the Utility Industry

edited by

Fereidoon P. Sioshansi
President, Menlo Energy Economics
USA

AMSTERDAM • BOSTON • HEIDELBERG • LONDON
NEW YORK • OXFORD • PARIS • SAN DIEGO
SAN FRANCISCO • SYDNEY • TOKYO
Academic Press is an imprint of Elsevier

Academic Press is an imprint of Elsevier
The Boulevard, Langford Lane, Kidlington, Oxford OX5 1GB, UK
Radarweg 29, PO Box 211, 1000 AE Amsterdam, The Netherlands
225 Wyman Street, Waltham, MA 02451, USA
525 B Street, Suite 1800, San Diego, CA 92101-4495, USA

Notice
Knowledge and best practice in this field are constantly changing. As new research and experience
broaden our understanding, changes in research methods, professional practices, or medical
treatment may become necessary.

Practitioners and researchers must always rely on their own experience and knowledge in evaluating
and using any information, methods, compounds, or experiments described herein. In using such
information or methods they should be mindful of their own safety and the safety of others, including
parties for whom they have a professional responsibility.

To the fullest extent of the law, neither the Publisher nor the authors, contributors, or editors, assume
any liability for any injury and/or damage to persons or property as a matter of products liability,
negligence or otherwise, or from any use or operation of any methods, products, instructions, or
ideas contained in the material herein.

Library of Congress Cataloging-in-Publication Data
Sioshansi, Fereidoon P. (Fereidoon Perry), author.
 Distributed generation and its implications for the utility industry / Fereidoon P.
Sioshansi. – First edition.
    pages cm
 Includes index.
 ISBN 978-0-12-800240-7
1. Distributed generation of electric power. 2. Electric utilities. I. Title.
 TK1006.S56 2014
 333.793′2–dc23

                                                                   2014012837

British Library Cataloguing in Publication Data
A catalogue record for this book is available from the British Library

ISBN: 978-0-12-800240-7

For information on all Academic Press publications
visit our web site at store.elsevier.com

This book has been manufactured using Print On Demand technology. Each copy is produced to
order and is limited to black ink. The online version of this book will show color figures where
appropriate.

Working together
to grow libraries in
developing countries

www.elsevier.com • www.bookaid.org

Contents

Part II
Implications and Industry/Regulatory Response 163

Part III
What Future? 357

Author Biographies

Heather Bailey is the executive director of Energy Strategy and Electric Utility Development for the city of Boulder, where she helps manage Boulder's Energy Future project by providing direction for the city's short- and long-term strategies and the municipalization exploration efforts.

With over 30 years of utility operations, finance, accounting, and auditing experience, Ms. Bailey has worked as a regulator, utility executive, and consultant. Her experience includes consulting on transmission, financial, regulatory, and strategic issues including managing one of the largest public power organizations in the country. Her consulting experience includes developing long-term utility transmission and generation investment strategy, representing solar and wind developers, management of operations and cost reductions, rate cases, and regulatory support.

Ms. Bailey has a master's in business administration from the University of Texas and a bachelor's in business administration from Sam Houston State University. She is a Certified Public Accountant.

Liam Blanckenberg is an energy economist of Frontier Economics, a microeconomics consultancy firm based in Melbourne, Australia.

Liam provides both quantitative and qualitative economic advice across a range of issues to clients who predominately operate in Australia's gas and electricity markets. Liam's specific areas of expertise include network regulation and investment, tariff design and regulation, wholesale electricity market forecasting, commercial advisory, and electricity and gas market data analysis.

Liam holds an honors degree in economics from the University of Melbourne and is a member of the Chartered Financial Analyst (CFA) Institute.

Jonathan Blansfield is a third-year JD candidate and a research associate in the Institute for Energy and the Environment at Vermont Law School where he is managing editor of the Vermont Journal of Environmental Law and a member of the Moot Court Advisory Board. He has worked as a legal and policy intern at the US Department of Energy's Office of Electricity Delivery and Energy Reliability and a law clerk in the Federal Energy Regulatory Commission's Office of General Counsel.

Jonathan's legal education is focused on energy law, with an emphasis on electricity policy and regulation, natural gas regulation, and renewable energy resources.

Jonathan holds a BA in anthropology and political science from the University of Connecticut.

Timothy Brennan is a senior fellow at Resources for the Future and professor of public policy and economics at the University of Maryland, Baltimore County. He has been an economist in the Antitrust Division of the US Department of Justice and a staff consultant in the Federal Trade Commission's Bureau of Economics. He served as senior economist for the Council of Economic Advisers and held the TD MacDonald Chair in Industrial Economics at the Canadian Competition Bureau. He has advised competition and regulatory authorities in numerous countries.

His expertise covers topics in antitrust and regulation and how they interrelate. His utility work has focused on telecommunications and electricity. Recent electricity-related research looks at energy efficiency, outage liability, and the role of consumer error.

Dr. Brennan has a BA in mathematics from the University of Maryland and a PhD in economics from the University of Wisconsin.

Christoph Burger is a member of the faculty at the European School of Management and Technology and managing director of ESMT Customized Solutions GmbH. Christoph has taught microeconomics at the University of Michigan, was project leader at Otto Versand, worked as a strategy consultant at Arthur D. Little and as an independent consultant, and was vice president at the Bertelsmann Buch AG.

His research focus is on long-term industry development, specifically in the energy sector, and decision making and negotiations. Christoph has published several articles, case studies, and book chapters/books in the field of customer management, business plan assessment, industry development, and decision making/negotiations. In 2013, he published the book *The Decentralized Energy Revolution: Business Strategies for a New Paradigm.*

He studied business administration and economics at the Saarland University in Saarbrücken (Germany), the University St. Gallen (Switzerland), and the University of Michigan, Ann Arbor (the United States).

Ralph Cavanagh is a senior attorney and codirector of NRDC's energy program, which he joined in 1979. Ralph has been a visiting professor of law at Stanford University and University of California, Berkeley; a lecturer on law at the Harvard Law School; and a faculty member for the University of Idaho's Public Utility Executives Course and served on the US Secretary of Energy Advisory Board from 1993 to 2003.

His focus at NRDC has been mobilizing the utility industry to invest in clean energy solutions, led by cost-effective energy efficiency, and ensuring that progress is not blocked unintentionally by regulatory policies. Ralph has received the Heinz Award for Public Policy, the National Association of Regulatory

Utility Commissioners' Mary Kilmarx Award, the Yale Law School's Preiskel-Silverman Fellowship, the Bonneville Power Administration's Award for Exceptional Public Service.

He is a graduate of Yale College and the Yale Law School.

Edward G. Cazalet is the founder of TeMIX Inc., MegaWatt Storage Farms Inc., and the Cazalet Group and served as a governor of the California Independent System Operator. His prior professional activities included the founding of Automated Power Exchange Inc. and Decision Focus Inc. He has been active in the design and implementation of markets for electricity and development of smart grid standards, in transmission, generation, storage, and demand management.

Dr. Cazalet has successfully promoted storage legislation both in California and at the federal level and has advocated new electricity market designs and the use of price-responsive demand and storage to support high penetration of variable renewables and efficient grid operation and investment. He is cochair of the OASIS Energy Market Information Exchange (EMIX) on Energy Interoperation and Scheduling.

Dr. Cazalet holds a PhD from Stanford University in economics, decision analysis, and power system planning and engineering degrees from the University of Washington.

Kelly Crandall is an energy strategy coordinator for the city of Boulder, where she is a key member of the city's Municipalization Exploration Project. Her responsibilities span policy, technical, regulatory, and project management. She recently implemented a statistical decision analysis process to enhance the city's municipalization feasibility study, and as part of the Energy Future team, Ms. Crandall is helping develop a vision for an innovative electric utility of the future in Boulder.

She also collaborates with the staff of other local governments to address their regulatory needs. Prior to joining the city of Boulder, she was a smart grid analyst at the National Renewable Energy Laboratory involved in consumer protection issues.

Ms. Crandall received a JD from the University of Colorado School of Law and a BA from the University of Florida. She is active on the board of Women in Sustainable Energy.

Ahmad Faruqui is a principal at the Brattle Group where he is focused on smart grid policy, dynamic pricing, demand response, advanced metering infrastructure, energy efficiency, and demand forecasting. He has worked with numerous utilities in the United States and abroad on a variety of engagements, with FERC in developing the National Action Plan on Demand Response and National Assessment of Demand Response Potential, and assisted the California Energy Commission and the Saudi Electricity Regulator.

Dr. Faruqui coauthored EPRI's national assessment of the potential for energy efficiency, EEI's report on quantifying the benefits of dynamic pricing for the New York Independent System Operator and economic demand response for the Midwest ISO and ISO New England and the PJM Interconnection.

Dr. Faruqui holds a PhD in economics from the University of California, Davis, and has published four books and more than a hundred articles on energy economics.

Frank A. Felder is director of the Center for Energy, Economic and Environmental Policy and associate research professor at the Edward J. Bloustein School of Planning and Public Policy at Rutgers University.

Frank directs applied energy and environmental research. His ongoing and recent projects include studies on energy efficiency evaluation and economic impact of renewable portfolio standards and power system and economic modeling of state energy plans. He is also an expert on restructured electricity markets. He has published widely in professional and academic journals on market power and mitigation, wholesale market design, reliability, transmission planning, market power, and rate design issues. He was a nuclear engineer and submarine officer in the US Navy.

Dr. Felder holds undergraduate degrees from Columbia College and the Harvard School of Engineering and Applied Sciences and a master's and doctorate from Massachusetts Institute of Technology (MIT) in Technology, Management, and Policy.

Clark W. Gellings is a fellow at the Electric Power Research Institute, responsible for technology strategy in areas concerning energy efficiency, demand response, renewable energy resources, and other clean technologies.

Mr. Gellings joined EPRI in 1982 progressing through a series of technical management and executive positions including seven vice president positions. Prior to joining EPRI, he spent 14 years with Public Service Electric and Gas Company in NJ. He is a recipient of numerous awards; has served on numerous boards and advisory committees, both in the United States and internationally; and is a frequent speaker at industry conferences.

Gellings has a bachelor of science in electrical engineering from Newark College of Engineering in New Jersey, a master of science degree in mechanical engineering from New Jersey Institute of Technology, and a master's in management science from the Wesley J. Howe School of Technology Management at Stevens Institute of Technology.

Yael Gichon is an energy sustainability coordinator for the city of Boulder, Colorado, and a member of the city of Boulder's Energy Future and the Municipalization Exploration Project team. With background in renewable energy, energy efficiency, sustainable development, and ecology, she leads projects that include technical analysis, policy development, and public engagement.

She managed the financial analysis, which led to a passing ballot initiative to continue exploring a municipal electric utility. She continues efforts to explore alternative resource options to increase the contribution of renewables in Boulder's energy mix.

Yael developed the nation's first performance-based rental housing policy, implemented in 2011, which has proved successful in increasing energy efficiency in rental housing. An award-winning residential energy efficiency program, it has reached a milestone of 10,000 participants.

Yael holds a master's in interdisciplinary ecology from the University of Florida.

Joel Gilmore is the principal of renewable energy and climate policy at ROAM Consulting, a leading provider of expert services in energy market modeling. With experience in energy market modeling and analysis, Joel leads a team to model and advise on topics related to the impacts of climate policy and renewable energy on electricity systems, such as renewable energy target schemes, carbon pricing, and the integration of wind and solar energy.

Dr. Gilmore's accomplishments include work for market participants in the National Electricity Market (NEM) and South West Interconnected System (SWIS) in Australia. He is interested in solar technologies and their integration into electricity grids.

Dr. Gilmore has a bachelor of science degree and a PhD in physics from the University of Queensland, Australia.

Koen Groot is a researcher at the Clingendael International Energy Programme (www.clingendaelenergy.com) in The Hague, the Netherlands, where he focuses on industry and regional market analysis.

His research interests are in the European electricity market, in the European transportation market, and in global LNG markets.

Koen holds a master of science degree in business studies and in international relations, both from the University of Amsterdam.

George Grozev is a principal research scientist and a research group leader, "Buildings, Utilities and Infrastructure," in the Urban Systems Program, CSIRO Ecosystem Sciences, Australia. His background is in operations research, and he has extensive experience in energy market modeling and simulation, network theory and algorithms, and software development.

During his tenure at CSIRO, he has carried out research and development in the areas of high-performance infrastructure, sustainable development, and electricity and gas markets. He is a member of the Australian Society for Operations Research.

George has a PhD from Institute of Engineering Cybernetics and Robotics, Bulgarian Academy of Sciences, and bachelor's in engineering physics from St. Kliment Ohridski University, Sofia, Bulgaria.

Dian Grueneich is the founder and principal of Dian Grueneich Consulting LLC where she assists clients on a full range of energy issues, focusing on regulatory systems affecting sustainable energy investments.

Dian is a nationally and internationally recognized energy expert, with 35 years' experience. Dian served as a commissioner on the California Public Utilities Commission from 2005 to 2010 and led its efforts on energy efficiency, developing the *California Long-Term Energy Efficiency Strategic Plan* and overseeing a 40% expansion of California's energy efficiency funding. Dian streamlined California's transmission siting process and initiated the California Renewable Energy Transmission Initiative. She currently serves on the US Department of Energy (DOE)'s Electricity Advisory Committee, the DOE-EPA State Energy Efficiency Action Plan Leadership Group, the Leadership Council of the China-US Energy Efficiency Alliance, and the Advisory Council of Stanford University's Precourt Energy Institute, among others.

Dian is a graduate of Stanford University and holds a JD from Georgetown University.

Philip Hanser is a principal of the *Brattle Group* and has over thirty years of consulting and litigation experience in the energy industry. He has appeared as an expert witness before the US Federal Energy Regulatory Commission and numerous state public utility commissions, environmental agencies, utility boards, arbitration panels, and federal and state courts. He served for six years on the American Statistical Association's advisory committee to the Energy Information Administration.

Prior to joining the *Brattle Group*, he held teaching positions at the University of the Pacific, University of California at Davis, and Columbia University and served as a guest lecturer at the Massachusetts Institute of Technology, Stanford University, and the University of Chicago. He currently is also a senior associate in the Mossavar-Rahmani Center for business and government at John F. Kennedy School of Government, Harvard University. He was manager of the demand-side management program at the Electric Power Research Institute.

Philip's undergraduate training was in economics and mathematics from the Florida State University, and his doctoral studies were in economics and mathematical statistics at Columbia University.

Andrew Higgins is a principal research scientist at CSIRO where he leads a systems engineering and dynamics group specialized in computational analysis of complex landscape problems in urban planning, natural hazards, agriculture, and bushfires.

He is an expert in mathematical forecasting of technology uptake given demographic and socioeconomic variables relevant to the decision-making process. He has published extensively on topics such as electric vehicles, solar PVs, water heaters, and air conditioners. Andrew has previously developed methods to optimize supply chain problems found in freight, agriculture, and natural resource management.

Andrew received his PhD from the School of Mathematics and School of Civil Engineering at Queensland University of Technology.

Magnus Hindsberger is a specialist at the Australian Energy Market Operator (AEMO) working on strategic projects in the area of long-term planning of electricity and gas networks. Previously, he worked at Transpower New Zealand Limited, the Nordic consulting company ECON, and Elkraft System, now merged into Energinet.dk, which is a Danish transmission system operator.

He has a wide experience from more than 15 years in the energy industry covering three different jurisdictions. He has a particular strong background in electricity and gas market modeling and its application for long-term scenario planning, forecasting market outcomes, cost-benefit assessments, and policy analysis.

Dr. Hindsberger holds MS and PhD degrees in engineering (operations research) from the Technical University of Denmark.

Delphine Hou is a senior market design and policy specialist at California Independent System Operator (CAISO) engaged in development of market design policies. She has led stakeholder initiatives focused on transmission reliability issues ensuring that CAISO has sufficient reserves to recover from a contingency and to improve the ISO's modeling of the Western Interconnection to enhance reliability and operate an energy imbalance market. She has also been involved in policy efforts to increase retail demand response participation in the wholesale market.

Before her current job, she was an associate at the Brattle Group focused on electric transmission planning and cost allocation, utility strategy, and scenario-based planning. She has advised utilities, transmission and generation companies, regulatory bodies, financial investors, industry associations, independent system operators, and regional transmission organizations.

Ms. Hou holds an MA in electric transmission policy from the Fletcher School at Tufts University.

John W. Jimison is the managing director of the Energy Future Coalition, a nonpartisan public policy initiative in Washington, DC. In that role, he manages campaigns to improve energy efficiency, modernize the electricity sector, and increase the role of clean energy. Prior to his present post, he served as senior counsel to the Committee on Energy and Commerce of the US House of Representatives, responsible for energy efficiency, natural gas policy, energy markets, and grid modernization.

He practiced law in federal and state jurisdictions, emphasizing natural gas regulation and distributed generation, and was principal administrator at the International Energy Agency in Paris, France. Earlier, he held positions with the Committee on Energy and Commerce, the Congressional Research Service, and the US Senate Committee on Commerce.

He has degrees from Georgetown University Law Center and the College of Wooster and is a former US Peace Corps volunteer in Somalia.

Kevin B. Jones is the deputy director and senior fellow for energy technology and policy of the Institute for Energy and the Environment at Vermont Law School where he leads the Smart Grid Project. Previously, he was the director for power market policy for the Long Island Power Authority, associate director of Navigant Consulting Inc., and director of energy policy for the city of New York.

Dr. Jones has been at the center of the transformation of the electric power industry in the Northeast United States and while at LIPA collaborated on energy policy with both the Large Public Power Council and the New York Transmission Owners.

Kevin has a doctorate from the Lally School of Management and Technology at Rensselaer Polytechnic Institute; a master of public affairs from the LBJ School of Public Affairs, University of Texas at Austin; and a BS from the University of Vermont.

Malcolm Keay is a senior research fellow at the Oxford Institute for Energy Studies where he works on issues connected with electricity and the transition to a low-carbon energy system.

He has had a wide-ranging career in the energy sector, including energy policy development (he was director of energy policy at the UK Department of Trade and Industry in 1996-99), international energy affairs, energy regulation, and energy consultancy. He has acted as an adviser on many energy studies and was special adviser to the House of Lords' Committee of Inquiry into energy security in Europe and director of the Energy and Climate Change Study Group for the World Energy Council.

Malcolm has an MA degree from University of Cambridge.

Urban Keussen is senior vice president of Technology and Innovation at the E. ON group, Germany's biggest energy company, based in Dusseldorf. In this capacity, he leads and is responsible for all technology and innovation activities within the company.

The focus on new technologies and related innovations plays an increasingly important role for strengthening the future competitive position of the company within its home and external markets. Prior to his present position, he worked in several roles within the E.ON group, became a member of the executive board of E.ON Netz and the power transmission operator in 2003, and promoted to his present job in 2010.

Urban studied physics and gained his diploma and PhD in physics in 1993.

Chris King is global chief regulatory officer, Siemens Smart Grid, a leading provider of smart grid and smart meter software. He is responsible for strategic activities, including market analysis and product strategy and policy, working with regulators and legislators worldwide.

Earlier, he was rate design director at Pacific Gas and Electric Company, VP of sales and marketing at CellNet Data Systems Inc., CEO at an electricity retailer utility, and chief strategy officer at a smart meter software provider eMeter. Chris' research includes smart grid economics, regulatory policy, technology, and consumer benefits. He chairs the European Smart Energy Demand Coalition and is a director of the Demand Response and Smart Grid Coalition, Association for Demand Response and Smart Grid, and Smart Grid Consumer Collaborative. He has testified before the US Congress and other state and international regulatory bodies.

Mr. King holds bachelor's and master's degrees in biological sciences from Stanford University, a master's in management science from Stanford Graduate School of Business, and a doctorate in law from Concord Law School.

Steven Kline is managing director of Sustainability and Policy Consulting where he advises organizations on energy-related policy and regulatory issues. Prior to this, he worked as VP of Regulation at Pacific Gas and Electric Company and VP for Corporate Environmental and Federal Affairs at PG&E Corporation and was PG&E's first chief sustainability officer.

Mr. Kline has testified before federal and state regulators and the US House of Representatives and Senate. He serves on the board of the China-US Energy Efficiency Alliance and the Executive Advisory Board of the EDGE at Duke University's Fuqua School of Business.

He has a BA in political science and economics from Coe College and an MA in diplomacy from the Patterson School of Diplomacy and International Commerce and is a graduate of the University of Michigan's Executive Business Program.

Jonathan Koehn is the regional sustainability coordinator for the city of Boulder, where he oversees aspects of the city's sustainability agenda, specifically in relation to climate action and waste reduction. He has been focused on various aspects of Boulder's energy efforts, primarily, the city's Municipalization Exploration Project.

Jonathan came to Boulder with over 10 years' experience working with state, regional, and local governments and their constituencies domestically and internationally to develop strategic and tactical solutions to energy, economic, and climate challenges. Prior to his current role, he was the environmental affairs manager for the city.

Jonathan holds degrees in environmental science from Northern Arizona University and marine sciences and biology from the University of Florida.

Lorenzo Kristov is principal specializing in market and infrastructure policy at California Independent System Operator (CAISO) where he develops CAISO policy in market design, transmission planning, new generator interconnection, and distributed resource integration to support California's clean energy policy goals.

His current focus is on alternative ways to manage the operational and wholesale market impacts of the accelerating proliferation of distribution-connected resources. He has been instrumental in redesigning the transmission planning process, integrating generator interconnection procedures with transmission planning, and designing CAISO's LMP-based market structure. Prior to CAISO, he worked at the California Energy Commission, developing the rules for retail direct access, and prior to that, he was a Fulbright scholar in Indonesia working on a commercial and regulatory framework for private power development.

He received a BS in mathematics from Manhattan College, an MS in statistics from North Carolina State University, and a PhD in economics from the University of California at Davis.

Iain MacGill is an associate professor in the School of Electrical Engineering and Telecommunications at the University of New South Wales and joint director (engineering) for the University's Centre for Energy and Environmental Markets (CEEM). Iain's teaching and research interests include electricity industry restructuring and the Australian National Electricity Market, sustainable energy technologies, and energy and climate policy.

CEEM undertakes interdisciplinary research in the monitoring, analysis, and design of energy and environmental markets and their associated policy frameworks. It brings together UNSW researchers from the faculties of engineering, business, science, law, and arts and social sciences. Dr. MacGill leads work in two of CEEM's three research areas: *Sustainable Energy Transformation* including energy technology assessment and renewable energy integration and *Distributed Energy Systems* including "smart grids" and "smart" homes, distributed generation, and demand-side participation. He has published and consulted widely in these and related areas.

Dr. MacGill has a bachelor of engineering and a masters of engineering science from the University of Melbourne and a PhD on electricity market modeling from UNSW.

Judith McNeill is an economist and senior research fellow in the Institute for Rural Futures at the University of New England in Armidale, New South Wales, Australia. Prior to academic life, Judith held research and advisory positions to members of parliament in the Australian and Northern Territory Treasuries, the Parliamentary Library of the Department of Parliamentary Services, and the Office of National Assessments in Canberra.

Judith's research interests include environmental policy, climate change adaptation, infrastructure finance, and ecological economics. Judith also teaches The Sustainable Business at the Queensland University of Technology's executive MBA program in Brisbane.

She has a master's in economics (DH Drummond Dissertation Prize) and a PhD (economics) from the University of New England, Armidale.

William C. Miller works for Lawrence Berkeley National Laboratory where he is assigned to the Energy Efficiency and Renewable Energy Program in the US Department of Energy. Prior to his current position, he managed the strategic regulatory issues group for the Customer Energy Efficiency Department at Pacific Gas and Electric Company (PG&E) where he oversaw the company's Zero Net Energy Pilot Program. Before joining PG&E, he spent 10 years in academia, engaged in teaching and research.

He has worked for over 25 years in the areas of energy forecasting, strategic market analysis, energy efficiency planning, measurement, policy, and litigation. From 1990 to early 2010, he managed the evaluation, policy, and regulatory activities for the Customer Energy Efficiency Department at PG&E.

Dr. Miller received a bachelor of science degree in economics from Stanford University and a PhD in economics from the University of Minnesota.

Bruce Mountain is the founder of Carbon and Energy Markets, an energy economics consultancy based in Melbourne, Victoria, engaged in the economic regulation of networks and in the design of electricity markets and renewable energy policies. Bruce has been an independent energy economics consultant since 2000 and before that worked in the strategy and policy team at PricewaterhouseCoopers in London; for Ofgem, on secondment to Electricite de France; and for Eskom in South Africa.

His consultancy provides advice to energy regulators, government departments, industry and their associations, energy users and their associations, multilateral lenders, investment banks, other consultancies, and various nongovernmental organizations.

He has a master's in electrical engineering from the University of Cape Town, is qualified as a chartered management accountant in England, and is writing a PhD thesis on the political economy of energy regulation in Australia.

Tim Nelson is the head of Economics, Policy and Sustainability at AGL Energy—one of Australia's largest energy utilities—where he is responsible for AGL's sustainability strategy; greenhouse accounting and reporting; economic research; corporate citizenship program, Energy for Life; and energy and greenhouse policy. Before joining AGL Energy, Tim was an economic adviser to the NSW Department of Premier and Cabinet and the Reserve Bank of Australia.

Tim has advised governments and utilities on energy and climate change policy for around fifteen years. He has had several papers published in Australia and internationally and has presented at conferences in Australia and throughout Asia and Europe.

Tim is an adjunct research fellow at the University of New England, holds a first-class honors degree in economics, and is a chartered secretary. He is currently completing a PhD in energy economics.

Paul H. L. Nillesen is a partner with PwC based in Amsterdam. He is part of the firm's global energy, utilities, and mining practice and the coleader of PwC's global renewables practice. He is also a member of the International Energy Agency's Renewable Energy Working Party.

Paul is specialized in regulatory economics, strategic analyses, advanced market and demand analyses, business simulation, and modeling primarily focused on energy and utilities. He has advised companies, governments, regulators, and national and international institutions and has published numerous articles in the field of energy and regulatory economics.

He holds first-class master's in economics from Universities of Edinburgh and Oxford and a PhD in economics from Tilburg University.

Phillip Paevere is the leader of the Urban Systems Program in CSIRO, where he directs a portfolio of multidisciplinary research spanning the domains of energy, climate change, and the built urban environment.

Phillip leads CSIRO's research on modeling impacts of new technologies, such as electric vehicles, and on the electricity grids and homes of the future and is also the lead researcher on technology trials related to the development of vehicle-to-grid and vehicle-to-house smart grid technologies. He has published widely in the peer-reviewed academic literature across several disciplines including energy, sustainable building, numerical simulation, and technology forecasting.

Dr. Paevere holds a bachelor of engineering (first-class honors) and a PhD (engineering), both from the University of Melbourne, Australia.

Glenn Platt leads the Local Energy Systems theme within CSIRO's Energy Transformed Flagship, developing technologies for dramatically reducing the carbon emissions and increasing the uptake of renewable energy around the world. The theme's work ranges from solar cooling, electric vehicles, smart grids, and the integration of large-scale solar systems to understanding people's response and uptake of particular low-carbon energy options.

Prior to CSIRO, Glenn worked in Denmark with Nokia on the standardization and application of cutting-edge mobile communications technology, and prior to that, he was employed in an engineering capacity for various Australian engineering consultancies, working on industrial automation and control projects.

Glenn holds PhD, MBA, and electrical engineering degrees from the University of Newcastle, Australia, and is an adjunct professor at the University of Technology, Sydney.

Michael G. Pollitt is university reader in business economics at Judge Business School, University of Cambridge. He is the assistant director of the Energy

Policy Research Group (EPRG) at the University of Cambridge and a fellow and director of studies in economics and management at Sidney Sussex College, Cambridge. He was coleader of the Cambridge-MIT Electricity Project from 2001 to 2005 and served as founding executive director of the EPRG in 2005 and 2006.

Dr. Pollitt is an economist with particular interests in the efficiency and regulation of network utilities. He has published 9 books and over 50 refereed journal articles on efficiency analysis, energy policy, and business ethics. He is a founding coeditor of the Economics of Energy and Environmental Policy.

He holds an MA in economics from the University of Cambridge and an MPhil and DPhil in economics from the University of Oxford.

John Rhys is a senior research fellow at the Oxford Institute for Energy Studies, where he works on issues related to energy policy, with particular reference to the electricity sector, climate change, and market reforms. He currently acts as secretary to the British Institute of Energy Economics (BIEE)'s Climate Change Policy Group.

Dr. Rhys is a former managing director of NERA Economic Consulting and a former chief economist at the UK Electricity Council. At NERA, he was closely involved in UK electricity restructuring and privatization, led the firm's international energy consulting work, and was personally involved in a number of energy policy and power sector reform studies in Eastern Europe and Asia, carried out for national governments under the auspices of the World Bank, ADB, and other agencies.

He has a mathematics degree from the University of Oxford and an MS and PhD from the London School of Economics and Political Science.

Chris Riedy is an associate professor at the Institute for Sustainable Futures, University of Technology, Sydney. He is also the president of the Climate Action Network Australia and cochair of the Australasian Node of the Millennium Project, a global futures think tank working on sustainability issues. Prior to his present post, he worked as an environmental consultant with CH2M Hill in Australia and the United Kingdom.

Dr. Riedy's interests include the social and cultural dimensions of climate change response, community engagement, sustainability communications, behavior change, and public participation processes. He makes use of applied foresight frameworks and methods and was recently appointed to the editorial board of Futures.

Dr. Riedy has a PhD in Sustainable Futures from the University of Technology, Sydney.

Jenny Riesz is a research associate at the Centre for Energy and Environmental Markets, at the University of New South Wales in Sydney, Australia. Prior to this, she was a senior consultant in Strategic Energy Advisory Services at

AECOM and the principal of renewable energy and climate policy at ROAM Consulting. Her work focuses on the integration of renewable technologies into electricity markets for clients such as the Australian Energy Market Commission and the Independent Market Operator of Western Australia.

Dr. Riesz is interested in multidisciplinary research at the nexus between technical and economic factors. She is an active member of the CIGRE C5-11 international working group, contributing to research on market design for large-scale integration of intermittent renewable energy sources and demand-side management.

Dr. Riesz has a PhD in physics from the University of Queensland, Australia, and is a chartered engineer with the Institution of Engineering and Technology (IET).

Roland J. Risser is the director for the US Department of Energy (DOE)'s Building Technologies Office. By working with the private sector, state and local governments, national laboratories, and universities, the Building Technologies Office improves the efficiency of buildings, and the equipment and systems within them, as well as how they can transact with the grid at scale. He is currently chairman for the ISO Technical Committee on Energy Management, ISO 50001.

Before joining DOE, Roland was director of Customer Energy Efficiency Department at the Pacific Gas and Electric Company. In this role, he was responsible for developing and implementing energy efficiency and demand response programs. Over his career, he also managed tariffs, electric and natural gas vehicles programs, a building and appliance codes and standards program, and the Pacific Energy Center, Energy Training Center, and Food Service Technology Center.

Roland received a bachelor of science degree from the University of California, Irvine, and a master of science degree from the California Polytechnic State University in San Luis Obispo. He also graduated from the Haas School of Business, Executive Program, at the University of California, Berkeley.

David Robinson is a senior research fellow at the Oxford Institute for Energy Studies. He specializes in the field of public policy and corporate strategy in the energy sector, with a particular interest in the role of markets and regulation in meeting the challenges of climate change. He has worked for 25 years as a consultant for governments, regulators, energy companies, and international organizations. He is based in Madrid, where he runs his own consulting firm.

He earned his DPhil in economics at the University of Oxford.

Paul Simshauser is the chief economist at AGL Energy Ltd and has overall responsibility for regulated pricing, economic policy and sustainability, energy regulation, government affairs, media, and corporate communications. He is also a professor of economics at Griffith University's.

Professor Simshauser has over 20 years' experience in the energy industry in the areas of energy trading, power station development, capital markets and project finance, retail pricing, and energy policy.

Paul holds a bachelor's in economics and in commerce, a master's in accounting and finance, and a PhD in economics from the University of Queensland. He is an FCPA, an AFMA-accredited dealer, and a fellow of the Australian Institute of Company Directors.

Fereidoon Sioshansi is president of Menlo Energy Economics, a consulting firm, and the editor and publisher of *EEnergy Informer*, a newsletter with international circulation. His professional experience includes working at Southern California Edison, the Electric Power Research Institute, National Economic Research Associates, and Global Energy Decisions (GED), acquired by ABB.

His interests include climate change and sustainability, energy efficiency, renewable energy technologies, smart grid, dynamic pricing, regulatory policy, and integrated resource planning. He has edited 7 books prior to this, *Electricity Market Reform: An International Perspective*, with W. Pfaffenberger (2006); *Competitive Electricity Markets: Design, Implementation, Performance* (2008); *Generating Electricity in a Carbon-Constrained World* (2009); *Energy, Sustainability and the Environment: Technology, Incentives, Behavior* (2011); *Smart Grid: Integrating Renewable, Distributed & Efficient Energy* (2011); *Energy Efficiency: Towards the End of Demand Growth* (2013); and *Evolution of Global Electricity Markets* (2013).

He has degrees in engineering and economics, including an MS and PhD in economics from Purdue University.

Robert Smith is the manager of economic policy and strategy at Ausgrid, the largest electricity distribution company in Australia based in Sydney. He has over 25 years' experience as an economist including a dozen years of working in electricity market design, regulation, energy efficiency, and demand management. In his current capacity, he has been involved in creating two energy efficiency centers, full retail competition for electricity, web-based energy efficiency and e-commerce tools, Australia's first mass market CFL giveaway, electric vehicles, and the Smart Home Family project.

Mr. Smith's interests include cost-benefit analysis and understanding how economics, technology, incentives, regulation, and customers' behavior interact to create change. He is a regular speaker and presenter on energy efficiency, electric vehicles, and economic issues.

Robert has a graduate degree in econometrics, a master's in economics from the University of NSW, and postgraduate qualifications in finance from the Securities Institute of Australia.

Rajat Sood is a member of the energy practice within Frontier Economics, a microeconomics consulting firm based in Melbourne, Australia.

Rajat has advised policy makers, regulators, businesses, consumer groups, and law firms operating in the Australasian energy sector for over a dozen years. His clients have included the Australian Energy Market Commission, the Australian Energy Regulator, the New Zealand Electricity Authority, the Singapore Energy Market Authority, and most Australian state governments and energy businesses. Rajat's key areas of expertise are the restructuring and reform of state-owned utilities, market design, network pricing and regulation, and trade practice (antitrust) economics. Prior to consulting, he was a solicitor with Freehills.

Rajat has honors degrees in economics and law from the University of Melbourne.

Paul Szuster is an energy analyst and founder of TrueDemand, a consultancy providing analysis and advice of renewable energy policies and the impact on the utility business model. Paul previously worked in roles of structural integrity, facilities, and decommissioning engineering at Esso Australia, ExxonMobil Corporation.

His consultancy provides clients with analysis and advice in strategy, policy, and economics in the energy industry. His interests are in the commercial development of renewable energy and its effects on the utility business model.

He has a first-class honors degree in civil/structural engineering from the University of Adelaide, currently completing his master of engineering degree in sustainable energy at RMIT University, Melbourne, Australia.

Dustin R. Thaler is program associate at the Energy Future Coalition, where he coordinates, researches, and develops content for initiatives on high-voltage transmission, utility reform, building efficiency, and aromatic hydrocarbons. Prior to his present job, he worked as the chief operations officer for Floridanomics, an economic consulting startup website, and as policy fellow for Groundswell, a DC-based energy efficiency nonprofit, and interned for the International Carbon Bank and Exchange.

His interests include energy production technologies, environmental and resource economics, carbon management and finance, energy law and policy, and understanding the constraints on the US electric grid. He has worked on transmission line cost allocation, energy intensity, and efficiency in low-income households.

Dustin graduated from the University of Florida with an honors bachelor's in economics and a minor in sustainability studies and is currently attending Johns Hopkins University on a master's in energy policy and climate.

Kai E. Van Horn is a PhD candidate in electrical and computer engineering at the University of Illinois at Urbana-Champaign and a consultant for the Brattle Group. His work focuses on the grid integration of renewable energy sources and policy and market issues related to grid planning, demand response, renewable energy, and energy storage.

Kai received his BS in multidisciplinary engineering from Purdue University in 2007 and his MS in electrical and computer engineering from University of Illinois at Urbana-Champaign in 2012.

Carl Weinberg served as manager of R&D at Pacific Gas and Electric Company for 19 years prior to his retirement. During his tenure, he was involved in photovoltaic utility system application in Advanced Customer Technology Test for maximizing energy savings in buildings, was instrumental in building the Modular Generation Test Facility to study the impact of distributed technologies, and contributed to Distributed Utility Valuation Project in collaboration with EPRI and NREL. He served on EPRI committees and the National Academies of Science and is the chairman of the board of the Regulatory Assistance Project and a board member of the Center for Resource Solutions.

Carl's main interests are distributed resources and the integration of renewables. He has also an involvement in regulatory policy and the impact of disruptive technologies on utilities.

Carl holds a BS and MS in civil engineering from University of California, Berkeley, and an MS in physics from Vanderbilt University.

Jens Weinmann is program director at the practice group Telecommunication, Transport, and Utilities of ESMT Customized Solutions GmbH at the European School of Management and Technology. Prior to that, he worked in a government-sponsored research project on electric mobility and at the economic consultancy ESMT Competition Analysis.

Jens' research focus lies in the analysis of decision making in regulation, competition policy, and innovation, with a special interest in energy and transport. In 2013, he published the book *The Decentralized Energy Revolution: Business Strategies for a New Paradigm* together with Christoph Burger.

He graduated in energy engineering at the Technical University of Berlin and received his PhD in decision sciences from London Business School. He also held fellowships at the John F. Kennedy School of Government, Harvard University, and the Florence School of Regulation, European University Institute.

Jon Wellinghoff is a partner at the law firm Stoel Rives LLP and cochair of the firm's energy team. Prior to his present post, he was the longest-serving chairman of the Federal Energy Regulatory Commission. He served as general counsel at the Nevada Public Utilities Commission and Nevada's first consumer advocate, among others.

During his FERC tenure, he led efforts to make the US power grid cleaner and more efficient by promoting and integrating renewable energy, demand response, energy efficiency, and storage. He championed FERC's landmark Order 1000, which required grid operators to integrate solar and wind resources into the network, created FERC's Office of Energy Policy and Innovation to promote new efficient technologies, and oversaw development of the National Action Plan on Demand Response.

Wellinghoff received an MAT in mathematics from Howard University and JD from Antioch School of Law. He has written and lectured extensively on numerous subjects related to energy policy and practice.

Eva M. Witteler is a manager in the finance and regulation energy group of PwC in Düsseldorf, Germany, and has also worked within the energy, utilities, and mining practice of PwC Amsterdam, the Netherlands.

Ms. Witteler's work is primarily focused on the energy sector, in particular the electricity and district heating market, conventional and renewable power, heat generation, and applications of energy sector laws. Eva has advised several national and international clients including large companies, SMEs, and industrial enterprises regarding feasibility studies, make-or-buy decisions, financial modeling of different energy projects, and optimization of energy costs.

Eva holds a master's in electrical engineering and economics from University of Braunschweig—Institute of Technology. She performed part of her studies at Pontifícia Universidade Católica do Rio de Janeiro.

This volume, at its core, is about technological change and its impact on utility business model. In this context, it is worth noting that the progress of technology is both rapid and uncertain. Political, environmental, technological, and competition factors are all driving uptake of new technologies in the energy sector, and companies such as E.ON are at the cutting edge of these developments.

These technologies could result in not only an energy system that is transformed over the lifetime of utilities' asset base, creating opportunities for both new and adapted businesses, but also threats where there is reluctance or inability to change.

As a number of chapters in this volume point out, failure to monitor both incremental and disruptive changes could result in poor investments in new or existing technologies or missed future business opportunities. These are the issues that technology and innovation activities need to address.

Historically, technology development in utilities has focused on incremental improvement of existing assets and capabilities. However, in the face of system transformation pressures and the pace of technological developments, this approach is no longer appropriate, a fact noted by a number of contributing authors.

At E.ON, we believe that the response needs to be a group-wide technology and innovation strategy to provide direction in accessing the capabilities required to deliver the business strategy.

Our analysis of the global environment has identified energy system trends that we believe will occur regardless of utilities business activities:

- Significant increase in "at-scale" renewable generation, leading to
  - need for more flexible generation, energy storage, and/or grid enhancement and
  - decoupling of supply and demand that increases the value of demand-side response services.
- Increasing deployment of information systems technologies and distributed energy technologies, leading to
  - bidirectional energy and information flow and
  - more active participation of our customers in the energy system.

However, there is a high level of uncertainty about possible transition pathways to potential future situations. For example,

- there are many potential "branching points" in the transition to a decarbonized energy world, and directions taken at these points will significantly alter the speed of transition and technology mix at the end;
- there is uncertainty over how rapidly a transition to a more decentralized energy system will unfold and how the trend will be sustained;
- current sovereign debt crisis and economic downturn may drive a near-term focus towards adaptation to climate change, meaning that the technology choices of a world of adaptation may be very similar to today.

Analyzing these trends has led to core beliefs about the key technology success factors. They form the basis of our approach to technology and innovation investment at E.ON, which include the following:

- Green energy is a global theme, while there is uncertainty over political will, consumer demand, and energy prices.
- Increasing role for distributed generation, storage, and demand-side management.
- Technology breakthroughs will become increasingly rapid.
- Information technologies will form the backbone of the future energy system.

These issues are among the topics covered in this volume, a timely contribution to a topical and relevant subject.

Urban Keussen
Senior Vice President Technology and Innovation
E.ON

Preface

The San Francisco Earthquake struck me as a reasonable metaphor for the kind of changes the energy utilities have experienced in this country in the least several years, in effect turning their world upside down . . . The basic proposition of this symposium is that the dramatic underlying changes in the economics of energy in the United States occasion a corresponding examination of the very important role of our energy utilities.[1]

Neighborhood electric cooperatives; rooftop systems integrated into a community storage and backup system; utilities as mere dispatching agents or maintenance crews – these are a few of the visions of the urban future. Overnight we have witnessed a dramatic change in the assumptions underlying our electric generation technologies and regulatory procedures.[2]

The electricity industry is a classic example of a market ripe for breakout disruptive technologies. The grid was a good idea 100 years ago. So good that we were able to graft a much more resilient network – the Internet – on top of it. It's time for the grid to learn what the Net already knows.[3]

Although these all sound like contemporary discourse, characteristic of much that follows in this volume, the actual publication dates were 1980, 1983, and 2003, respectively. The quotations are a reminder that urgent rhetorical challenges to the utility business model have a rich history, along with formulaic claims that technology innovation will shortly relieve people of any need for further reliance on their hometown utility's generation, transmission, and/or distribution systems, with dire consequences for shareholders, bondholders, and everyone else connected to ubiquitous but imminently obsolete ventures. I remember sitting in a hearing room packed with utility managers when a regulator in the nation's most populous state declared electricity distribution monopolies obsolete and admonished them to prepare for mass institutional funerals. That was twenty years ago. That regulator is now a widely respected CEO for a major electricity distribution monopoly.[4]

The resilience of the utility model reflects durable economies of scale in grid management, reliability assurance, and resource procurement and integration

---

1. California PUC, Energy Efficiency and the Utilities: New Directions, p. 1 (July 1980)
2. Howard R. Brown (ed.), Decentralizing Electric Power Production, p. 55 (1983)
3. Steve Silverman, *Taming the Electricity Beast,* Wired, November 2003, p. 036
4. The regulator was then California PUC Commissioner Jessie Knight. He can be found today at the San Diego Gas & Electric Company, adroitly navigating what has proved to be stubbornly persistent utility regulation.

(but not generation ownership). Grid bypass remains unappealing to almost everyone with the option to connect or stay connected, notwithstanding every advance in the composition and management of rooftops; as EF Lindsey observed thirty years ago: "Until you've walked into a totally dark room with a flashlight in one hand and a toolbox in the other, you haven't had a firsthand experience with onsite power."[5] Inventors and venture capitalists have learned that it is far more productive to treat utilities as partners than adversaries. For those committed to a clean energy future, utilities remain the most important investors. Technological progress in the electricity sector has been and remains much more about opportunities for grid enhancement than grid displacement.

This is not to say that regulatory models don't need to change; this volume is an eloquent rebuttal to any such contention, and I personally have devoted a lifetime to the reform project and don't intend to stop now. US electricity sales aren't collapsing, but since 2000, the rate of growth has lagged well behind population growth. The traditional commodity vision of the business won't work, as Edison knew from the beginning. This yields two crucial questions: given declining growth in commodity sales, how do utilities secure the reasonable revenue certainty required to make enduring provision for reliable and affordable services, and how can regulators allocate the costs of those services equitably among all who use them? The simple but entirely wrong answer to both questions would be to stop charging for electricity service based on electricity use and to make all or most of an electricity bill independent of consumption. The "all you can eat" pricing model may work well for some businesses, but none have environmental and equity dimensions comparable to electrical generation.

The real solutions begin with revenue decoupling, which makes utilities indifferent to retail energy sales without abandoning the tradition of volumetric pricing and its incentives for customers to use energy efficiently. More than half the states have now adopted this approach for at least one electric or natural gas utility, and a comprehensive June 2013 order by the Washington Utilities and Transportation Commission is the latest primer on how to do it effectively, using modest annual true-ups in rates that few if any customers even notice.[6]

This book addresses additional aspects of utility business model reform and rate design that are also critical to a clean energy transition. "Net metering" programs in wide use across the United States have helped valuable "distributed" technologies such as solar power gain traction and improve performance, but additional approaches are needed now. Although such generation can reduce a grid's needs for central station generation and other infrastructure, it typically does not eliminate its owners' needs for grid services. When they use distribution

---

5. E.F. Lindsley, Planning Practically for a Decentralized Electrical System, in Howard R. Brown (ed.), note 2 above, p. 245
6. Washington Utilities and Transportation Commission, In the Matter of the Petition of Puget Sound Energy and the Northwest Energy Coalition, UE 121697 and UG 121705 (June 25, 2013)

and transmission systems to import and export electricity, owners and operators of on-site/"distributed" generation should provide reasonable cost-based compensation for the utility services they use while also being compensated fairly for the services they provide.

Customers deserve the opportunity to interconnect distributed generation to the grid quickly and easily. For their part, utilities deserve assurances that recovery of their authorized nonfuel costs will not vary with fluctuations in electricity use. This does not require rate designs that reduce rewards to customers for using less electricity by instituting or raising fixed charges that are independent of peak or average use. Alternatives include minimum bills that convert to volumetric charges if the customer exceeds a monthly consumption threshold and variable demand charges that take advantage of digital meter capabilities.[7]

Finally, investor-owned utilities' earnings opportunities should include performance-based incentives tied to benefits delivered to their customers by cost-effective initiatives to improve energy efficiency, integrate clean energy generation, and improve grids. In general, business models should include profit opportunities linked to utilities' performance in delivering safe, reliable, and affordable energy services. This would represent a decisive and overdue break with the regulatory tradition of tying utilities' earnings opportunities to tonnages of capital invested in their own physical assets.

Elements of this agenda appear repeatedly in the pages that follow, with elaborations and analyses that will help everyone involved in shaping the future of the electricity sector. I salute the authors and their indomitable editor for the most constructive engagement on issues in which all humanity shares a stake.

Ralph Cavanagh
Energy Program Codirector
Natural Resources Defense Council

---

7. For a discussion of these concepts in the context of a specific utility's rate design proceedings, see http://switchboard.nrdc.org/blogs/rcavanagh/changing_how_utilities_do_busi.html (September 2013).

The Rise of Decentralized Energy

Fereidoon P. Sioshansi

Menlo Energy Economics

At a news conference in mid November 2013, Peter Terium, CEO of RWE, Germany's second-biggest utility, said, "Our traditional business model is collapsing under us." He was, of course, referring to flat demand for electricity and continued growth of renewable generation leading to excess capacity resulting in steady erosion of wholesale electricity prices, evaporation of profits, and declining share prices.

While German—and by extension European—utilities' financial woes appear to be more pressing than those afflicting utilities in other parts of the world, the fundamentals are strikingly similar. In a number of mature economies, utilities are discovering that their traditional business model is rapidly *eroding* if not collapsing. As further described in the following chapters, the signs of change are initially appearing in selected countries with low or nonexisting electricity demand growth, high and rising retail tariffs, ambitious renewable targets, and supportive policies that encourage decentralized generation. With the passage of time, however, similar conditions are likely to apply to an increasing number of countries.

HISTORICAL PERSPECTIVE: CENTRALIZED GENERATION AND GROWING CONSUMPTION

With minor variations, the electricity supply industry (ESI) evolved in different parts of the world propelled by the growing demand for electricity and the wonderful services it provides, on the one hand, and the remarkable and persistent decline in prices accompanied by improved service quality and reliability, on the other.

From the beginning, the industry's growth was accompanied by pervasive regulations that ensured sufficient financing would be available to the growing industry to invest in and maintain the capital-intensive infrastructure. During its first century, the ESI not only succeeded to keep up with the continued growth in demand but also managed to do so at substantially falling per-unit costs over extended periods of time. The persistent falling prices, in turn, encouraged continued demand growth, bringing prosperity, higher standards of living, and comfort to billions of customers around the world.

Moreover, throughout its history, the ESI was able to invest untold billions in a long value chain that typically stretches from coal mines, natural gas fields, hydro reservoirs, nuclear power stations, and other *upstream* facilities all the way down to the ubiquitous outlets in every home, office, shop, farm, or factory. Consumers did not need to know how and where the electricity came from and how it was generated, transmitted, and distributed to the outlet on the wall. Until recently, the per-unit costs of electricity were consistently falling in real terms. It was a win-win arrangement: As customers consumed more, the ESI generated and delivered more at lower costs. During the industry's so-called golden years, some utilities in fact actively encouraged *increased* consumption by offering *declining* block rates.

As described elsewhere in the book,[1] the industry's business model gradually evolved around two basic principles:

- Centralized generation, delivering the output of large plants through a massive transmission and distribution network designed to deliver power to consumers while adjusting supply to match the load in real time
- A simple mechanism to collect revenues from consumers based on a flat tariff applied to volumetric consumption[2]

The centralized generation model allowed a few massive power plants to be built, usually far from major load centers—partly because nobody likes to have power plants in their backyard[3] and partly because thermal plants had to be located near cooling water or coal mines—with their output transported over long distances. Customers preferred not to see, smell, hear, or otherwise be aware of where the power was coming from or how it was produced or transported.

The flat tariff applied to volumetric consumption meant that for a customer using 100 kilowatt-hours a month with 10 cents/kWh, calculating the bill would be straightforward.[4] In many cases, such as in California, residential consumers were not charged a fixed connection or service fee, which meant that if they consumed no kilowatt-hours, their bill would be zero. As will be pointed out in numerous chapters of this book, the ESI's overwhelming reliance on volumetric consumption—which was predicated on continued demand growth

---

1. Refer to Chapter 7 by Smith and MacGill.
2. For examples, refer to J. C. Bonbright, Principles of Public Utility Rates, Columbia Univ. Press, 1961.
3. This has given rise to *NIMBY*, not in my backyard.
4. Over time, many variations evolved including fixed monthly fee based on the capacity, but for the most part, the revenue recovery mechanism was straightforward.

and falling retail rates—has become a liability in an environment where the reverse may be true.

The flat tariff scheme was simple not only in terms of accounting but also in terms of what had to be measured, recorded, and billed. A primitive, inexpensive, and hardy spinning meter was all that was required to measure volumetric consumption. Before the advent of digital meters, also called smart meters, it was not easy or feasible to measure other attributes of service, such as the pattern of consumption.[5]

All that was necessary to produce a bill was a single number from a simple meter. For decades, this was done (it is hard to believe today) by sending a person—a meter reader—to visually *read* the dials on the meter at every consumers' meter, every month, repeated 12 times a year. In the early days, this was done manually, recorded in a ledger as the meter reader walked from meter to meter, chased by unfriendly neighborhood dogs. The data in the ledger would subsequently be transcribed by another person into whatever accounting and billing "system" was in place to produce customer bills. In some countries, meters were read less frequently—say quarterly—or in some cases, readings were estimated and reconciled once a year. These variations aside, it took one single meter read per customer per month (or quarter) to produce a bill.[6]

How the flat fee was determined could get complicated. In essence, however, this was set by a regulatory agency or government ministry to allow a *reasonable* return on investment on the infrastructure required to supply the juice plus allowance for the recovery of reasonable operating and maintenance expenses deemed necessary to keep the kilowatt-hours flowing. For example, labor costs, fuel costs, and operating expenses were typically allowed to pass through. Investments in infrastructure—steel in the ground, poles, and wires—were allowed a regulated rate of return, which encouraged more investment and more consumption. The term energy efficiency not only was alien to the industry but also ran against the business principles.

It is a slight exaggeration to say that the ESI essentially operated as a cost-plus business.[7] It was allowed to collect sufficient revenues through the volumetric sales to cover the financial costs and operational expenses. The rates would be periodically adjusted to keep the business viable while making sure it was not overly profitable. This, as it turned out, was a fine balancing act left to regulators in many parts of the world where the main stakeholders—the

---

5. Sophisticated billing options such as dynamic pricing or transactive energy, described in the following chapters, were not feasible and some would require further development before they can be implemented even today.

6. This points to another ESI historical legacy: the near total lack of communication or interface with customers, who were treated as *passive* consumers in a centralized, unidirectional world where electrons were delivered in bulk and billed accordingly.

7. Other competitive, capital-intensive businesses have a hard time understanding how this arrangement has lasted as long as it has.

generators, transmission and distribution owners/operators, and retailers—were privately owned and often vertically integrated.

The centralized business paradigm was not limited to generation but expanded to all functions. In many places, the ESI grew into large vertically integrated enterprises that spanned generation, transmission, distribution, and customer services such as billing and service restoration. In some cases, these vertically integrated utilities expanded their reach to include coal mining, LNG importing terminals, natural gas transportation, storage, and other enterprises.

To keep the enterprise intact—there are variations in different parts of the world depending on the form and organization of the ESI and how it is organized and regulated—it was decided that consumers would buy *all* their electric needs from the "grid," the umbilical cord that connects them to those upstream investments. And until recently, there were no feasible and/or compelling reasons to do otherwise. Except for a few large energy-intensive industries, few consumers had any reason to bypass the "grid" or the "utility" that served them or to generate some or all of their needs.[8]

WHAT HAS CHANGED?[9]

Since the early 1970s, a number of developments have begun to change the fundamentals of how the ESI operated and, more importantly, recovered sufficient revenues to stay viable.

The first major change began to unravel in the 1970s with the advent of modern and efficient gas turbines that could produce power at reasonable cost and on small to medium scale. This and other technological developments led to the realization that the generation segment of the industry need not be a regulated monopoly. Private generators not affiliated with transmission, distribution, or retiling could be viable enterprises and effectively compete with the incumbent vertically integrated utilities. Small or smallish entities could finance, build, and operate independently owned power plants and sell the output to incumbent "utilities" and/or large customers, provided the prevailing regulations were modified to allow such transactions.

The passage of the Public Utility Regulatory Policy Act of 1978 (PURPA), a seminal piece of legislation in the United States, allowed a fledgling industry, known as independent power producers (IPPs), to flourish. Power generation was no longer limited to large vertically integrated utilities—globally.

In retrospect, this was a major game changer for the established industry, which, for the first time, was confronted with competition from nimble IPPs,

---

8. The exceptions were/still are remote facilities that are too far or economic to be connected to the grid or in cases where, for example, extensive demand for process steam made it cost-effective to generate power from the same heat source.

9. The first part of the book covers the same topic in more detail and from different perspectives.

who in some cases were more efficient and could generate power at lower cost than they could.

Not surprisingly, the incumbents did not welcome the newcomers and did their best to frustrate their growth. With the support of the regulators, however, the IPPs eventually prevailed. The entrenched incumbents with vested interests and significant investment in infrastructure have a habit of resisting change even when the business fundamentals render the status quo outdated or obsolete. This explains why the established players, broadly speaking, do not welcome distributed generation on the customer-side of the meter, such as rooftop solar PVs, because it competes with their core business, which is centralized generation.

This major development was followed by market reforms in the late 1980s and 1990s that eventually led to the privatization, liberalization, and/or restructuring of monopoly, vertically integrated utilities in many parts of the world, in some cases introducing competition not only in generation but also in retailing and supply. These developments—it is a long story—go beyond the scope of the present volume.[10]

With the successful implementation of restructured or reformed markets in selected countries, notably England and Wales in the 1990s, it became apparent that the traditional "centralized" business paradigm that had led to vertically integrated "utilities" was not necessarily efficient, nor central to the operation of the ESI. Restructured markets in many parts of the world demonstrated the benefits of decentralization, vertical disintegration, and competition applied to portions of the ESI.

In the United States, Europe, Australia, and a number of other places, policy makers introduced centralized wholesale markets and market operators encouraging trading, competitive bidding, bilateral contracts, and other forms of transactions that compelled the stakeholders to compete for generation. On the retail end, an increasing number of countries and states in the United States allow retailers to compete to serve customers. These developments resulted in significant operational efficiencies and cost savings, allowing customers—for the first time—to select a local supplier or retailer from among a number of competitors willing to offer lower prices and improved and/or more customized services.

Once again, the existing stakeholders did not welcome the changes, resorting to legal challenges and other means to prevent, frustrate, or impede anything that would undermine the status quo. In the United States, for example, incumbent vertically integrated companies claimed that the introduction of restructured markets would result in billions of dollars of stranded costs, bargaining

---

10. For examples, refer to Electricity Market Reform: An International Perspective (with W. Pfaffenberger), 2006; Competitive Electricity Markets: Design, Implementation, Performance, 2008; and Evolution of Global Electricity Markets, New Paradigms, New Challenges, New Approaches, 2013.

for bailouts in the form of cost recovery mechanisms.[11] Many parts of the United States—and most other countries—still operate as monopoly, vertically integrated utilities earning regulated rate of return.[12]

More recently, a number of other developments began to impact the business fundamentals:

- End of the industry's *golden age* and the realization that investing in bigger central plants no longer results in lower average costs.[13]
- Environmental restrictions on power generation further increased costs.[14]
- Broadly rising cost of fossil fuels and network maintenance began to impact retail electricity costs.[15]
- The realization that investing in energy efficiency could be a preferable alternative to increased investment in infrastructure—that is, negawatts are superior to megawatts.[16]
- Concerns about climate change and sustainability increased interest in low-carbon generation.[17]

These powerful forces have brought many of time-tested assumptions about the "traditional" utility business model into question. Notably, the ESI's continued reliance on volumetric sales, which made sense when demand was growing and per-unit costs were falling, makes little sense in an age where electricity consumption in many advanced economies is barely growing while retail rates are rising.[18]

---

11. In California, for example, the 1996 legislation allowed the incumbent utilities to recover stranded costs through a mechanism called competition transition charge or CTC, which appeared on customer's electric bills as a separate item.

12. Politically powerful utilities in southeastern United States, for example, have succeeded to maintain the status quo to this day.

13. The ESI had essentially run out of the economies of scale. This reality was particularly acute in the case of nuclear power plants. As larger nuclear plants became more sophisticated and safety requirements became more stringent, building larger reactors resulted in *increased* per unit costs, not the other way around.

14. Restrictions started with NOx and SOx emissions; have been extended to include harmful particulates, once-through water cooling, and other restrictions; and are now being expanded to cover greenhouse gas emissions in some places.

15. The particulars vary from place to place, of course. In the United States, for example, retail electricity tariffs have been stable in the recent past partly due to the economic recession and soft demand and partly due to the bounty of domestic shale gas at unprecedented prices. In Australia, on the other hand, retail tariffs have nearly doubled in the past 5 years—as described in the chapter by Mountain and Szuster mostly because of rising network costs.

16. A recent edited volume by the author, Energy efficiency: Towards the end of demand growth, Elsevier, 2013, covers this topic in more detail.

17. Two edited volumes by the author, *Generating electricity in a carbon-constrained world* and *Energy and environmental sustainability*, cover efforts to decarbonize electric generation and sustainability, respectively.

18. The chapter by Faruqui and Grueneich examines these issues.

Flat tariffs have another fundamental problem. They give the ESI perverse incentives to encourage increased consumption, all else being equal, since this allows more revenues and keeps per-unit costs low. For some time, policy makers have been trying to tinker with the revenue collection mechanism to eliminate the incentives to sell more, rather than less.

Others have proposed the need for a new definition of service for the industry, not simply as purveyors of kilowatt-hours but as enterprises who are in the business of meeting customers' energy service needs at the lowest possible cost, where costs include societal and environmental costs. RWE, among the largest European energy companies, has changed its future business focus from "volume to value," acknowledging the new realities of the market.[19] Surely under such a definition of service, a "utility" should not only be allowed but in fact be encouraged to invest in customers' premises to reduce energy consumption if in doing so, the customer, the society, and the "utility" are better off. Several chapters in the book examine the fundamental issues surrounding the basic definition of *service*.[20]

In this case, an increased number of consumers are likely to be attracted to anyone who can meet their basic *energy service* needs—the proverbial hot shower and the cold beer—at the lowest possible cost.[21] The time has arrived to revisit the idea that a viable business model can be made for utilities who are *energy service providers* rather than providers of a commodity, kilowatt-hours.

THE RISE OF DISRUPTIVE TECHNOLOGIES

As the preceding discussion explains, the traditional "utility" business model has been under increasing pressure on a number of fronts for some time. But the urgency of coming up with an alternative business model did not seem urgent or pressing until the arrival of disruptive technologies in the last few years.[22]

These powerful forces, now progressing at accelerating speed, require a fundamental rethinking of the traditional "utility" business model, redefining the nature, types, and value of services delivered and how they will be paid for. At stake, as described by Nillesen et al., in chapter 2 is who *delivers* and who *captures* value in the evolving environment.

The fundamental changes—many are overlapping and/or interrelated—include

---

19. Refer to Chapter 7 in this volume.
20. For examples, see the chapter by Crandall.
21. This simple concept, much talked about for a very long time, may be ripe for reexamination. Duke Energy's Save-A-Watt, for example, came close to redefining basic "utility" service, but did not take off for a variety of reasons.
22. Refer to Chapter 7 by Sioshansi and Weinberg.

- low and falling electricity demand growth, most pronounced in mature economies—further described in the chapter by Faruqui and Grueneich;
- generally rising retail rates,[23] mostly as a result of rising network costs—further described in the chapter by Nelson et al.;
- ambitious renewable targets and/or financial incentives for decentralized generation as described by Mountain and Szuster for Australia and Burger and Weinmann and Groot in Europe;
- the growth of distributed generation, notably rooftop solar PVs, driven by rising retail rates and generous net energy metering (NEM) laws;
- rising investment in energy efficiency through appliance efficiency standards, more stringent building codes, requirements such as zero net energy (ZNE) or passive buildings;
- behavior and demographic changes resulting in flat or possibly lower per capita electricity consumption rates—setting aside the potential takeoff of EVs as described in the chapter by Platt et al.;
- growing interest in microgrids, semi-independent communities, or biovillages and interest in higher levels of service reliability[24];
- the rise of *prosumers*—consumers who generate some or all of their needs from on-site generation; and
- increased levels of consumer engagement assisted with the proliferation of inexpensive and easy to use/programmable devices that manage/control usage and host of others.

CHANGING THE FUNDAMENTALS

The main premise of this volume is that the time has arrived to reexamine the industry's business model for at least two reasons, and possibly others:

- First, the traditional model no longer applies to certain segments of the business in jurisdictions where "utilities" have already been vertically disaggregated. For example, in places where distribution companies operate as "stand-alone" poles and wires enterprises—for example, Australia or Texas—charging consumers on a volumetric basis makes little or no sense since virtually, all the costs are fixed. Moreover, given the developments

---

23. Rising retail tariffs are a growing phenomenon, notwithstanding the United States, currently blessed with low-cost natural gas. The United Kingdom's National Audit Office (NAO), for example, recently reported that energy (gas and electricity) increased 44% between 2002 and 2011 with projections of 18% increase by 2030. The 44% rise occurred while low-income consumers' wages fell by 11%. Fuel poverty—defined as consumers who spend more than 15% of their disposable income on energy—now afflicts 2.4 million households or 11% of the population. Similar figures apply in many other parts of the world.
24. An edited volume by the author of the chapter on smart grid covers developments that may address some of these concerns.

already mentioned, these enterprises are likely to encounter increased costs—say from the increased stresses imposed by *prosumers*—while at the same time facing flat or declining volumetric sales. These types of scenarios are not difficult to envisage, and the situation is likely to get progressively worse.

- Second, if one accepts the projections of flat or possibly declining consumption over time, accompanied with rising costs, retail tariffs will have to rise to keep the incumbent stakeholders viable. Rising rates are likely to result in even lower demand growth, more investment in energy efficiency, and more distributed generation—leading to a vicious cycle.[25]

Take the case of an ultraefficient building, residential or commercial, with ultraefficient appliances and energy-using devices and significant investment in on-site generation, say, rooftop PVs, a solar hot water system, ground source heat pumps, fuel cells, and CHP. Add one or more EVs with large batteries programmed to charge only when prices are ultralow, say, during early morning hours when demand is low and wind generation is high, resulting in virtually free—and carbon-free—electricity. How would a stand-alone distribution company remain viable on a flat tariff applied to volumetric consumption in such a scenario?

Even more important is the new definition of *service* for prosumers. As further described in Chapter 7, the most valued "service" for prosumers is likely to be *connectivity* to the grid and the ability to feed into it or take energy out of it depending on demand and on-site generation/storage. In an extreme case, such a customer may use few if any *net* kilowatt-hours while putting enormous stress on the distribution network for having to constantly balance the variable supply and demand.

An analogy with wireless communication business may be appropriate, even if not perfect. What today's consumers crave the most from the mobile network is *connectivity* or *service availability* as they roam the city, the country, and the world and ample *bandwidth* that allows them to download and upload growing volumes of data, files, videos, pictures, and applications on the go. Paying on a per-minute basis does not make much sense in a world where few consumers are actually talking and those who do use essentially free services such as those offered by Skype or companies who offer low-cost mobile service via prepaid cards.

This suggests the need for a redefinition of *service* for electricity, especially for an increasing number of prosumers, ultralow usage consumers, consumers with EVs who charge only when rates are ultralow, consumers with smart programmable devices that essentially do as EV batteries do, or an increasing number of consumers or communities that operate in a grid-parallel or grid-assisted

---

25. Nelson et al. and others describe the "death spiral" scenario.

mode. If storage technology evolves to the point where it is cost-effective, many observers believe it will be "game over" for the centralized business model as we know it.

Just as mobile network providers have adjusted their business model to set significant fixed charges based on bandwidth, speed, signal strength, and service ubiquity—moving away from per-minute charges—utilities need to make a similar transition to charge for the network's ability to balance customers' load and on-site generation.

ORGANIZATION OF THE BOOK

This collected volume, which includes contributions from a number of scholars, academics, experts, and practitioners from different parts of the world with differing perspectives, is focused on examining the future of the ESI in view of the fundamental changes that are challenging the ESI's historical business model. The fundamental questions are as follows:

- Are these developments real and imminent?
- What may be their consequences? (a central concern is the potential erosion of volumetric consumption of grid-supplied electricity or the traditional bread and butter of the industry)
- What are potential solutions to the challenges facing the ESI?

As the reader will find out, not all contributors believe in the doom and gloom or the death spiral scenarios. Moreover, since different authors have different expertise and background, a rich variety of topics are examined and different perspectives are offered, not necessarily in full agreement with one another.

The book is divided into 3 parts:

Part I: What Is Changing? examines the changes taking place within and without the ESI and sets the tone for the remaining chapters.

In Chapter 1, **Decentralized Energy: Is It as Imminent or Serious as Claimed?**, **Fereidoon Sioshansi** describes the fundamental drivers of change with significant impact on the ESI, including the flattening of demand, the rapid rise of renewables, and the recent growth of decentralized energy resources (DERs). The author argues that while DER is currently concentrated in a handful of markets where retail tariffs are high and/or generous incentives for DER exist, the same can be expected to apply elsewhere.

Rapid advances in technology, declining costs combined with broadly rising retail tariffs, are likely to propel DER into a significant phenomenon in the coming years. The evidence in places such as California, Australia, Germany, and Japan, to name a few, is compelling and is impacting retail tariffs for non-DER customers.

The chapter's main contribution is to examine the likely impacts of disruptive technology on the traditional stakeholders whose revenue model is still largely dependent on a flat tariff applied to volumetric consumption.

In Chapter 2, **New Utility Business Model: A Global View**, **Paul Nillesen**, **Michael Pollitt**, and **Eva Witteler** examine the response to the surge in distributed energy resources (DERs) and prosumers by traditional utilities. Using the results from a survey of CEOs of major utilities around the world, conclusions on the major threats and opportunities are drawn.

Although the creation of new markets, new niches, and new entrants might suggest the demise of the big and established utilities, the authors suggest that they might be able to "fast second" by scaling up and exploiting the new business opportunities—provided they make the right moves in time. Otherwise, the emergence of "platform" markets might lead to intermediaries capturing the value of interactions among the stakeholders before the utilities do. The critical skill for the traditional players will therefore be not in creating new business models but in identifying and cultivating them in mass-market scale. The authors point out some of the new entrants already appearing to challenge the established players.

The chapter's main contribution is to synthesize the discussions around new utility business models and the different views on how to respond to the challenge of DERs.

In Chapter 3, **Germany's Decentralized Energy Revolution**, **Christoph Burger** and **Jens Weinmann** examine the effects of the generous German feed-in policies for renewable and distributed energy sources confronting grid operators, utilities, and politicians with unprecedented challenges. While operators of rural distribution systems struggle to ensure grid stability, utilities are experiencing a significant erosion of revenues because renewables drive down wholesale market prices and rooftop PVs lead to loss of sales.

The authors point out that the success of the feed-in policy comes at a cost to society. Regulatory changes to incentivize backup by conventional power plants are under discussion, including capacity payments, subsidies for flexible supply, and demand-side management.

The chapter's main contribution is a critical assessment of Germany's unfolding distributed energy revolution, which allows an increasing number of consumers, businesses, and entire communities to virtually abandon purchasing power from the grid while relying on enhanced energy efficiency and self-generation.

In Chapter 4, **Australia's Million Solar Roofs: Disruption on the Fringes or the Beginning of a New Order?**, **Bruce Mountain** and **Paul Szuster** describe the stunning growth of rooftop solar PVs in Australia from a negligible base at the start of 2010 to more than 1 million at the end of 2012.

The authors point out that market penetration—more than one in eight household rooftops—is now the highest in the world. Households spent more than $9 billion and will receive subsidies over the life of their systems of $8 billion. The average cost of electricity from the rooftop systems, around $160/MWh, is about half the average household electricity prices.

The chapter concludes that, excluding feed-in tariffs, claims of cross-subsidy between energy users that do not have PV and those that do are not conclusive. PV households are paying $252 million per year less to monopoly network service providers, but they are avoiding future network expansion.

In Chapter 5, **As the Role of the Distributor Changes, so Will the Need for New Technology, Clark Gellings** points out that the "distributor" or the electric distribution utility will potentially face the greatest technology needs and business environment changes of all of the ESI stakeholders. In addition, information technology (IT) is becoming an increasingly integral part of the electric infrastructure, impacting transmission and distribution owners and operators and consumers alike. However, it is in the distribution system that the implications of IT will be most profound.

The author suggests that in tomorrow's distribution system, the proliferation of sensors, controls, intelligent electronic devices (IEDs), and the resultant "marriage" of customers to the utility system will lead to an increased complexity in distribution system operations. As efficiency accelerates, distributed generation proliferates, and demand response and new electric technologies are added to the mix, the distribution utility will need technologies to manage an increasingly complex system.

The chapter's main contribution is to explore these new technologies and their potential implications for the distribution system of the future.

In Chapter 6, **The Impact of Distributed Generation on European Power Utilities, Koen Groot** highlights how the growth of distributed generation in the EU influences the corporate strategies of major European utilities.

Driven to a large extent by renewable support policies, the proportion of distributed power has increased in the EU—to the detriment of the share of electricity supplied by incumbent utilities that are not only selling less electricity but also changing market conditions that have led to lower prices for electricity sold. The uptake of distributed renewables adds to the already troublesome outlook for traditional utilities, further eroding their profitability. In response, these firms engage in strategic restructuring, cost-cutting, and the expansion of activities in growth markets outside the EU.

The chapter examines the impact of a rapidly growing share of distributed power in the EU and the incumbents' initial response to this new dynamic in their European home markets.

In Chapter 7, **Lessons from Other Industries Facing Disruptive Technology, Fereidoon Sioshansi** with contribution from **Carl Weinberg** examines how other industries confronting sudden technological change and/or disruptive technologies have reinvented their core business models.

The authors point out that while many parallels exist and useful lessons can be learned, not all industries facing rapid technological change have made a successful and timely transition to a new pricing or business paradigm that incorporates the underlying disruptive technology.

The chapter's main message is that the ESI needs to be cognizant of the tectonic changes taking place, and instead of focusing on strategies that attempt to stop or impede the advances in disruptive technologies, it should embrace them in ways that are win-win-win, for consumers, for the industry, and for purveyors of the new distributed energy resources. One promising approach may be to expand the traditional utility boundaries to include DERs.

Part II: Implications and Industry/Regulatory Response examines the implications of developments and the industry and regulators' initial response to these developments.

In Chapter 8, **Electricity Markets and Pricing for the Distributed Generation Era**, **Malcolm Keay**, **John Rhys,** and **David Robinson** point out that for the most part, wholesale and retail electricity markets have changed little over the past century, with flat kilowatt-hour pricing applied to volumetric consumption still the norm in many parts of the world—the United States, for example. This outdated business model will have to change to meet the future challenges facing the ESI.

The authors explain that present market structures are ill-adapted to the cost and operating characteristics of a decarbonized supply side especially in view of technological innovations in distributed energy resources—they give no useful signals for intermittent sources like wind and solar, and their volatile prices add to investment risk. Meanwhile, consumers are either sheltered from the impact of this volatility or faced with confusing or perverse price signals.

The chapter considers various options for a pricing structure linking supply and demand coherently, including a reconceptualization of electricity into two products: *when available* and *on demand* power. The implications of these developments for electricity business models and investment are examined.

In Chapter 9, **Transactive Energy: Linking Supply and Demand through Price Signals**, **Chris King** defines "transactive" energy as the mechanism(s) that links electricity customers, who consume, produce, and/or store energy, with resources—generation, transmission, and distribution—via price signals that reflect the availability of grid resources in real time.

The authors shows that by responding to prices, customers can lower their costs while increasing grid efficiency and reliability. Consumption, generation, and storage decisions are most efficient when based on wholesale prices, which vary by time and location, especially with the rapid penetration of distributed energy resources and self-generation. The ultimate example of transactive is "prices-to-devices" where real-time prices are automatically delivered to smart devices.

The chapter's main contribution is to describe the various mechanisms for turning this basic concept into reality by using case studies from the United States and abroad that demonstrate the potential of transactive energy in the real world.

In Chapter 10, **Transactive Energy: Interoperable Transactive Retail Tariffs**, **Edward Cazalet** says it is becoming increasingly apparent that existing retail tariffs will be inadequate in a rapidly evolving environment where massive amounts of intermittent generation and potentially equally large amounts of decentralized generation, storage, semi-independent microgrids, and smart energy use devices are joining the existing network, changing both the quantity and the direction of power flows, and impacting wholesale and retail prices in unexpected ways.

The author examines the ramifications of these developments and proposes an elegant framework for addressing the many operational, investment, and economic complexities resulting from the decentralized revolution now sweeping the industry.

The chapter's main contribution, interoperable transactive retail tariffs, addresses the myriad of transactions taking place as consumers become prosumers, as decentralized generation, storage, and microgrids offer new opportunities to coexist and transact in new ways with the existing network.

In Chapter 11, **The Next Evolution of the Distribution Utility**, **Philip Hanser** and **Kai van Horn** suggest that electric distribution utilities (EDUs) must radically redefine their core mission to confront the widespread impact of DERs.

The authors argue that the twofold potential impact of DERs—on their livelihood and on how they interact with wholesale markets—cannot be addressed by mere tariff redesign, only a redefinition of their fundamental purpose and their structure can. No longer will the EDU be a mere conveyor of electricity, but the future EDU will also facilitate transactions among customers and between customers and wholesale markets.

The chapter's main focus is to describe the repurposed EDU's potential alternative business models. Those models vary from a membership model in which the EDU's tariff is proportional to the customer's use mode to an EDU model in which tariffs are proportional to the customer's desired level of reliability.

In Chapter 12, **An Expanded Distribution Utility Business Model: Win-Win, or Win-Maybe?**, **Timothy Brennan** suggests that important policies based on quarantining regulated monopolies from competitive markets are being forgotten. The rationale is neither a lack of entrepreneurship nor difficulties in changing to a business model predicated on reducing electricity use.

Rather, the author argues, utility diversification into competitive services such as distributed generation or energy management creates incentives to discriminate against rivals in those markets, primarily in access to regulated services. It may also lead to cross-subsidization, charging competitive services costs to the regulated sector. These can distort competition in unregulated markets and raise prices of utility services. These ideas justified the breaking-up of AT&T in the 1980s and FERC policies to separate generation ownership from control of transmission.

The chapter also examines economic and legal factors that could mitigate these concerns noting that utilities may be unduly vulnerable to entry from distributed generation because of pricing regimes that allow DG customers to get backup electricity capacity essentially for free.

In Chapter 13, **From Throughput to Access Fees: The Future of Network and Retail Tariffs**, **Tim Nelson**, **Judith McNeill**, and **Paul Simshauser** examine how utilities need to adjust their tariff structures to compete with new forms of energy production and point out that the new tariffs will be a critical determinant of the their success or failure in the future.

The authors assess existing and emerging tariff designs to address the emergence of substitutes such as distributed generation to grid-supplied service. The economic efficiency and equity implications of changing tariff designs are examined based on existing and emerging technologies.

The chapter concludes that utilities must be cautious in rapidly redesigning their business models and rebalancing their tariffs in an effort to recover foregone revenue previously obtained through rising volumetric consumption. Importantly, adjusting tariffs to recover revenues in the short-term may hasten the adoption of energy storage technologies, further undermining the financial stability of utilities in the long term.

In Chapter 14, **Industry Response to Revenue Erosion from Solar PVs**, **Jonathan Blansfield** and **Kevin Jones** examine early examples of industry response to challenges posed by distributed generation, notably from solar PVs, and offer insights on possible approaches to help mitigate cost shift and revenue erosion.

The authors examine case studies of American utilities' initial approaches to address the revenue erosion problem, covering a range of strategies in the solar PV integration and acceptance spectrum including the regulatory policy landscape and the ways selected utilities have successfully or unsuccessfully attempted to adapt to an ever-growing fleet of solar PV within their service territory.

The chapter concludes by offering, through a comparative analysis, strategies likely to be successful in embracing this disruptive technology, such that utilities can maintain sufficient revenue streams by working cooperatively with their customers and state regulators, rather than against the tide of growing solar PV penetration.

In Chapter 15, **Making the Most of the No Load Growth Business Environment**, **Ahmad Faruqui** and **Dian Grueneich** examine the implications of the slowdown in sales growth for the ESI. The chapter begins by reviewing five major reasons for the slowdown: the weak economic recovery from the recession, utility energy efficiency programs, governmental codes and standards, the rise of distributed generation, and fuel switching to natural gas. The slowdown is expected to persist in the indefinite future.

The authors presents different strategies and tactics that utilities can pursue to deal with the slowdown, including a transition to simply become a wires company or provider of distributed generation. The tactics include redesigning electric rates, revamping load forecasting efforts, and reinventing load and market research.

The chapter lays out a decision-making process that utilities can use to pick the winning strategy that comports with their long-term vision of themselves and that is grounded in their core competency.

In Chapter 16, **Regulatory Policies for the Transition to The New Business Paradigm**, **William Miller, Roland Risser**, and **Steve Kline** identify new regulatory principles and pricing options that will allow utilities to move toward sustainable business models. The authors note that, at least in the United States, the industry's historical revenue recovery mechanism can only change as state-level regulators allow it to change. Regulatory oversight differs elsewhere in the world, but in all cases, some authority sets boundaries and determines the ultimate prices customers pay.

The chapter describes traditional regulatory principles underlying the current basis of utility tariffs including the determination of components of fixed and variable costs commonly deployed in determining rates. The authors examine the potential application of these principles as near-term solutions in view of the changes taking place due to the rapid rise of distributed energy resources.

The chapter's main conclusion is a discussion of possible new regulatory principles facilitating the transition to new utility business models.

In Chapter 17, **Electric Vehicles: New Problem or Distributed Energy Asset?**, **Glenn Platt, Phillip Paevere, Andrew Higgins**, and **George Grozev** examine the likely uptake of different types of electric vehicles over time under different scenarios, a prerequisite to a better understanding of the potential impact of EVs on the distribution network. It is not enough to know that, say, 20% of the new vehicle fleet in 2020 will be EVs, but rather an understanding of how many EVs will be in a particular street, on a particular segment of the distribution network, and the specifics of when and how fast they are being charged.

Using empirical data, the authors develop a modeling methodology to predict the impact EVs may have at a particular distribution circuit. The results show that the EV phenomenon—along with the myriad of others examined in this volume—requires a fundamental rethinking of the utility business model, particularly its distribution component.

The chapter's main contribution is a modeling methodology that predicts the local impact of EVs on the distribution system, including an examination of the potential role of "vehicle-to-grid" (V2G) concept, where EVs can actually *benefit* the distribution system by acting as storage medium during periods of excessive distributed self-generation, feeding some of the stored energy back into the grid during peak demand periods.

Part III: What Future? examines alternative views of how these developments will evolve over time and what the industry may look like in the future under different scenarios.

In Chapter 18, **Rethinking the Transmission-Distribution Interface in a Distributed Energy Future**, **Lorenzo Kristov** and **Delphine Hou** show how the proliferation of distribution-connected resources calls into question the prevailing model of power markets, where an ISO or RTO delivers energy to transmission grid takeout points and utility distribution companies (UDCs) move the energy radially from the grid to the end users.

The California Independent System Operator (CAISO) is seeing increasing interest from distribution-level resources to participate in the wholesale markets, while customers exhibit a growing desire to adopt new technologies and self-optimize their energy use. Reliable system operation in this new world requires fundamental rethinking of ISO/RTO and UDC roles around the transmission-distribution interface.

The chapter explores two potential approaches. One involves extending the ISO/RTO operational and market functions to include distribution-level resources. The other redefines the UDC as a distribution system operator (DSO) responsible for real-time balancing at the distribution level.

In Chapter 19, **Decentralized Generation in Australia's National Electricity Market? No Problem**, **Rajat Sood** and **Liam Blanckenberg** examine how participants in Australia's National Electricity Market (NEM) are responding to the twin challenges of low or falling demand growth and rising renewable and distributed generation.

The authors note how the competitive landscape of the NEM has been altered in recent years by technological developments and policy imperatives, including an obligation on retailers to procure renewable energy. These factors have helped increase the pace of vertical integration over the past decade as independent generators and retailers have morphed into *gentailers*.

The chapter's main finding is that participants have responded in a manner to be expected given the energy-only design of the market in an environment of falling demand and rising renewables generation. While Australia has not experienced utility-scale solar plants, there is no fundamental reason to believe these developments would undermine the efficient functioning of the market.

In Chapter 20, **What Future for the Grid Operator?**, **Frank Felder** examines whether and how the fundamental changes that are affecting electric utilities in mature economies will affect organized wholesale electricity markets.

The author examines two different scenarios. In the first, "incremental changes," regional transmission organizations and independent system operators (RTOs/ISOs) continue their present practices and even expand their central role in administering electricity markets, operating the power system, and planning transmission. They develop new wholesale electricity products to address the intermittency of wind and solar resources, extend their dispatch and unit commitment to include retail supply and demand resources, and increase the numbers of their membership. In the second, "decentralization dominates," RTOs/ISOs are relegated to a primarily balancing function of a minimalist grid operator, balancing, when needed, numerous microgrids that infrequently, if at all, rely on a centralized power system.

The chapter's main contribution is an examination of the implications of these opposing scenarios.

In Chapter 21, **Utility 2.0: Maryland's Pilot Design**, **Dustin Thaler** and **John Jimison** report on potential areas in which the utility can experiment, on a pilot basis, to preserve for itself a robust role in the inevitably decentralized future.

The authors describe the Energy Future Coalition's rough sketch for a pilot project, pieces of which are starting to be implemented in Maryland with lessons that are certainly applicable elsewhere. The design, which incorporates input from a number of stakeholders, attempts to move from the abstraction to concreteness by recounting practical steps that can be taken by the utility to successfully transition to the future. These practical steps include exploring performance-based ratemaking, smart grid services, nontraditional utility investments, microgrids, and electric vehicles as storage assets.

The chapter's main contribution is to illustrate how utilities may carve out a greater role and ensure a creative transition toward the future rather than a destructive one.

In Chapter 22, **Turning a Vision to Reality: Boulder's Utility of the Future, Kelly Crandall, Heather Bailey, Yael Gichon** and **Jonathan Koehn** describe the unique attributes of the city of Boulder, Colorado, and how these traits have contributed to a new vision for energy localization.

The authors describe the community's vision, which is to ensure that Boulder's residents, businesses, and institutions have access to reliable energy that is increasingly clean and competitively priced. This underpins a utility that delivers energy as a service rather than a commodity. Boulder is currently engaged in a process to determine the path toward this vision through the formation of a municipal utility or the development of a new partnership with its current electricity provider.

The chapter explains the opportunities inherent in this process and points out that regardless of the selected path, localized renewable energy generation and community-based decision-making will be central to implementing the vision.

In Chapter 23, **Perfect Storm or Perfect Opportunity? Future Scenarios for the Electricity Sector, Jenny Riesz, Magnus Hindsberger, Joel Gilmore**, and **Chris Riedy** use futures thinking to explore possible scenarios that electric utilities may face in the coming decades.

The chapter applies a top-down approach to identify the key drivers that could influence business models. It describes three possible futures. Firstly, the "centralized" future moves toward decarbonization but retains the centralized model present in most power systems today. In contrast, the "decentralized" future moves toward greater decentralization while retaining a significant role for the grid. The "disconnected" future moves to complete decentralization, with most customers disconnecting from the grid entirely.

The chapter concludes that all three scenarios are possible and will have important implications for electric utilities. Wise businesses will adopt a risk management approach.

In Chapter 24, **Revolution, Evolution, or Back to the Future? Lessons from the Electricity Supply Industry's Formative Days, Robert Smith** and **Iain MacGill** begin with the invention of the light bulb by Thomas Edison that gave birth to today's ESI. What is, however, often forgotten is that during its formative years, the future shape and eventual success of the industry were far from certain.

The authors describe the fledgling industry's struggle to find an effective business model, the war between AC and DC currents, central versus distributed generation, patent disputes, and regulatory uncertainties due to rapid technological change. However, history also cautions that current arrangements and business models have been remarkably adaptive for over a century and may well prove more resilient than imagined.

The chapter's main contribution is that the industry is facing a similar set of challenges as it responds to current changes and to expect a bumpy path ahead for some stakeholders.

In closing, **Jon Wellinghoff** puts the insights gained from the preceding chapters in to the book's Epilogue.

What is Changing?

Chapter 1

Decentralized Energy: Is It as Imminent or Serious as Claimed?

Fereidoon P. Sioshansi

ABSTRACT

The recent rise of decentralized energy resources or distributed energy resources (DERs), which has caught the attention of incumbent stakeholders, is the culmination of three drivers: slowing demand growth, high and/or rising retail tariffs, and ambitious renewable targets plus supportive policies favoring decentralized generation. These powerful forces, currently manifested in a number of advanced economies, are likely to spread to other regions over time, pushed by rising retail tariffs, pulled by falling costs of DERs, and propelled by *disruptive* nature of technology. As with any disruptive technology, there will be winners and losers.

Keywords: Distributed generation, Utility business model, Disruptive technology, Renewable energy, Utility strategy

1 INTRODUCTION

Up to now, the rise of decentralized or distributed energy resources (DERs) has been limited to a handful of countries or regions where one or more of the following conditions prevail:

- High and/or rising retail tariffs
- Ambitious renewable targets or mandates
- Supportive policies favoring DERs

One might ask, "where do these conditions currently apply?" The answer includes Germany, Japan, California, and Australia, among others. The speculation is that with passage of time, a growing number of mature economies will follow.

This chapter examines some of the fundamental drivers responsible for the rise of DERs and their implications for the electricity supply industry (ESI). At the outset, it must be noted that the term *distributed* or *decentralized* energy resources or DERs refers to both energy efficiency (EE)—using electricity

Distributed Generation and its Implications for the Utility Industry. http://dx.doi.org/10.1016/B978-0-12-800240-7.00001-1
3

more judiciously and sparingly—and distributed generation or DG. The latter is extensively covered in a prior volume edited by the author;[1] the present volume is primarily focused on DG, broadly defined as generation on customer side of the meter.[2]

Another caveat, reinforced in various chapters of the book, is that at least at this stage, few are prepared to write the eulogy of the incumbent utilities. While some (e.g., owners of thermal generation) may face challenging times in markets with excess capacity and depressed wholesale prices, others (e.g., owners of critical transmission and distribution networks) may in fact thrive in a decentralized future with little or no demand growth once they make the necessary adjustments in how they deliver and—more important—capture value from vital balancing and reliability services provided by the "grid." As with any disruptive technology, there will be winners and losers in this game.

The chapter is organized as follows: Section 2 examines some of the fundamental reasons for the observed slowing demand growth. Section 3 briefly examines the rise of renewables and the push towards low-carbon energy. Section 4 examines the drivers of DER growth. Section 5 questions the seriousness and immediacy of the threats—or opportunities—implied by DERs, followed by the chapter's conclusions.

2   WHAT IS BEHIND THE SLOWING DEMAND GROWTH?

The rise of DERs would not be as noticeable if the underlying demand for electricity was growing at historical rates. In other words, if the total size of the electricity demand pie were growing, the growing DER slice would not be as worrisome. The stagnant growth—recently experienced in a number of countries—makes the growth of renewables and DERs so much more pronounced and painful for incumbent generators while impacting grid operators[3] and owners of transmission and distribution networks.

While there are many reasons and interpretations for the tepid or nonexistent demand growth among mature OECD economies—the reasons vary from place to place and are not necessarily applicable to rapidly growing global economies—the following stand out:

- Structural change in the composition of mature economies
- Demand saturation
- Rising retail tariffs
- Negawatts cheaper than megawatts
- More efficient appliances

---

1. F.P. Sioshansi (Ed.), Energy Efficiency: Towards the end of demand growth, Academic Press, 2013.
2. The exponential growth of utility-scale renewable generation, while critical to the debate, is not the main focus.
3. Refer to Chapter 18 by Kristov and Hou.

- Codes and standards
- Regulatory fiat

The first is the irrefutable structural change taking place imperceptibly among the mature OECD economies. As economies advance, they slowly move away from energy-intensive industries to services, finance, education, innovation, and other knowledge-based, high-value added sectors where relatively little energy is needed to generate large amounts of output—measured in terms of energy intensity or similar metrics.

Figure 1 shows the long-term relationship between economic growth and electricity consumption for the United States for the period 1950 to present with projections to 2040. The basic message—similar trends are observed within most OECD countries—is that mature economies can sustain reasonable economic growth rates with relatively little growth in energy/electricity consumption. As described by Faruqui and Grueneich, there are many who believe that the phenomenon of flat, or possibly declining, electricity demand growth may be permanent.

According to the International Energy Agency, the total energy consumption within OECD in 2012 was the same as in 2002, while GDP of the rich countries over the same period grew by 22%.[4] As illustrated in Figure 2, projections

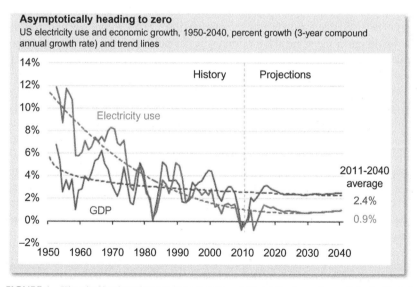

FIGURE 1   Historical/projected trends in economic and electricity demand growth in the United States. *Source: U.S. Energy Information Administration, Annual Energy Outlook 2013 Early Release.*

---

4. Total energy consumption within OECD fell in 4 of the past 5 years.

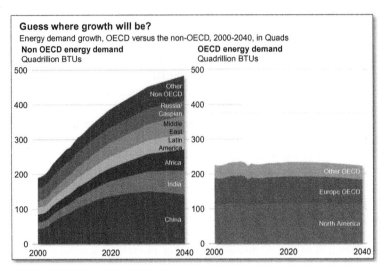

Guess where growth will be?
Energy demand growth, OECD versus the non-OECD, 2000-2040, in Quads

FIGURE 2   Total energy demand is projected to remain virtually flat or slightly declining over time within OECD block (right), while it continues to grow for some time in developing countries (left), in quadrillion BTUs. *Source: The outlook for energy: A view to 2040, ExxonMobil, Dec 2012.*

for total demand growth within OECD are virtually flat, while economic output is projected to grow. It is, of course, a different story for the developing countries. The critical question is how long before the rapidly growing economies of the world reach the same stage of maturity. In case of China, for example, the peak is projected in mid-2020s, according to ExxonMobil's The Outlook for Energy: A View to 2040, published in Dec 2012.[5]

Another explanation for the stagnant demand growth—not just for electricity but also for many goods and services—is the phenomenon of demand saturation. The basic explanation is that consumers in rich countries can afford and already use as much of the basic commodities and services they need and want. This applies to soda consumption in the United States[6] (Figure 3) or pasta consumption in Italy.[7] In the former case, after growing for decades, Americans are now moving away from sweetened fuzzy drinks to bottled water, juices, coffee, and other beverages driven by concerns about obesity and diabetes. In the latter case, Italians appear to have had enough pasta and are consuming more vegetables and other foods.

---

5. Exxon's updated 2014 outlook, released in Dec 2013, reinforces these trends.
6. It is hard to believe, but Mexico's per capita soda consumption is 40% higher than that of the United States, prompting the government to introduce a new tax to discourage consumption due to concerns about obesity and diabetes.
7. According to an article by M. Mesco titled "Italians lose their taste for pasta" in the Wall Street Journal (12 Oct 2013), Italy's per family pasta consumption reached a peak of 40 kilos (88 lbs.) around 2009 and has fallen to 31 kilos (71 lbs.) now—roughly equal to the US level, which is still rising.

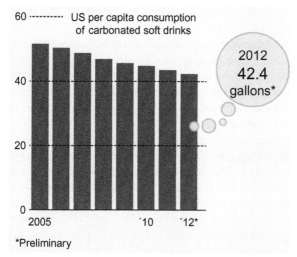

FIGURE 3   U.S. per capita soda consumption, 2005-2012, in gallons/person/year. *Source: The Wall Street Journal, 19 Jan 2013.*

Demand saturation phenomenon, of course, is manifested in numerous forms in many rich countries. Consider, for example, per capita car ownership where the United States, Germany, and Japan have already reached peak levels (Figure 4). In the case of the Untied States, there are now more cars than licensed drivers. In Japan, car ownership is constrained by lack of parking space in urban areas. Moreover, in many congested cities in developing countries, car ownership does not automatically translate into increased mobility or comfort

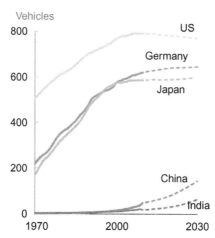

FIGURE 4   Car ownership in selected countries per 1000 residents. *Source: BP Energy Outlook 2030, Jan 2012.*

due to traffic snarl as anyone who has visited Mumbai, Jakarta, Beijing, Hong Kong, or Bangkok can attest.

A variation of the saturation phenomenon is the number of miles traveled by U.S. drivers, which, after growing steadily since mass production of cars in the 1920s, appears to have reached a peak in 2007 at around 3 trillion miles and show signs of dropping. The explanation to this is that Americans already drive as much as they want or need. As the U.S. population ages, fuel prices remain high, roads get more congested, and mass transit opportunities expand, the demand for driving will taper off.

Looking at the top global electricity consumers on per capita basis, it is hard to imagine top-ranked Icelanders[8] consuming more than they currently do even if it were offered at no cost (Figure 5). Ditto for second-ranked Norway, enjoying some of the lowest prices in the world due to the abundance of hydro. Canada's per capita consumption fell between 1993 and 2010, while Sweden and the United States appear to have reached a plateau. Over this period, per capita consumption only grew in developing countries and/or in places where electricity prices are heavily subsidized, such as Kuwait, Qatar, and the UAE.

The speculation is that consumers in many mature OECD countries are approaching or are already beyond demand saturation levels when it comes to electricity consumption. That is to say, when consumers reach high per capita levels of consumption, be it pasta, soda, miles driven, or electricity, there is little or no gain from consuming more. Think of calories and obesity, now reaching epidemic levels in many parts of the world.

Beyond demand saturation are the high and/or rising retail tariffs currently plaguing a number of markets. Consumers confronted with high and rising

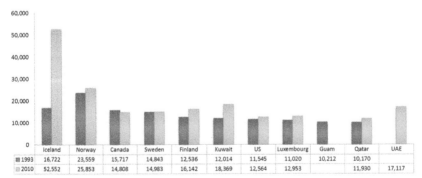

FIGURE 5   Per capita electricity consumption in selected countries, 1993 and 2010, in kilowatt-hours per person/year. *Source: The Wall Street Journal, 12 Nov 2013. http://online.wsj.com/news/articles/SB10001424052702303382004579129873603534290.*

---

8. In the case of a country with small population, such as Iceland, per capita electricity consumption can be overly influenced by the introduction of energy-intensive industries such as smelting or processing.

tariffs tend to cut down on consumption—not by sitting in the dark but rather by using energy more efficiently and sparingly.[9]

Retail tariffs are already high in many parts of the world and continue to rise. The reasons are varied: In Europe, the main driver is high energy taxes and levies to support the growth of renewables. In Australia,[10] it is due to the rapid rise of network fees. In other places such as Japan or Hawaii, virtually all energy is imported and is expensive. In California, it is due to the rising residential tiered pricing,[11] further described in Section 4.

Needless to say, as retail tariffs rise, for whatever reason, EE becomes more compelling.[12] Numerous studies in different parts of the world have concluded that, on the margin, a kilowatt-hour saved is by far cheaper than a kilowatt-hour generated as illustrated in Figure 6 and further described in Chapter 15 by Faruqui and Grueneich.

Moreover, electricity-consuming devices are getting more efficient, propelled by continuously improved technology that allows more energy services for a unit of energy and assisted by higher appliance efficiency standards. Take the case of the ubiquitous refrigerator. As illustrated in Figure 7, the average

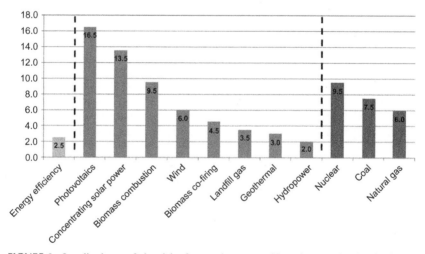

FIGURE 6 Levelized cost of electricity from various renewable and conventional technologies and energy efficiency, in cent/kilowatt-hour. Includes current federal- and state-level incentives; natural gas price is assumed at $4.50/MMBTU. *Source: U.S. Renewable Energy Quarterly Report, ACORE, Oct 2010.*

9. Consumers are interested in lighting, heating, and comfort, not energy per se. Many of these services can be had more economically by investing in efficient delivery of the services, for example, more efficient LEDs for lighting or a better-insulated home for increased comfort.

10. See Chapter 13 by Nelson et al.

11. Will California get its tariffs right? EEnergy Informer, Nov 2013.

12. Refer to *Energy Efficiency: Towards the End of Demand Growth* on this topic.

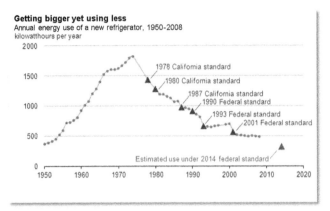

FIGURE 7   Annual electricity consumption of a new refrigerator sold in America, 1950-2014, in kilowatt-hour/year. *Source: U.S. Department of Energy, Office of Energy Efficiency and Renewable Energy, Building Technologies Office.*

fridge sold in the United States in 2014 will consume about the same number of kilowatt-hours as in 1950 despite the fact that today's units are, on average, three times as big and come with multitudes of advanced features. The same applies to virtually all electricity-consuming appliances, lighting devices, and other electronic gadgets.

Aside from appliance efficiency standards and programs funded by consumers, EE can be encouraged through regulatory intervention such as building codes. Regulation and policy *does* matter especially when applied persistently and comprehensively over time as is apparent by divergence of per capita electricity consumption in California versus America as a whole (Figure 8). Truly, there are many reasons behind the divergence of trends,[13] but most experts attribute part of California's success to stringent building codes, appliance EE standards, and consumer-funded EE programs.[14]

A case in point may be the adoption of zero net energy (ZNE) code applicable to *new* residential buildings in California starting in 2020 and to *new* commercial buildings starting in 2030.[15] The only way for a ZNE-like concept to be practical is to reduce on-site consumption to the point where limited on-site generation can be sufficient to meet overall demand, the definition of ZNE (Figure 9). Policy makers in other parts of the world, notably in Europe, are also focusing on similar schemes using terms such as passive buildings. New single-family residential

---

13. What's behind Rosenfeld's curve? EEnergy Informer, Oct 2013.
14. Not everyone, of course, favors more regulations and mandatory codes and standards, but when supported by rigorous cost-benefit analysis, a case can be made for standards where the end justifies the means.
15. Refer to chapters in Energy Efficiency: Towards end of demand growth, F. Sioshansi (ed), Academic Press, 2013.

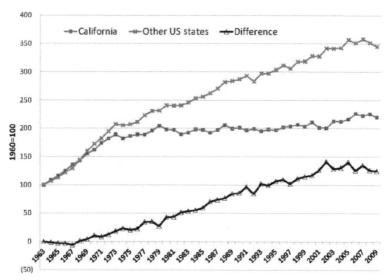

FIGURE 8   Per capita electricity consumption in California compared to other states, 1963-2010, 1960 = 100. *Source: Lucas Davis, University of California, Berkeley blog posted on 5 Aug 2013. http://energyathaas.wordpress.com/2013/08/05/deconstructing-the-rosenfeld-curve/*

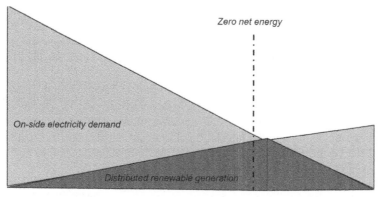

FIGURE 9   As buildings use less and generate more over time, they eventually reach zero net energy, where distributed generation equals or exceeds on-site consumption.

buildings built in Japan in post-Fukushima era are super efficient, but 80% come with solar PVs and 50% reportedly have fuel cells.[16]

The net result of the phenomena outlined earlier can be seen in flat or declining electricity demand in countries as far apart and different as Australia, New Zealand, and the United Kingdom (Figure 10) or in parts of the United States.

---

16. P. Landers & M. Negishi, in post-tsunami Japan, homeowners pull away from grid, Wall Street Journal, 18 Sept 2013.

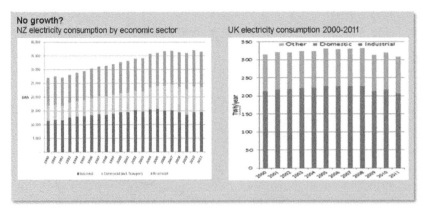

FIGURE 10 Flattening of electricity demand in New Zealand, 1990-2011 in gigawatt-hours (left), and the United Kingdom, 2000-2011 in terawatt-hours (right). *Source: for NZ (left), energy data file from the Ministry of Business Innovation and Economy (MBIE) at http://www.med.govt.nz/sec tors-indsutries/energy/pdf-docs-library/energy-data-and-modelling/publications/energy-data-file/ energydatafile-2011.pdf; for the United Kingdom, UK Department of Energy & Climate Change.*

In the case of Australia, the Australian Energy Market Operator reports declining electricity consumption in the populous eastern part of the country (Figure 11). There is an ongoing debate if this is permanent or temporary.[17] Similar developments are being observed in a number of other mature, slow-growth economies. In October 2013, the market operator in New England, ISO New England, projected an essentially flat demand curve for the region for the next decade.[18]

Moving forward, it is not *inconceivable* to project flat or possibly declining electricity demand in mature economies. In fact, such an outcome appears *inevitable* for all the reasons mentioned. Moreover, future demand growth *can* be further curtailed through the imposition of more stringent codes and standards plus EE investments. Pursuing such a flat or declining demand strategy not only is possible but also, by most accounts, will be cost-effective.[19]

In the case of the United States, currently the biggest electricity-consuming nation (Figure 12), future demand can be managed depending on which path is pursued. Under a business-as-usual scenario, U.S. electricity consumption is projected to grow at a modest rate over time[20] (top line in Figure 13). But the

---

17. Refer to Robert Smith, Crouching demand, hidden peaks: What's driving electricity consumption in Sydney? in Energy efficiency: Towards the end of demand growth in F. Sioshansi (Ed.) Academic Press, 2013.

18. End Of Demand Growth: Coming To A Theatre Near You, EEnergy Informer, Dec 2013.

19. Amory Lovins makes a strong case for the cost-effectiveness of energy efficiency in *Reinventing Fire*, 2012., where he claims that the United States can drastically cut the use of fossil fuels by converting to a more efficient and mostly renewable future at considerable cost saving.

20. IEE white paper, May 2011.

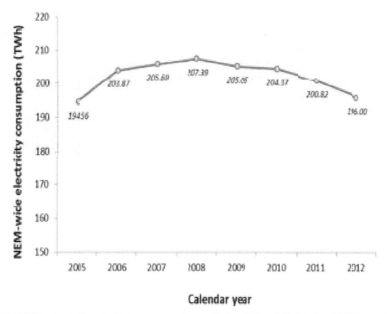

FIGURE 11   Australian electricity consumption on National Electricity Market (NEM) covering the eastern interconnected network, 2005-2012, in terawatt-hours. *Source: AEMO.*

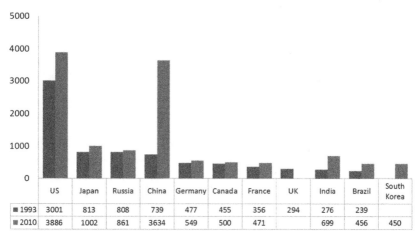

FIGURE 12   Top global electricity-consuming countries, 1993 and 2010, in billion kilowatt-hours. *Source: The Wall Street Journal, 12 Nov 2013.*

future is not preordained. Demand growth can be modified by applying more stringent codes and standards and by increasing EE investments—decisions that can be made by policy makers at state and/or federal level. Figure 13 shows two alternative scenarios, both technically feasible and cost-effective in the sense that net consumer costs would be lower than the business-as-usual case.

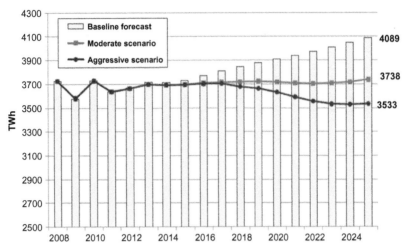

FIGURE 13 Baseline projection of U.S. electricity demand growth with alternative scenarios based on moderate or aggressive energy efficiency standards, 2008-2025, in terawatt-hours. *Source: IEE white paper, May 2011.*

The preceding discussion suggests that

- electricity demand growth among mature global economies will be anemic at best and
- even this anemic growth can be modified further downward with supportive policies, which, by all measure, will be cost-justified.

For the ESI, a future with little or no demand growth is unprecedented and painful given the industry's traditional dependence on continued growth, investment, and rising revenues.

3   RISE OF RENEWABLES

A second recurring theme in this volume is the growth of renewable generation,[21] assisted by subsidies and/or mandatory targets in a growing number of countries. A number of countries have already set ambitious targets to move toward a low-carbon energy future (Table 1). As time goes on, technology improves, and prices continue to fall, more countries can be expected to follow. More important, renewables are no longer limited to developed economies; both India and China[22] are aggressively expanding renewable generation domestically while targeting the all-important export market for wind turbines and solar PV modules.

---

21. In this volume, renewables usually refer to large-scale, mandatory, or financially assisted technologies such as wind, solar, geothermal, biomass, and small hydro. Small-scale self-generation such as rooftop solar, fuel cells, and solar water heaters is referred to as self-generation or DG, usually on the customer side of the meter.

22. By 2035, renewable generation in China is expected to exceed those of the United States, EU, and Japan combined, according to the World Energy Outlook 2013, the International Energy Agency, Nov 2013.

TABLE 1 Renewable Targets for Selected Countries

Country	Current	2020 Renewable Target % of Total Generation	2050 Renewable Target % of Total Generation
Denmark	22.2	50	100
Sweden	47.9	49	100
Spain	13.8	20	100
Germany	11	35	85
California[a]	11.6	33	80
Australia	9.6	22.5	85
UK	7.4	30	No target

[a]*California is not a country, yet, but it has the world's tenth largest economy if it were.*
Source: New Scientist, 22 June 2013.

The drivers vary from place to place but usually include concerns about climate change, interest to increase reliance on energy sources whose price does not fluctuate with fossil fuel prices, local or national self-reliance and/or energy-security issues plus the promise of jobs, and export potential that comes from technological leadership and investment in renewable generation.

In the United States, the prime driver of renewable growth is *Renewable Portfolio Standards* or RPS[23] (Figure 14). Different countries apply different schemes for supporting renewables. Europeans, for example, have historically relied on feed-in-tariffs (FiTs), while Australia has established a renewable energy target.[24]

The RPS schemes are driving the renewable demand in the United States. In the past few years, renewables account for a growing percent of all *new* capacities coming online (Figure 15). This is even more amazing given the plentiful supply of natural gas at historically low prices.[25] The rise of renewables and the abundance of natural gas combined with sluggish demand and environmental restrictions have resulted in depressed wholesale prices in parts of America and early retirements in coal-fired generation.

A similar, except more pronounced, outcome is apparent in Germany under its *Energiewende*, or *energy turnaround*, which includes phasing out all remaining nuclear plants by 2022 and replacing them, mostly, with renewable generation.

---

23. For the latest data, refer to Database of State Incentives for Renewable Energy (DSIRE) at http://www.dsireusa.org/documents/summarymaps/RPS_map.pdf.
24. Australia's RET may be scaled back under its current government as described in chapters by Mountain and Szuster and Sood and Blanckenberg.
25. According to the data released by the Federal Energy Regulatory Commission (FERC), 99.3% of new capacity added in the month of October 2013 in the United States was from renewable resources.

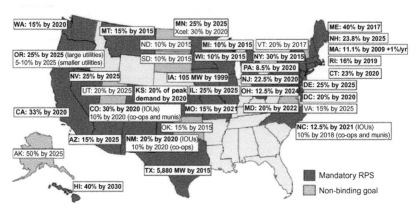

FIGURE 14   U.S. Renewable Portfolio Standards in effect. *Source: 2011 wind technologies market report. DOE.*

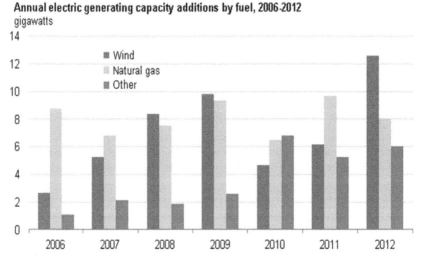

FIGURE 15   U.S. new capacity additions, 2006-2012 in gigawatt. *Source: Energy Information Administration, Annual Electric Generator Report, Form EIA-860.*

The result has been a literal flood of renewable capacity (Figure 16), depressing wholesale prices ironically accompanied by significant rise in retail tariffs.

The predictable outcome, as further described in Chapter 3 by Burger and Weinmann, is likely to lead to excess capacity, albeit not of the dependable or dispatchable kind.[26] According to one estimate, Germany may end up with

---

26. The rapid rise of renewables poses a number of other challenges to grid operators, including integration of so much intermittent generation into the network further described in the chapters by Sood and Blanckenberg and Kristov and Hou.

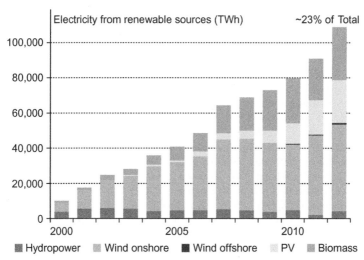

FIGURE 16   Germany: renewables flood. *Source: Macroeconomics of German Energiewende, Prof. Georg Erdmann presentation at IAEE Conf. In Dusseldorf, Germany, 20 Aug 2013.*

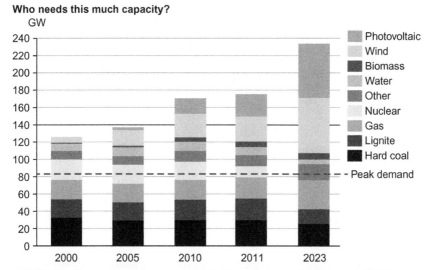

FIGURE 17   German generation capacity by fuel/technology type for the period 2000-2023, in gigawatt. *Source: Bundesnetzagentur.*

as much as 240 GW of capacity by 2023, while peak demand is projected to remain more or less stagnant, around 80 GW (Figure 17). While these projections are subject to change based on recent government efforts to address the overcapacity and unsustainable wholesale prices, the fundamental issues are likely to persist (Box 1).

Box 1 [27] How to Lose Half a Trillion Euros

An article in the 12 Oct 2013 issue of *The Economist* asks how to lose half a trillion Euros. The article describes a breezy, sunny Sunday when renewable generation in Germany topped 50% when demand was low. Between 2pm and 3pm on 16 June 2013, renewable capacity feeding the grid reached 28.9 GW. Despite the network operators' best attempts to turn down all flexible thermal units to a level as low as they could go, they were faced with 51 GW of capacity and barely 45 GW of demand. To keep the network from collapsing, wholesale prices plunged to minus €100/mwh— paying customers, traders, or whoever a hefty reward for taking the unwanted capacity.

Had there been millions of electric vehicles or other storage devices, the excess capacity could have been easily absorbed, but today's networks in Germany and elsewhere are not yet ready for such episodes. And such episodes are becoming more common, not only in Germany but also in Texas, in California, and in a number of other places where renewables are a growing share of total generation.

The reverse problem, when renewable production falls off suddenly and/or unpredictably, such as when prevailing winds die unexpectedly or when cloud cover reduces solar output, requires thermal generators to quickly make up the difference. This is becoming an issue since many thermal plants are not dispatched frequently enough and/or for long enough hours to remain viable, especially when wholesale prices remain depressed, as they have been in Europe, the United States, and Australia, to name a few.

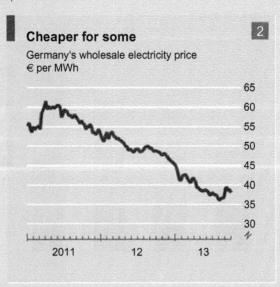

Cheaper for some

Germany's wholesale electricity price
€ per MWh

Source: The Economist, 12 Oct 2013. http://www.economist.com/news/briefing/21587782-europes-electricity-providers-face-existential-threat-how-lose-half-trillion-euros.

---

27. Excerpted from Dec 2013 issue of *EEnergy Informer*.

> **Box 1 How to Lose Half a Trillion Euros—cont'd**
>
> The Economist reckons that Europe's top 20 utilities have lost roughly half their market value, around half a trillion Euros, since 2008. Declining wholesale prices, notably in Germany, get most of the blame. As the share of renewables grows to 35% by 2020 and 80% by 2050, the problem will only get worse. Germany's renewables accounted for 23.5% of the generation in 2012, up from 15% in 2007.
>
> The rapid growth of solar PVs, now at or approaching grid parity, is adding to utilities' woes. As a growing number of consumers become *prosumers,* they will contribute little to utility coffers while relying on the critical services provided by the grid to balance their variable consumption and DG. As The Economist notes,
>
> *In such a world, the old fashioned utilities play two vital roles. They will be the electricity generator of last resort, ensuring the lights say on when wind and solar generators run out of puff. And they will be providers of investment to help build the grand new grid. It is not clear that utilities are in good enough shape to do either of these things.*

Regardless of the specifics, the continued growth of renewables is likely to result in

- excess capacity in a number of key markets, accompanied by depressed wholesale prices;
- ironically higher retail prices due to the various subsidies and support schemes; and
- new challenges for grid and network operators who must maintain system reliability.

4   THE RISE OF DERs

As already mentioned, decentralized energy resources or DERs is a two-sided coin: one side is EE, the other decentralized or DG. While the book is mostly focused on the latter, the former plays a significant role. The combination of the two is likely to become more pronounced with passage of time: consumers using less while generating more—the definition of ZNE as illustrated in Figure 9.

Section 2 described a number of powerful forces encouraging consumers to use *less.* The two most important drivers, of course, are rising retail tariffs and improved efficiency of electricity-consuming devices. They work like hand and glove. Consumers in high-retail-tariff regions, such as in Europe, Australia, or California[28] (Table 2), increasingly find investment in EE to be cost-justified. Modern light-emitting diodes (LEDs), while still expensive, are becoming

---

28. For the latest details, consult utility websites under residential tiered tariffs.

TABLE 2 California Tiered Residential Electricity Prices Rise with Increased Consumption, in cents/kilowatt-hour

Tier	Volume of Use	PG&E	SCE	SDG&E[a]
Tier 1	Within baseline	13	13	14
Tier 2	101-130%	15	16	16
Tier 3	131-200%	30	24	24
Tier 4	201-300	34	28	31
Tier 5	>300%[b]	34	31	NA

[a]SDG&E has slightly different rates for summer and winter, making it more complicated for consumers.
[b]PG&E shows five tiers but the price for the top two tiers is the same.
Source: Utility websites, updated periodically.

popular due to their frugal electricity consumption and long life. As consumers demand more LEDs, costs decline due to mass manufacturing. Lighting accounts for roughly a quarter of electricity consumption within most buildings; hence, efficient lighting combined with natural lighting can make a big dent in consumption.

Using the latest technologies and designs, overall energy consumption of modern buildings can be easily reduced by 85% or more with a marginal increase in up-front investment.[29] In high-retail-tariff areas, the increased investment can be recovered in the first few years of the building's long life. A growing number of new buildings incorporate ultra EE features in their design such as better lighting and higher insolation. Many upscale new homes come with solar roofs as standard equipment.[30] It does not take much for an ultra-efficient building to go to the next step by becoming ZNE.

The EE side of the DER coin is, however, old news. What is new and potentially more powerful is the other side of the coin, namely, the compelling economics of DG. As the cost of DG technologies continues to fall and their performance and reliability improve, the day of reckoning—grid parity—may not be far-off.

The typical scenario is depicted in Figure 18: rising retail tariffs accompanied by continuously falling costs of DG—in this case, electricity generated from solar PVs in Germany. The argument goes that as long as the two trends continue to move in the directions they are headed—retail tariffs *rising* and DG costs *falling*—there is bound to be a crossover at some point. In Denmark, Germany, Italy, Japan, Australia, Hawaii, and California—to name a few—the crossover is near if not already here. Other parts of the world will reach

---

29. Another new normal: Eco housing moving mainstream, EEnergy Informer, June 2013.
30. The Wall Street Journal, 3 May 2013.

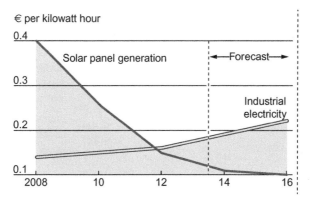

FIGURE 18   Cost of grid-supplied electricity and electricity supplied from solar PVs in Germany, 2008-2016, in €/kilowatt-hour. *Source: Renewables: A rising power, Financial Times, 8 Aug 2013 based on data from IHS Solar Demand Tracker.*

the crossover at some point in the future, some sooner than others, especially if the costs of environmental externalities are included and energy subsidies are phased out.

A study by Navigant Research concluded that solar grid parity is imminent and is likely to apply everywhere except for only a handful of low-retail-price areas in the United States before the end of the decade.[31] Another study[32] shows the potential impact of solar grid parity across the United States based on the current PV costs and retail rates (Figure 19). In the state of Hawaii, where average retail rate is around 40 cents/kWh, the economics of solar—and for that matter all kinds of renewables from wind to geothermal to biomass—are compelling. The surprising thing is why has DG not caught on more than it already has.[33]

The growth of PVs has been short of miraculous in a number of countries, including sunny Australia, which has experienced a gravity-defying rise since 2009, propelled by overgenerous FiTs.[34] According to the Clean Energy Council, Australia passed the 3 GW solar PV milestone in late 2013.[35]

Solar PVs, already exceeding 35 GW in Germany, are expected to grow globally for the foreseeable future (Figure 20). Other forms of DG, including solar hot water heaters, CHP, fuel cells, ground-source heat pumps, and other devices, are expected to continue to drop in price while improving in performance. The bottom line is that consumers in high-retail areas are likely to *consume less* while *generating more*.

31. Solar PV market forecast, 3rd Qtr. 2013, Navigant Research.
32. John Farrell, Renewable Energy World.com, 9 July 2013.
33. Hawaii has recently adopted a 40% renewable target for 2030.
34. These have been scaled back recently, as described in Chapter 4 by Mountain and Szuster.
35. Clean Energy Council press release dated 5 Dec 2013.

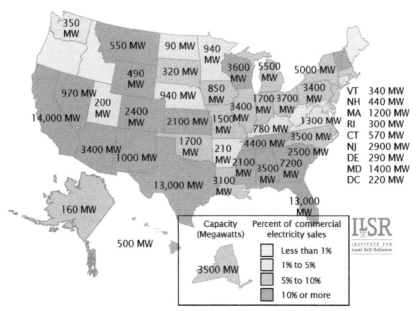

FIGURE 19   Projected potential for solar PV installations in the United States based on prevailing retail tariffs, solar insulation, and other variables. *Source: Commercial Rooftop Revolution, Institute for Local Self-Reliance (ILSR), Dec 2012.*

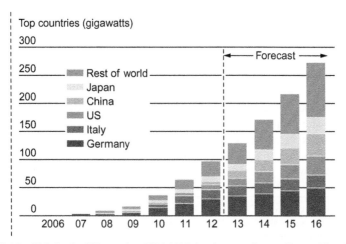

FIGURE 20   Global solar PV capacity, 2006-2016, in gigawatt. *Source: Renewables: A rising power, Financial Times, 8 Aug 2013 based on data from IHS Solar Demand Tracker.*

Over time, this suggests that a growing number of consumers will become *prosumers*—a term that refers to the fact that they will produce some, if not most, of what they consume. This, as explained in the following section, has significant implications for the ESI.

5   HOW SERIOUS A THREAT—OR OPPORTUNITY?

The preceding discussion suggests that the ESI may be approaching a critical *tipping point* where consumers, for the first time in history, may be able to generate electricity at costs that may be cheaper than the grid-supplied variety. Such a statement, of course, must be prefaced by a couple of important caveats:

- First, such an outcome is currently applicable only to a number of places facing high retail tariff—that is, is not universally true.
- Second, the necessity to remain connected to the grid and the services provided by the upstream infrastructure will likely remain critical for the foreseeable future[36]—some would say it becomes far more valuable[37]— even for *prosumers* whose net consumption may negligible. In other words, even assuming grid parity, the great majority of consumers would have compelling reasons to remain dependent on the network.

This suggests that *prosumers* and the ESI are presently intertwined in a bitter love-hate relationship. At least until storage, fuel cell technology and micro-grids develop to the point where *prosumers* no longer need the reliability provided by the grid; the two will have to find a way to coexist. Neither can live without the other. This, ironically, may entail higher connection or fixed fees for prosumers to pay for the upkeep and upgrading of the grid as suggested in a number of chapters, including Chapter 7.

The first problem, further described in the chapters by Nelson et al., Keay et al., and others, is that as their numbers grow, prosumers will have a detrimental impact on ESI revenues, which are primarily, if not exclusively, recovered through volumetric tariffs today. As/if volumetric consumption flattens or potentially falls, it is bad news for virtually all stakeholders upstream of the meter since their costs are mostly fixed.

The second problem, further described by Gellings and others, is that prosumers continue to rely on a host of critical services provided by the grid—the transmission and distribution network and thermal generators that will increasingly be needed to provide balancing service—while contributing relatively little to their upkeep under the prevailing flat tariffs.

---

36. Refer to Chapter 21 by Thaler and Jimison.
37. See, for example, Chapter 5 by Gellings.

The rise of DREs, in other words, leads to a growing *missing money* problem. The ESI's costs are predominantly fixed and growing, while the prevailing tariffs are mostly collected on volume, which may be falling.

Referring to a growing number of virtually ZNE communities, such as the University of California, Davis West Village, Jesse Berst of SmartGridNews observed,

With projects like these ... we can expect more and more net zero subdivisions, office parks, industrial parks, campuses, etc. Does your utility have a plan in place for a world where new neighborhoods mean more costs – new wires, poles, transformers, etc. – without more sales?

Blog posted on 5 Dec 2013, www.smartgridnews.com

How big is the missing money problem? Estimates vary, and there are many arguments about how to measure the gains and the losses in trying to come up with the net costs or the true revenue loss.[38]

It is fair to say that the scale of the missing money problem is significant and is likely to grow until a reasonable and equitable solution is reached. King, Cazalet, Hanser and van Horn, Keay et al., Miller et al., and Brennan, for example, offer possible solutions, as do a number of other contributors in this volume.

As far as the longer-term potential impact of the threats—and opportunities—presented by the rise of decentralized energy revolution is concerned, opinions also vary and this is reflected in the chapters that follow. A slim but widely quoted report, *Disruptive challenges: Financial implications and strategic responses to a changing retail electricity business*, published by the Edison Electric Institute in Jan 2013, warned of grave consequences,[39] stating in part:

Recent technological and economic changes are expected to challenge and transform the electric utility industry. These changes (or "disruptive challenges") arise due to a convergence of factors, including: falling costs of distributed generation and other distributed energy resources (DER); an enhanced focus on development of new DER technologies; increasing customer, regulatory, and political interest in demand-side management technologies (DSM); government programs to incentivize selected technologies; the declining price of natural gas; slowing economic growth trends; and rising electricity prices in certain areas of the country.

Similarly, a 2013 report by Eurelectric, the European association of the electricity industry, presented a rather grim prognosis for the incumbent stakeholders with the sobering message that the time has arrived for the ESI to "evolve or die" (Box 2).

---

38. For an example, refer to California NEM draft cost-effectiveness evaluation, prepared for CPUC by E3, 26 Sept 2013, reported in Net Metering: Guilty as charged, EEnergy Informer, Nov 2013.

39. Disruptive challenges: Financial implications and strategic response to a changing retail electric business, Edison Electric Institute, Jan 2013Ref 2013 EE report.

Box 2 [40]Industry Prognosis: Evolve or Die

A report published by Eurelectric in May 2013 starts by acknowledging that "It is a tough time to be an electricity utility in Europe," adding, "The EU power sector is going through one of the *most profound changes in its history*" (emphasis added). The punch line, buried much later in the report, is that electric utilities may be facing a stark new reality: they must *evolve* or *die*.

What is causing such dire prognosis? Eurelectric spells out 4 major trends that it claims are reshaping the power sector:

- Rapid rise of renewables
- Equally rapid rise of DG
- Evolution of the smart grid
- Emergence of retail competition and new services

The report points out that "Since 2000, renewable energy sources have accounted for the large majority of new capacity additions in Europe. Many nations have already reached high renewable shares, including 31% of electricity in Denmark, 22% in Germany, and 21% in Spain and Portugal combined."

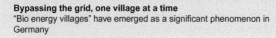

Bypassing the grid, one village at a time
"Bio energy villages" have emerged as a significant phenomenon in Germany

Number of bio-villages in Germany

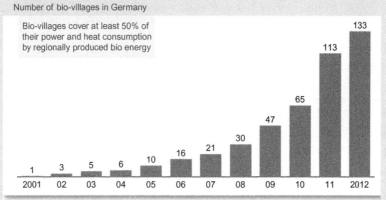

Source: Utilities: Powerhouses of innovation, Eurelectric, May 2013.

It adds, "The variable nature of wind and solar energy is putting new pressures on the operation of conventional generators and electricity systems. These all present new cost drivers for owners of conventional generation assets, even as declining wholesale power prices are reducing operating revenues"—the current reality in Germany.

The report also notes that "While much of the recent growth in renewable energy capacity consists of large, centralized wind farms, the European power sector is also moving in parallel towards a more decentralized system of electricity generation.
Continued

---

40. Excerpted from Oct 2013 issue of EEnergy Informer.

Box 2 Industry Prognosis: Evolve or Die—cont'd

Small-scale power producers are giving rise to new ownership structures and business models, some of which directly compete with conventional utilities," adding that more than 3 million households have started producing their own electricity with solar PV, while 133 "bio-villages" have emerged in Germany since 2000, generating more than 50% of their electricity and heat from bioenergy resources (accompanying graph).

The third driver of change is the European Commission's third Electricity Directive, which requires member states to equip at least 80% of consumers with intelligent metering systems by 2020 and notes that roughly 50 million smart meters have already been installed across Europe as of the end of 2012 with European utilities taking steps towards a more intelligent and controllable distribution grid.

The final driver is the emergence of competitive retail electricity markets in EU, offering customers a proactive role in managing their power consumption. According to the report, more than 510 million customers in Europe reside in markets with liberalized supply and retail markets,[41] and it notes that in 2011, some 6% of households exercised their new options by switching to a different supplier.

Eurelectric says that the confluence of these factors will "erode traditional utility market shares" and send profits into a "free fall." Moreover, while these issues may currently be confined to European markets, similar trends "may simply preface broader revolutionary forces under way across much of the global electric power sector."

Together, these "profound changes" are hitting European utilities right where it hurts: the traditional focus of their core business and historical source of the majority of their profits, "conventional generation assets."

What is striking about the Eurelectric report is that this is not a question of a "what-if" in a distant future but one that is actually happening here and now. The market value of Germany's two biggest utilities, E.ON SE and RWE AG, has fallen to 76% since 2008. E.ON's earnings for the first half of 2013 were nearly half as much as the corresponding period in 2012, while profits are down to 40%, according to Bloomberg.

Ominous trends
The cost of renewable technologies is expected to decrease, by as much as 60% for solar PVs by 2020

EUR/MWh[1]

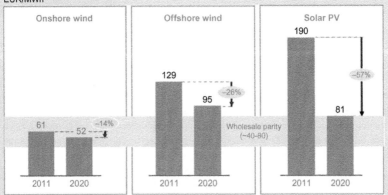

1 Cost refers to the levelised cost of energy attainable using leading technology in favourable conditions. Assumptions include (2011, 2020): Onshore wind capex in EUR/kW (1188, 1108), load factor (36%, 39%), WACC 9%, Offshore wind capex in EUR/kW (3772, 2830), load factor (51%, 54%), WACC 10%; Solar PV capex in EUR/kW (2162, 927), load factor (16%, 16%), WACC 7%.

Source: Utilities: Powerhouses of innovation, Eurelectric, May 2013.

---

41. The corresponding figure for the United States is 110 million and roughly 20 million for Australia and New Zealand combined.

Box 2 Industry Prognosis: Evolve or Die—cont'd

"A significant part of our business model is now facing new challenges," RWE Chief Financial Officer Bernhard Guenther recently told Bloomberg. "Whatever we do in terms of cost and Capex-cutting won't fully compensate the profit loss we see in conventional power generation."

Eurelectric, like others who have examined the industry's plight, blames the low demand growth, relentless gains in EE, and the continued growth of renewable energy and customer self-generation among the main culprits leading to depressed wholesale prices (accompanying graph).

These forces may spell the end of the traditional source of utility profits in Europe. "Even with possible changes such as the introduction of capacity payments, higher commodity prices, or large-scale industry consolidation, the 2020 profit pool appears unlikely to grow larger than that seen in 2012," concludes Eurelectric.

The sober conclusion is that

"With market shares and profitability of conventional centralized generation assets declining, electricity demand growth stagnating, and emerging technologies enabling new ways to meet consumer demands, electric utilities may be facing a stark new reality: they must evolve or die." (emphasis added)

Having delivered the bad news, Eurelectric switches to positive, declaring that utilities must now become "powerhouses of innovation," the title of the report.

How can this be good for business?
The home of the future may feature smart meters, micro-generation, and a host of new services and appliances

The home of the future: example of a new single house

Decrease in energy needed from the grid by the HotF 2020, indexed, 100 = kWh of household consumption

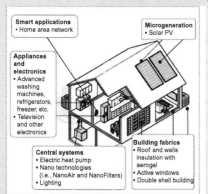

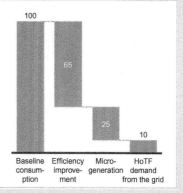

Source: Utilities: Powerhouses of innovation, Eurelectric, May 2013.

Even though the prognosis is grim, Eurelectric's report has an upbeat title and does its best to portray that a turnaround can, somehow, be achieved. But for anyone who knows the industry, this may be wishful thinking. The ESI is not known for taking risks, for being innovative, or for thinking outside the box.

That explains why David Crane, the CEO of NRG famously, called the U.S. utility industry "an industry of Neanderthals," adding that it has "been the least

Continued

Box 2 Industry Prognosis: Evolve or Die—cont'd

innovative industry in America, maybe the world." Crane certainly does not have much hope for industry innovation from *within* or an imminent turnaround.

The problem is not unique to the ESI. In nearly all industries, most of the innovation often comes from *outside* the large established businesses. Tesla's successful electric car, for example, was not developed within GM, MBZ, Fiat, or Toyota.

The Eurelectric report highlights how the costs of alternatives to traditional grid-supplied generation such as solar PVs, renewables, storage batteries, microgrids, and a host of other disruptive technologies are expected to continue to decline over time (accompanying graph). Simultaneously, homes, offices, appliances, lights, air conditioners, computers, and everything else that uses electricity are becoming more efficient, thus reducing the demand. It is a real puzzle how Eurelectric, presenting such sobering evidence, can still come up with an upbeat assessment of a resurrected ESI.

How's this blueprint going to save the industry?

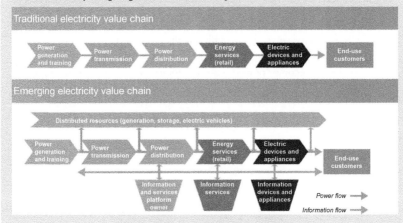

Source: Utilities: Powerhouses of innovation, Eurelectric, May 2013.

The accompanying graph, for example, shows that home of the future—whatever that means—can be expected to need a mere 10% of the energy conventional homes receive from the grid today. That means 90% *fewer* kilowatt-hours needed from the grid. How can the incumbents respond to such an eventuality? No wonder 60% of the respondents to a survey, mentioned in the Eurelectric report, said they expect a negative impact on the "conventional utilities."

Setting these inconsistencies aside, the report suggests a different value chain for the ESI (accompanying graph). Looking at the schematic presented, however, the future looks identical to today's except that distributed resources are added on top and services are somehow enhanced by information and/or added intelligence on the bottom.

While Eurelectric is focused on the plight of European utilities, similar trends are impacting the ESI in other mature OECD economies, including the United States, Australia, and Japan, to name a few. The drivers are the same, growing renewables and increasing efficiency and rapid uptake of self-generation, all leading to stagnant or declining electricity demand.

Major utilities in affected areas—for example, Germany, California,[42] and Australia—are already taking steps in response to the challenges of the evolving market as described in the chapter by Groot. RWE, Germany's second largest utility, has announced a new business strategy (Box 3), while E.ON, the biggest utility in Germany, is acknowledging the need for radical departure from the status quo.

Box 3 [43]RWE: From Volume to Value

The past couple of years have been harsh on utilities in Europe, notably in Germany. With wholesale prices depressed due to the rapid proliferation of renewable capacity flooding the market and rising retail tariffs due to a variety of taxes and levies to pay for them, the Big 4, E.ON, RWE, EnBW, and Vattenfall, have suffered drops in their stock prices and profitability. Under the prevailing conditions, the future appears bleak and the business as usual is not a sustainable strategy.

Following discussions at the company's board meeting in September 2013, RWE announced in October 2013 that it was changing its business strategy from large-scale thermal power production to become an *enabler* in what remains as the only promising growth segments within the beleaguered industry. Henceforth, RWE will focus on creating *value* rather than *volume*.

Not good for investors
Share performance of Germany's two largest utilities since 2010 speaks volumes about the unsustainability of the traditional utility business model

Source: The Wall Street Journal, 9 Sept 2013 based on data from FactSet.

RWE's new strategic roadmap says the company, which traditionally relied heavily on coal-fired and nuclear power generation, will undergo a radical change to

Continued

---

42. California legislature passed a bill in late 2013 authorizing the state's regulatory agency additional flexibility to modify residential tariffs and adjust the net energy metering laws mostly in recognition of the changing dynamics of PVs and their revenue impact on incumbent utilities.
43. Excerpted from EEnergy Informer Dec 2013.

Box 3 RWE: From Volume to Value—cont'd

survive under Germany's *Energiewende* or energy turnaround and EU's energy policies. "The massive erosion of wholesale prices caused by the growth of German photovoltaics constitutes a serious problem for RWE which may even threaten the company's survival," according to the strategic roadmap.

Moreover, RWE has decided not to play a leading role in the new growth sector of decentralized, subsidized power production, which it sees as "the only growth segment in the European power generation market" for the foreseeable future. "In a low-interest environment, it will not be possible for RWE to generate sufficient return within this subsidized industry. Our cost of capital will not be competitive against funding from private and institutional equity investors," it said.

What will RWE do then? The company will leverage its skills and resources to become an *enabler*, *operator*, and *system integrator* in the growing renewable energy sector while pursuing a capital-light approach largely relying on third parties for financing.

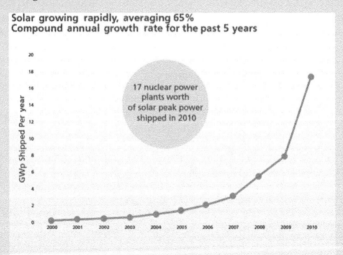

Source: PV industry growth, Navigant Research, 2013.

Noting that the retail European energy sector is "about to undergo a massive transformation in the coming years," RWE sees its 25 million customer base as valuable cash cows, with plans to retain and expand by "applying energy supply capabilities and information technologies intelligently." The company intends to do so through

- *customer centricity*, which "will become a critical capability which we have not sufficiently addressed yet";
- "developing an innovative and profitable prosumer business model ... alongside the traditional value chain"; and
- "develop a culture, structures and processes that allow us to develop new business models, which go beyond incremental improvement of the existing value chain."

Box 3 RWE: From Volume to Value—cont'd

RWE says, "In a highly uncertain and volatile business environment, we need to transform ourselves from a 'risk mitigation' to an 'uncertainty management' company." Additionally, RWE plans to rely on two other businesses areas in which it already excels: energy trading and its indispensible transmission and distribution network.

As for RWE's conventional generation, the company says that demand for *reliable* thermal capacity—in contrast to intermittent renewables—will not decline significantly any time soon. RWE's sphere of interest, the region covering Central and Western Europe, requires roughly 260 GW of *reliable* capacity today. RWE figures that the same would be needed in 2030. "We have to live with our assets and make the best of them," it says.

A major point of contention for owners of *reliable* thermal generators like RWE is that under current market rules, they do not get paid adequately for providing backup capacity when intermittent renewables are not available. RWE, like its counterparts, believes that this issue will have to be rectified, sooner rather than later, perhaps through market redesign and/or some form of capacity payment scheme that keeps thermal units viable as renewables continue to grow.

The stakes are high, and the time is of the essence for Germany's second biggest utility, based in Essen and serving the industrial heart of Europe's biggest economy, the Ruhr Valley. The company, whose share prices have been on the decline for some time, reported net loss of $500 million in the quarter ending in Sept 2013. Speaking at a conference in mid November 2013, its CEO Peter Terium acknowledged more trouble ahead, with operating profit projected to fall 24% in 2014. The company, which employed 72,000 in 2011, has already slashed 6200 jobs; another 6750 cuts are expected over the next 4 years—unprecedented in company's long history.

Describing the industry's woes—depressed wholesale prices, declining profits, and evaporating share prices—Mr. Terium said, "Our traditional business model is collapsing under us."

Mr. Terium, who has been at the helm of RWE since July 2012, is planning to turn the company's culture and institutions around—and it is clear that he is under no illusion of the challenges ahead. RWE's new mission is to "create value by leading the transition to the future energy world." According to the documents released, RWE aspires to be "the most efficient, integrated European energy utility in Europe by 2018" by achieving three targets: "successfully shape Europe's energy industry, regain our financial fitness and transform RWE into a high-performing organization."

Realizing the political realities of *Energiewende* and EU's myriad of energy and environmental policies, Mr. Terium appears conciliatory. At least in its public statements, the company says that it does *not* merely support Germany's energy turnaround, but it *actively wants to lead it*. According to the strategy paper, the company's goal is that in five years' time, "RWE will have been vital in shaping the energy industry across Europe."

In an interview with *Energy Post's* Brussels correspondent, Sonja van Renssen, published on 5 Dec 2013, E.ON's CEO Johannes Teyssen said,

I think everyone understands at this point that value creation is moving downstream, to the customer, from the commodity side to the solution side. So Eon is striving hard in Europe to go to those parts of the value chain where we are close to our 25 million customers, where we enable them and can share with them the advantages of efficiency.

Teyssen, who is the head of Europe's second largest utility and the current president of Eurelectric, says E.ON is now focused on customers and their needs: "We are there to be on the side of our customers. What's right for them is right for us!"

For the most part, industry observers—including the majority of those contributing to this volume—are of the opinion that the ESI will not go away any time soon nor will its assets be stranded overnight. While some assets, for example, inefficient and polluting coal-fired plants, may be forced to retire prematurely, efficient and flexible thermal plants and all forms of energy storage are likely to become more valuable in the future where intermittent renewables and DG are dominant.

A number of contributors including Thaler and Jimison predict incremental change taking place gradually as the industry responds to the changing environment. Others have ideas on how such a transition might take place and when.

Others believe that the ESI may be facing a serious and potentially devastating threat, which requires more immediate and radical responses—both from within and from without—such as changes in regulations particularly retail pricing and revenue collection.

Yet others see the decentralized energy revolution as a disruptive technology—further described in Chapter 7—which suggests that the old model and the old delivery network may be under threat as consumers initially become prosumers but may gradually find ways to operate in grid-assisted, grid-parallel, and eventually off-grid modes. These experts point to the fate of other industries facing disruptive technologies and draw doom and gloom scenarios. Several chapters examine alternative future scenarios, including Felder, Kristov and Hou, Riesz et al., and Smith and MacGill.

6   CONCLUSIONS

The rapid pace of change facing the incumbent players is serious and is accelerating. The industry must rapidly evolve if it is to make a successful transition into a future where new players, new services, new technologies, and new means of delivering service are replacing traditional ones. The ESI has many strengths. It needs to reinvent itself—and the regulators and policy makers must have the wisdom to allow such transition to take place.

Chapter 2

New Utility Business Model: A Global View

Paul Nillesen, Michael Pollitt and Eva Witteler

ABSTRACT

This chapter examines the response to the surge in distributed energy resources (DER) and prosumers by traditional utilities. Using the results of a comprehensive survey of senior management of major utilities around the world, conclusions on the major threats and opportunities are drawn. Although the creation of new markets, new niches, and new entrants might suggest the demise of big and established utilities, this chapter suggests that they might be able to survive and thrive by scaling up and exploiting the new business opportunities—provided that they make the right moves at the right time. Otherwise, the emergence of "platform" markets might lead to intermediaries capturing the value of interactions among the stakeholders before the utilities do. The critical skill for the traditional players will therefore not be in creating new business models, but rather in identifying and cultivating them in mass-market scale. The authors point out some of the new entrants already appearing to challenge the established players.

Keywords: Utility business model, Distributed energy resources, Disruptive technology, Management perspective, Platform markets

1 INTRODUCTION

"We have to rethink what is our role, and our place in the energy sector," Frank Mastiaux—EnBW's boss—was quoted as a saying in The Economist recently (12 October, 2013) in their briefing on European utilities.[1] RWE's board recently announced that the company's guiding principle will change from volume to value and that it intends to reposition itself to a project enabler, operator, and system integrator of renewables.[2] The energy sector is undergoing a rapid shift from central large-scale production to small-scale local production, from

---

1. The Economist, 12 October 2013, page 24.
2. Energy Post, 21 October 2013a, Exclusive: RWE sheds old business model, embraces transition (www.energypost.eu/index.php/exclusive-rwe-sheds-old-business-model-embraces-energy-transition). See also Sioshansi's chapter in this book.

Distributed Generation and its Implications for the Utility Industry. http://dx.doi.org/10.1016/B978-0-12-800240-7.00002-3

thermodynamics to electronics, and from market prices to subsidies and government intervention. Several factors—new renewable technologies, IT solutions,
customer demand, and government carbon and renewable policies—have created the perfect storm for traditional utilities, putting pressure on traditional
business models and decentralized investment decision-making and revenue-
generating activities. As described by Sioshansi in the introduction of this book,
the traditional business model formula—revenue collection based on volumetric tariffs—has already been under pressure for the last decades. However, the
urgency of creating new business models did not seem as pressing until the
recent surge in DERs and the steady rise of prosumers. Although these developments are currently limited to a number of more mature markets, it is plausible to assume that this trend will spread to other developing markets.

The focus of this chapter is to highlight what the companies themselves see
as major challenges and opportunities, discuss some new and innovative emerging business models, and question the ability of traditional companies to
become successful in this new world. Although it may not seem obvious, these
big established companies could become what Markides and Geroski (2005) of
the London Business School define as "fast second": traditional players that are
not the ones to create new markets, but are the ones best-positioned to take these
new niche markets and to scale them up into mass markets.[3]

The traditional "utility" value chain was relatively straightforward: centralized large-scale electricity generation was transmitted to large industrial users
and further distributed to households. The role of the network infrastructure was
to ensure balance in the grid, ample capacity, and long-term planning to meet
demand growth. The production of electricity was focused on providing an efficient portfolio of base load and flexible generation to meet the load profile of
demand. Electricity suppliers (or retailers) bundled these products into a single
invoice, where (household) customers paid for the volume consumed and additional fixed charges for the infrastructure. Figure 1 demonstrates the traditional
value chain in electricity.

With the rise of DER in the electricity grid, the former top-down structure
changes into a multidirectional system. Instead of just consuming electricity,
households (or other larger users) can now also produce electricity—either lowering their demand or supplying their excess production to the grid. At the same
time, with the increase in IT and new technology, such as smart grids and smart
metering, there are more information and control over the timing and amount of
consumption—changing the consumption pattern. Figure 2 shows prosumers
supplying part of their electricity demand themselves while feeding in residual
electricity production into the low-voltage grids. At the same time, prosumers
are still (partially) dependent on the generation upstream to meet peak or emergency energy requirements (see also Kristov and Hou and Cazalet in this book).

---

3. Markides, C.C. & Geroski, P.A. (2005), Fast Second: How smart companies bypass radical innovation to enter and dominate new markets, John Wiley & Sons.

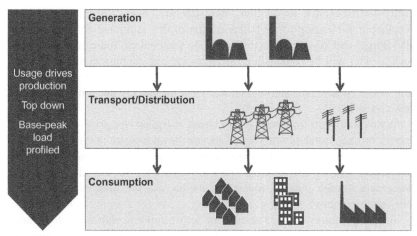

FIGURE 1 The traditional value chain. *Source: PwC, 2013a,b.*

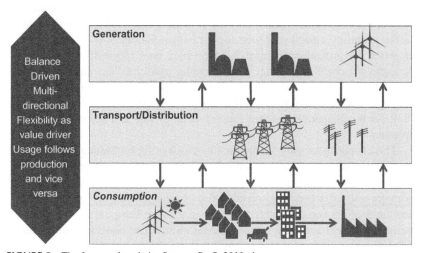

FIGURE 2 The future value chain. *Source: PwC, 2013a,b.*

In this evolving market, production is more nebulous and scattered through the system, rather than centralized and controllable. New constraints may appear further down in the grid, which has been dimensioned for the one-way flow of electricity. Balancing new sources of supply—such as intermittent solar PV, small-scale wind, and domestic gas-fired electricity generation—with more intelligent usage, such as energy consumption monitoring and steering, pooling of demand between households, and storage, will become increasingly important. The customer in essence becomes a microcosm of the traditional utility—with energy production, infrastructure, energy management, and consumption—leading to the balkanization of the traditional value chain (see also Burger and Weinmann in this book). In this context, some questions arise, such as who will be responsible for balancing and running this new system? Who is the supplier of last resort? Who pays what and to whom? What is the role of the government and regulators? How do you tax and subsidize in this new system?

A structural shift from transaction-based, marginal cost pricing to fee-based service business models often accompanies the emergence of "platform" markets, i.e., multisided markets where an intermediary captures the value of the interaction between user groups. The many examples include telecommunications, data storage, cinema, music and media, and the automobile industry.[4] Why not electricity?

In Europe, for example, a number of "crowdfunding platforms" for renewables exist, enabling individuals and organizations—from private consumers to small business and community organizations—to invest in renewable energy with minimum stake.[5] In return, for a certain fee from the project owner, the platform operator collects the money and organizes the payouts. Even though the funds are still small compared with the total cost of transitioning to a low-carbon economy, chances are seen as promising in Europe where crowdfunding increased by around 65% from 2011 to 2012 to €735 million.[6] Examples throughout Europe show that these funding platforms are able to provide money in a comparatively short period. And this development also impacts the larger traditional utilities. RWE recently announced that their cost of capital "will not be competitive against funding from private and institutional equity investors."[7]

This chapter's focus is on the possible shape of future business models, and the potential traditional utilities have to not only survive but also actually thrive in this new world. Section 2 provides an overview of what senior management of major utilities around the world thinks—where answers are split by region. Section 3 offers some insights into new innovative business models that are starting to appear and what role traditional utilities can play. The final section provides some conclusions and suggestions for the future.

---

4. Weiller, C.M. & Pollitt, M.G. (2013), Platform markets and energy services, EPRG Working Paper (forthcoming).

5. Energy Post, 29 October 2013b, Crowdfunding renewables: game-changer for the energy sector? (www.energypost.eu/crowdfunding-renewables-game-changer-energy-sector/).   See   Cazalet's chapter in this book where contracting between parties takes place.

6. *Ibid.*

7. Energy Post, 21 October 2013a, *op. cit.* An example of alternative funding options is the financing option provided by SolarCity—discussed later in the chapter.

2   WHAT IS THE GLOBAL PERSPECTIVE? VIEWS FROM THE SENIOR MANAGEMENT OF MAJOR GLOBAL UTILITIES

In Europe, where the share of energy from renewable sources is around 13%[8] of gross final energy consumption, few expected the intensity of current market dynamics to force operators of highly efficient and flexible new installed generation to a standstill.[9] However, the effects of the changing energy value chain are visible in other parts of the world as well, albeit with different degrees of strength and in different parts of the value chain. Within the complex interactive system of the market, new technology, and sector regulation, it is difficult to predict where the worldwide changes will in the end lead to. In the 2013 PwC Global Power and Utilities Survey, senior management from 53 power and gas utilities in 35 countries were surveyed to get their view on the future.[10] Almost 40% expect the existing power utility business model in their market to transform or even be unrecognizable in the period between now and 2030.

The survey is based on research conducted between April and July 2013. The 35 countries were from Europe (including Russia), North and South America, Asia Pacific, Middle East, and Africa (MEA). The objective of the survey was to map out "the direction of some of the forces shaping" the future utilities' business model from the CEO's perspective. It is based on the participants' answers to a range of questions and their assessment of a series of future scenarios. The survey addresses the following topics: transformation, disruption, technology, supply, companies, and regulators. The following highlights some of the key findings related to DER and business model transformation.

The survey points out that the expectation of a transformation of the utilities' business model is highest in Asia (69%). In contrast, in South America and MEA, all, or at least 70%, of the companies expect similar business models as today. At the same time, less than 10% worldwide expect the business models of the future to remain more or less the same as today. According to the majority of the survey participants (67%), the future energy market will most likely be a mixture of large-scale centralized and smaller-scale distributed generation. Only South America and MEA are ambivalent in their expectation. Half of

---

8. Numbers for EU27 in 2011 (Eurostat, 2013); renewable energy sources cover solar thermal and photovoltaic energy; hydro (including tide, wave, and ocean energy), wind, and geothermal energy; and biomass (including biological waste and liquid biofuels). The contribution of renewable energy from heat pumps is also covered for the member states for which this information was available. The renewable energy delivered to final consumers (industry, transport, households, services including public services, agriculture, forestry, and fisheries) is the numerator of the Europe 2020 target. The denominator, the gross final energy consumption of all energy sources, covers total energy delivered for energy purposes to final consumers and the transmission and distribution losses for electricity and heat.

9. See, for example, The Economist, *op. cit.* or PwC (2013a), Financial and economic impact of a changing energy market (Netherlands) for the Dutch Energy Association (www.energie-nederland.nl).

10. PwC (2013b), Energy transformation – The impact on the power sector business model, 13th PwC Annual Global Power & Utilities Survey (www.pwc.com).

Which energy market transformation vision most closely matches your expectation for your market?

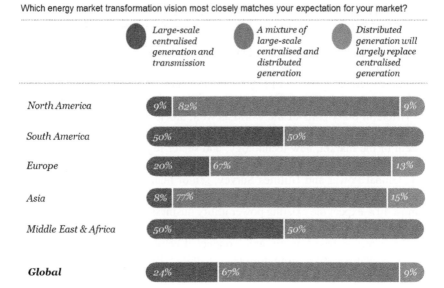

FIGURE 3 Energy market vision by region. *Source: PwC Annual Global Power & Utilities Survey (2013).*

Percentage of respondents saying it is likely or highly likely that increasing levels of distributed generation will force utilities to significantly change their business models

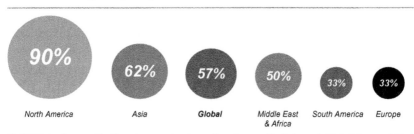

FIGURE 4 Impact distributed generation by region. *Source: PwC Annual Global Power & Utilities Survey (2013).*

the participants envisage a mixed market, and the other half expect a market of large-scale centralized generation and transmission (Figure 3).

When asked if the increasing levels of distributed generation will force utilities to significantly change their business models, a greater regional spread can be observed (Figure 4). While nearly all participants from North America (90%) think that the effect from distributed generation exists, in South America and, surprisingly, in Europe, only one-third (33%) share this opinion. While the survey states that "perhaps European participants see such changes as already underway," a further reason might be seen in the correlation between the given answer and what part of parts of the value chain the respondent operates in (e.g., network and nonnetwork).

Meanwhile, the majority of the participants see decentralized generation as an opportunity rather than a threat (80% compared with 20%). The strategy, which is seen as the most successful in a distributed generation market, is to provide distributed generation services (67%). Second- and third-rated successful strategies are the role as an efficiency contractor to support customer energy savings (60%) and the role as an intelligent grid operator to support prosumer energy share (56%), respectively. Further, about half of the participants believe that becoming an "energy partner" rather than an "energy supplier" is a successful strategy in a distributed generation market.

In connection with improvements in customer relations and services, it is worth mentioning that a significant proportion of the survey participants (41%) foresee a rise of a new type of active energy customer in ten years' time, who is more engaged in energy saving and energy generation than today, e.g., the so-called prosumer (see Sioshansi on RWE's strategy in his chapter and also Blansfield et al.). Especially participants in Asia (46%), North America (50%), and Europe (60%) share this impression. However, most survey participants agree that there is still a long way to go before being able to compete for those demanding new customers. Three-fifths (61%) say that there is "high" or "very high" scope for improvement in customer relations and services.

In addition to the change to offensive strategies and preparation for the new prosumer, a significant number of the survey participants (65%) see potential to reduce their cost base and improve efficiency by more than 10%. Europe leads in this area with a share of over 90%. And still one-third (31%) of all participants even consider a potential improvement of more than 20%, led again by Europe (58%) (Figure 5).

Worldwide, nearly three-quarters (73%) see the scope for performance improvement in one of their core activities—asset performance management. Bigger utilities like RWE and its rival E.ON have embarked on significant programs to restructure their portfolios and target cost reductions in response to the current market environment and the prospects going forward (see, e.g., the chapter by Groot).

Even if cost cutting is realized, decentralized generation is already eating into the margins of traditional utilities today. As the German experience shows and the Californian ISO warns, the impact of decentralized generation shaves peak demand leading to lower prices in the peak, lower utilization throughout the year, and inevitable impacts to margins and profitability.

Therefore, most survey participants do not expect centralized generation and transmission to play a lead role in the future. It is mainly envisaged that future demand growth will be met by a mixture of centralized and distributed generation. As a result, not only a decline in revenue but also simultaneously an increase in costs from increased demand for transmission, distribution, and balancing services is expected.

What is the scope for power utility companies to reduce the cost base and improve efficiency?*

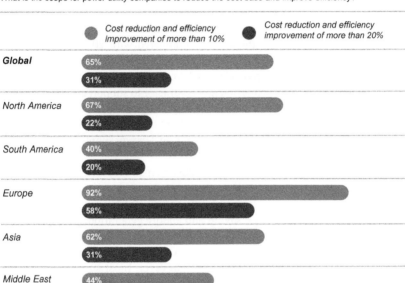

* % of respondents

FIGURE 5   Scope for efficiency and cost reductions by region. *Source: PwC Annual Global Power & Utilities Survey (2013).*

Percentage of respondents rating the following strategies as likely or highly likely to be successful in a distributed generation market

Global

* 'prosumers' refers to customers that generate their own electricity.

FIGURE 6   Strategies for dealing with distributed generation. *Source: PwC Annual Global Power & Utilities Survey (2013).*

With respect to future customer relation and revenues, more than two-thirds of the participants see a medium or high probability for a scenario in which power utility companies need to become much more tariff-savvy to attract and retain customers by innovative pricing schemes (Figure 6). On this point, utilities can perhaps learn from other industries such as telecommunications with their bundling and "free allowance" packages. The question is as follows: how will utilities adjust to compete and ensure revenues in times of increasing DER and decreasing consumption (see also Nelson et al)?

When asked what regulators can do to enable system operators to balance an increasingly intermittent system, survey participants worldwide state that the introduction of demand response and demand-side management markets is the most effective (67%). Approximately the same number of participants worldwide sees the need to curtail intermittent generation during low-demand periods as a preferable regulatory measure (66%).

Although there may be some regional differences, there is overall consensus among the senior management that the fundamentals under their traditional business model are being upset by distributed generation and the transition to renewable energy. Most companies are optimistic that this change offers them opportunities to adjust their value proposition with the provision of distributed energy services, supporting customers with energy savings or as an intelligent grid operator (see, for example, Keay et al. on some of the new concepts being developed).

3   WHAT BUSINESS MODELS ARE ACTUALLY BEING EMPLOYED AND WHO WILL BE SUCCESSFUL IN THIS NEW MARKET

As the survey clearly demonstrates, senior managers expect the traditional value chain to change and that traditional companies will need to adapt, as customer preferences and their behavior changes and as new entrants and new markets evolve. A recent newspaper article in Die Zeit in Germany stated that traditional utilities like E.ON and RWE were facing competition from more than 1 million renewable energy power plants, mostly household scale.[11] For decades, energy utilities have been used to capital-intensive investments with predictable long-term returns—partially regulated and partially unregulated. These companies are now struggling in order not to miss the shift in their underlying business. These shifts are not uncommon where new technologies disrupt traditional business models—suggesting that regulation, which is pervasive in energy markets, also needs to act (however unlikely) as an enabler and not an inhibitor.[12] It seems that the energy giants in Europe also see the regulatory changes coming and thus want to actively coshape the regulatory future. RWE announced the need "to offer its expertise in order to contribute to the political opinion forming-process in a credible, trustworthy but also self-confident way."[13]

---

11. Die Zeit, Riesen taumeln im wind, 5 September 2013.
12. See, for example, Pollitt, M.G. (2010), Does electricity (and heat) network regulation have anything to learn from fixed line telecoms regulation? Energy Policy (38), pp. 1360–71.
13. Energy Post, 21 October 2013a, www.energypost.eu/index.php/exclusive-rwe-sheds-old-business-model-embraces-energy-transition.

The main question facing the sector is: Where is the value and who captures it? And is this value temporal only to dissipate as new technology or solutions enters? The electricity retail market, for example, is emerging with elements of a platform market, where an intermediary is needed to match supply and demand (much like a stock market, or eBay). In the case of electricity, there is a need for one or more "matchmaking" intermediaries between suppliers who cannot predict their generation and consumers who start participating in active energy demand management. The value added is then in matching supply and demand, not in meeting demand itself. Several chapters in this book (e.g., Mackeay) explore the increasing role of intermediaries.

But value can only be derived from this service if there are some barriers to entry in the form of, e.g., infrastructure (storage, flexible generation, etc.) or know-how (algorithms, systems, etc.). The rapid development and disruptive nature of this change seem to suggest that barriers are likely to be low and disappear over time. The challenge for the market is coming not only from traditional competitors or from the "energy community" but also from energy and energy services in particular that are attracting new players from adjacent and historically separated markets, such as retail, furniture, automotive, and multimedia.

In a recent article in The Guardian, furniture maker IKEA announced that it will be offering "solar panel packages" at all of its 17 British stores starting in 2014.[14] From IKEA's point of view, this is a logical extension of their value proposition but is clearly a new threat to the traditional utilities. The success will need to be seen, although IKEA itself has deep experience with rooftop solar production for its own facilities. SolarCity on the other hand offers a different challenge for utilities. In the United States, SolarCity leases solar panels to homeowners, allowing them to pay for them monthly over 20 years, avoiding large initial investments. This establishes a longer-term relationship with a customer and weakens the traditional relationship with the utility. The company is now "securitizing" these PV leases and offering them to investors as banks do with mortgages.

In the Die Zeit article, E.ON Connecting Energies indicates that traditional generation will no longer be the main revenue contributor, but that the focus will be on assisting customers to produce and consume their own electricity in the most efficient manner—a market that they estimate to be worth over 100 billion euros.

This view is supported by a recent report by the Lawrence Berkeley National Laboratory, which finds that the growth of energy service companies (ESCOs) over the period 2009-11 was greater than US GDP growth.[15] The growth rate over the period was 9% per year, leading the authors of the report to project an annual market of more than US$15 billion in 2020 for the United States alone, as shown in Figure 7. This is not substantial when compared to the total energy

---

14. The Guardian, IKEA to sell solar panels in UK stores, 30 September 2013.
15. Lawrence Berkeley National Laboratory (2013), Current Size and Remaining Market Potential of the U.S. Energy Service Company Industry.

Current size and remaining market potential of the U.S Energy Service Company industry

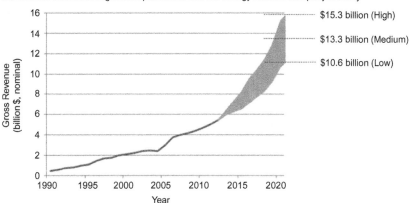

FIGURE 7   Projected growth in ESCO revenues in the United States. *Source: Lawrence Berkeley National Laboratory (2013, page 3), op.cit.*

market, but a sizeable opportunity that could grow further and eat into the more traditional business models of utilities.

The dilemma of this development is also clearly demonstrated by the report. This new market offers opportunities for revenue growth, but the ESCOs are actually eating into the traditional revenue streams of utilities by helping customers save energy (roughly US$4 billion per year). The trend of ESCOs in the larger commercial and industrial market is also starting to appear in the retail market, albeit with mixed success so far. Generating companies such as NRG are acquiring energy aggregators because they recognize that the future is not about having MWs but rather a combination of MWs and "negawatts" that balances supply/demand and meets customers' needs.[16]

The classic example of a player that uses new internet-based technology to create new niche markets is Google. In 2009, Google launched Google PowerMeter—a free energy monitoring tool to raise awareness and encourage energy efficiency. The project was terminated just two years later after not generating enough users around the world. Google, however, has continued to invest in smart home solutions and even has an energy subsidiary, Google Energy. The use of sophisticated algorithms, like those used by Amazon, Google, and other Internet companies, could spur a wave of innovation in household-level ESCOs. As other chapters point out (for example, Keay et al. and Kristov and Hou), moving downstream seems feasible with smart meters and players such as credit card companies and Telcos, who are used to collecting and manipulating lots of atomized data.

Some new players are also entering the market for demand response and balancing possibilities. REstore is an example of a small new player offering

16. See, for example, EEnergy Informer, October 2013, http://www.eenergyinformer.com.

demand response aggregation services. It has a two-sided market, offering payments to industrial consumers willing to curtail their consumption and offering a load-shedding option to transmission operators. These types of products are also referred to in the utility survey results and have been successful in markets like PJM in the United States—where the capacity market has seen an increase in not only committed generation but also demand-side response and even energy efficiency solutions.

Within their research on the application of a two-sided retail electricity market platform and related pricing strategies, Weiller and Pollitt (2013) saw the benefit of new entrants like information and communication technology companies in the stimulation of innovation and new capital investment, saying that platform services still could be provided by incumbent utilities and ESCOs.[17] They fulfill the role as an intermediary between supply companies and end users. Connecting the dots between the small-scale (domestic) renewable energy productions in particular is also a route that some new entrants are taking. In the German market, Next Kraftwerke has interconnected via remote access hundreds of small-scale renewable energy production units and bundles these for resale in the energy spot market. In 2012 alone, they sold 1TWh of production as a virtual power plant.

These developments provide some hope that there are new profitable business models that can be exploited—but the exact form of these markets remains unclear for now. The question is whether the traditional incumbent utility players can successfully adapt and maneuver into this new value chain and exploit the opportunities that these new markets bring. Or to put it more bluntly, as the European electricity industry association (Eurelectric) states, they will need to *"Evolve or Die."*[18]

It is rare to find business model innovations that originated from established players. For example, Markides (2008) demonstrated that (i) the majority of business model innovations are introduced by newcomers; (ii) incumbents find it extremely difficult to respond appropriately; (iii) most respond by imitating the innovation rather than neutralizing it, for example; and (iv) the majority of responses fail because they cannot manage two competing business models at the same time—the original business model and the new model.[19]

The evidence also shows that the new dominant design is likely to emerge in a disorganized and chaotic way, with lots of candidate designs championed by lots of new entrants, usually appearing in niches of the existing market. These niches are likely to be populated by consumers who are innovators or early adopters of the new technology... Finally, we know that the champions of the old design – the market leaders of today

17. Weiller, C.M. & Pollitt, M.G. (2013), Platform markets and energy services, EPRG Working Paper (forthcoming).
18. Eurelectric (2013), Utilities: Powerhouses of innovation.
19. Markides, C.C. (2008), Game-changing strategies, John Wiley & Sons.

and the first-movers of yesteryear – will be among those who are least willing to see change occur and least willing to participate in the change process.

(Markides & Geroski, 2005, Page 162)

The main thesis of Markides' work is that traditional firms should not attempt to develop new business models, but rather outsource this to a community of new entrants and smaller firms. The traditional player then nurtures these innovations and helps scale up the emerging dominant products or services. This may sound radical, but this "business model" is common in industries where corporate survival depends solely on the continuous stream of creative new products, such as movies, theaters, art galleries, book, and music publishing.

Markides and Geroski provide a number of examples of markets where the original innovator in an industry was not the innovator that created the mass market for that product or service (Table 1).

TABLE 1 Colonists Versus Consolidators: The Inventors of New Business Models and the Creators of Mass Markets

Industry	Innovator that Came up with the Idea	Innovator that Created the Mass Market
35mm cameras	Leica	Canon
ATMs	De La Rue	IBM/NCR
Diapers	Chicopee mills (J&J)	P&G
Personal computers	Osborne/Apple	IBM
Online bookselling	Charles Stack	Amazon
Online brokerage	Net investor	Schwab
VCRs	Ampex	JVC
Copiers	(Haloid) Xerox	Canon
CAT scanners	EMI	GE
Videogames	Magnavox/Atari	Nintendo
Operating systems	Digital Research	Microsoft
Pocket calculators	Bowmar	TI
Mainframes	Atanasoff-Berry Computer (ABC)	IBM

Source: Markides & Geroski (2005)

This theory could also apply to the traditional utilities, with access to large balance sheets and a broad (mostly) captive customer base. Rather than fighting the newcomers, traditional utilities could embrace and nurture them and transform their own business along the way. Here, the question remains whether the incumbent players will get an opportunity to create the platform that all market participants want to use or the service standard everybody wants to ascribe to and whether consumers actually see a role for the incumbents when smart metering, new IT, and new entrants really appear.

4   CONCLUSION

This chapter has provided a brief overview of the major shifts in the traditional value chain for electricity. As Sioshansi points out in "Introduction," the rise of distributed generation and the transformation of the traditional consumer into a prosumer are changing the way the revenue flows in the value chain and changing the value chain itself. The move from the traditional utility to utility 2.0 is challenging and a development that most of the major utilities around the world see unfolding—as supported by the survey results. This chapter discussed some of the new business models being adapted in the market—from a retail approach, such as IKEA, to technology-led aggregation of small-scale renewables into a virtual power plant. Most of the names that appear do not belong to the traditional companies however.

Tomorrow's energy company will be very different than today's, and the question is whether traditional utilities are able to make a successful transformation in time. The answer is yes, but a strongly qualified yes. Moving first into a new market does not mean that a company will survive over the longer term, but moving quickly enough before things have settled is critical. Timing is therefore everything. With the sector under tremendous pressure, will management be able to resist the pressure and get the timing right?

REFERENCES

Die Zeit, September 5, 2013. Riesen taumeln im wind.
EEnergy Informer, October 2013. http://www.eenergyinformer.com.
Energy Post, October 21, 2013a. Exclusive: RWE sheds old business model, embraces transition www.energypost.eu/index.php/exclusive-rwe-sheds-old-business-model-embraces-energy-transition.
Energy Post, October 29, 2013b. Crowdfunding renewables: game-changer for the energy sector? www.energypost.eu/crowdfunding-renewables-game-changer-energy-sector.
Eurelectric, 2013. Utilities: Powerhouses of innovation.
Lawrence Berkeley National Laboratory, 2013. Current Size and Remaining Market Potential of the U.S. Energy Service Company Industry.
Markides, C.C., 2008. Game-changing strategies. John Wiley & Sons, San Francisco.
Markides, C.C., Geroski, P.A., 2005. Fast Second: How Smart Companies Bypass Radical Innovation to Enter and Dominate New Markets. John Wiley & Sons, San Francisco.

Pollitt, M.G., 2010. Does electricity (and heat) network regulation have anything to learn from fixed line telecoms regulation? Energy Policy. 38, 1360–1371.

PwC, 2013a. Financial and economic impact of a changing energy market (Netherlands) for the Dutch Energy Association www.energie-nederland.nl.

PwC, 2013b. Energy transformation—The impact on the power sector business model, 13th PwC Annual Global Power & Utilities Survey (www.pwc.com).

The Economist, October 12, 2013. page 24.

The Guardian, September 30, 2013. IKEA to sell solar panels in UK stores.

Weiller, C.M., Pollitt, M.G., 2013. Platform markets and energy services, Cambridge University, EPRG Working Paper (forthcoming).

Chapter 3

Germany's Decentralized Energy Revolution

Christoph Burger and Jens Weinmann

ABSTRACT

This chapter examines the effects of the German feed-in policy for renewable and distributed energy sources. The success of the instrument confronts grid operators, utilities, and politicians with unprecedented challenges. While grid operators struggle to ensure stability in the network, utilities experience a significant erosion of revenues, because renewables and self-generation drive down wholesale market prices and lead to loss of sales. The feed-in policy has led to a major financial legacy. Final customers are billed a rising levy to compensate the owners of renewable energy installations for their investments. However, decentralized supply has also triggered a change in the perception of energy, with bioenergy villages, energy cooperatives, and municipal utilities discovering the value of locally produced power and heat. Germany's unfolding decentralized energy revolution suffers from multiple unresolved issues, but may serve as a role model for other countries and regions if it succeeds.

Keywords: Feed-in tariff, Decentralized energy generation, German Energiewende, Prosumer, Energy emotionalization

1 INTRODUCTION

The German energy turnaround, or "Energiewende," is considered one of the country's most ambitious and far-reaching experiments of economic transformation (Carrington, 2012). By 2050, the German government plans to obtain 60% of country's gross final energy consumption from renewable sources and "a minimum of 80% of the electricity supply is to be generated from renewables," with an intermediate goal of 35% in 2020; power from cogeneration is supposed to reach 25% in the same year (Federal Government, 2012). Decentralized energy generation is a major contributor to these two targets.

While some elements of the strategy to reduce greenhouse gas emissions lag behind the schedule, in particular in transportation and energy efficiency, the success of the policy instruments to promote renewable and distributed energy supply has exceeded even the most optimistic forecasts: in 2012, more than a

Distributed Generation and its Implications for the Utility Industry. http://dx.doi.org/10.1016/B978-0-12-800240-7.00003-5

million independent self-producers with renewable energy installations fed their electricity into the German grid (Renewable Energies Agency, 2013). The momentum fundamentally questions the current market design and the role of utilities. After a focus on sustainability and the transition to a low-carbon supply system, the energy policy triangle may bounce back to security of supply and affordability, as the decentralized energy revolution unfolds.

The following sections of this chapter are structured as follows: Section 2 covers how regulation and industrial policy triggered the success of decentralized energy in Germany. Section 3 provides an overview of the momentum that decentralized energy has gained and links it to two related concepts, empowerment and emotionalization of energy. Section 4 highlights major challenges that Germany encounters in implementing decentralized energy, including key unresolved questions in financing and redistribution, technologies lagging behind the momentum, ensuring security of supply, and counterstrategies of German energy utilities. The chapter's conclusions appear in Section 5.

2   WHAT IS DRIVING THE GERMAN DECENTRALIZED ENERGY REVOLUTION?

Germany's move toward distributed energy resources is the result of a long-term sustainability strategy of the government, even predating the liberalization of the electricity market in the late 1990s. The history of its practical implementation is reviewed in Section 2.1. Germany's overarching objective to become both lead market and lead supplier—the rationale underlying the policy—is discussed in Section 2.2.

2.1   German Feed-In Regulation for Renewable Energies and Cogeneration

Germany is a country with the modest endowment of "classical" renewable energy sources. Until the early 1990s, the dominant renewable primary energy was hydropower, which contributed only 3.1% to the overall electricity generation; power produced from biomass was negligible (Renewable Energies Agency, 2013). The move toward a larger share of renewable energy supply started in 1991 with a first law that allowed for the feed-in of renewable sources. Energy utilities were obliged to compensate the feed-in of electricity generated from hydropower, biomass, and waste with 75% of the average retail electricity price, while wind and photovoltaic (PV) power received 90% of the retail price.

This first law gave a boost to wind power—installations increased from 56 MW at the beginning of the 1990s to 4400 MW in 1999 (BMU, 2009). At the end of the 1990s, the liberalization of the electricity sector led to declining retail prices, which became an impediment for the promotion of renewable technologies. The governing coalition of social democrats and the green party decided to overhaul the existing law and decouple it from electricity prices.

In 2000, the coalition ratified the Renewable Energy Sources Act with the objective of creating more stable investment conditions. Differentiated according to generation technology and the size of the installation, the law determined the financial compensation of each kilowatt hour fed into the grid with predefined rates over a time span of 20 years. In addition, it imposed priority feed-in for renewable energies. The installation of PV panels gained particular momentum with the "100,000 Roofs Program," which offered favorable loans to private homeowner, freelancers, or small- and medium-sized enterprises. In 2003, the program's objective of reaching 300 MW installed solar capacity was achieved.

Subsequently, German feed-in legislation underwent three modifications in 2004, 2009, and 2012 to align the scheme with actual costs of generation technologies. Its compensation mechanism remained practically the same, though.

The program's overall success can be measured in the amount of electricity produced from renewable energies, which rose roughly eightfold from 17 to 136 TWh per year between 1990 and 2012. While hydropower remained approximately constant, wind power increased from virtually nil to 46 TWh, biomass to 41 TWh, and PV power to 28 TWh. In total, renewable energies contributed 23% of German electricity supply in 2012 (Renewable Energies Agency, 2013). Figure 1 shows the evolution of renewable energies between 1990 and 2012.

In 2012, around 71.4 GW of renewable energy installations benefitted from subsidies linked to the Renewable Energy Sources Act (50hertz et al., 2013). In 2013, 9.2 GW was added under the scheme. Figure 2 depicts all active renewable energy installations that were financially supported by the scheme in 2013, as measured at the point of delivery (ibid.).

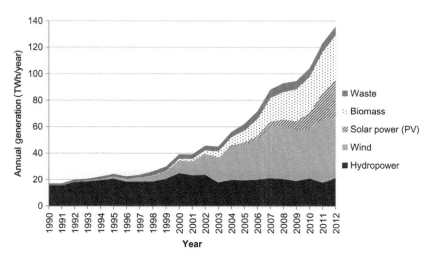

FIGURE 1   Annual contribution of renewable energies in electricity generation in Germany. *Source: BMU (2013).*

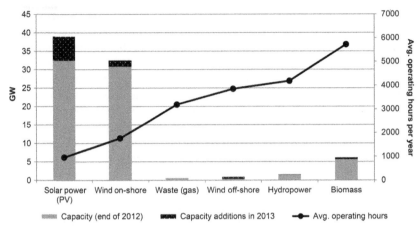

FIGURE 2   Renewable energy installations in Germany supported by feed-in regulation. *Source: 50hertz et al. (2013).*

Cogeneration is promoted in a parallel scheme that is also based on feed-in subsidies. However, small-scale combined heat and power (CHP) plants have not seen a market rollout comparable to renewable energies. Wünsch (2013) estimated that CHP plants smaller than 1 MW generated 3.5 TWh electricity in 2011—less than 1% of the overall production. In 2012, Germany had around 1250 units smaller than 2 kW, around 35,000 between 2 and 50 kW, and around 700 between 50 and 1 MW. Despite the subsidies, the prospects for cogeneration plants are highly uncertain and vary between around 260,000 plants smaller than 2 MW in 2020 and a complete stagnation at 2012 levels (ibid.). Figure 3

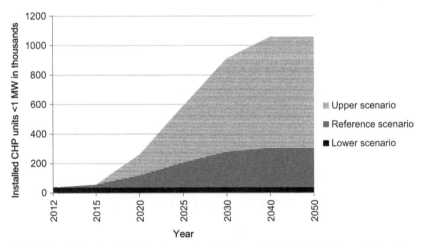

FIGURE 3   Market size in 2012 and growth projections until 2050 of CHP plants with less than 1 MW$_{elec}$ capacity. *Source: Wünsch (2013).*

shows the range of current estimates, with a reference scenario reaching saturation at around 300,000 units in 2040.

Internationally, the feed-in instrument became a widespread success. According to the Federal Ministry for the Environment, more than 40 countries adopted some type of feed-in regulation based on principles like the German Renewable Energy Sources Act (BMU, 2009, p. 14).

2.2   Industrial Policy: Lead Market and Lead Supplier

Although the rationale underlying German feed-in regulation was to mitigate climate change, it was also intended to give a boost to the domestic diffusion of these technologies and to promote the German manufacturing sector—a dual strategy to become both "lead market" and "lead supplier." As Chancellor Angela Merkel pointed out in 2010: "The structural shift toward a low-carbon economy must be pursued consistently. This will of course also have added economic value, for introducing this structural change in Europe early on will lead to significant competitive advantages for German industry in global competition" (Federal Government, 2012, p. 115).

Manufacturing of PV panels evolved into an important pillar of economic development and employment in the former communist states in the eastern part of the country. As an allusion to the Silicon Valley in California, a region with a high density of manufacturers of solar cells was coined "Solar Valley." The dual strategy worked well, at least for some time: Germany claimed to be the world leader in solar PV installation and manufacturing with a global market share of 46% in 2008 (Arnold et al., 2009), while the amount of PV installations on German rooftops by far exceeded the installations in any other country. In 2012, almost a third of the worldwide PV capacity was located in Germany, more than four times the amount installed, for example, in the United States (BP, 2013). Another intention when establishing the instrument—achieving economies of scale in the manufacturing of PV panels—also materialized: the unit costs decreased by more than 60% between 2006 and 2012 (Bundesverband Solarwirtschaft, 2013). According to the estimates of the Federal Ministry for the Environment, in 2012, almost 380,000 people were employed in the renewable energies sector, and €19.5 billion was invested in the installation of renewable energies, of which more than half was directed toward solar energy (Renewable Energies Agency, 2013; Figure 4).

From a "lead market" perspective, the strategy was a success. It came at high costs though, considering the ratio of having financed 380,000 jobs with cumulative investment of around €54 billion by 2012 (Renewable Energies Agency, 2013). Similarly, curbing greenhouse gas emissions could have been achieved with significantly lower investments—the marginal abatement cost curve typically identifies building efficiency measures as the least-cost option to reduce the carbon footprint (see, e.g., McKinsey, 2007).

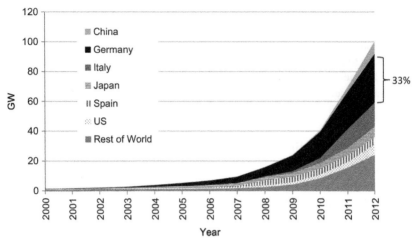

FIGURE 4　Cumulative installed photovoltaic (PV) power in selected countries. *Source: BP (2013).*

The success of "lead supplier" strategy was short-lived, with the global market share of German manufacturers of PV panels declining below 7% by 2012. Due to increased competitive pressure from Chinese PV manufacturers, a number of major domestic firms like Q-Cells, Solon, or Conergy filed for bankruptcy between 2011 and 2013.

Nonetheless, German companies maintained market shares of around 20% and 35% in silicon and inverter production, respectively (Haselhuhn, 2012). In addition, decentralized energy may ensure local employment: In 2011, the share of the German PV industry in the value creation of a PV system installed in German was almost 70% (ibid.). Hirschl (2012) estimated that in the same year, decentralized energies contributed more than €10 billion to communal value creation.

3　PROSUMER EMPOWERMENT: A GENUINE BOTTOM-UP MOVEMENT

While the major German utilities have only recently started recognizing the importance and impact of distributed energy on the overall electricity and heat supply, the agents of change came from the midst of society: Section 3.1 highlights the role of private residents and farmers in the rollout and Section 3.2 focuses on bioenergy villages as triggers of communal self-supply. In Section 3.3, strategies of the municipal utilities and regional suppliers are discussed.

3.1　Who Drives the Revolution?

Decentralized energy has changed the electricity market more drastically than liberalization in the late 1990s, because it turned out to be a genuine bottom-up movement, a grassroots revolution fueled by private investors—often with

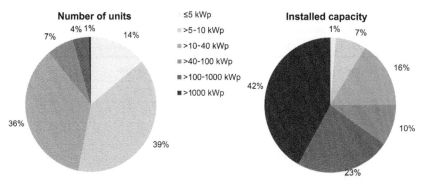

FIGURE 5   PV additions in Germany in 2012 by the number of units and installed capacity. *Source: Bundesverband Solarwirtschaft (2013).*

small installations on their rooftops. Those residential self-producers, or prosumers,[1] were the main driver in the first years of the PV boom. Until the end of 2007, Germany had approximately 430,000 solar installations with a total capacity of 3.8 GW_{peak}. Hence, on average, an installation accounted for around 8.8 kW_{peak}. Later, more commercially oriented investors entered the market. In 2012, around 190,000 installations with an aggregate capacity of 7.6 GW_{peak} were added, which corresponds to an average capacity of 40 kW_{peak} per new installation (Bundesverband Solarwirtschaft, 2013, and previous annual statistics; Figure 5).

Power plants and CHP plants fired with biomass experienced a similar development like solar installations: over time, their size expanded. While the average biomass plant had 60 kW_{el} in 1999, over the five subsequent years, it doubled its size (Bruns et al., 2010, p. 45). In 2011, more than 7200 CHP plants operated with biogas, and around 560 and 260 plants were fueled with liquid and solid biomass, respectively (Renewable Energies Agency, 2013).

In 2012, private residents and farmers still constituted almost half of the owners of renewable energy installations in Germany. By contrast, the "Big Four" energy utilities EnBW, E.ON, RWE, and Vattenfall only held 5% of the installations in terms of capacity (ibid.). Table 1 shows the ownership according to investor groups in 2012.

The actual amount that renewable energies contribute to the overall electricity supply can rise up to more than half of German electricity consumption on a sunny day. For example, on 18 April 2013, a record feed-in of 35.9 GW was registered, with overall electricity demand hovering around 70 GW (Photovoltaik.org., 2013).

---

1. For a discussion on the role of prosumers, see the contribution by Blansfield and Jones.

TABLE 1 Ownership Distribution of Installed Renewable Energy Capacity for Electricity Production (in Total 72.9 GW_{elec}, 2012)

Owners	Percent
Private individuals	35%
Project firms	14%
Industry	14%
Investment funds/banks	13%
Farmers	11%
Other energy suppliers	7%
The "Big Four" energy suppliers	5%
Others	1%

Source: Renewable Energies Agency (2013).

3.2   Empowerment: Bioenergy Villages and Cooperatives

Decentralized energies are a trigger for and expression of empowerment; that means the quest for more self-determination. This not only encompasses the ownership of assets but also entails nonmonetary values: "Empowerment is a multi-dimensional social process that helps people gain control over their own lives. It is a process that fosters power (that is, the capacity to implement) in people, for use in their own lives, their communities, and in their society, by acting on issues that they define as important" (Page and Czuba, 1999). Whereas the liberalization of energy markets allowed final consumers to choose their electricity and gas supplier, decentralized energy turns them into outright producers. Two movements of collective empowerment will be briefly presented, namely, bioenergy villages and cooperatives.

Decentralized energies enjoy particular support in rural areas of Germany—for obvious reasons: the natural resource endowment is more appropriate, including the availability of biomass, a low population density, and, consequently, sufficient surface area for solar panels, wind, and agricultural products to coexist. Bioenergy villages are a prominent example of communal energy autonomy. Typically, they are a cooperative ventures with villagers as co-owners and collective decision-makers. By definition, a bioenergy village covers at least 50% of its power and heat consumption by regionally produced energy. In mid-2013, there were more than 130 bioenergy villages operational or in the process of setting up their local energy supply system, scattered across all regions of the country (Figure 6).

In most cases, the villagers use local biomass in cogeneration plants. They keep the connection to the main grid and leave it up to each resident whether to

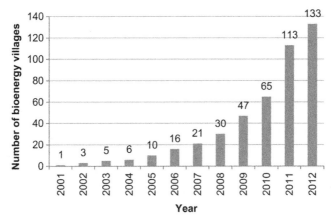

FIGURE 6   Number of bioenergy villages in Germany. *Source: Eurelectric (2013).*

join the autonomous energy supply or to stick with the incumbent regional supplier. The financial viability is ensured by domestic feed-in subsidies for the use of biomass in energy production. In some states like Baden-Württemberg, the initial investment is complemented by direct subsidies.

Jühnde, the first German bioenergy village located slightly north of the country's center, was chosen by the nearby University of Göttingen among several contenders to establish local energy supply. The university withdrew from the engagement relatively soon after the facilities became functional, and the villagers took over the full responsibility of running the system. In the first 10 years of the bioenergy supply's existence, more than two-thirds of the local population joined the heat supply (Burger and Weinmann, 2013). According to Eckhard Fangmeier, the manager of the system, the communal effort could only succeed because there was enough communal support for such an endeavor. The villagers organized themselves in a cooperative, with collective decision mechanisms that triggered a high degree of identification and involvement (ibid.).

The number of energy cooperatives increased almost tenfold from 66 in 2001 to 656 in 2012 (Renewable Energies Agency, 2013). These cooperatives share the characteristic that they have some type of collective ownership in energy assets, many of them finance, for example, solar and wind farms or biomass plants. The main economic activity is the generation of solar power, biomass, wind power, and hydropower, as shown in Table 2.

The Renewable Energies Agency estimates that the 130,000 members of German energy cooperatives invested around €1.2 billion in the so-called citizens' power plants, or *Bürgerkraftwerke*. Even with donations less than €100, citizens are often able to join these cooperatives (ibid.). Their membership ranges from 3 to over 1000 persons, project volume ranges from €50,000 to more than €20 million, and average equity and investments were around €1 and €3 million, respectively (Bellmann, 2012). On aggregate, they were

TABLE 2 German Energy Cooperatives' Main
Economic Activities (2012)

Main economic activity	Percent
Solar	43%
Biomass, wind energy, hydropower	19%
Cogeneration	14%
Grid operations	12%
Distribution	6%
Other services (e.g., consulting)	5%
Construction	1%

Source: Renewable Energies Agency (2013).

able to supply 160,000 households with electricity (Renewable Energies Agency, 2013).

3.3   Municipal Utilities and the Emotionalization of Energy

Municipalities and their local utilities have been active agents in deploying distributed energy resources, especially cogeneration, in many countries (see, e.g., the contribution of Krandall et al. on the city of Boulder, Colorado). In Germany, more than 800 regional distribution companies and utilities operate on the local distribution grid. The smaller ones do not own any generation capacity, but larger operators often produce part of their energy by themselves. Before liberalization, many of these utilities were integrated across different municipal functions including public transport, waste, and water services. Heat and power supplies were typically one of the major cash cows of the enterprises and cross-subsidized deficits in, say, public transport or the operation of swimming pools.

After liberalization, the fate of the municipal utilities seemed unclear, because final customers, especially in the lucrative commercial and industrial segments, started to opt out of old contracts, shop among cheaper entrants, or use the freedom to renegotiate contracts. Some of the local and regional utilities formed alliances, especially to increase their market power in purchasing the energy they needed, but many other grid operating firms remained independent and prospered under the regulation of the German Federal Network Agency, which secured reasonable and risk-free returns from operating the distribution network. Between 2005 and 2013, around 70 new municipal utilities were founded, and major German cities consider reacquiring the distribution network from private operators (Berlo and Wagner, 2013). For example, citizens of

Hamburg voted for repurchasing the electricity and natural gas grids from utilities Vattenfall and E.ON, respectively.

Many municipalities have discovered that decentralized, renewable energy in their home territory increases acceptance and emotionally compensates for comparatively high tariffs, because final consumers can identify with their product offers. They are able to build upon proximity to their clientele by, say, supporting local sports teams or cultural events and combine it with marketing campaigns for carbon-free, locally produced energy, for example, by solar panels on top of public buildings like sports facilities, town halls, or schools. Municipal utilities are particularly active in cogeneration, around 43% of their own generation facilities are based on CHP, while the share of renewable energies increased from 8.7% to 11.4% between 2011 and 2012 (VKU, 2013).

Some municipal utilities may free ride on the green, local branding that they convey. However, in 2012, around 1400 MW of cogeneration units and 800 MW of renewable energy installations were in approval proceedings or under construction (ibid.). Intelligent marketing is likely to create a positive association with final energy consumption and transforms the indistinguishable commodity energy into a communal and emotionalized good of self-determination (Burger and Weinmann, 2013).

4   CHALLENGES AND UNFORESEEN CONSEQUENCES OF THE FEED-IN SCHEME

The momentum that decentralized energy created led to a number of yet unresolved conflicts of interest and regulatory shortcomings in Germany. This section focuses on four dimensions: (1) open questions in financing and redistribution, (2) technologies lagging behind the momentum, (3) ensuring supply, and (4) counterstrategies of Germany's energy incumbents.

4.1   Open Questions in Financing and Redistribution

While the Federal Ministry for the Environment claimed that renewable energy intake led to benefits of the order of €2.8 billion in 2010 because of the so-called merit-order effect, lower wholesale prices—the result of the merit-order effect—turned into a major policy obstacle, especially with respect to acceptance of PV power generation: final electricity consumers have to compensate the owners of the panels by paying the difference between the wholesale tariff and the fixed feed-in subsidy. If wholesale prices decline, the gap becomes larger, and the financial burden for consumers consequently rises. From 2012 to 2013, the levy increased by 47%/kWh. For 2014, the transmission operators expect more than €22 billion to be distributed among the owners of renewable energy installations, of which more than half will be directed toward PV and around 25% to wind power and biomass, respectively (50hertz et al., 2013).

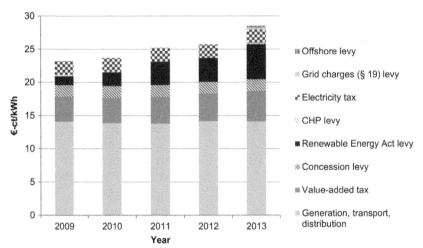

FIGURE 7   The development of residential electricity prices, 2009-2013. *Source: Götz and Lenck (2013).*

These figures dwarf subsidy schemes in other countries. For example, Australia's Climate Change Authority estimates *aggregate* renewable electricity subsidies for large-scale renewable installations at around \$28 billion between 2012 and 2030 (Mountain and Szuster, Chapter 4). For final customers, the German levy is expected to reach 6.2 €ct/kWh (Figure 7).

By contrast, the levy for small cogeneration plants never reached half a euro cent per kilowatt hour over the observation period and even decreased to almost zero in 2012 because of decreasing subsidies for existing installations.

In 2013, around 50% of the average tariff of a residential electricity consumer consisted of taxes and levies, including two new levies that were introduced in 2013, the grid charges (Chapter 19) levy and the offshore levy, which were intended to ease the burden of energy-intensive industries and bear the risk of connecting offshore wind installations, respectively. Self-producers and prosumers can bypass all taxes and levies except the value-added tax (Bode and Groscurth, 2013).

As a response to public criticism of the financial burden imposed on final customers, the government lowered the compensation for new PV panels, putting the profitability in the range of around 4.5%, as opposed to 15% or more, as before; in October 2013, for the first time, compensation fell below 9 €ct/kWh (Schultz, 2013). However, the subsidy scheme for PV installations that are already in operation will continue with the agreed rates. Conservative forecasts estimate that the financial burden induced by feed-in compensation will increase from around €18.5 billion in 2013 to €25.7 billion in 2017 (IWR, 2012). In comparison, cogeneration plants in the parallel subsidy scheme called KWK-G were only supported by €220 million in 2011 (50hertz et al.,

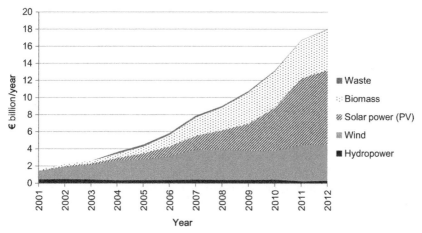

FIGURE 8 Transfer payments related to the Renewable Energy Sources Act in Germany. *Source: BMU (2013).*

2013). In the public discussion about regulatory changes, this legacy of the regime serves as one of the reasons for claims that the entire feed-in system should be abandoned and replaced by a quota system or auctions (see, e.g., Monopolkommission, 2013, for a discussion on the topic; Figure 8).

Similar to the debate on net energy metering in the United States, the discussion about the compensation of owners of PV panels has encountered a social dimension. It has become the trigger of a double controversy, between poor and rich parts of the population on the one hand and between wealthy and less wealthy states in the other hand: Typical prosumers reside in single-family houses with rooftops, they tend to be more affluent or middle class, while residents with rented apartments, which constitute more than half of Germany's population (Destatis, 2012), most of them urban dwellers, cross-subsidize them. For example, final customers in North Rhine-Westphalia, the most populous German state suffering from structural problems because of the decline of the heavy metal and coal industry, sent a net sum of €1.8 billion to prosumers in other German states in 2012. By contrast, the wealthy state of Bavaria in the south of the country received a net €1.2 billion from other states for its renewables plants. One year earlier, in 2011, net transfers into Bavaria were around €200 million lower. The amount redistributed among consumers and prosumers reached €18 billion in 2012—more than double the conventional money transfers between wealthy states like Bavaria and poorer states like North Rhine-Westphalia (Hoffmann, 2013). The grand coalition between conservatives and social democrats, which was elected in the autumn of 2013, planned a fundamental reform of the feed-in legislation to increase public approval of the energy turnaround and decrease the risk of energy poverty of the poorer parts of the population.

With the reform, the exemptions of energy-intensive industries would become more selective and will only be granted if a competitive disadvantage vis-à-vis foreign competitors can be detected.

On the distribution level, an unresolved topic in the regulatory debate is the cost for connecting individual producers of renewable energy to the grid. In residential neighborhoods, the distribution grid was designed to satisfy the typical demand of a household. If homeowners decide to install PV panels on their rooftop, the production may exceed the capacity of the cables. Under current regulation, distribution grid operators are obliged to reinforce or outright upgrade the network. This rule stems from the preliberalization times when electricity supply was a unidirectional public service obligation and all consumers had to be connected, irrespective of their location. With the rise of prosumers, who financially benefit from feeding their electricity into the grid, it becomes an equity issue, whether part of the reinforcement or replacement costs of the network should be borne by the solar customers. Distribution grid operators in rural areas in the south of Germany, where PV intake is particularly high, are latently disadvantaged compared to, say, municipal utilities with a high urban density and few self-producers (Burger and Weinmann, 2013).

Unresolved conflicts about redistribution are not limited to residential customers: Commercial and industrial consumers are exempted from paying the feed-in taxes if they consume more than 1 GWh per year and electricity costs contribute to gross value added above 14%. In 2012, more than 2000 firms applied for exemption, while in 2013, the number further increased by another 300 firms, which in total corresponds to approximately 50% of Germany's industrial electricity consumption or around one-sixth of the country's total electricity consumption. Especially small and medium enterprises that do not meet the conditions tend to suffer from financial disadvantages. For every kilowatt hour they consume, they have to pay 5.3 €ct/kWh, instead of 3.8 €ct/kWh if no exemptions at all were allowed (Reuter, 2013).

Hence, it becomes increasingly attractive for industrial and commercial customers to produce at least parts of their electricity requirements by themselves, since this enables them to avoid paying up to 43% of their electricity prices related to feed-in taxes or the generic energy tax (BDU, 2013). In the 2012 industry survey of the Association of German Chambers of Commerce and Industry (DIHK) among German companies, 13% of the respondents had already implemented self-generation, and a further 16% considered some type of self-generation as a future option (Sorge, 2012).

German energy consumers have already paid more than €200 billion for the energy transition, according to research by the Centre for Solar Energy and Hydrogen Research Baden-Württemberg. By 2025, the overall costs of *Energiewende* are expected to reach €300 billion (Staiß, 2013). However, after 2025, the investments will eventually pay off and lead to substantial savings compared to the status quo (ibid.).

4.2   Technologies Lagging Behind the Momentum

The decentralized energy revolution in Germany has been very successful with respect to the investments in solar, wind, and biomass feed-in installations. It has been far less successful in other areas, in particular building efficiency and microcogeneration (CHP) plants, although the government supports CHP plants from $1.8 €ct/kWh_{el}$ for plants larger than 2 MW to $5.41 €ct/kWh_{el}$ for plants smaller than 50 kW. The subsidies for cogeneration plants with biomass intake rise from $6 €ct/kWh_{el}$ for plants larger than 5 MW to $14.3 €ct/kWh_{el}$ for plants smaller than 150 kW to (Wünsch, 2013). According to industry experts, one of the obstacles for the mass rollout of micro-CHPs is the lack of qualified staff among the craft firms (Egger, 2012).

Among the "Big Four," E.ON launched the most decisive strategy to integrate small cogeneration units in its portfolio. In 2012, the company founded a new subsidiary called "E.ON Connecting Energies." In the first 6 months of 2013, it started operating 35 new cogeneration plants, and further 70 units were under construction until the end of 2013. The company has installed around 6000 units worldwide, of which two-thirds are located in Germany (Flauger, 2013).

Despite numerous efforts of the German government to promote building efficiency, including tax breaks, direct subsidies, and credits from KfW, the German development Bank, the statistics of Germany's largest online database for renovations in buildings show that the overwhelming majority of home-owners change boilers and the heating system only at the end of its functional lifetime—and not as a result of government incentives (Burger and Weinmann, 2013). This reluctance is certainly related to known market failures affecting building efficiency, like the principal-agent dilemma, split incentives, or incomplete contracts. Like in the case of micro-CHPs, another reason may be found in the fragmented market of Germany's around 300,000 craft firms. In Germany, 70% of craft firms in plumbing, gas, and water supply had annual revenues of less than €0.5 million per year (BVR/IFO Institut, 2013). Often, those mini-enterprises do not have the capacity to obtain information on how to improve a homeowner's heating system. Consequently, systemic innovations that manufacturers of, say, heat pumps present at trade fairs do not trickle down into the working practice and skill set of a typical plumber. Hence, final customers do not get to know the full range of options available for a renovation, because the craftsmen are not aware of them. With the digital society and a widespread access to the Internet, this information deficit is likely to be reduced, though.[2]

The market for energy performance contracting is slowly developing. Energy incumbents like E.ON and RWE have traditionally provided energy management services for larger commercial and industrial customers from

---

2. See, for example, Burger and Weinmann (2013) for an extensive discussion.

within their regional subsidiaries. RWE bundled its know-how in a separate entity. Apart from the big players, new entrants like Kofler Energies or Argentus Energie experiment with new concepts like a less engineering-oriented approach, following the principle that 80% of the desired reduction in consumption can be achieved with 20% of typical savings measures, including changes in customer behavior. Other companies, like start-up MeteoViva from Aachen, offer intelligent software solutions that use local meteorologic forecasts to optimize heat requirements over a time span of 72 h. However, the potential of that market is far from being successfully exploited.

As opposed to, say, heat pumps, the sales of micro-CHPs have not experienced any substantial boom in sales since 2010—despite the fact that the German government intends to increase the electricity produced by cogeneration to 25% by 2020 and provides financial incentives and tax breaks. Electricity retailer LichtBlick and incumbent car manufacturer Volkswagen allied for an ambitious project to set up virtual power plants with residential homeowners who have a micro-CHP device installed in their basement. While Volkswagen had been experimenting with micro-CHPs since the 1980s, LichtBlick was founded in the aftermath of liberalization as a vendor of renewable energies and became one of the largest players in the German retail market with over half a million customers. The company's strategy to promote micro-CHPs was based on the idea that a typical cogeneration plant requires a constant heat sink and an operating time of, say, 4000-6000 h per year to run profitably. Since heat demand of single-family or two-family houses is affected by strong seasonal variations, the profitability of the scheme could only be ensured by the strategy to let the plant run only in times when electricity wholesale prices were high. Then, the revenues would allow for operating times in a much lower range to reach profitability. The heat could be stored in a residential low-cost storage device for later use.

LichtBlick announced a target of selling 100,000 home-generating units. However, consumer acceptance turned out to be less enthusiastic than anticipated, and up to June 2013, LichtBlick was able to install around 700 units in the market.

With declining electricity wholesale prices, the idea of connecting multiple micro-CHPs and selling the aggregated electricity on the wholesale market became equally under threat as the feed-in of conventional thermal power plants, as described in the succeeding text. In 2013, LichtBlick changed its strategy and used a 2012 amendment to the Renewable Energy Sources Act, which rewards zero carbon and flexible supply technologies, to ensure profitability of its units: instead of conventional natural gas, the micro-CHPs use biogas that is fed into the gas grid whenever the electricity system requires stabilization. The home-generating CHPs then produce electricity for the homeowners' own consumption requirements and receive a subsidy for making biogas-generated electricity available in times of deficits of fluctuating other renewable energies like solar or wind (Stahl, 2013).

Apart from LichtBlick and VW, other companies like Viessmann and Senertec try to enter the market with Stirling motors. German heating device manufacturer Vaillant offers a micro-CHP based on a solid oxide fuel cell, of which 100 units have been part of a field test financed by the EU. Until mid-2013, overall customer acceptance of micro-CHPs remained sluggish, though.

4.3 Ensuring Security of Supply

Will decentralized energy supply threaten long-term system stability and the financial viability of utilities? The German experience suggests that system stability is less affected than the survival of established incumbents.

Wind and solar power are erratic by nature, but forecasting becomes increasingly sophisticated. Nonetheless, the interventions of Germany's largest transmission grid operator TenneT rose from 2 interventions in 2003 to 290 in 2010 to 970 in 2012 (Hübner, 2013), mainly due to redispatch measures linked to a divergence between predicted and actual renewable energy supply. Solar peaks like the one on 18 April 2013, which was already mentioned earlier, will become more frequent in the near future and are likely to affect operation of the distribution grid. A number of pilot projects in the so-called lighthouse regions in Germany are under way, which experiment with smart grid components, including complex control systems for balancing intermittent renewable energies, sensors, and "intelligent" transformer stations. If parts of the distribution grid become more autonomous or experience a high intake of decentralized renewable energy, local storage solutions may be more cost-efficient to buffer excess feed-in than to reinforce the grid. For example, distribution grid operator EnBW ODR in southern Germany has entered a collaboration with battery manufacturer Varta to test stationary lithium-ion batteries in 2012.

By government decree, the nuclear phaseout is scheduled to be completed by 2022. Many of the nuclear reactors are located near the industrial hubs in southern Germany. Meanwhile, offshore wind parks in the North Sea and Baltic Sea are supposed to replace the missing electricity from the nuclear plants. Hence, major investments in AC or DC high-voltage lines connecting the northern wind farms to the southern load centers are planned. In April 2013, the German government ratified the German energy agency's recommendations to newly build 2800 km and reinforce 2900 km of transmission lines with investments of €21 billion by 2023 (Gartmair, 2013). However, if decentralized energy leads to largely autonomous island systems that will develop into local markets with trading platforms for aggregators of renewable energy, demand-side management, and regional brokers, the ambitious extensions of the transmission grid may not be necessary.

While transmission and distribution are regulated businesses under the current market design, generation is still a competitive element in the electricity value chain. Not only the major German energy utilities but also some municipal utilities with their own fossil generation portfolio started to experience the

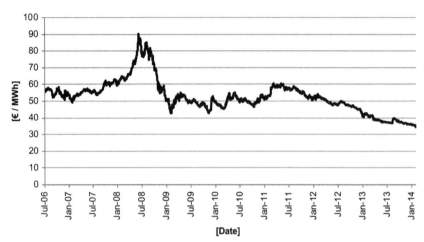

FIGURE 9 Prices for Phelix Baseload Year Futures (Cal-14) on the European Electricity Exchange, as of 19 December 2013. *Source: EEX (2013).*

impact of decentralized energy on their revenue streams in 2011. Between that year and mid-2013, average wholesale prices for baseload declined by 28.9% on the electricity spot market (IWR, 2013; Figure 9).

Decentralized generation is considered one of the underlying reasons: the peak of solar radiation coincides with the midday demand peak that used to guarantee reasonable returns for the energy utilities. According to estimates by the German Association of Energy and Water Industries (BDEW), from 2011 to 2012, wholesale spot market prices decreased by around 8 €/MWh between April and September, which the association calls the "level effect." During the midday hours, when radiation is highest, a further 4 €/MWh reduction in the price can be attributed to PV feed-in (Figure 10).

Since regulation imposed that PV energy and other renewable energies had a priority feed-in vis-à-vis thermal or nuclear plants and they bid at zero marginal cost, the dispatch of more expensive plants occurs less frequently. In particular, natural gas plants, which have comparatively high marginal costs because of their primary fuel, are crowded out of the wholesale market. The owners of several highly efficient gas plants that were recently built threatened to mothball some of their assets, because they were not running for a sufficiently large amount of hours per year to justify their standby mode, let alone recoup the investment. For the owners, it became cheaper to shut the plants outright down.

This crowding-out effect on the wholesale market led to a more intense use of coal and lignite plants with lower operating costs but higher specific carbon dioxide emissions than natural gas plants. From an environmental perspective, natural gas plants as a "bridge technology" would be preferable to complement volatile renewable energies, but the business case for new gas-fired capacity has been substantially dampened due to the combination of low wholesale prices and low operating hours. Paradoxically, the success of the feed-in regulation

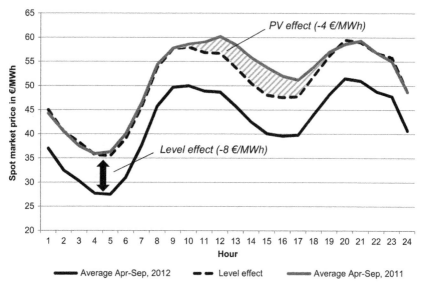

FIGURE 10   Change in the hourly wholesale prices, averages April to September 2011 and 2012. *Source: Hille (2013).*

has thus led to the paradoxical situation that Germany's carbon dioxide emissions increased by 2% from 2011 to 2012—despite further extension of renewable energies in the overall generation portfolio (Umweltbundesamt, 2013).

The German regulatory agency entered individual negotiations with the owners if the plants were considered system-relevant; that means a shutdown could have consequences for the overall grid stability. For example, E.ON reached a deal with the regulatory agency in April 2013 to receive a double-digit million euro compensation per block per year to keep one of the most efficient CCGT plants in Germany on standby for 3 years (Hack et al., 2013). The costs for this additional reserve capacity are borne by final consumers. With the new government in place, a market for capacity will most likely replace the current reserve mechanism.

4.4 Counterstrategies of Energy Utilities

Since 2011, the revenues of all major German utilities have substantially decreased. Share prices of the two incumbent German energy companies E.ON and RWE plummeted by around 70% between 2008 and early 2013. Deutsche Bank downgraded integrated German utilities E.ON and RWE from *Hold* to *Sell* in January 2013 (Sanz De Madrid Grosse, 2013). The energy turnaround leaves the big German energy utilities confronted with a major strategic challenge, which Peter Terium, CEO of Germany's second largest utility RWE, coined "the biggest sectoral crisis of all times" (Student, 2013).

One reason for the erosion of income and gloomy prospects for the utilities was the government's post-Fukushima decision to phase out nuclear power.

The "Big Four" were obliged to shut down and disconnect some of their units, whereas beforehand, nuclear power benefitted from cheap primary fuel and the fact that investment costs had been written off for most of the reactors.

However, the crisis of the major German utilities cannot solely be associated with the phaseout decision: "During the 2000s, European utilities overinvested in generating capacity from fossil fuels, boosting it by 16% in Europe [. . .]. The market for electricity did not grow by nearly that amount, even in good times; then the financial crisis hit demand," comments the Economist (2013).

Most importantly, the revenues of the energy incumbents were directly affected by distributed energy resources: on the German electricity exchange EEX, the mass adoption of decentralized energy technologies has led to an increasing amount of hours with very low or even negative spot market prices (Table 3).

The comparison between the first 6 months of 2012 with the same period 1 year later shows that hours with spot prices below 10 €/MWh increased almost fourfold. According to Mayer et al. (2013, p. 9), high PV and wind power intake drove prices down in the spring and early summer of 2013 (Figure 11).

Distributed energy offers opportunities for a range of new competitors. Apart from companies such as Volkswagen, LichtBlick, Honda, or Vaillant, which target the micro-CHP segment, new entrants in energy performance contracting and building efficiency like Kofler, Argentus, or MeteoViva nibble into

TABLE 3 Number of Hours with Negative Spot Prices at EPEX Spot Market

Year	2008	2009	2010	2011	2012	January to June 2013
Number of hours	15	71	12	15	56	36

Sources: Monopolkommission (2013; data for 2008-2012); Mayer et al. (2013; data for 2013).

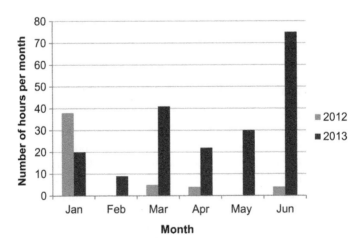

FIGURE 11   Number of hours with spot prices below 10 €/MWh at EPEX spot market. *Source: Mayer et al. (2013).*

the market share of energy incumbents. Start-ups like digitalSTROM offer intelligent and easy ways to connect all electric household appliances, while companies like family-owned "Hidden Champions" Mennekes develop plugs and charging stations for electric vehicles.

German energy incumbents do not only face challenges from competitors *within* the energy and manufacturing sector. Incumbents and start-ups from the Internet and communication technologies sector have discovered the potential value associated with the smart grid, local area networks, and grid optimization. Among them, leading players like Deutsche Telekom, Cisco, or SAP are still in the orientation phase and have not yet decided whether to target the electricity sector in outright competition to the energy incumbents or to enter the market in strategic alliances.

The following "nibble" chart provides an overview of companies that are likely to offer or have already started to offer products and services along the electricity value chain with the objective of nibbling away the business from incumbents (Figure 12).

The strategy of German incumbent energy suppliers vis-à-vis the momentum that decentralized energy has gained is to use their financial and technical expertise to build and operate large-scale renewables, in particular offshore and onshore wind farms. All of them expand complementary services, especially in energy performance contracting and the optimization of procurement, self-production of commercial and industrial customers, and products and services related to the grid, in terms of both hardware and software.

As mentioned in the previous section, incumbent utility E.ON already installed 4000 smaller cogeneration units in Germany. Its strategy is to provide CHP to large customers like wholesale market chain Metro. In 2013, decentralized generation contributed €400 million to the company's overall EBITDA (Flauger, 2013). In addition, it allied with Entelios, a start-up company based in Munich that specializes in demand-side management, while southwestern

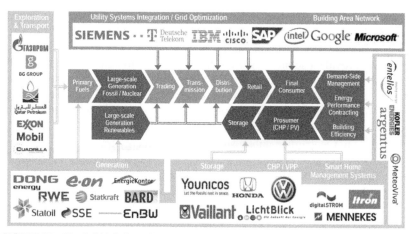

FIGURE 12   The "nibble" chart. *Source: Own analysis.*

utility EnBW aims to increase its EBITDA in "sales" and as "decentralized service provider" from €200 million in 2012 to €500 million in 2020. RWE targets the market with innovative products like Smarthome devices or public charging station equipment for electric vehicles. The company intends to strengthen its role as a "holistic energy service provider" and shifts its strategic objectives from generation to retail and grid operations.

German utilities have realized that distributed energy poses a serious threat to their business models. All of them explore how the decreasing revenue streams from centralized power plants can be offset by new products and services. Smart meter rollout, smart grid, and metering services may compensate to some extent the decreasing importance of traditional generation, but maintaining the workforce at current numbers seems unlikely.

5   CONCLUSIONS

The unprecedented growth and mass adoption of distributed energy resources in Germany have paved the way into a future of low-carbon, decentralized power supply—much faster and more fundamental than political and corporate decision-makers anticipated. Germany experiences all advantages and disadvantages of a first-mover: The value proposition of a clean and green country ignited initial enthusiasm and found widespread public support. It was accompanied by a mushrooming, innovative industry active in all parts of the value chain, which became a nonnegligible factor in the economy. Promoted by a generous regulatory regime, private residents, farmers, local associations, and communities started investing in distributed energy technologies. They were attracted not only by guaranteed financial returns but also by the promise of empowerment and self-determination.

The scheme became a victim of its own success—a phase of disillusionment, public controversy, and prominent bankruptcies followed. It became clear that the switch to decentralized energy supply can be achieved, albeit at substantial costs to society. The focus on sustainability led to dilemmas with the other two dimensions, affordability and security of supply. Incumbent energy companies still struggle to redefine their strategies and business models. The state has yet to provide a sound regulatory framework to strengthen systemic resiliency while intermittent energy production continues to grow. How can investment at controllable and predictable levels be incentivized? How should the burden of risk and reward be shared between prosumers and consumers? How could an appropriate compensation and redistribution scheme be designed, which does not jeopardize public support for the energy transition?

Germany has now entered a third phase of consolidation and incremental rather than radical transformation. Its trajectory toward decentralized energy will be decelerated by the envisaged downscaling of financial support for new installations. However, the country has all the preconditions to make the shift to distributed energy resources a lasting success: Its population is sufficiently wealthy to finance the transition, its industry provides technological expertise to master the challenges of a less reliable and stable power supply, and innovative newcomers experiment with business models that exploit the largely untapped potential of demand-side management, building efficiency, and energy performance contracting.

At the moment, the grid infrastructure is highly reliable, there has not yet been any major blackout, and backup power from conventional power plants is abundant. The Energiewende will face its real baptism of fire only in 5-10 years when the hardware of the existing system will incrementally retire.

If global electricity demand will increase by 70% between today and 2035, as the International Energy Agency predicts (IEA, 2012), political and corporate decision-makers all across the globe will closely monitor how Germany manages the transformation of its electricity sector. It may serve as a blueprint for a low-carbon trajectory for industrialized and developing countries. Departing from the German experience, they may be able to avoid mistakes, accelerate the transition, and leapfrog to the next stage of energy supply—decentralized, renewable, and secure.

REFERENCES

50hertz, Amprion, Tennet, Transnet BW. 2013. EEG/KWK-G Portal [Online]. Available: http://www.eeg-kwk.net (accessed 22 October 2013).

Arnold, Z., Jenkins, J., Lin, A., 2009. Case Studies in American Innovation – A New Look at Government Involvement in Technological Development. Breakthrough Institute, Oakland, CA.

BDU, 2013. Energiewende im Mittelstand. Bundesverband Deutscher Unternehmensberater, Berlin.

Bellmann, K., 2012. Energiegenossenschaften: Der Bürger gestaltet die Energiewende – neue Kooperationen vor Ort. In: BDEW-Symposium "Energiewende braucht Investitionen", Berlin.

Berlo, K., Wagner, O., 2013. Stadtwerke-Neugründungen und Rekommunalisierungen. Wuppertal Institut für Klima, Umwelt, Energie, Wupeertal.

BMU, 2009. In: Dürrschmidt, W. (Ed.), Strom aus erneuerbaren Energien – Zukunftsinvestition mit Perspektiven. Bundesministerium für Umwelt, Naturschutz und Reaktorsicherheit, Berlin.

BMU, 2013. Zeitreihen zur Entwicklung der erneuerbaren Energien in Deutschland. Bundesministerium für Umwelt, Naturschutz und Reaktorsicherheit, Bonn/Berlin.

Bode, S., Groscurth, H.-M., 2013. Zur vermeintlichen "Grid Parity" von Photovoltaik-Anlagen. Energiewirtschaftliche Tagesfragen 63, 7.

BP, 2013. Statistical Review of World Energy 2013, London.

Bruns, E., Ohlhorst, D., Wenzel, B., 2010. 20 Jahre Förderung von Strom aus Erneuerbaren Energien in Deutschland – eine Erfolgsgeschichte, Renews Spezial.

Bundesverband Solarwirtschaft, 2013. Statistische Zahlen der deutschen Solarstrombranche (Photovoltaik).

Burger, C., Weinmann, J., 2013. The Decentralized Energy Revolution – Business Strategies for a New Paradigm. Palgrave Macmillan, Basingstoke, UK.

BVR/IFO Institut, 2013. Klempner, Gas- und Wasserinstallateure, VR Branchen Special.

Carrington, D., 2012. Germany's renewable energy revolution leaves UK in the shade. The Guardian, 30 May.

Destatis, 2012. Wohnen 2010: mehr Wohnungen, mehr Wohneigentum, Pressemitteilung Nr. 093 ed.

Economist, 2013. How to lose half a trillion euros. The Economist, 12 October.

EEX, 2013. Phelix Baseload Year Futures (Cal-14) [Online]. European Electricity Exchange, Leipzig, Available at http://www.eex.com/ (accessed 12 November 2013).

Egger, R., 2012. Ein Beitrag zur Energiewende: Mini- und Mikro-KWK-Technologie. Heizungsjournal. 9.

Eurelectric, 2013. Utilities: Powerhouses of Innovation. Eurelectric, Brussels.

Federal Government, 2012. National Sustainable Development Strategy – 2012 Progress Report. Federal Government, Berlin.

Flauger, J., 2013. Abschied vom Gigantismus. Handelsblatt. 20.

Gartmair, H., 2013. Stand des Netzausbaus – Bedarf, Planung und Umsetzung, FfE-Tagung Energieeffizienz und Erneuerbare Energien im Wettbewerb, TenneT.

Götz, P., Lenck, T., 2013. Kompensieren sinkende Beschaffungskosten den Anstieg der EEG-Umlage für Haushaltskunden? Energy Brainpool, Berlin.

Hack, J., Kaeckenhoff, T., Steitz, C., 2013. E.ON says reaches deal to keep Irsching plant open. Reuters.

Haselhuhn, R., 2012. Aktuelle Situation der Photovoltaik in Deutschland. Jahreshauptversammlung DGS – Landesverband Berlin Brandenburg, Berlin.

Hille, M., 2013. Alte Märkte, neue Märkte, überhaupt noch Märkte? Ideen für die Zukunft des Energiemarkts. Berliner Energietage, Berlin.

Hirschl, B., 2012. Kommunale Wertschöpfung und Beschäftigung durch erneuerbare Energien. In: Energiesymposium Handwerk, Stuttgart.

Hoffmann, K.P., 2013. Bayern streicht ein, Berlin zahlt drauf. Tagesspiegel.

Hübner, C., 2013. Herausforderung Energiewende aus der Sicht eines Übertragungsnetzbetreibers, TenneT.

IEA, 2012. Agency, I.E. (Ed.), World Energy Outlook 2012. OECD, Paris.

IWR, 2012. EEG-Prognosen: Alles auf Nummer sicher? Available from: http://www.iwr.de/ (accessed 1 October 2013).

IWR, 2013. Börsen-Strompreis fällt im ersten Halbjahr 2013 um 12,6 Prozent.

Mayer, J.N., Kreifels, N., Burger, B., 2013. Kohleverstromung zu Zeiten niedriger Börsenstrompreise. Fraunhofer-Institut für Solare Energiesysteme, Freiburg.

Mckinsey, 2007. Costs and potentials of greenhouse gas abatement in Germany. McKinsey, Berlin.

Monopolkommission, 2013. Energie 2013: Wettbewerb in Zeiten der Energiewende.

Page, N., Czuba, C.E., 1999. Empowerment: what is it? J. Extension. 37.

Photovoltaik.org., 2013. Rekord bei der Einspeisung von Solarstrom und Windkraft. Available from: http://www.photovoltaik.org (accessed 1 October 2013).

Renewable Energies Agency, 2013. Charts and Data [Online]. Available, http://www.unendlich-viel-energie.de/ (accessed 1 October 2013).

Reuter, B., 2013. EEG-Umlage: Gefährden Industrie-Ausnahmen die Energiewende? Available from: http://green.wiwo.de/ (accessed 1 October 2013).

Madrid Grosse, V., 2013. Utilities 2013 Outlook. Deutsche Bank.

Schultz, S., 2013. Erneuerbare Energien: Markt für Solarkraftwerke kollabiert. Spiegel Online.

Sorge, N.-V., 2012. Großkunden laufen RWE und Eon davon. Manager Magazin.

Stahl, L.-F., 2013. LichtBlick ZuhauseKraftwerke wechseln vom KWKG zum EEG. Available from: http://www.bhkw-infothek.de (accessed 1 October 2013).

Staiß, F., 2013. Henning, H.-M. (Ed.), Example Germany: The Cost of the Energy Transition. Fraunhofer ISE, Gleisdorf.

Student, D., 2013. RWE-Chef Terium hält Gehaltskürzungen für möglich. Manager-Magazin.

Umweltbundesamt, 2013. Treibhausgasausstoss in Deutschland 2012 – vorläufige Zahlen aufgrund erster Berechnungen und Schätzungen des Umweltbundesamtes.

VKU, 2013. Verband Kommunaler Unternehmen, (Ed.), Deutsche Stadtwerke liefern wichtigen Beitrag zur Energiewende, Berlin.

Wünsch, M., 2013. Maßnahmen zur nachhaltigen Integration von Systemen zur gekoppelten Strom- und Wärmebereitstellung in das neue Energieversorgungssystem. Prognos, Berlin.

Chapter 4

Australia's Million Solar Roofs: Disruption on the Fringes or the Beginning of a New Order?

Bruce Mountain and Paul Szuster

ABSTRACT

Between the start of 2010 and the end 2012, 900,000 rooftop photovoltaic (PV) systems were installed in Australia. Market penetration—more than one in eight household rooftops—is now the highest in the world. Households spent more than $9 billion and will receive subsidies over the life of their systems of $8 billion. The average cost of electricity from these rooftop systems (around $160/MWh) is about half the average household electricity prices.

Excluding feed-in tariffs, claims of cross-subsidy between energy users that do not have PV and those that do are not conclusive. PV households are paying $252 million per year less to monopoly network service providers, but they are avoiding future network expansion. The latter seem to be worth less than the former, but the gap may not be large and the question remains whether monopoly network service providers should recover income lost to competitors, from their remaining captive customers.

Keywords: Solar PVs, Australia, Cost-effectiveness, PV subsidies, PV penetration

1 INTRODUCTION

At the end of 2012, 12% of Australia's detached and semidetached houses—a little over 1 million homes—had photovoltaic (PV) systems on their roofs. Per household, Australia is, by a wide margin, the global leader in PV installation. This outcome was achieved from a negligible base in 3 years from the start of 2010. Such rapid expansion was widely unexpected and has involved substantial capital and production subsides.

This chapter examines first whether those households that installed rooftop PV have had a windfall gain. It then examines the rapid rise of PV from the perspective of other energy users: has PV on others' rooftops been all cost and no benefit to them?

Distributed Generation and its Implications for the Utility Industry. http://dx.doi.org/10.1016/B978-0-12-800240-7.00004-7

The analysis in this chapter is not a traditional cost/benefit analysis. Rather, it contributes to the issues discussed in other chapters of this book by evaluating the profitability of PV investment to households that installed PV, by valuing the subsidies rooftop PV has received, and by presenting preliminary estimates of the impact of rooftop PV on wholesale electricity markets and on networks. Other outcomes on which PV support policy was predicated—industry development, job creation, and reduction in greenhouse gas emissions—are not examined.

The chapter is organized as follows: Section 2 provides context, focusing mainly on the factors that promoted rapid PV uptake. Section 3 describes the methodology; Section 4, the results; and Section 5, a discussion. A concluding section draws out the main points.

2   CONTEXT

Between the end of 2007 and start of 2013, the number of PV systems installed on the roofs of homes in Australia increased from 8000 to 1.1 million as shown in Figure 1.

The analysis in this chapter focuses on the period from 1 January 2010 to 31 December 2012 during which the major expansion in PV capacity occurred. In this period, 899,014 rooftop PV systems were installed. These systems sit on the roofs of 12% of Australia's detached or semidetached dwellings. The density varies in the different parts of Australia from 21% of households in the state of South Australia to 5% of households in the state of Tasmania. The total

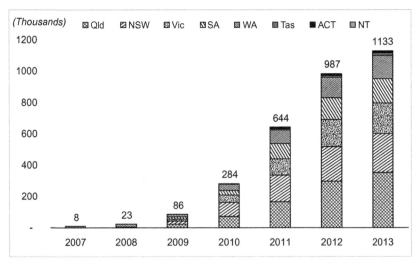

FIGURE 1   Australian cumulative solar PV installations, 2007 to October 2013 (Clean Energy Regulator, 2013c).

capacity of these 899,014 PV systems is estimated to be 2.3 GW and they are expected to produce 3.4 TWh during 2013.

Average PV system size in Australia is small by global standards. For example, Germany, the global leader in PV, had an average system size per installation at the end of 2012 of 40 kW_{peak}.[1] The comparative figure in Australia is just 3 kW_{peak}.

PV system size in Australia has progressively increased from 1.1 kW for systems installed in 2009 to 3 kW for systems installed in 2012. The small system size reflects the dominance of household rooftop PV installation. The design of PV subsidies—discussed in the succeeding text—has also affected system size. In addition, the very high level of private home ownership[2] in Australia is likely to have affected the propensity of households to invest in PV.

The rapid expansion of PV in the period 2010-2012 seems to be explained by three factors: rising electricity prices, capital and production subsidies, and declining PV system costs, described in the rest of this section.

2.1   Rising Household Electricity Prices

Australian electricity prices rose to 90%—adjusted for changes in the consumer price index—between 2007 and 2012 as shown in Figure 2. Rising network costs have been the main contributor to these increases (see Productivity Commission (2013a) for a full discussion on this).

Household electricity prices in Australia, at purchasing power parity exchange rates, are higher than average prices in the European Union and Japan and much higher than average prices in Canada or the United States (Mountain, 2012a). In 2013, average household electricity prices on regulated reference tariffs in the different jurisdictions of Australia range between AUD280/MWh and AUD380/MWh, with a national average of AUD320/MWh.[3] Some households that have chosen tariffs offered in the contestable market have however been able to achieve price reductions of around 10-15% from regulated reference tariff prices.

2.2   Subsidies

From 2010 to 2012, significant capital and production subsidies were available to promote the uptake of PV. Capital subsidies were paid through a mandatory renewable energy certificate scheme (Clean Energy Regulator, 2013a).

---

1. See Chapter 3.
2. Eighty-eight percent of all detached or semidetached dwellings on whose roofs the vast majority of PV systems have been installed are privately owned.
3. CME analysis based on tariff data in jurisdictional tariff schedules is identified in ActewAGL, 2013b; Aurora, 2013; EnergyAustralia, 2013; Essential Services Commission, 2013; Government of South Australia, 2013; Queensland Competition Authority (QCA), 2013b; Synergy, 2013.

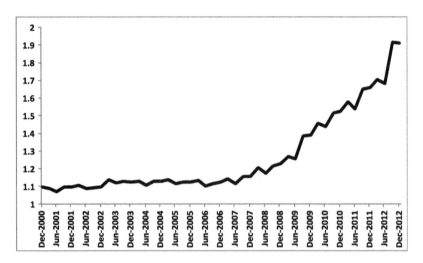

FIGURE 2   Australian household electricity price index, adjusted for changes in the consumer price index (Australian Bureau of Statistics, 2013).

Production subsidies were paid through jurisdictional government-determined feed-in tariffs (FiTs).[4]

2.2.1   Capital Subsidies

A nationwide mandatory renewable certificate scheme (the renewable energy target) funded capital subsidies. Under this scheme, a number of certificates can be created once a PV system has been commissioned. The number of certificates per installed system is based on the expected—"deemed"—production for 15 years. Over the period from 2009 to 2013, a multiplier for the deemed production affected the certificate entitlement per installation for the first 1.5 kW of installed capacity. The multiplier reduced from 5 in July 2009 to 1 from the start of 2013.

Electricity retailers have an obligation to surrender a number of certificates to the Clean Energy Regulator each year. The regulator calculates the obligation each year as a proportion of the retailers' sales volumes. Through this, the cost of certificates incurred by the retailers is recovered from energy users. Large emission-intensive and trade-exposed energy users are partially exempt from certificate obligations (Commonwealth of Australia (Climate Change Authority), 2012).

The capital subsidy arrangements were fundamentally changed from the start of 2011. Before 2011, the certificates created by PV were fungible with the certificates from other certificate-eligible renewable sources. At this time, certificate prices were determined in a market that reflected an annual

---

4. For example in the NSW jurisdiction, see Independent Pricing and Regulatory Tribunal (IPART, 2013).

mandatory demand for certificates but a variant supply of certificates from PV and other competing eligible renewable energy technologies. After 1 January 2011, certificates from PV became part of a subsidy scheme with unlimited certificate volume and the Australian government became a buyer of the last resort offering a fixed price of $40 per certificate. Again, charges are passed through to users through an annual obligation determined by the regulator based on the number of certificates created.

The majority of PV owners sold their certificates to their installers as part of their PV purchase. PV installers would typically offer to purchase certificates from their customers for $25 to $30 per certificate, and they would then typically sell these certificates to retailers directly or through brokers for around 25% more than they paid (Martin and Rice, 2013).[5]

2.2.2   Production Subsidies

Production subsidies were paid through jurisdictional government-determined—state and territory government—FiTs that set a price for electricity fed back into the electricity grid. The FiT prices varied significantly in each jurisdiction over the period from 2008 to 2013, scaled back as solar PV installations increased, as shown in Figure 3.

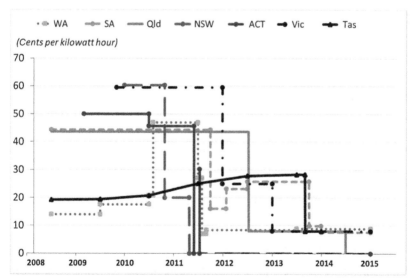

FIGURE 3   Jurisdiction government-specific feed-in tariffs, 2008-2015.[6]

5. The currency in this chapter is Australian dollar, roughly USD $0.92 at the time of writing.
6. CME analysis using data in sources identified in ACIL Tasman, 2013b; ACT Government, 2009, 2011a, 2011b; Collier, 2010; Gallagher, 2011; Government of Western Australia, 2013a; Green, 2013; NSW Government, 2013; Queensland Government, 2013; State Government of Victoria, 2013a, 2013b, 2013c, 2013d; Tasmanian Government, 2013.

In addition to jurisdictional government-determined FiTs, energy retailers typically also offered payments to households for electricity exported to the electricity grids from their PV systems. In all jurisdictions other than New South Wales (NSW) and the Australian Capital Territory (ACT), the FiT and retailer payments are based on the net-metered generation—that is, PV production less electricity used at the point of installation. In NSW and the ACT, FiT payments were based on total PV production.[7]

The retailers in turn recover the aggregate mandatory feed-in payments by charging this to the regulated network service providers, except in Western Australia. The regulated network service providers in turn recover the FiT payments from electricity consumers through regulated charges to the consumers that they supply. In the case of one jurisdiction, NSW, the retailers recover the FiT payments from the jurisdictional government who then in turn recovers this through a levy on all energy users. The NSW government, uniquely, also requires retailers to pay to the government a set amount[8] for the electricity that they procure from PV owners.

In addition to FiTs, PV owners typically also receive payment from their energy retailers for the electricity production exported to the grid (which the retailers then on-sell to their other customers). In four jurisdictions (Victoria (VIC), South Australia (SA), Western Australia (WA), and Tasmania (TAS)), retailers were required to pay PV owners a specified price for the electricity exported to the grid, while three other jurisdictions (Queensland (QLD), NSW, and ACT) do not mandate the price that retailers are required to pay PV owners for electricity exported to the grid.

2.3   PV System Costs

The installed cost of PV on household roofs declined from an estimated $12 per watt in 2008 to $3 per watt in 2012 (Flannery and Sahajwalla, 2013; Eadie and Elliott, 2013). The decline in costs over this period is attributable to three main factors: appreciation of the Australian dollar, sharp reductions in the price of solar panels manufactured in China, and greater competition among system installers (Martin, 2011; Martin and Rice, 2013).

3   METHODOLOGY

Conventional project investment analysis methods are applied to evaluate the net benefit to the roughly 900,000 households that installed PV in the period from 1 January 2010 to 31 December 2012. Discounted present values of the capital outlay are set against the discounted present value of the capital

---

7. "Gross-metered" production.
8. For FY 2013-2014, this was deemed 6.6 cent/kWh.

subsidies, production subsidies, income from retailers, and avoided electricity purchases to derive the internal rate of return (IRR).[9]

In period from 2010 to 2012, there were significant changes in production and capital subsidies and also in electricity prices and the capital costs of PV systems. For this reason, the relevant inputs in the analysis—the number of installations, FiTs, capital subsidies, and system costs—are specified monthly and per jurisdiction.[10]

Data on production subsidies per year were created based on PV installation data by jurisdiction per month multiplied by the applicable monthly jurisdiction-specific certificate eligibility formulas.

The calculation of avoided purchases reflects data on jurisdiction-specific average residential load profiles and average demand, jurisdictions-specific solar radiation profiles, and PV system sizes. The calculation reflects the progressive degradation in PV output over time.

Production subsidies—FiT payments—once started, decline at 6% per year to account for average household tenure in Australia of 16.7 years (Bloxham et al., 2010) (production subsidies terminate once the recipient of the FiT sells their home). The electricity produced from rooftop PV over the assumed 25-year installation life declines by 0.5% per year, reflecting the gradual degradation in system output over time (Jordan and Kurtz, 2013; Peng et al., 2013). Household electricity prices for the next 25 years, having risen significantly over the last 5 years, are assumed to return to their trend behavior for most of the last 40 years—no change in the constant currency.

4   RESULTS

4.1   IRR of Investment in Rooftop PV

Figure 4 shows the Australian-wide aggregate net present value (NPV) of investment in PV by 899,000 households over the period from 1 January 2010 to 31 December 2012, discounted at a rate of 9.8% (the internal rate of return—the discount factor at which the NPV is zero).

In this figure,

- capital outlay is the present value of the outlay that households made at the time of the PV system's installation,

---

9. This is the rate of return at which the net present value is equal to zero.

10. A notable feature of an analysis of PV in Australia is the absence of measured data on PV production or grid exports. For this reason, the calculation of annual PV production is based on installation data and the Clean Energy Council's Grid-Connected Solar PV Systems Design Guidelines of production per kilowatt of solar capacity in Australian capital cities Clean Energy Council (2012). In addition, for calculations that require hourly and monthly PV production, data from the ANZSES (2006). Australian Solar Radiation Data Handbook (ASRDH) was used for hourly irradiance at capital cities on a north-facing plane inclined at the capital city's angle of latitude.

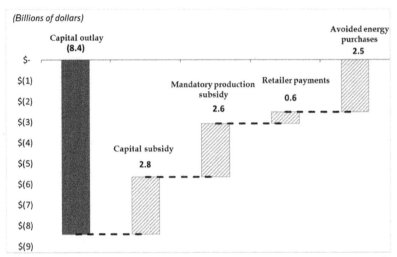

FIGURE 4   Discounted present value (to households) of their investment in PV ($ billion).[11]

- capital subsidy is the present value of the renewable energy certificates received by households at the time of installation,
- the mandatory production subsidy is the present value of the jurisdictional government-mandated FiTs,
- retailer payments are the present value of the payments received by households from energy retailers for electricity fed back into the electricity grid,
- avoided energy purchases are the present value of the energy produced by PV and consumed at the household thereby avoiding purchases from the grid.

The Australian-wide IRR is 9.8%. However, the IRR varies for different jurisdictions: SA (11.5%), NSW (10.7%), ACT (10.5%), QLD (9.7%), WA (9.7%), VIC (7.7%), and TAS (5.5%). These calculations pertain to the national and jurisdictional aggregates. Individual households may have obtained a higher or lower return on their investment depending on the price they paid, their household electricity consumption, the size of the PV system, their location, and so on.

11. CME analysis using data in sources identified in ActewAGL, 2010, 2011, 2012, 2013a; AECOM, 2010; AGL Energy Limited, 2013; Aurora, 2009, 2010, 2011, 2012; Business Spectator, 2013; Clean Energy Regulator, 2013a, 2013c; Commonwealth of Australia, 2013; Deloitte, 2011; Economic Regulation Authority, 2012; EnergyAustralia, 2009, 2010, 2011, 2012; Essential Services Commission of South Australia (ESCOSA), 2013; Essential Services Commission Victoria, 2013; Government of South Australia, 2009, 2011, 2012; Government of Western Australia, 2011, 2013b, 2013c; Green Energy Markets, 2013; Independent Competition and Regulatory Commission, 2009; OTTER, 2013; Queensland Competition Authority (QCA), 2009, 2010, 2011, 2012, 2013a; Red Energy, 2010; Victorian Competition and Efficiency Commission (VCEC), 2012.

Also, these calculations pertain to the costs and benefits of PV received by the initial owners of the PV systems, that is, those households who incurred the outlay to install PV—"original PV owners"—and consequently received the capital and production subsidies. Many of these homes will be sold before the assumed 25-year life of the PV. When the house is sold, mandatory production subsidies end.

Subsequent owners of the home on whose roof the PV system is attached—"subsequent PV owners"—do not make the capital outlay and neither do they receive production or capital subsidies. Over the remaining life of the installed PVs, in addition to the amounts shown in Figure 4, retailer payments of $534 million and avoided energy purchases from the grid of $1.6 billion will accrue to subsequent PV owners.

The net benefit that subsequent PV owners receive from the PV system on the roof of the house they have acquired will depend on whether the price they paid for the house reflects a premium that reflects the existence of the PV system. If we assume that no premium is factored into the price of PV-equipped houses, then subsequent PV owners are in aggregate $2.1 billion better off (retailer payments of $534 million plus avoided energy purchases of $1.6 billion).

This calculation reveals the sensitivity of the calculation of the return on investment to original PV owners and to assumptions about the extent to which subsequent PV owners value the existence of a PV system on the roof of the houses they acquire. The results shown in Figure 4 assume that subsequent PV owners do not pay a premium for PV systems on the houses they acquire, and hence, they, rather than the original PV owners, capture all of the benefit from the remaining retailer payments and avoided energy purchases over the life of the PV.[12]

4.2   Valuation of Capital and Production Subsidies

This subsection examines the value of the production and consumption subsidies borne by energy users and then expresses the aggregate per megawatt-hour consumed. Energy users—including households that installed PV—bore this cost, although some emission-intensive trade-exposed energy users obtained partial relief from exposure to the recovery of capital subsidies.

4.2.1   Capital Subsidies

In the period from 1 January 2010 to 31 December 2012, around 122 million solar PV renewable energy certificates were created. The estimated total undiscounted cost of these certificates is $3.8 billion. As shown in Figure 4, the estimated

12. An alternative assumption is that subsequent PV owners recognize the value of the remaining retailer payments and avoided energy purchases that they will receive and factor some or all of this into the price of the house that they buy. In this case, the return on investment for the original PV owner will be higher than as shown in Figure 4.

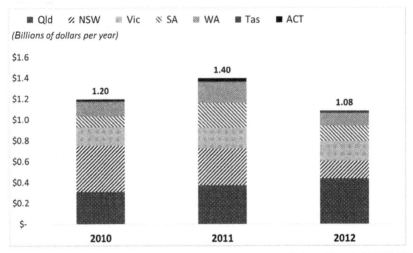

FIGURE 5　Renewable energy certificates created through the installation of household PV $ billion (2013).[13]

present value of the production subsidy—discounting at the IRR—provided through certificates is $2.8 billion. The discount factor and the portion of the certificate subsidy that has been captured by PV installers, electricity retailers, certificate brokers, traders, and aggregators explain the difference.

Figure 5 shows that the value of capital subsidies peaked in 2011 so that despite sharply higher PV installation rates in 2012 (see Figure 1), the decline in the multiplier meant that the total certificate creation started to decline.

4.2.2　Production Subsidies

Figure 4 shows that the present value—discounting at the IRR—of the mandatory production subsidy (the FiT) is estimated to be $2.6 billion, varying by jurisdiction. The undiscounted total FiT is $4.5 billion, and the highest is NSW ($1.35 billion) then QLD ($1.3 billion), VIC ($0.8 billion), SA ($0.6 billion), WA ($0.24 billion), ACT ($131 million), and TAS ($11 million). Figure 6 shows the profile of the payment of FiTs to 2030, at which point they will all have been terminated. The large hump for NSW reflects the use in NSW of a "gross" FiT.

The sum of the capital subsidies ($3.7 billion) and expected mandatory production subsidies ($4.5 billion) paid by energy users per megawatt-hour of electricity delivered between 2010 and 2030 is calculated for each jurisdiction per megawatt-hour expected to be sold and is shown in Figure 7. The peak in 2011

---

13. CME analysis using data in sources identified in References.

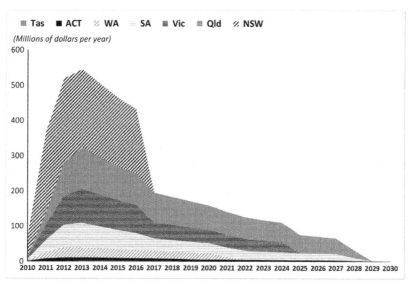

FIGURE 6   Jurisdictional government-mandated feed-in tariff (FiT), 2010-2030, $ millions (2013).[14]

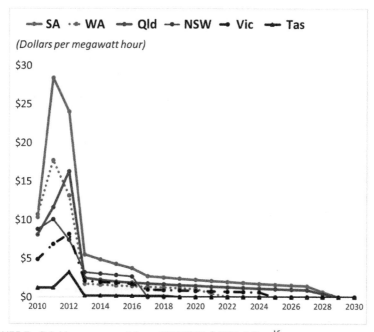

FIGURE 7   Subsidy per megawatt-hour, 2010-2030, $ 2013 dollars.[15]

14. CME analysis using data in sources identified in NEMREVIEW6, 2013.
15. CME analysis using data in sources identified in References.

and 2012 drops rapidly after that. This reflects the recovery of capital subsidies in or close to the year in which they were incurred.

It should be noted that in Figure 7, as for Figure 6, the data show the incidence/payment of subsidies, not the profile of their recovery from other energy users. The recovery, determined regulators, lags the incidence. As a result, the recovery of subsidies from users will be somewhat more smoothed than their incidence as shown in Figures 6 and 7.

5   WINNERS, LOSERS, AND CONSEQUENT QUESTIONS

The results presented in the previous section provide insights into relevant issues in Australia and internationally. In the format of rhetorical questions, this section discusses the results and elaborates on relevant issues.

5.1   Have PV Owners Had a Windfall Gain?

The previous section explained the estimated IRR of investment in PV by c. 900,000 homes between the start of 2010 and end of 2012, discounting at the rate of 9.8%. It might also be argued that households obtain other benefits from installing PV such as greater control over their electricity supply or meeting personal environmental objectives.[16] Households also have tax advantages for their investment in PV.[17] Therefore, it might reasonably be argued that households would have been willing to accept a lower rate of return than 9.8%. In this sense, it might be argued that households that installed PV have profited unreasonably from their investment in rooftop PV. However, the gap is not large enough to support a conclusion of windfall gain.

5.2   Is It All Cost and No Benefit for Other Energy Users?

In total, around $8.2 billion will be paid in capital and production subsidy for the 900,000 rooftop PV systems. As a present value—discounting at the IRR—per megawatt of PV installed, the total subsidy is $3.7 million/MW. Per megawatt-hour produced over the life of the PV, the subsidy is $108/MWh. Energy users—including the households with rooftop PV—are bearing this subsidy.

While the subsidy can be valued with reasonable certainty, it is more difficult to be certain about the benefits that all users would share, as a result of PV's impact on the wholesale electricity market and on networks. A preliminary discussion follows.

---

16. See Chapter 3 for an excellent discussion of this in Germany.
17. Income from PV produced in households is not taxed—Australian Taxation Office (ATO, 2013).

5.2.1   Energy Market Impact

What impact has household PV had in reducing wholesale electricity prices? Australia's renewable energy subsidy[18]—which the Climate Change Authority's advisors estimated would cost $28 billion between 2012 and 2030— essentially pays for itself through the impact that the additional renewable generation has in reducing wholesale electricity prices (Climate Change Authority, 2012). In Chapter 19, Sood and Blanckenberg suggested that renewable generation entering the market in quantities that exceeds demand growth has been "one of the key drivers of the reduction in spot prices experienced in recent years."

The analysis presented in this chapter estimates that the 900,000 PV roofs will produce around 3.4 TWh per annum. This is just 1.6% of Australia's centrally dispatched electricity production. Prima facie, a 1.6% reduction in average demand is unlikely to have a significant or lasting impact on prices.

However, 90% of PV production occurs in just 6 h from 10 a.m. to 4 p.m., so PV's share of consumption is more appropriately stated as a percentage of consumption during this time, in which case PV's share of the market (half-hourly demand) rises to around 5%. This is just a national average, and again, the assessment is made more meaningful by looking at individual regions.

In South Australia, where PV is installed on 21% of households, PV's share of South Australia's centrally dispatched electricity production between 10 a.m. and 4 p.m. is around 15%. Furthermore, this PV production is at the time when the supply cost curve is most likely to be relatively steep, which implies that effective reduction in residual demand—total demand less PV production— can be expected to have a reasonably significant impact on wholesale market prices. Furthermore, this calculation is just for the PV systems installed between 2010 and 2012. Including the pre-2010 and post-2012 installations, PV's market share will be even higher and the impact on prices more noticeable.

The conclusion from this preliminary analysis is that household PV can reasonably be expected to have an impact on wholesale electricity prices, at least in some of Australia's regional electricity markets. The benefit of this should be calculated and counted as part of the assessment of PV subsidies. This area is amenable to rigorous analysis and it would be valuable to undertake such analysis to quantify the possible value of such wholesale market price effects. A good example of such analysis in California can be found in the Energy and Environmental Economics, Inc. (2013). Retrospective modeling of the merit-order effect[19] on wholesale prices from PV in the Australian energy market by McConnell et al. (2013) found that for 5 GW of capacity,[20] the reduction in wholesale prices would have been worth in excess of AUD$1.8 billion over

---

18. With respect to large-scale renewable generation.
19. The wholesale price reduction of €2.8 billion in 2010 due to the merit-order effect was calculated by the German Federal Ministry for the Environment (see Chapter 3).
20. Present per capita installation of PV in Germany.

2009 and 2010. Wholesale energy market investment effects such as those covered in Burke (2013) also merit deeper analysis.

5.2.2 Network Impact

The beneficial impact of household PV in reducing network losses and by effectively augmenting capacity by pushing electricity back into the grid in the opposite direction to the predominant flows is likely to be significant. For example, in Western Australia, the Independent Market Operator estimates that PV systems in Perth are producing around 27% of their installed capacity at the time of regional peak demand (Independent Market Operator, 2013a). The analysis in this chapter for South Australia identifies that the average summer PV output is around 35% of the installed capacity at the time of peak daily demand in South Australia. The Australian Energy Market Operator (2012) suggested that PV's contribution at the typical time of regional peak demand ranged between 28% (QLD), 29% (NSW), 35% (VIC), and 38% (SA).

Assuming that 30% of the installed capacity of the 900,000 systems coincides with the time of peak demands in the regional electricity markets of Australia, this amounts to an aggregate average demand reduction of around 690 MW[21] at the time of system peak demand.

Network costs in Australia have been extraordinarily high by international standards (Mountain and Littlechild, 2010). Regulated network service providers in Australia have been provided with regulated expenditure allowances that average $3 million per megawatt of capacity added on distribution networks and $1.3 million per megawatt of capacity added on transmission networks (Mountain, 2012a, 2012b). On the assumption that the 900,000 PV units were effective in avoiding augmentation on both distribution and transmission networks, the value of the avoided expenditure might therefore be estimated at between $0.9 billion[22] and $2.1 billion.[23]

However, the extent to which PV avoids the need for augmentation of much of the distribution network is not clear. Residential maximum demands occur around or after sunset and so the ability of PV to defer augmentation of residential feeders and other bits of the network dominated by residential load is likely to be small. However, the typology of the network is likely to be important including the extent of meshing and load control across circuits.

Load flow and statistical studies would be useful in understanding with greater certainty how asset utilization is affected by PV at or near the times of peak demand at different points on the network, from the local reticulation lines all the way to zone and bulk supply substations, subtransmission, and the extra-high-voltage transmission lines.

---

21. Thirty percent of 2286 MW.
22. 690 MW multiplied by $1.3 million per megawatt.
23. 690 MW multiplied by $3 million per megawatt.

5.3 Should Tariff Structures Be Changed?

The Australian Energy Market Commission—the Australian energy market policy advisor and industry rule maker—said that its stakeholders identified tariff issues related to rooftop PV as a key priority (Australian Energy Market Commission, 2013). Several Chapters 8,12–15,19—Keay et al., Brennan, Nelson et al., Blansfield and Jones, Faruqui and Grueneich, and Sood and Blanckenberg—also draw attention to tariff issues arising from distributed generation.

A consultancy report[24] to the Energy Supply Association of Australia—an association that represents the chief executives of electricity producers and distributors in Australia—suggested that households with PV should compensate the incumbent centrally dispatched electricity producers and retailers and also the monopoly network service providers. Their argument is that, by meeting their own needs for electricity through their rooftop PV, "they avoid paying for electricity (from the grid) they do not need."

This claim is not credible with respect to compensation to grid-based electricity producers for their loss of market share to nongrid competitors. However, in the case of "compensation" to monopoly network service providers—or, more precisely, to their other captive customers—the arguments are more complex.

The argument for compensation is that networks were designed to meet expected demands—in the case of households around 20 kVA[25] per household. Households with PV are still drawing similar peak demands from the grid, but as a result of their rooftop PV, their annual consumption from the grid is much lower and so their contribution to the recovery of the fixed costs of the grid is lower.

The commonly accepted view in Australia is that the current regulatory arrangement grants monopoly rights to network service providers and with this an implicit right to recover revenue lost from one group of customers, from the remaining customers. Network service providers do this, in due course, by raising their prices so that they recover their regulated revenue entitlement.[26] In this sense, households with PV, like other distributed generators or indeed other forms of demand reduction, are able to impose higher prices on others if they reduce their purchases from the grid. Consequently, it is argued that households with PV are being subsidized by other electricity users.[27]

On the other hand, it is also necessary to have regard to positive externalities of rooftop PV that those households with PV are not being currently compensated for.

---

24. ACIL Tasman (2013a).
25. Energy Matters (2013).
26. For many network service providers' prices, revenues are regulated. This brings additional complications although the essential point remains.
27. This analysis rests on the assumption that network service providers have an unquestionable right to recover lost revenue by raising prices to other consumers. But should the owners of the network service providers be insulated from disruptive changes that offer alternatives to their services? How are networks to be held accountable for excessive or wasteful investments they have made? In the United States, the concept of revenue recovery of "used and useful" expenditure is well accepted in the regulation of networks. Should similar concepts apply in Australia?

In Australia, leaving the FiTs to one side, households with PV are paid 6-8 cent/ kWh for electricity exported to the grid. This is determined by regulators based only on the energy value of their grid exports. But the discussion earlier presented preliminary estimates that the 900,000 PV systems avoid network augmentation that could be valued at between $0.9 billion and $2.1 billion or between $72 million and $168 million annualized. While more needs to be done to confirm the magnitude with greater certainty, the impact is likely to be nontrivial.

The 900,000 PV systems produce around 3.4 TWh per year, of which 1.8 TWh displaces electricity in those households that would otherwise be supplied by the grid. This results in around $252 million of income that monopoly distribution network service providers have lost from households with PV.[28] This loss is likely to be more than the network augmentation that PV avoids, although a more rigorous assessment may find that in some parts of Australia, the benefit exceeds the disbenefit.

If we assume that network service providers have a right to recover lost income, then some form of tariff adjustment is needed to adjust for the difference between the disbenefit of lost income and the benefit of network augmentation avoidance. However, it is hard to see that this difference will be large, and the question remains whether network service providers should have the right to recover all of the income that they have lost, effectively to competitors, from their remaining captive customers.

The question of the appropriate way to change tariffs to households with rooftop PV is also contentious. A peculiar feature of the contemporary Australian tariff debate is the view, seemingly widely held, that cost-reflective tariffs for households with PV will not be politically acceptable, and therefore, a "crude way"[29] to fix the problem—of the network disbenefit—is to raise the fixed charges that households with PV pay for their usage of the network.

However, households with PV are generally connected to half-hourly meters and have typically been placed on diurnal tariffs. Tariffs for households with PV that are more cost-reflective—that is, greater temporal and locational differentiation—would surely encounter little (reasonable) opposition.

In addition, in considering changes to tariff structures, it is essential to be clear on the existing situation. Network and energy charges are bundled in the tariffs that retailers charge to households. There are no regulatory controls

---

28. This lost income is 0.7% of the $10.4 billion total annual regulated income that will be recovered by distribution network service providers in Australia in 2013. Hence, to recover the $252 million that has been lost, average network prices will need to rise by 2.4%. Since distribution network prices are 20% to 40% of the final electricity bill for the average consumer—network charges vary considerably in different jurisdictions of Australia—this means electricity prices paid by end users would need to rise by between 0.5% and 1%.

29. See, for example, comments by Keiran Donoghue, Energy Supply Association of Australia Policy Manager at http://www.businessspectator.com.au/article/2013/7/26/solar-energy/taking-heat-solar-charge.

on the way that network charges are reflected in the tariffs that households actually pay, and the split between fixed and variable elements of both network and retail tariffs varies widely across Australia.[30]

An acceptable resolution of the tariff debate in Australia will inevitably reflect not only economic arguments but also perceptions of fairness and equity. Again, there are conflicting views. On the one hand, the popular perception that overly generous FiTs have led to windfall gains for PV owners—a misconception this chapter concludes—will sway public opinion. On the other hand, households with rooftop PV will smart at the suggestion that they should pay higher network charges because they are using the network less, when one of the reasons for such high network charges is that households that installed large air conditioners failed to face the full cost of the consequential network expansion and from which network service providers have profited handsomely.[31]

Like the death spiral debate, the tariff debate in Australia has a long way to run. The least that may be concluded from the discussion here is that an acceptable resolution requires the careful construction of economic arguments and also a detailed understanding of the actual circumstances and data and a fair assessment of equity concerns.

5.4 Is This the Beginning of a Death Spiral for the Centrally Dispatched Model?

The uptake of PV in Australia, even as subsidies have declined, has sparked debate on whether the economics of PV, installed at the point of use, is such that the established centrally dispatched electricity model is at the beginning of a death spiral.

The average cost of electricity produced by the 900,000 PV systems installed between 2010 and 2012 is $162/MWh.[32] This compares to the nationwide average household electricity price (see Section 2.1) of $320/MWh. In other words, for households in Australia that installed PV during the period from 2010 to 2012, it costs, on average, half as much to produce electricity from

---

30. For example, in Victoria, the fixed element of the network service provider charges for household tariffs makes up about 5% of the charge, while for the retail tariff (which is the tariff the householder sees), the fixed element is around 15% of the average bill. The opposite applies in Queensland, where up to 30% of the network charge is fixed, but just 10% of the charge that households actually pay is fixed.

31. Households running a 2-kWe reverse cycle air conditioner, during peak times, receive an implicit subsidy equivalent of $350 per year from consumers who do not use air conditioners during this period—Productivity Commission (2013b). Electricity Network Regulatory Frameworks. Canberra. p. 349-352.

32. This is a simple calculation of cost divided by expected lifetime production (not discounted future production). It is meant to reflect the type of calculation that typical households can be expected to do in comparing PV to purchasing electricity from the grid.

a PV system than it does to purchase electricity from the electricity grid. This is obviously a large gap. And since this analysis covers a period during which PV costs were typically considerably higher than they are now, the conclusion would be that the gap between grid and PV is now even wider.

This suggests significant potential for further expansion of PV, and if battery storage can be supplied for less than around $160/MWh, then it would seem to be more economic for households who have the opportunity to delink from the national grid altogether. Platt et al. in Chapter 17 describe the integration of electric vehicles and the grid with solar PV. Perhaps, this and other developments may nonetheless extend the usefulness of the grid, despite its apparently increasingly weak comparative economics.

Prima facie therefore, it seems reasonable to think that the traditional centrally dispatched industry model is facing a serious competitive threat. This is not to suggest that its demise is certain or will happen quickly. Frictions in the further augmentation of solar capacity and changes in tariff structures—for example, a greater proportion of fixed rather than variable charges—could nonetheless delay the transition. The incumbents are also unlikely to wish to facilitate their loss of market share.[33] Blansfield and Jones in Chapter 14 and Nelson et al. in Chapter 13 however made the point that utilities should engage constructively to ensure that their needs align with the realities of a modern and engaged consumer.

This struggle has a long way to play out. It will be shaped by changes in regulatory policy, tariff structures, technology and market developments, and of course relative costs. Considering the implications of such a major shift for producers and consumers, this is another area where careful analysis of the public interest would be valuable.

6   CONCLUSIONS

Despite the generous subsidies this chapter finds, contrary to popular perception, households that invested in PV have not received windfall gains.

The average cost of electricity supplied by the 900,000 rooftop PV systems installed between 2010 and 2012 averages $162/MWh. This compares to the average residential electricity price of $320/MWh. While many factors may delay the further uptake of PV, the size of this gap suggests that the traditional centrally dispatched industry model is facing a serious competitive threat in Australia, as it has already in Germany.

---

33. For example, South Australian Power Networks, the monopoly provider of distribution network services in South Australia, is proposing changes to the regulatory arrangements so that if households add battery storage capacity to their PV systems, they will no longer be eligible to receive feed-in tariff payments for electricity exported to the grid (see Morris, 2013. The South's solar storage feed-in gaffe).

This chapter also finds that arguments of cross-subsidy between energy users that do not have PV and those that do are not clear. The 900,000 households are contributing around $250 million per year less income to monopoly network service providers. But on the other hand, they do not seem to be compensated for their contribution to the reduction of future network expansion. While the latter seem to be worth less than the former, the gap may not be large, and the question remains whether network service providers should have the right to recover revenue effectively lost to competitors, from its remaining captive customers.

Finally, as a general theme of this chapter, much more could and should have been done by Australia's governments, regulators, consumers, and academies to understand the public interest impact of distributed generation in general and rooftop PV in particular. Careful construction of economic and equity arguments and a good understanding of the actual circumstances and data are vital.

ACKNOWLEDGMENTS

This chapter improved considerably from the helpful comments made by David Young, Darryl Biggar, Jamie Carstairs, Damien Moyse, Craig Memery, Michael Reid, Darren Gladman, Paul Troughton, Brian Green, Paul Smith, and Ann Whitfield.

REFERENCES

ACIL Tasman, 2013a. Distributed generation: implications for Australian electricity markets. Prepared for the Energy Supply Association of Australia.

ACIL Tasman, 2013b. The fair and reasonable value of exported PV output. Prepared for the Essential Services Commission of South Australia. ACIL Tasman.

ACT Government, 2009. ACT Electricity Feed-in Tariff Scheme.

ACT Government, 2011a. ACT Electricity Feed-in Tariff Scheme.

ACT Government, 2011b. Feed-in Tariff Re-opened.

ActewAGL, 2010. Standard Retail Electricity Supply—Schedule of Charges from 1 July 2010.

ActewAGL, 2011. Standard Retail Electricity Supply—Schedule of Charges from 1 July 2011.

ActewAGL, 2012. ACT Standard Retail Electricity Supply—Schedule of Charges from 1 July 2012.

ActewAGL, 2013a. ActewAGL Solar Buyback scheme [Online]. Available: http://www.actewagl. com.au/Product-and-services/Offers-and-prices/Prices/Residential/ACT/Feed-in-schemes/ ActewAGL-Solar-buyback-scheme.aspx (Accessed 15 August 2013).

ActewAGL, 2013b. Out ACT Electricity Prices—Schedule of Charges from 1 July 2013.

AECOM, 2010. Solar Bonus Scheme. AECOM Australia Pty Ltd., Sydney, NSW.

AGL Energy Limited, 2013. Annual report.

ANZSES, 2006. Australian Solar Radiation Data Handbook (ASRDH), fourth ed. Australian and New Zealand Solar Energy Society, Sydney.

Aurora, 2009. Aurora's Approved Electricity Tariffs for 2009-10.

Aurora, 2010. Aurora's Approved Electricity Tariffs from 1 July 2010.

Aurora, 2011. Aurora's Approved Electricity Tariffs from 1 July 2011.

Aurora, 2012. Aurora's Approved Electricity Tariffs from 1 July 2012.

Aurora, 2013. Aurora's Approved Electricity Tariffs from 1 July 2013.

Australian Bureau of Statistics, 2013. Consumer Price Index 6401.0.

Australian Energy Market Commission, 2013. Strategic Priorities for Energy Market Development 2013, Sydney.

Australian Energy Market Operator, 2012. Rooftop PV Information Paper.

Australian Taxation Office (ATO), 2013. Authorisation Number 90083—private ruling for "Are you assessable on energy credits generated from your domestic solar power grid?" [Online]. Available: http://www.ato.gov.au/rba/content/?ffi=/misc/rba/content/90083.htm (Accessed 10 November 2013).

Bloxham, P., Mcgregor, D., Rankin, E., 2010. Housing Turnover and Firs Home Buyers. Reserve Bank of Australia.

Burke, K.B., 2013. The Reliability of Distributed Solar in Critical Peak Demand: A Capital Value Assessment. Physics Department, University of Newcastle.

Business Spectator, 2013. Tas Gov't slashes feed-in tariff by two-thirds [Online]. Available: http://www.businessspectator.com.au/news/2013/8/19/tas-govt-slashes-feed-tariff-two-thirds (Accessed 20 September 2013).

Clean Energy Council, 2012. Consumer guide to buying household solar panels (photovoltaic panels).

Clean Energy Regulator, 2013a. About the Renewable Energy Target.

Clean Energy Regulator, 2013b. How to Create STCs for Small-Scale Solar, Wind, and Hydro Systems [Online]. Available: http://ret.cleanenergyregulator.gov.au/Certificates/Small-scale-Technology-Certificates/Creating-STCs—solar-panels/creating-sgu (Accessed 08 October 2013).

Clean Energy Regulator, 2013c. List of SGU/SWH Installations by Postcode [Online]. Australian Government Clean Energy Regulator. Available: http://ret.cleanenergyregulator.gov.au/REC-Registry/Data-reports (Accessed 03 October 2013).

Clean Energy Regulator, 2013d. Solar credits phased out [Online]. Available: http://ret.clean energyregulator.gov.au/Latest-Updates/2012/November/3 (Accessed 08 October 2013).

Climate Change Authority, 2012. Renewable energy target review, final report.

Collier, P., 2010. Feed-in Tariff Scheme Provides Incentive [Online]. Government of Western Australia. Available: http://www.mediastatements.wa.gov.au/pages/StatementDetails.aspx?listName=StatementsBarnett&StatId=2665 (Accessed 26 November 2013).

Commonwealth of Australia, 2013. In: Australian Bureau of Statistics, (Ed.), 4670.0—Household Energy Consumption Survey, Australia: Summary of Results, 2012. ACT, Canberra.

Commonwealth of Australia (Climate Change Authority), 2012. Renewable energy target review—final report—Chapter 6. Liability and Exemption Framework.

Deloitte, 2011. Advanced metering infrastructure customer impacts study (final report). Report for the Department of Primary Industries (DPI).

Eadie, L., Elliott, C., 2013. Going Solar: Renewing Australia's Electricity Options. Centre for Policy Development.

Economic Regulation Authority, 2012. Synergy's costs and electricity tariffs—final report. Perth, Western Australia.

Energy Matters, 2013. Energy matters calls for real electricity pricing reform [Online]. Available: http://www.energymatters.com.au/index.php?main_page=news_article&article_id=3984 (Accessed 13 November 2013).

EnergyAustralia, 2009. Residential Customer Price List—Regulated Retail and Green Energy Tariffs—Effective from 1 July 2009.

EnergyAustralia, 2010. Residential Customer Price List—Regulated Retail and Green Energy Tariffs—Effective from 1 July 2010.

EnergyAustralia, 2011. Residential Customer Price List—Regulated Retail and Green Energy Tariffs—Effective from 1 July 2011.

EnergyAustralia, 2012. Residential Customer Price List—Regulated Retail Tariffs—Effective from 1 July 2012.

EnergyAustralia, 2013. Residential Customer Price List—Regulated Retail Tariffs—Effective from 1 July 2013.

Essential Services Commission, 2013. Standing Offer Tariffs 1994-95 to 2013.

Essential Services Commission of South Australia (ESCOSA), 2013. Variation to the 2012 Determination of the Solar Feed-in Tariff Premium. Adelaide, SA.

Essential Services Commission Victoria, 2013. Minimum Electricity Feed-in Tariffs—For Application from 1 January 2014 to 31 December 2014—Final Decision.

Flannery, T., Sahajwalla, V., 2013. The Critical Decade: Australia's Future—Solar Energy.

Gallagher, K., 2011. Feed-in Tariff Scheme Closes. ACT Government.

Government of South Australia, 2009. Supplementary Gazette—The South Australian Government Gazette—AGL SA Electricity Standing and Default Contract Prices.

Government of South Australia, 2011. The South Australian Gazette—AGL SA Electricity Standing and Default Contract Prices.

Government of South Australia, 2012. The South Australian Government Gazette—AGL SA Electricity Standing and Default Contract Prices.

Government of South Australia, 2013. The South Australian Government Gazette—AGL South Australia Pty Ltd—Standing and Default Contract Prices for Small Customers.

Government of Western Australia, 2011. Tariff and Concession Framework Review—Issues Paper.

Government of Western Australia, 2013a. Feed-in tariff rates [Online]. Available: http://www.finance.wa.gov.au/cms/content.aspx?id=14714 (Accessed 12 August 2013).

Government of Western Australia, 2013b. Feed-in Tariff Scheme Changes.

Government of Western Australia, 2013c. Renewable Energy Buyback Scheme [Online]. Department of Finance. Available: http://www.finance.wa.gov.au/cms/content.aspx?id=14271 (Accessed 13 August 2013).

Green, B., 2013. Solar Feed-in Tariffs Guaranteed for Five Years [Online]. Premier of Tasmania, Hobart (Accessed 30 September 2013).

Green Energy Markets, 2013. Small-scale technology certificates data modelling for 2013 to 2015—report to the clean energy regulator. Hawthorn, VIC.

Independent Competition and Regulatory Commission, 2009. ActewAGL Retail—Schedule of Charges for Standard Retail Electricity Supply 2009–2010.

Independent Market Operator, 2013. Electricity Statement of Opportunities. Perth.

(IPART)Independent Pricing and Regulatory Tribuna, 2013. Solar Feed-in Tariffs—The Subsidy-Free Value of Electricity from Small-Scale Solar PV Units from 1 July 2013, NSW.

Jordan, D.C., Kurtz, S.R., 2013. Photovoltaic degradation rates—an analytical review. Prog. Photovoltaics Res. Appl. 21, 12–29.

Martin, J., 2011. Solar PV System Prices Reaching Unprecedents Lows. Why? And How Long Will It Last? [Online]. Solar Choice. Available: http://www.solarchoice.net.au/blog/solar-pv-system-prices-reach-unprecedented-lows-in-australia (Accessed 24 October 2013).

Martin, N., Rice, J., 2013. The solar photovoltaic feed-in tariff scheme in New South Wales, Australia. Energy Policy. 61, 697–706.

McConnell, D., Hearps, P., Eales, D., Sandiford, M., Dunn, R., Wright, M., Bateman, L., 2013. Retrospective modeling of the merit-order effect on wholesale electricity prices from distributed photovoltaic generation in the Australian National Electricity Market. Energy Policy. 58, 17–27.

Morris, N., 2013. The South's Solar Storage Feed-in Gaffe. Climate, Spectator [Online].

Mountain, B.R., 2012a. Reducing electricity costs through Demand Response in the National Electricity Market. A report for ENERNoC, Melbourne.

Mountain, B.R., 2012b. Electricity prices in Australia: an International Comparison. A report for the Energy Users Association of Australia, Melbourne.

Mountain, B.R., Littlechild, S.C., 2010. Comparing electricity distribution network revenues and costs in New South Wales, Great Britain and Victoria. Energy Policy. 38, 5770–5782.

NEMREVIEW6, 2013. NEM-Review [Online]. Available: http://v6.nem-review.info/what/index.aspx (Accessed 20 October 2013).

NSW Government, 2013. Solar Bonus Scheme [Online]. Trade & Investment. Available: http://www.energy.nsw.gov.au/sustainable/renewable/solar/solar-scheme/solar-bonus-scheme#Solar-Bonus-Scheme-parameters (Accessed 08 August 2013).

OTTER, 2013. Regulated Feed-in Tariff for Tasmanian Small Customers. Office of the Tasmanian Economic Regulator, Hobart, Tasmania.

Peng, J., Lu, L., Yang, H., 2013. Review on life cycle assessment of energy payback and greenhouse gas emission of solar photovoltaic systems. Renew. Sust. Energy Rev. 19, 255–274.

Productivity Commission, 2013. Electricity Network Regulatory Frameworks, Canberra.

Queensland Competition Authority (QCA), 2009. Queensland Government Gazette No. 41.

Queensland Competition Authority (QCA), 2010. Queensland Government Gazette No. 41.

Queensland Competition Authority (QCA), 2011. Queensland Government Gazette.

Queensland Competition Authority (QCA), 2012. Queensland Government Gazette.

Queensland Competition Authority (QCA), 2013a. Estimating a Fair and Reasonable Solar Feed-in Tariff for Queensland. Brisbane, Queensland.

Queensland Competition Authority (QCA), 2013. Queensland Government Gazette No. 23.

Queensland Government, 2013. Solar Bonus Scheme—Current Eligibility for the Scheme [Online]. Department of Energy and Water Supply. Available: http://www.dews.qld.gov.au/energy-water-home/electricity/solar-bonus-scheme/current-eligibility (Accessed 08 August 2013).

Red Energy, 2010. Default Contract Prices for Small Customers.

State Government of Victoria, 2013a. Feed-in Tariff for New Applicants [Online]. Department of State Development Business and Innovation (Accessed 11 August 2013).

State Government of Victoria, 2013b. Premium Feed-in Tariff [Online]. Department of State Development Business and Innovation. Available: http://www.energyandresources.vic.gov.au/energy/environment-and-community/victorian-feed-in-tariff-schemes/closed-schemes/premium-feed-in-tariff (Accessed 10 August 2013).

State Government of Victoria, 2013c. Standard Feed-in Tariff [Online]. Department of State Development Business and Innovation. Available: http://www.energyandresources.vic.gov.au/energy/environment-and-community/victorian-feed-in-tariff-schemes/closed-schemes/standard-feed-in-tariff (Accessed 10 August 2013).

State Government of Victoria, 2013d. Transitional Feed-in Tariff [Online]. Department of State Development Business and Innovation. Available: http://www.energyandresources.vic.gov.au/energy/environment-and-community/victorian-feed-in-tariff-schemes/closed-schemes/transitional-feed-in-tariff (Accessed 10 August 2013).

Synergy, 2013. Standard Electricity Prices and Charges—South West Interconnected System Effective 1 July 2013.

Tasmanian Government, 2013. Feed-in Tariffs: Transition to Full Retail Competition Issues Paper. Department of Treasury and Finance, Hobart, Tasmania.

Victorian Competition and Efficiency Commission (VCEC), 2012. Power from the People: Inquiry into Distributed Generation.

Chapter 5

As the Role of the Distributor Changes, so Will the Need for New Technology

Clark W. Gellings

ABSTRACT

As elucidated in several other chapters, the grid provides substantial value that goes beyond the delivery of electricity to consumers who produce their energy locally. This chapter asserts that implementing technologies that will modernize the grid will enhance its value to all consumers and enable higher penetration of local energy resources and that grid connectivity provides the overall least-cost approach to society with reliability and high quality with increasing penetration of distributed generation. The chapter's primarily goal is to elucidate how the need for new technology will change as the role of the distributor changes.

Keywords: Local generation, Distributed generation, Integrated grid, Photovoltaic (PV) generation, Smart grid

1 INTRODUCTION

Broadly speaking, in the debate surrounding the changes in consumer demands, the industry—utilities, consumers, and regulators alike—are underestimating the value of grid connectivity to consumers who elect to generate their own electricity locally. This is the case for those that partially meet their electricity requirements from on-site generation and those that seek to self-generate all of their needs. Without the grid, a distributed generation (DG) consumer would suffer greatly reduced reliability and, without the grid or without substantial investment in supporting technologies, would not have the ability to power their building or premise when the sun does not shine or have the electrical support to start a residential air conditioner. Generating energy locally is referred to in this chapter as local generation or customer generation.[1]

---

1. Local generation is more often referred to as DG. Photovoltaic (PV) power generation installed by customers is a form of DG. Most practitioners also include DG in a basket of technologies called distributed energy resources or DERs that include DG, responsive loads, and electric energy storage.

Distributed Generation and its Implications for the Utility Industry. http://dx.doi.org/10.1016/B978-0-12-800240-7.00005-9

This chapter makes the assertion that from a societal perspective, the overall lowest-cost approach to meeting future consumer needs including use of local or DG is to enable connectivity to the grid. To embrace wide-scale connectivity, the industry needs to modernize the grid. This will require additional technologies. The power delivery system provides benefits beyond simply delivering energy, and those benefits become more pronounced for consumers who generate some of their own electricity.

This chapter consists of five sections in addition to the introduction. Section 2 provides an overview of the value the grid provides to consumers who have local generation. Section 3 describes the advantages of accommodating local generation. Section 4 elucidates the distribution system technologies needed to accommodate local generation. Section 5 offers an overall conclusion.

2 VALUE OF THE GRID

At present, the value of connectivity to the grid for customers who choose to generate electricity locally is understated and not well understood. Without connectivity to the grid, the reliability and capability of the consumer's power system would be diminished unless substantial investments were made in the local power system to achieve equivalence of service quality to what the grid provides. The grid provides electrical support to consumer installations.

Connection to the grid, called interconnection or grid connectivity, provides substantial improvements in power quality, reliability, and generator operating economics as well as in voltage and frequency regulation, harmonic distortion, efficiency, and operating costs. Utility standby rates, which are charged to recover the cost of maintaining T&D standby capacity and not for the support value to the local generation, are usually but not always less than the cost of a parallel utility system connection.

Figure 1 illustrates the primary benefits of grid connectivity for consumers with local generation. From an operating perspective, this support enables the robustness of the electric infrastructure to provide the following:

2.1 Reliability (Balancing and Backup Resource)

The grid backs up disruptions in the performance of local generation systems and provides a balancing function. This includes compensation for variable generation photovoltaic and wind power generation. When the sun does not shine and the wind does not blow, the grid provides backup. In this role, the grid is acting as bulk storage. This paradigm will change as storage costs decline potentially allowing for storage to be located where it is most optimal. This has the grid providing balancing in all time frames—hours, minutes, and

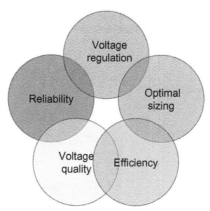

FIGURE 1   Primary benefits of grid connectivity.

seconds—to balance the variability and intermittency of the distributed resources.[2]

The average reliability provided by the grid for the United States as a whole, assuming the consumer is served by the distribution system, is 99.9% including major events. Figure 2 illustrates the 10-year average of system average interruption duration index, called SAIDI.

Figure 3 illustrates the variability of PV load generation systems. A New Jersey location is used for illustrative purposes. The variability needs to be supplemented by either the grid or the addition of technologies in order not to impact the reliability of power quality and availability.

Figure 4 illustrates the reliability inherent in a 1 MW power system. In order for a local generation system to achieve the same reliability as the grid, additional sources (generators are used in this example) need to be added.

2.2   Voltage Regulation

The grid provides start-up service or adequate "inrush" current for consumer appliances and devices such as compressors, air conditioners, and welders. This is so they can start reliably without severe voltage dips and equipment damage.

Figure 5 illustrates inrush current for a typical motor. Note that inrush currents are often five times or more greater than steady-state power draw. A local generation system without batteries or other technologies cannot provide adequate "energy" to start such motors.

---

2. The conventional measure of reliability is the value of unserved energy or the so-called lost load, which uses customers' value of electricity to set the value of service reliability. In the case of total grid disconnection, it is more logical to ascertain what would be required for a customer to disconnect from the grid but retain the same level of reliability. This will emphasize the implicit value of the grid connection, especially because the disconnect cost is likely to require an annual cost (amortized) of $1000 or more and the embedded distribution cost is probably between $150 and 250/ customer/year (about 15-20% of the residential customers total bill).

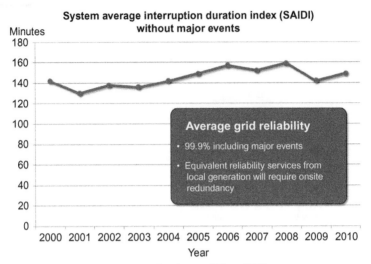

FIGURE 2   Grid provides reliability service. *Source: Wilson (2013).*

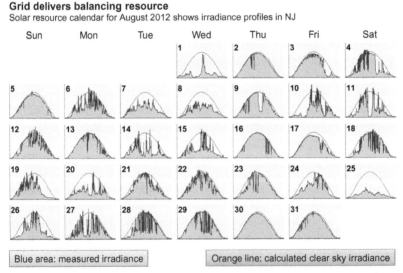

FIGURE 3   Solar system output—typical month for a PV load generation system. *Source: Trueblood et al. (2013).*

2.3   Voltage Quality

The robustness of the grid provides some mitigation against certain power quality events. In particular, transients, voltage flickers, sags, and swells as well as harmonic distortion created within customer facilities are dampened. The electrical stiffness of the grid mitigates the impacts harmonics created

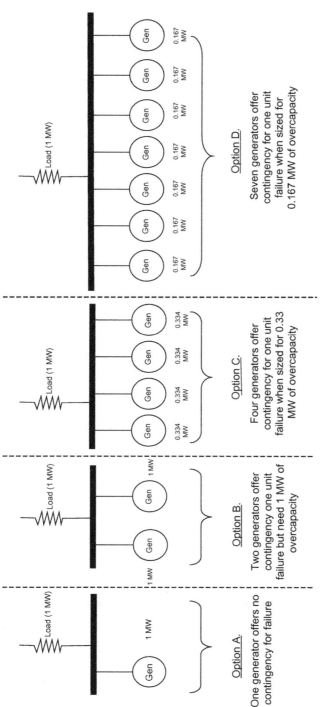

FIGURE 4 Multiple generation sources increases reliability. *Source: EPRI Report 1001668: "Benefits Provided to Distributed Generation by a Parallel Utility System Connection," 2013.*

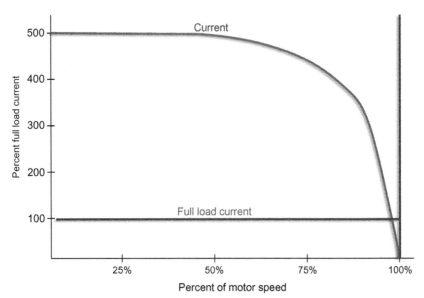

FIGURE 5 Inrush current for a typical motor.

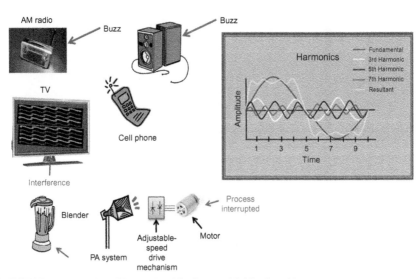

FIGURE 6 Illustration of harmonic mitigation provided by the grid.

from nonlinear loads can have as an independent power system. This is illustrated in Figure 6. Connectivity to the grid also allows synchronization that keeps analog clocks in tune with national time standards.

Figure 7 illustrates the difference between voltage quality delivered by the grid and that from a local generation system, referred to here as DG.

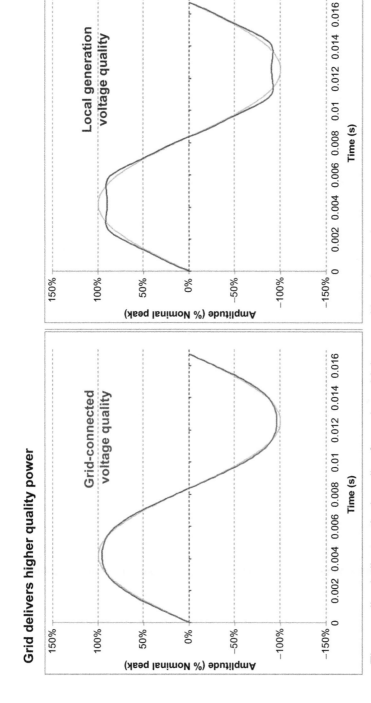

The grey line indicates the baseline for a sinusoidal wave with no harmonics

FIGURE 7   Illustration of grid-connected versus local generation power quality.

Harmonics have a deleterious effect on electrical devices—both consumer end-use devices and utility distribution equipment such as transformers. Harmonics cause heating in many electrical components that affects dielectric strength and deteriorates the life of equipment. The economic impacts of reduced life are hard to quantify and not obvious to consumers.

In addition, operation of generation systems in parallel with the utility system allows the local generation system to focus primarily on real power generation and not provide reactive support. This depends on the utility's interconnection requirements. Without interconnections, each premise has to manage real power needs with live voltage. In future system development, local generation systems may be compensated for providing reactive power support on demand.[3]

2.4   Efficiency

Operating microturbines or combined heat and power with the grid will allow internal combustion engine (ICE)-driven generators to run 10-20% more efficient. In addition, for all local generation systems, there will be a reduction in transmission and distribution losses.

Overall, lower harmonics also reduce hysteresis losses in distribution transformers. Reduced losses ultimately translate to reduced costs for the utility. This reduction is ultimately passed on to consumers in some form, often reduced revenue requirements for the utility (Kefalas et al., 2010).

2.5   Optimal Sizing

Parallel connections with the grid allow the owner/operator of local generation systems to design the optimal capacity into the local generation system that allows best use of local generation in assets. For example, an interconnected system could allow for dispatching with the objective of minimizing the impact of power production on the environment.

The additional benefit of a local generation system is in the ultimate flexibility it provides toward the grid of the future. Referred to as Grid 3.0, it would enable full integration of customer resources, end uses, and electrical and thermal storage with grid components. Referred to as the ElectriNet[SM], it enables the "flexible load shape" concept first envisioned in the initial construct of demand-side management (DSM), as is illustrated in Figure 8.[4]

DSM is the planning and implementation of those utility activities designed to influence customer use of electricity in ways that will produce desired

3. For additional explanation of these electrical engineering terms, the reader may wish to refer to The Electric Power Engineering Handbook, L.L. Grigsby, editor, ISBN 0-8493-2509-9, CRC Press, 1998.
4. How did the flexible load shape concept originate? The flexible load shape concept originated from the foundations of demand-side management (DSM) as envisioned in the late 1970s and early 1980s.

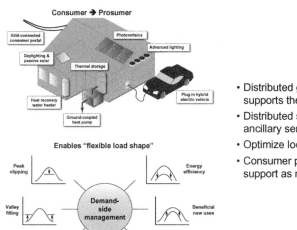

• Distributed generation
  supports the grid
• Distributed system provides
  ancillary services
• Optimize location of storage
• Consumer pays for grid
  support as needed

FIGURE 8   Connectivity enables the ElectriNet^SM. *Based on EPRI EA/EM 3597 (EPRI 1984).*

changes in the utility's load shape—i.e., changes in the time pattern and magnitude of a utility's load. Utility programs falling under the umbrella of DSM include load management, new uses, energy efficiency, electrification, customer generation, and adjustments in market share.

In DSM planning, six load shapes were envisioned as representative of optimal strategies for given power systems. These include peak clipping or the reduction of system peak loads; valley filling or building off-peak loads; load shifting, which involves shifting load from on-peak to off-peak periods; energy efficiency, which is the load shape change that results from reducing end-use consumption; electrification or strategic load growth, which is the load shape change that refers to a general increase in sales beyond valley filling; and flexible load shape.

The ElectriNet^SM embodies the value of support provided to the local generator by a parallel utility system connection. The support value that the power system offers is substantial and can make local generation viable in applications where if it were stand-alone, i.e., not grid-connected, the performance would be marginal.[5]

---

5. Why is the grid so strong? The utility system is a strong source and stable from a frequency perspective. It represents a vast interconnected network of thousands of megawatts of generation capacity, and any single distribution feeder load is tiny in comparison and has no significant impact on the bulk generation dispatch needs or operating efficiency. By comparison, the DER system is weak, and the starting of large motors and other loads can create severe power quality problems. Typically, reliability of utility system power is in the range of 99-99.999% (national average is about 99.97%). This compares to a single DER that is on the order of 94-97% available.

3   THE ADVANTAGES OF ACCOMMODATING LOCAL GENERATION

As the use of local generation proliferates, so must the architecture of the power system. The power system must transform into ElectriNet^{SM}. The ElectriNet^{SM} is a combination of local energy networks (LENs).[6] LENs are combinations of interconnected distributed end-use, local generation, storage, and utility technologies at the building, community, or distribution level that will increase the independence, flexibility, and intelligence for the optimization of energy use and energy management at the local level and then integrate LENs with the bulk power system to form the ElectriNet^{SM}.

The enabling infrastructure for LENs is the transmission and distribution system. Most of today's LENs are fairly simple in design, consisting of a single generator supplying a dedicated load or of multiple identical generating units ganged paralleled to operate much like a single unit. Compared to traditional power systems, LENs offer the potential for improvements in energy delivery and efficiency, power quality, reliability, and cost of operation for very concentrated and localized loads. To fully realize these benefits, however, it will be necessary to coordinate the control of complex LENs potentially composed of several different kinds of local generation and storage systems, which may also be geographically dispersed.

Groups of interconnected loads and DERs within clearly defined electrical boundaries that act as a single controlled entity can be referred to as microgrids[7] (Smith and Ton, 2013). Other microgrid benefits can include the following:

- Enabling grid modernization and the integration of multiple smart grid technologies.
- Enhancing and easing the integration of distributed and renewable energy sources that help reduce peak load and also reduce losses by locating generation near demand.
- Meeting end-user needs by ensuring energy supply for critical loads, controlling power quality and reliability at the local level, and promoting customer participation through DSM and community involvement in electricity supply.
- In addition, in some jurisdictions and on some systems, local generation can achieve benefits from avoided renewable portfolio standards (RPS) purchases. Depending on the generation mix, utilities may also be able to use local generation resources as a hedge against volatility in natural gas prices.

---

6. This concept of a network of LENs was first described by EPRI as the ElectriNet. See Gellings and Zhang (2010).
7. The reader should note that a collection of microgrids is not a sufficient construct to establish an ElectriNet^{SM}. An ElectriNet^{SM} enables flexible load shape and, as such, an entirely transactive topology where consumer uses, consumer generation, and storage devices and utility assets are monitored and potentially controlled in real time.

- Supporting the macrogrid by handling sensitive loads and the variability of renewables locally and supplying ancillary services to the bulk power system.

LENs, energy sources, and a power distribution infrastructure are first integrated at the local level. This could be an industrial facility, a commercial building, a campus of buildings, or a residential neighborhood (refer to Figures 9–11). Local area networks are interconnected with different localized systems to take advantage of power generation and storage through the smart grid enabling complete integration of the power system across wide areas. Localized energy networks can accommodate increasing consumer demands for independence, convenience, appearance, environmentally friendly service, and cost control.

The ElectriNet^{SM} enables the "flexible load shape" concept that also allows utilities to operate their distribution system in concert with consumer's power systems for a variety of other potential benefits:

- To optimize the energy consumption of the consumer's end-use equipment in balance with enhancing the efficiency of the grid.
- To utilize on-site generation, possibly in conjunction with energy storage, to avoid peak energy costs and even create revenue streams for the consumer by selling energy back to the grid once price signals justify it economically.

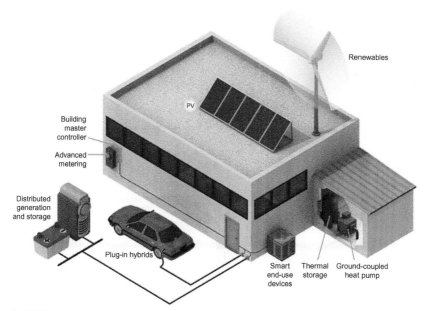

FIGURE 9 Building-level local energy network (LEN). The building-level LEN is the simplest network that involves distributed technologies deployed at the building level. *Source: Gellings and Zhang (2010).*

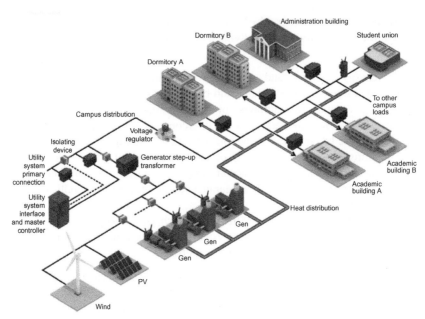

FIGURE 10 Campus-level LEN. The campus-level LEN is a network composed of building-level LENs coupled with other distributed technologies at the campus level. *Source: Gellings and Zhang (2010).*

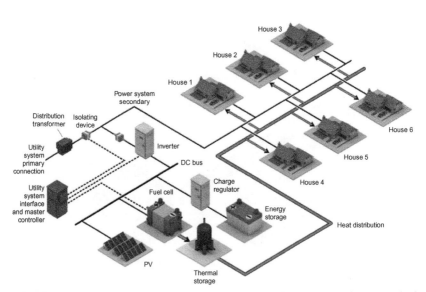

FIGURE 11 Community-level LEN. The community-level LEN is a network composed of building-level LENs coupled with other distributed technologies at the community level. *Source: Gellings and Zhang (2010).*

- To facilitate enrollment in demand-response programs, this can generate revenue by reducing load on the grid. Demand response can be provided by both self-generation and end-use loads.
- While grid energy transactions and fuel costs dominate the economics of a power system, participation in capacity and ancillary services markets by local generators can also be important incremental revenue drivers.
- The reliability improvements obtained through islanding capability can be quite valuable, depending on the mission of the facility and the critical load served during islanded operations.
- Local generation installations can also be viewed as a means of creating zero-net-energy communities and meeting other environmental goals established by states or regulatory agencies.

In addition, LENs allow for the following:

- The optimization of energy availability across a larger variety of energy sources, resulting in improved economics
- The creation of an infrastructure for more optimum management of overall energy requirements including heating, cooling, and power
- The control and management of reliability at the local level

The interconnection methods that have been developed for conventional parallel operation of DG include guidelines based on the concept that the DG will have a minimal impact on the bulk supply and only a small impact on the local distribution system. The reason for this is that the distribution system is small compared to the capacity of the bulk system connection and the distribution system. Some key characteristics of this type of interconnection approach are the following:

- The generator operates in a voltage-following mode, which means it is not attempting to directly regulate voltage or hold voltage to a set point.
- The generator has anti-islanding protection that disconnects the generator if the generator becomes islanded with any portion of the distribution system or voltage conditions leave the acceptable window of operation.
- The generator does not attempt to regulate frequency, but simply follows system frequency.

Of course, these operational characteristics are the antithesis of what is needed in the LEN when it changes to an islanded state. During islanded operation, generator tripping during island detection is not desirable. So protection needs to be established that allows generators to operate in parallel mode during conditions associated with normal utility system connection and under the guidelines of traditional local generation interconnection requirements but to transition to islanded control during that mode.

Inherently, LENs can operate somewhat independently, but their value is maximized when they are nested with each other and with the bulk power system. This nesting concept also allows for increased overall stability within the

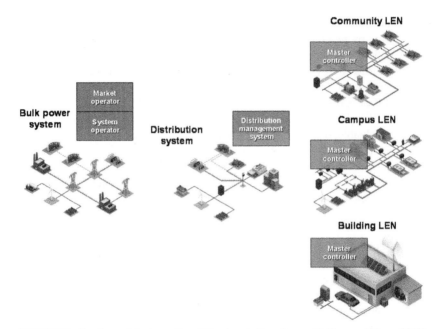

FIGURE 12 Creating architecture with multilevel controllers. *Source: Gellings and Zhang (2010).*

power system. Both energy storage and power electronics can be employed at all levels of the power system to reduce interdependencies between system components and to make the system relatively immune to temporary disturbances.

Nesting of LENs to form the ElectriNet^{SM} is illustrated in Figure 12 where a "master controller" is embedded into each LEN. These controllers are not all the same—their complexity increases for the LENs that have more components. The master controllers and the LENs they control will require local intelligence and infrastructure. In turn, integration of LENs will require higher levels of integration involving more significant infrastructure transformation in communications and control, as well as in the overall power delivery infrastructure.

It is often advantageous to use several generating units as is the case with a centralized supply system instead of one large unit for several reasons[8]:

- *Increased reliability*. The system can continue to operate when individual units are forced out of service or taken out for maintenance.
- *Greater efficiency*. Unneeded units can be disconnected during periods of light system load, allowing the remaining units to operate near their optimum power output level.

---

8. EPRI report #1001675, "Compatibility Tests for Dissimilar Types of Distributed Generation Powering a Microgrid," 2008.

- *Greater diversity of energy supply.* Units powered by renewable resources can be used with units driven by fossil fuels to reduce costs and environmental impacts while providing reliable power. Units that provide thermal energy for industrial processes may be replaced by other forms of generation when thermal energy is not needed.
- *Increased power quality.* By using a combination of generator types, the strengths of one type can be used to balance the weaknesses of others.
- *Expandability.* Systems can be expanded as the load grows, reducing the costs of construction; it may also be possible to use improved technology or different technology in the later stages of the project.

While the availability of local generation enhances reliability of the power system, systems with local generation can also play a vital role in recovering from high-impact, low-frequency events that may disrupt service from the larger grid. For example, after a major storm, utilities take steps to restore service as quickly as possible, but some outages are inevitable. At some time in the future when utility-connected local generation has been established throughout the electrical system, storms will still cause damage to electrical components, but outages will be contained, with fewer customers losing power.

To enable this architecture, the power delivery system must evolve by incorporating a variety of innovations including

- energy storage and DG to enable stand-alone installations and microgrids;
- systems to enable utilities to integrate local generation, storage, and end-use devices including energy management systems and distribution management systems (DMS);
- topologies, which encourage electrification based on their societal benefits;
- an expanded infrastructure, which includes sensors, data analytics, and communications as well as the means to manage the future infrastructure;
- technologies, which integrate the consumer into the power delivery system through services, social media, and "apps";
- leveraging the advantages of combined infrastructure investment including communications and electricity;
- expansion of business models to enable flexible load shape where consumers support the grid and the grid supports consumers.

Further research is needed to determine some of the technical issues and solutions associated with the operation of a hybrid power system with multiple generation units such as an ElectriNet^{SM}. Control of voltage and frequency will be key factors in the successful operation of the system. The wide variety of generation devices and configurations available means that research will be required to properly design and operate these systems.

In addition, the use of uniformly DG in the ElectriNet^{SM} facilitates the ability to build distribution systems that do not need any high-voltage elements—they are predominantly medium voltage. This medium-voltage approach has potential for significant cost savings and power quality/reliability improvements and can provide improved safety benefits as well. However, special controls and generator protection are required to facilitate proper operation of the ElectriNet^{SM}. Control methods currently under development for conventional interconnection of local generation are not necessarily suitable for the ElectriNet^{SM}.[9]

There are three value streams that customers with LENs can benefit from: (1) providing ancillary and other services, (2) participating in interruptible tariff options, and (3) participating in demand-response programs.

Depending on how the system is designed, a customer with an LEN can provide the utility with various ancillary services, including operating reserves, regulation and load balancing, and voltage support. While the value of these kinds of services can be significant, it is highly variable (King, 2006).

3.1   Enabling LENs

A LEN could be enabled simply by the use of technologies such as the intelligent universal transformer (IUT) as illustrated in Figure 13. The "old mode" in the figure simply illustrates traditional consumer connectivity to the grid. The grid supplies alternating current (AC) power and energizes an "AC bus."

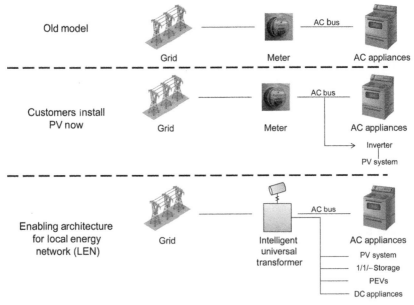

FIGURE 13   The evolution of customer-grid interface: tomorrow's LEN.

9. EPRI report #1018097, "Program on Technology Innovation: Distributed Photovoltaic Power Applications for Utilities," 2009.

As customers install local generation today—using PVs, as an example—they are installing PV along with the necessary inverter to convert the direct current (DC) output of the PV system to AC and to then feed the AC bus.

The ideal evolution of LENs would move toward increasing use of DC. The enabling technology to facilitate this is the IUT. The IUT is an electronic-based device that converts medium-voltage AC (usually 4-13 kV) to low voltage (120-480 V) using power electronics. Its topology takes the medium-voltage power and literally "chops" the waveform into small segments so as to literally create DC. It then takes the DC and reassembles it into an AC waveform at a specified voltage. The IUT has two "buses"—both AC and DC.

Using the enabling architecture as illustrated incorporates the IUT so as to create both an AC bus and a DC bus in the customer's building. A DC bus facilitates the integration of PV systems as the inverter is eliminated, reducing cost and losses. This same DC bus also positions the LEN to easily adapt batteries, plug-in electric vehicles, and DC appliances. Research and development in DC appliances is increasing.

As electronics are increasingly employed in such things as lighting and motors, the same concept of chopping and reassembling waveforms that are used in the IUT is used in lighting ballasts, electric motor drives, computers, and home entertainment devices. In essence, many seemingly AC appliances have "DC inside." The IUT could unleash a fundamental change in building electrical systems and facilitate the transactional environment that the flexible load shape concept embodies.

3.2   Costs and Benefits of the ElectriNet[SM]

Table 1 details the costs for the United States as a country based on EPRI research that would be incurred by consumers to build the smart grid needed by the ElectriNet[SM] and to provide the connectivity highlighted earlier.

These costs total between $338 and $476 billion for the United States and can yield benefits of between $1294 and $2028 billion.[10]

3.3   Examples of the Value of the ElectriNet[SM] and the Flexible Load Shape Concept

The ElectriNet[SM] and the flexibility load shape concept would enable true synergy between consumers and the grid. First, PV systems offer natural support for distribution systems. For example, as shown in Figure 14, the coincident demand from PV offsets some consumer demand as evidenced by a typical

---

10. Refer to EPRI report #1022519 for details: "Estimating the Costs and Benefits of the Smart Grid: A Preliminary Estimate of the Investment Requirement and the Resultant Benefits of a Fully Functioning Smart Grid," 2011b.

TABLE 1 Consumer Implications—Enabling Connectivity

	Smart Grid Cost to Consumers: Allocated by Annual kWh							
	$/Customer, Total cost		$/Customer-year, 10-year amortization		$/Customer-month, 10-year amortization		% Increase in monthly bill, 10-year amortization	
	Low	High	Low	High	Low	High	Low	High
Class	$/Customer	$/Customer	$/Customer/year	$/Customer/year	$/Customer/month	$/Customer/month	Low	High
Residential	$1033	$1455	$103	$145	$9	$12	8.4%	11.8%
Commercial	$7146	$10,064	$715	$1006	$60	$84	9.1%	12.8%
Industrial	$436,291	$151,877	$43,629	$15,188	$3636	$1266	0.1%	1.6%

Source: EPRI 1022519.

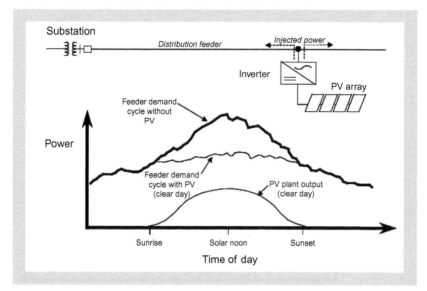

FIGURE 14   Grid support value of distributed PV. *Source: EPRI 1018096.*

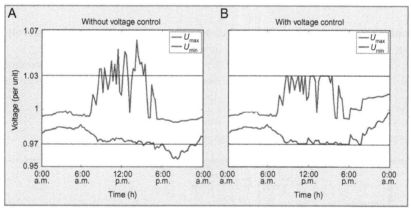

FIGURE 15   Minimum and maximum voltage magnitude (A) without and (B) with central voltage control. *Source: EPRI 1018096.*

distribution load profile. In some cases, the system can support utility needs and mitigate distribution system demand.

Figure 15 further illustrates the effectiveness of central voltage control-enabled by connectivity that interconnected local generation systems can yield. The figure on the left represents a system with local generation installed with no voltage control. The relative high voltage and fluctuations can interfere with the effective operation of customer and distribution equipment, leading to premature failure of components and appliances. The figure on the right

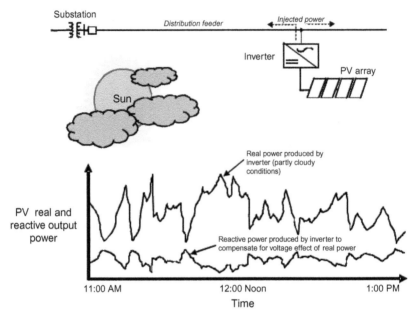

FIGURE 16   Advanced inverter for reactive compensation. *Source: EPRI 1018096.*

represents a system where the voltage can be controlled to mitigate the overvoltage problem and, hence, reduce the adverse costs that result.

Integrating local generation into the grid would enable coordinated control of high-penetration PV systems as illustrated in Figure 16. Here, advanced

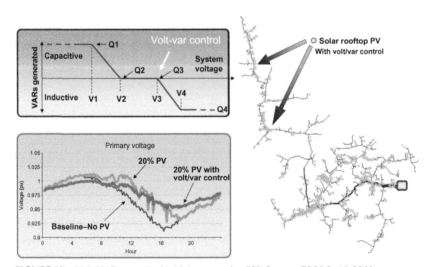

FIGURE 17   Volt-VAR control with high-penetration PV. *Source: EPRI Smith 2011.*

inverters are used to enhance voltage control and balance the ratio of real and reactive power needed to reduce losses and improve system stability.

Figure 17 illustrates another example of the value of integration. In this figure, a distribution system was analyzed to determine the feeder hosting capacity or the amount of local PV generation the feeders could support without substantial upgrading. As shown in the figure, the use of advanced inverters and control could substantially increase the feeders' capacity to accommodate PV. In this case, potential overvoltage situations can be mitigated, eliminating the need for extensive reconfiguring. The "y" axis represents the voltage in "per unit" or PU values. A PU of 1.0 would translate into a service voltage of a 120-V system of 120.

Overall, the ElectriNet[SM] could enable enormous value for both the power delivery system (a utility's ability to serve consumers) and the consumer itself.

4 DISTRIBUTION SYSTEM TECHNOLOGIES NEEDED

Utility distribution systems continue to be challenged by an aging infrastructure, conventional designs, and increased demands for power. Electricity distribution utilities are under pressure to improve reliability and system performance while dealing with the ongoing challenges of an aging infrastructure, conventional designs, and increasing customer demands for higher reliability and power quality. Budget and investment constraints require electric utilities to manage their distribution systems ever more efficiently. In addition, present expectations require that distribution systems support increased automation, new load types, increased DG and storage, increased demand-side controls, environmentally friendly (green) technologies, and so on.[11]

To enable the wide-scale deployment of DERs while meeting the increasing demand of all consumers, two major areas of technology development must be pursued: (1) technologies that improve system responsiveness, flexibility, and functionality and (2) technologies that specifically are focused on enabling the integration of DERs, further described later.

4.1 Distribution Planning for Flexible Load Shape

It is becoming more common that utility distribution planners are faced with accommodating a large penetration of local generation and storage on their power distribution circuits. In many states, the RPS and incentives from various sources have resulted in larger solar PV installations than previously experienced. Similar trends are taking place in many parts of the world as described in other chapters of the book.

---

11. For example, in collaboration with its members, EPRI has developed a strategic plan to articulate the objectives of EPRI research for the next 10 years and to ensure that the focus of its research is, and remains, aligned with those objectives. This appendix provides a high-level summary view of the strategic plan. The reader is encouraged to review the full distribution strategic plan (EPRI 2011a) for the details.

The distribution planning process has traditionally been focused on determining the least-cost alternative for meeting the peak load demand projected for some date in the future. The analysis is often simplified because loading patterns have been the same for many years, and there is much experience with dealing with these loading patterns such as for the rating of transformers. By looking at how the system behaves at one loading point, the planning engineer has a good idea of how it will respond at other times. The basic process can be summarized as follows:

- Define a distribution planning area and model it for power flow analysis.
- Develop a load forecast for a selected planning horizon.
- Determine when planning limits on voltages and current capacities will be violated based on the load forecast.
- Identify one or more alternatives for correcting the violations.
- Determine the least-cost alternative over the planning horizon using approved economic evaluation methods.

Research and development is needed to enable effective distribution planning.

4.2   System Responsiveness, Resiliency, Flexibility, and Customer Connectivity

The distribution system must become flexible, be able to respond to changing system conditions, be able to communicate real-time information, be able to easily and efficiently accommodate varied generation and load types, be able to easily accommodate the installation of new equipment and equipment types, and be able to support system and end-use energy efficiency and demand-response initiatives.

The following are key components for achieving this future state:

- Utilities will be prepared to accommodate DG, storage, and plug-in hybrid electric vehicles. Refer to Chapter 12, Electric Vehicles: New Headache or Asset?
- Intelligent distribution devices will support plug-and-play capability.
- The distribution system will be designed with devices to manually or automatically respond to meet changing system conditions (demand needs, fault conditions, and so on).
- Utilities will use the smart grid infrastructure to accomplish core distribution tasks.
- Gaps need to be addressed in distribution designs and protection schema that address anticipated increasing DG, storage, and PEVs.
- Uncertainty needs to be resolved regarding the capacity implications of new generation types.
- Technologies need to be deployed to allow operators to be able to control and dispatch DG.

- Distribution and generation integration planning needs to be improved.
- Distribution designs and protection that accommodate microgrids need to be developed.
- DG and storage need to be considered in the design of the smart grid.
- The design and regulation of DG and storage on the distribution system need to be refined.

Results can be used to help specify voltage-optimization systems as part of the overall smart distribution development, develop the business cases for voltage-optimization functions, and evaluate performance of systems being implemented.

These results will provide the resources for needs assessment, the development of business cases, and the specification of advanced system reconfiguration functions that could be implemented as part of a distribution management system, including performance assessment for advanced reconfiguration functions.

Many challenges are impeding the implementation of smart grid technologies, including (1) existing designs and conditions of distribution systems that are not yet ready to support smart applications, (2) the need for plans for an orderly application of smart technologies, (3) standards and protocols for the technologies, (4) the need for plans and methods for investment cost allocation, and (5) justification for capital investments by utilities. In addition, new load and generation types are being added to the grid, creating an urgent need for the dynamic system response enabled by smart applications.

4.3   Integration of Local Generation

The existing electric distribution system is ready to accommodate a reasonable demand for deployment of local generation. Practices to establish hosting capacities, procedures for making interconnections, and methods for mitigating problems are well established. Operating the distribution system with distributed renewable generation does not compromise system reliability, power quality, or safety. However, as the use of local generation proliferates, a number of key developments will be required.

The following are key components for achieving this future state:

- Distribution planners will have the data, experience, and needed modeling tools to plan for different renewable-deployment scenarios.
- Methods to determine feeder hosting capacities for variable DG will be available.
- Distribution operators will know what distributed resources are connected to their systems, will be able to forecast and manage variability, and will be able to employ mitigation strategies when needed.
- Distributed PV systems will include power and communication interfaces that are able to provide grid support.

- Communication protocols and interfaces will be available for DMS and will be able to take advantage of advanced functions available from electronic inverters.
- Effective business models for utility ownership and operation of distributed PV will be available.
- Utilities that own and operate distributed PV systems will have established practices to minimize operation costs and avoid downtime.
- Communication, metering, and power interfaces of distributed renewables to provide grid support, including the required communication protocols for integration.
- Smart technologies (such as inverters and energy-storage devices) and advanced measurements and controls (such as self-diagnosis and self-healing).
- Existing and emerging intelligent technologies and techniques to provide for proactive maintenance, optimal levels of asset utilization, and greater efficiencies.

Closing these R&D gaps will require that the industry invest resources in developing and implementing new processes, tools, techniques, and technologies.

5 CONCLUSION

Disruptive changes in the retail electric business are already under way. An integrated approach enables resources and technologies to be deployed operationally to realize all potential benefits. It requires a modern grid characterized by connectivity, rules enabling interconnection, and innovative rate structures that enhance the value of the power system to all consumers.[12] This approach allows society to enjoy the benefits of a blend of the most valuable generation, storage, power delivery, and end-use technologies tailored to meet local circumstances. It does not favor any one technology beyond its unique contribution. Instead, the approach includes objective and thorough analysis to identify optimum architectures considering end-user requirements, delivery system, grid operation, and available energy and storage resources.

REFERENCES

EPRI, 1984. Demand-Side Management – Volume 1: Overview of Key Issues. EPRI, Palo Alto, CA, EA/EM-3597.

EPRI, 2011a. Distribution Research Area Strategic Plan: September 2010. EPRI, Palo Alto, CA, 1022335.

---

12. Operational integration does not necessarily imply that utilities must control local generation. However, it does require monitoring and overall balance between resources.

EPRI, 2011b. Estimating the Costs and Benefits of the Smart Grid: A Preliminary Estimate of the Investment Requirement and the Resultant Benefits of a Fully Functioning Smart Grid. EPRI, Palo Alto, CA, 1022519.

Gellings, C., Zhang, P., 2010. The ElectriNet. Electra, Cigré.

Kefalas, T.D., et al., 2010. Grid voltage harmonics effect on distribution transformer operation. In: Proceedings of IEEE 7th Mediterranean Conference and Exhibition on Power Generation, Transmission, Distribution and Energy Conversion.

King, D.E., 2006. Electric Power Micro-grids: Opportunities and Challenges for an Emerging Distributed Energy Architecture. Dissertation, Carnegie Mellon University, Pittsburgh, PA.

Smith, M., Ton, D., 2013. Key connections. IEEE PES Magazine.

Trueblood, C., Coley, S., Key, T., Rogers, L., Ellis, A., Hansen, C., Philpot, E., 2013. PV measures up for fleet duty: data from a Tennessee plant are used to illustrate metrics that characterize plant performance. Power and Energy Magazine, IEEE. 11 (2), pp. 33,44.

Wilson, J., 2013. IEEE Distribution Reliability Working Group 2012 as reported. In: Wireless Connectivity for a Reliable Smart Grid, Electric Light and Power.

Chapter 6

The Impact of Distributed Generation on European Power Utilities ☆

Koen Groot

ABSTRACT

This chapter highlights how the growth of distributed power generation in the EU influences the corporate strategies of major European power utilities. Driven to a large extent by renewable support policies, the proportion of distributed power generation has increased in the EU. This has happened to the detriment of the share of electricity supplied by power utilities, which are not only selling less electricity but also facing lower prices for electricity sold, as a result of the changing market conditions. The uptake of distributed renewables adds to the already troublesome outlook for traditional utilities, further eroding their profitability. In response, these firms engage in strategic restructuring, cost-cutting, and the expansion of activities in growth markets outside the EU. The chapter examines the impact of a rapidly growing share of distributed power in the EU and the incumbents' initial response to this new dynamic in their European home markets.

Keywords: European power utilities, Distributed generation, Renewable energy, Utility business strategy, Corporate restructuring

1 INTRODUCTION

In recent years, the major EU power utilities[1] have been confronted with significant changes in their external environment seriously affecting their future viability. The economic crisis has translated in diminished demand for electricity; the subsidized

---

☆ This chapter draws from the Clingendael Energy Paper, "European Power Utilities under Pressure?" (May 2013).

1. In the power sector of the European Union, seven firms stand out in size when it comes to installed capacity, electricity production, and revenues. These are E.ON, EDF, Enel, GDF SUEZ, IBERDROLA, RWE, and Vattenfall. This chapter focuses on the Central Western European Market, although the portfolios of utilities active there include assets in adjacent markets.

rise of renewables caused more competition in a shrinking market, and the departure from nuclear generation in various member states has caused significant setbacks to firms operating nuclear power plants. The ongoing developments culminate in a dire outlook for many EU utilities. Profit margins of individual plants and entire generation portfolios are under pressure, and as a result, firms are retiring their generation assets—whether mothballing, wet-reserving, fully decommissioning, or even putting entire divisions up for sale.[2] Neither the current market conditions nor the prospects for market reform provide much relief.

The development with the most far-reaching ramifications for the traditional utility business model is the increase in distributed renewable energy sources (RES). In the EU, a significant growth in distributed generation is witnessed as installed capacity grew from 10 GW in 2010 to 70 GW in 2012 (Eurelectric, 2013b). The main driver of this trend is the EU push for renewable energy and associated support mechanisms.[3] To date, Germany and Italy are the largest markets for solar PV in the world (REN21, 2013; The Financial Times, 6 August 2013). For some time, the European Union has dominated both the total installed capacity and the new solar PV installations, although as of 2013, the latter is no longer the case (EPIA, 2013).

This upsurge of distributed generation—and especially its renewable nature—has a profound effect on the European electricity industry. In this already difficult market environment, the increase of distributed generation reduces the size of the market supplied by power utilities. The majority of electricity in the European Union is traditionally generated by a small number of companies, referred to here as the big 7—that is, E.ON, Enel, EDF, GDF SUEZ, RWE, IBER-DROLA, and Vattenfall. The generation portfolios of these companies consist of large, centralized power plants—mainly large hydro, nuclear, gas-fired, or coal-fired power plants and, as of late, some large-scale wind farms, concentrated solar power facilities, and ground-mounted solar panel fields. These are electricity generation assets, characterized by investments with rates of return based on predictable run-times and on sufficient exposure to peak demand. This is the so-called paradigm of centralized generation and volumetric consumption further described in the chapters by Sioshansi and others in this volume.

---

2. RWE discontinues the operation of 10 units of total 4.3 GW (RWE, 2013b); E.ON decommissions 30 units of total 11 GW by 2015 (E.ON, 2013); as GDF SUEZ continuously retires assets, the mothballing of another 1.4 GW was announced in 2013 adding to the already 8.6 GW of decommissioned or mothballed units since 2009 (GDF SUEZ, 2013a); EnBW and Statkraft take similar measures, shutting down, respectively, 0.5 GW of hard coal- and gas-fired capacity (Enerdata, 10 July 2013) and 2.2 GW of gas-fired generation capacity (Enerdata, 20 August 2013), while Vattenfall has announced to look for investors to offtake its share in non-Nordic European power production at a total of ~19.5 GW, of which nuclear capacity in Germany (Vattenfall, 23 July 2013).

3. By way of Renewable Energy Directive 2009/28/EC (European Commission, 2009), which has set targets for a share of 20% renewables in EU electricity production by 2020, the directive is translated into 27 National Renewable Energy Action Plans, which formulate the approach of the different member states toward attaining the EU goal.

The growth of distributed renewable generation harms the traditional utility business model, as it contributes to lower uptimes for conventional generators and lower margins for production.[4] As a result, large parts of the power generation portfolios of major European utilities are operating at a loss.[5] This changing market, where the electricity companies sell less electricity at lower wholesale prices, confronts the major utilities with a daunting picture. Over a short period of time, a new paradigm has emerged in which earlier premises of EU energy policies and corresponding corporate strategies no longer hold. Compelled by these challenging conditions, all major European power utilities are in a process of strategic reorientation.

This chapter examines the impact of these developments:

- What is the effect of the recent developments, especially of the growth in distributed RES generation in the EU electricity sector, on the profitability of the big 7?
- How are the big 7 responding to these challenges?

The chapter is organized into four sections including the introduction. Section 2 provides an overview of the developments in the European power sector. Section 3 describes the impact of this on utilities. Section 4 describes how firms are responding followed by the chapter's conclusions.

2   THE CHALLENGES FACING THE EU ELECTRICITY SECTOR

"European energy policy has run into the wall." These were the words used by the CEO of French utility GDF SUEZ during the press conference of an ad hoc group of ten European energy company CEOs[6] aimed at conveying the troublesome outlook for the EU electricity market. In its announcement, the group pleaded for a remuneration mechanism for owners of conventional generation capacity as well as for a stop to subsidies for renewables. This outcry from a group of industry representatives is emblematic of the situation in the European electricity sector. The operators of mainly centralized conventional capacity are faced with increasing competition from often subsidized distributed renewables. The question is how did it come this far?

---

4. Distributed renewable power generators, mainly solar panels although also small-scale wind power, absorb the market share of other power generation sources by displacing them in the merit order. Consequently, the increase of RES in the market flattens the merit order, leading to (wholesale) price decreases as well as to convergence of base load and peak load prices. This harms the business case for conventional power generation in the EU market, where especially gas- and coal-fired power plants are increasingly operating (even if) at a loss.

5. 110–130 GW of gas-fired power plants—about 60% of the total installed gas-fired generation in the EU—is currently unprofitable and at risk of closure in the next 3 years up (IHS, 2013); 30-40% of RWE's generation portfolio is operating at a loss (Management Team, 2013).

6. The so-called Magritte Group consists of the CEOs of France's GDF SUEZ, German RWE and E. ON, Spanish IBERDROLA, Swedish Vattenfall, Italian Enel and ENI, and Czech CEZ (Euractiv, 2013).

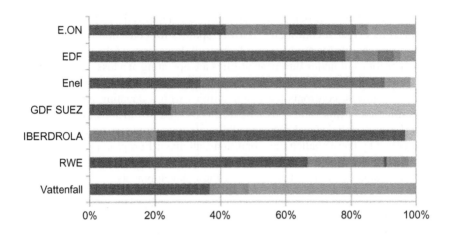

■ France ■ Germany ■ UK ■ Iberia ■ Italy ■ Benelux ■ Nordic ■ CEE ■ Other

FIGURE 1   Installed generation capacity of major EU power utilities in Europe. *Source: Clingendael Energy research (Annual reports and corporate websites of respective firms (The Financial Times (2013) Markets Data by Thomson Reuters)).*

Stimulated by EU policy aimed at the creation of an internal market for energy, various utilities have invested heavily in regional growth. Through mergers and acquisitions—mainly outside of their traditional home markets—as well as through substantial investments in new generation capacity, the firms became the largest players in the EU electricity market. Most of the electricity produced in the EU is generated by the big 7; most of this occurs in western Europe as illustrated in Figure 1.

The diversified portfolios of the major utilities traditionally provided a significant advantage, the ability to leverage market risk. In today's market, the diversification contributes to overexposure to the challenging developments in the sector. Most notably, the effects of the economic downturn, especially the stagnation of demand; the dwindling support for nuclear generation in various member states translating in nuclear phaseouts[7]; and the rise of distributed renewable generation that is cannibalizing their profit margins.

The latter especially has taken the industry by surprise. In the EU, the share of renewable generation is growing rapidly. Total installed solar PV and wind capacity rose from 13 GW in 2000 to over 175 GW in early 2013, accounting for ~19% of total EU generation capacity.[8] The contribution of these resources to electricity generation is more moderate as is reflected in Figure 2. This is due

---

7. In Germany, the government decided to phase out nuclear power generation before 2023; the Belgian government has decided to phase out nuclear power production completely by September 2025; in Switzerland, the last nuclear power generator will go offline by 2034; in France, discussion is ongoing to reduce the share of nuclear energy in the power mix from 75% to 50%.
8. 175 GW of wind and solar capacity out of the 930 GW of total generation capacity (EWEA, 2013a).

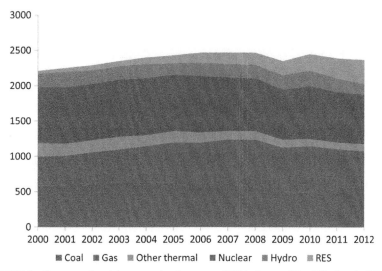

FIGURE 2   European electricity generation by source (TWh). *Source: Wood Mackenzie (2013).*

to relatively low load factors and the intermittency of these sources. In Germany, Spain, and Italy, three of the four largest European electricity markets, the shares of solar PV production amounted to between 3% and 6.5% of annual volumes in 2013,[9] although at times these volumes are much more substantial as described for Germany in the chapter by Burger and Weinmann.

In parallel with the increase of RES capacity, the share of distributed generation in Europe increased to almost 70 GW in 2013 (Eurelectric, 2013b). This share has been largely the result of the boom in distributed solar PVs. In the beginning of 2013, the installed base of distributed solar power in the EU reached 59 GW—about 85% of all installed distributed generation in Europe.[10] What is remarkable, but perhaps not surprising, is the lack of investment by major utilities. In the past decade, they have mainly invested in new thermal plants as reflected in Figure 3.

While the role of the big 7 in distributed renewables is limited, the effect of this capacity on them has been significant. First and foremost is the replacement effect: the growth in distributed renewables has resulted in more consumers generating their own electricity. This has reduced the need to consume electricity generated by utilities. The growth of self-generation by traditional

---

9. In Germany, from the 244.1 TWh produced in H1 of 2013, 14.3 TWh was solar energy, more than 5.8%, with peaks of up to 23.2 MW (Fraunhofer, 2013a). In Spain, solar PV accounted for 8.17 TWh out of 266.85 TWh consumed, fulfilling an approximate 3% of Spanish electricity demand (Red Eléctrica de España, 2013). In Italy, solar PV accounted for 6.4% of total electricity produced in Italy in 2012 (OECD, 2013), 18.86 TWh (GSE, 2013).

10. Other sources of distributed generation in the EU are mainly wind turbines and small-scale combined heat and power installations (Eurelectric, 2013b).

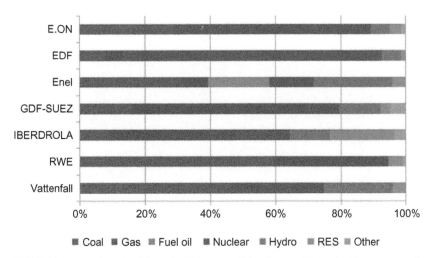

FIGURE 3   Production mix of the major EU power utilities. *Source: Clingendael Energy research (Annual reports and corporate websites of respective firms (The Financial Times (2013) Markets Data by Thomson Reuters)).*

consumers of electricity is also referred to as the rise of the "prosumer" described in other chapters including Blansfield and Jones. Besides providing a new source of competition to incumbents, distributed renewables affect the profit margins of power generation.

When large swaths of electricity produced by distributed RES come online, wholesale electricity prices get depressed. This is especially relevant as the peaks for solar power generation largely coincide with peak demand as illustrated for Germany in Figure 4. Lower uptimes and lower prices during uptimes significantly lower the margins on power generation. Investors in gas-fired generation—once expected to be the perfect partners to renewable generation because of the flexibility of combined-cycle gas turbines—had expected gas to play a prominent role in the fuel mix. This however, at least for now, has turned out otherwise. According to some estimations, an approximate 60%

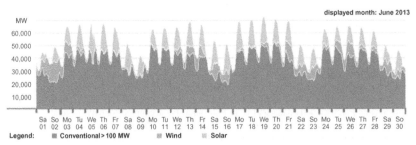

FIGURE 4   Electricity production in Germany (MW). *Source: Fraunhofer ISE (2013a).*

of total installed gas-fired capacity in the EU is currently unprofitable and at risk of closure (IHS, 2013). This is witnessed throughout Europe, where firms mothball and decommission gas plants from Germany to Spain and from Italy to the Netherlands, resulting in large-scale write-offs as described by Burger and Weinmann for Germany.

Meanwhile, in many European countries, the economic downturn has forced stringent austerity policy on governments. The growing cost burden associated with the growth of subsidized renewables—in combination with budgetary limitations resulting from the economic downturn—has resulted in several countries redrafting their renewable energy support schemes. In the Spanish electricity sector, this has resulted in lower income for utilities like IBERDROLA, Endesa, and ACCIONA (Platts, 5 August 2013). Besides affecting government budgets, RES support contributes to rising consumer costs.[11] At the same time, this provides consumers with new opportunities to invest in solar PV or other types of distributed generation to reduce their exposure to the grid-supplied electricity. In doing so, these "prosumers" also avoid potential hikes in retail electricity tariffs and associated levies. This phenomenon is referred to as the "utility death spiral" further described in other chapters.

The loss of customers during daylight hours, typically the most lucrative hours for generators, is a growing problem for power utilities in the EU. The sheer size of their portfolios and the lengthy payback periods for their investments drastically limit the ability of the firms to respond to this new market dynamic.

3   THE IMPACT ON MAJOR EU POWER UTILITIES

While total revenue of the big 7 has *increased* over this period, their combined net income has *declined* (Figure 5). This suggests that while there in fact is growth in revenues, costs are growing even faster, resulting in falling net income. The growth in turnover is partly the result of the integration of acquired business units and companies by the major power utilities over this period, while an increase in trading activities provides another explanation.

The financial outlook for major European utilities has deteriorated substantially, as the firms face negative margins on substantial parts of their generation portfolios, partly due to the growth in distributed renewables. The operating

---

11. In Germany, consumers are to face a 47% hike in the contribution to renewable support, which is part of their electricity bill; following an increase of 72% in 2010, the gross national surcharge is expected to be around 20.4 billion EUR in 2013, compared to 8.3 billion EUR in 2010. The reason for this is the vast expansion of RES generators in Germany 5/7 GW annually and the heavily subsidized industrial power rates, providing households with the bill (Petroleum Intelligence Weekly, 24 October 2013).

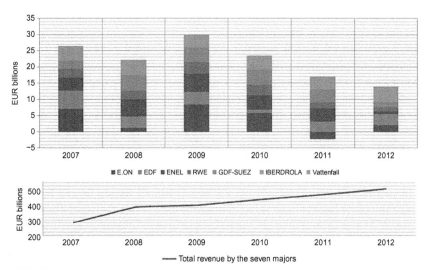

FIGURE 5   Net income and total revenue of the seven major EU power utilities. *Source: Clingendael Energy research (Annual reports and corporate websites of respective firms (The Financial Times (2013) Markets Data by Thomson Reuters)).*

income of the companies is further depressed by impairment charges and write-offs—related to specific events, such as the Atomausstieg[12]; to market conditions deteriorating the profitability of gas- and coal-fired generation assets[13]; and to the macroeconomic conditions in specific EU markets such as Spain.[14]

Falling profit margins have a significant impact on the stock market valuations of the firms. In the past five years, the 20 utilities captured in the MSCI European Utilities index have seen their combined share price halved, from an approximate trillion EUR (The Economist, 12 October 2013). Shareholders of EU power utilities are faced not only by these dramatic drops in value but also by cuts in dividend as announced by RWE.[15] The economic downturn

---

12. The write-offs on nuclear assets, caused by the nuclear phaseout in Germany, have forced E.ON, RWE, and Vattenfall to rebalance their portfolios, which resulted in the firms divesting from (non-core) assets (Reuters, 10 August 2012; The Wall Street Journal, 6 March 2013).

13. For example, GDF SUEZ wrote down €2 billion over 2012 (GDF SUEZ, 2012), while Vattenfall wrote down €4.6 billion, about half of which in the Netherlands, as part of an effort to restructure the company (Financial Times, 23 July 2013); similarly, RWE made impairment charges of €2.3 billion on power generation assets, mostly in the Netherlands (The Wall Street Journal, 5 March 2013).

14. Significant impairments and write-offs are made especially in Spain and Portugal, e.g., by Enel that made impairments of €2.58 billion over 2012 (Enel, 2013. Annual report 2012).

15. See RWE corporate website.

exacerbates the problematic financial situation for the firms in the European electricity market.[16] Recent downgrades in sovereign bond credit ratings have affected the credit ratings of Enel, Endesa, and IBERDROLA (Eurelectric, 2012), while the negative market outlook for the likes of RWE and E.ON has resulted in downward adjustment of their credit ratings (Bloomberg, 27 July 2012).

What makes the picture mentioned earlier even more pressing is that it complicates the ability of the firms to source capital for the refinancing of ongoing operations and, even more important, for investments in new activities. This is disadvantageous, as the companies will need to engage in new activities to overcome the current downward trend.

To deal with the problems in the short run, the firms are undergoing adjustments. Besides mothballing and idling unprofitable power plants, virtually all major utilities have started disposal programs aimed at selling noncore assets.[17] To improve capital positions, firms deleverage through divestitures while postponing or outright canceling investments in nonsubsidized generation activities in Europe. In addition to these divestitures, utilities have been forced to consider layoffs[18] and engage in gas contract renegotiations. In an attempt to overcome the negative margins on gas generation, various European power utilities have entered into contract renegotiation with their gas suppliers.[19] What else can the utilities do to overcome this pressing situation?

---

16. The France government is a majority shareholder in EDF and large shareholder in GDF SUEZ; the Swedish government fully owns Vattenfall; the Italian government is a shareholder in Enel; and German regional governments own stakes in RWE. Government ownership is common in the European electricity sector: The Danish government is the majority shareholder in DONG Energy, CEZ is for the majority owned by the government of the Czech Republic, and Fortum is owned by the Finnish state.

17. In order to execute its disposal program, RWE will need an approximate €5 billion before the end of 2013 (RWE, 2013a). In mid-2012, E.ON had already divested for €12.5 billion, to amount to €15 billion by end 2013 (E.ON, 2013). Vattenfall completed its envisioned divestment of noncore assets by 2012, after which it has imposed an additional reduction to be realized by the end of 2013 (Vattenfall, 2012). GDF SUEZ is looking to free up €11 billion (GDF SUEZ, 6 December 2012), while IBERDROLA is looking to make divestments to the order of €2-5 billion (IBERDROLA, 2012). IBERDROLA has in 2013 sold various assets in different countries, among others, its stake in Chilean electricity producer Empresa Eléctrica Lican and in Polish wind power projects (Bloomberg, 1 July 2013).

18. In an attempt to limit costs and reduce debt, Vattenfall and RWE have announced reorganizations, resulting in significant losses of jobs (Reuters, 10 August 2012; The Wall Street Journal, 6 March 2013).

19. E.ON successfully renegotiated suppliers Statoil and Gazprom and EDF subsidiary Edison with various suppliers including ENI, Statoil, and Sonatrach, while RWE has renegotiated the contractual terms of its gas offtake contracts with Statoil. The ICC arbitration court has ruled in favor of utilities in the cases of, among others, RWE versus Gazprom and EDF versus RasGas (World Gas Intelligence, 14 July 2012; 3 October 2012; 7 November 2012; 23 January 2013).

4   HOW ARE EUROPEAN POWER UTILITIES RESPONDING

Driven by the effects of a changing European power system, all major European power utilities are in a process of strategic reorientation. The earlier-mentioned measures will provide short-term relief at most. To improve the outlook for their business on the long run, the firms are taking other measures. In general, the long-term strategic restructuring of the power utilities follows three directions, or a combination thereof:

- Development of activities in growth markets inside and outside of Europe
- Development of large-scale renewables in the EU and globally
- Activities aimed at securing market share in their European home markets

In the latter case, where utilities chose to deal with the challenges in Europe, a general distinction can be observed between activities focused on maintaining the status quo and activities focused on establishing dominant positions in a changing market. This is a distinction between activities aimed at defending existing "distressed" assets, mainly thermal power plants in home markets, and activities aimed at developing new business around other existing business units such as the electricity grids, the trading floors, and marketing departments.

The major European power utilities that have the ability will restructure their assets to become less dependent on the mature economies in Europe and generate more future revenue in other markets. When feasible, the firms invest outside the EU, in power generation, thermal, nuclear, and renewable, or other activities, often upstream or LNG-related. The majority of growth in demand for electricity is expected to come from the Asia Pacific markets including China and India, the Middle East, and Latin America (ExxonMobil, 2013). GDF SUEZ, Enel, and IBERDROLA, for example, are already present in these markets, and in their corporate plans for the coming years, they convey to increase this substantially.[20] Others have more recently started investing outside of the EU; RWE's recent Brazilian and Turkish ventures are an example of this. The activities of the major power utilities outside of the EU are for many a major source of income as can be observed in Table 1. Besides diverting toward the far ends of the world, big 7 utilities invest in growth markets closer to home, for example, in eastern Europe. Utilities like E.ON, RWE, and Enel direct significant investments to developing activities in the Czech Republic, Slovakia, Hungary, Romania, Bulgaria, and Russia.

Whether the expansion outside the mature markets in Europe is temporary or permanent remains to be seen. Some firms are already able to offset the losses in net income of their European power generation portfolios against their

---

20. Of the 13 GW of new capacity under construction in GDF SUEZ's fleet, nearly 80% is located in growth markets in the Americas, the Middle East, and the Asian Pacific; for its total portfolio, this equals ~40% (GDF SUEZ, 2013b); Enel directs more than half of its CAPEX toward power generation activities in the Americas, with over 11 GW of projects in the pipeline (Enel, 2013).

TABLE 1 Major EU power utilities' core data

	Revenue (EUR billion)	Non-EU power sector revenue (% of total)	Net profit (EUR billion)	Capacity (GW)	Production (TWh)
E.ON	142.94	20%	2.18	70.00	271.20
EDF	72.73	4%	3.32	134.79	631.28
Enel	84.89	41%	0.87	97.34	291.09
GDF SUEZ	97.04	65%	1.55	117.31	465.00
IBERDROLA	34.75	24%	2.84	46.03	145.13
RWE	50.77	8%	1.31	49.24	205.70
Vattenfall	19.22	4%	1.98	35.85	153.70

Source: Clingendael Energy research (Revenue and Net Profit 2012 data, and others, estimation based on 2011 data. Annual reports and corporate websites of respective firms (The Financial Times (2013) Markets Data by Thomson Reuters)).

investments in other parts of the world. As long as the outlook for the European power sector—especially for utilities with large shares of conventional generation sources—does not improve, it is likely that firms will continue to strengthen their focus outside of the EU. Of course, investing in international growth markets is appealing and rational. International investments, however, are by no means a panacea to the problems suffered in home markets. Unless there is a strategic fit with the portfolio—be it in the firms' value chain or a link to other activities in the region—activities outside of the EU are unlikely to be sustainable. Developing global portfolios will therefore not be part of the solution for all major power utilities.

Back in the EU, the member states seem set to look for ways to entice investors in large-scale renewable capacity projects, as states seek to attain their shares of renewable energy laid down in the legally binding EU 2020 policy.[21] Under these circumstances, the development of renewable business will proceed, especially as renewable energy projects take on a much larger scale, certainly in the case of offshore wind.[22] Major utilities including E.ON, Vattenfall, and RWE, accompanied by Danish DONG, Norwegian Statkraft, and British firms SSE and Centrica, account for over 70% of the investments in European offshore wind project to date (EWEA, 2013b). The growth trend in offshore

---

21. For example, in the United Kingdom, the Netherlands, and France.
22. Virtually all major utilities have invested in large, multi-gigawatt wind power development projects (E.ON, 2012; EDF, 2013; Enel, 2013; GDF SUEZ, 2013b; IBERDROLA, 2012; RWE, 2012; Vattenfall, 2012).

wind is expected to continue, and as these projects are more akin to the major utilities' traditional expertise in large-scale centralized production, these firms can be expected to keep an active role. The type of role, however, might be changing, whereas the deteriorated financial performance of most major power utilities will make it harder to secure funding for the costly offshore wind projects.[23] Rather than being the main investors, the companies will seek to play the part of project developer and operators of these projects, as announced by RWE.

In a way, large-scale renewable generation projects can be compared to the international adventures the major European utilities have been branching out in. Not all utilities will be successful in developing these—some more than others. The realization of this has already occurred and is visible in the wave of divestments or withdrawals from various wind projects in the EU by the big 7 in 2013.[24] Activities outside of the mature markets in the EU and new large-scale investment projects are two ways of finding new income streams. To be sure, some firms will succeed in these endeavors; for others, this is less likely. So in what ways can these firms build strategies to last?

4.1   New Business Strategies for Europe

In the new approach to their home markets in Europe, many major utilities focus on maintaining the status quo. To do so, the firms have significantly increased their corporate communication; they focus more on marketing, aimed at appealing to consumers and on corporate political activity[25] aimed at influencing policy makers. Through marketing activities, the firms seek to maintain their customer base, by convincing electricity users of their vanguard position as they embrace change and offer new services. Such services have mostly been centered on business concepts that the major utilities had mostly neglected thus far. An example of this is the growth of solar PV services offered by large utilities that have as of recent started business units like RWE Solar, E.ON Solar, and Vattenfall Solar, further described in chapter by Burger and Weinmann.

In addition to convincing customers of the appeal of their products, the utilities focus on convincing policy makers of their critical services to public. Through corporate political activities, the major utilities attempt to influence EU policies in their favor. An example of this is the effort by the "Magritte Group," an ad hoc group of major power and gas companies calling for the

---

23. The retreat of RWE from the UK Atlantic Array project late 2013 might be regarded as an example of this change of direction.

24. In 2013, several major European utilities stopped and backed out of offshore wind projects; notable examples of this are the withdrawal of RWE from a large UK wind project, the Atlantic Array; the sale of IBERDROLA's Polish wind assets; and the withdrawal of IBERDROLA a large offshore project in Scotland, the Argyll Array.

25. Corporate political activity is a type of nonmarket strategy aimed at influencing the political and policy-making process, at both domestic and international levels, to achieve policy-based advantages.

introduction of capacity payments to owners of thermal generation capacity. The firms seek to maintain the viability of the thermal plants in their portfolios through payments for capacity in addition to the current market for electricity. In anticipation of market reforms toward such instruments, the utilities are planning to hang on to their thermal assets to the extent that they can, whether through mothballing, dry-reserving, or maintaining the assets operating at a loss. At the same time, they use the current slack in the market to shut down outdated plants to make way for the significant amount of new coal- and gas-fired capacity still under construction in mature EU markets.

In addition to the call for capacity payments, European utilities focus on policy support for renewables. On this topic, there is discord between the European utilities, with on the one hand the firms of the "Magritte Group" that oppose support for renewables and on the other hand the firms of the "Coalition of Progressive European Energy Companies" that are in favor of renewable support. The latter group consists of utilities that focus on renewables, like DONG, EDP, and SSE (SSE, 2013). For these firms, a continuation of the support for renewables in the EU will be beneficial, providing them with more options to expand their business. Indirectly, the firms also stand to win from a continuation of renewable support since this will further weaken their competitors.

Apart from the PR and corporate political lobbying, the firms are focused on establishing dominant positions in a changing market. Major utilities aim to attain a new role in the changing market by developing activities that will be important in a more decentralized and intermittent energy future. The general consensus is to move away from a supply-driven model toward more interaction with the customer. In the words of E.ON CEO Teyssen, "the upstream, commodity side is shrinking at high speed, we are trying to replace it with the distribution, retail side" (Energy Post, 3 December 2013).

Moreover, utilities will look to build new activities on the basis of their current strengths, such as their transmission and distribution networks, trading floors, and marketing departments. In a recent report by the European electricity industry association Eurelectric, these new downstream activities are expected to be the main source of growth in business toward 2020 (Eurelectric, 2013b; also elaborated upon in the chapter by Sioshansi).

While highly anticipated, most of these new activities are still in development and have yet to be proven successful. Depending on the fit between the capabilities of the different incumbent companies and the characteristics of the markets they seek to serve, the utilities take a variety of approaches to engage in new downstream activities.

RWE for one is set to develop toward becoming a *customer-centric* company, focusing more on the interplay with consumers at the front end of the business, moving away from a large-scale thermal power-based model, and not looking to play a role in the new growth sector of decentralized, subsidized business (Energy Post, 21 October 2013). Instead, it will focus on creating value in different segments of the electricity sector, mainly in an enabling role as a developer of energy projects, looking to match different types of customers, for example, institutional investors and households. Moreover, RWE seems determined to become a recognized brand name—crucial in a customer-driven world—by directing much attention to its retail activities. Thus far, electricity companies have mostly been generation-oriented, paying limited attention to their brand perception. If RWE succeeds in establishing itself as a true customer brand, it can obtain a serious lead on the competition.

EDF is one of the firms still heavily involved in transmission, through subsidiary RTE, and distribution, through subsidiary ERDF. Building on these activities, in combination with its central generation assets, EDF can become a provider of ancillary capacity in a market characterized by more intermittency. The firm is also focused on developing offshore wind projects, which provides synergy with its grid assets. EDF might also benefit from its distribution assets, by emphasizing the proximity to the consumers. In doing so, it might be able to market new energy services, for example, focused on energy efficiency and distributed generation. In the EU, transmission and distribution asset ownership by utilities are limited, especially on the scale of EDF's grid assets.[26] This might provide EDF with an advantage over the competition.

E.ON, in moves quite similar to RWE, has announced a strategic shift toward the distribution and retail business. It seeks to do so by focusing more on its retail business and distributions assets. Building on these assets, it wants to create value for its customers by offering services for smart grid development and integration of distributed generation and energy efficiency (Energy Post, 3 December 2013). While it is becoming more customer-oriented, E.ON will also look to find new ways to market natural gas, which is a significant part of its business. At the same time, E.ON seeks to become active in other markets by establishing and developing businesses in central and eastern Europe, Turkey, and Brazil.

While some of the major utilities are more focused on the development of business outside of the EU, they must at the same time develop sustainable and profitable operations in their home markets. IBERDROLA does so by investing heavily in large-scale offshore wind projects. In Europe, it has over 6 GW

---

26. Since the coming into force of the so-called Third Energy Package, there is a structural separation between transmission system operator activities on the one hand and generation, production, and supply activities on the other. As a result, many power utilities have unbundled or spun off their electricity transmission and distribution activities. This is especially the case for utilities with operations in Germany, while in France, Italy, and Spain, utilities still operate distribution and/or transmission systems. In Germany, most transmission assets have switched hands, from RWE to Amprion, from E.ON to TenneT, and from Vattenfall to Elia. EnBW is the exception in Germany as it still owns transmission capacity (EnBW, 2013; 2020). In France, all T&D assets are owned by French electricity giant EDF, and GDF SUEZ does own T&D assets, however in gas infrastructure rather than in electricity. In Spain, transmission assets formerly owned by Enel subsidiary Endesa were sold to Red Eléctrica de España. In Italy, Enel still owns grid operations on the distribution level. IBERDROLA also still owns and operates its distribution assets in the EU, mainly in Spain and the United Kingdom.

currently under construction in Germany, the United Kingdom, and France. This will be added to its global portfolio of wind generation capacity in Europe, Latin America, and the United States. These activities combined already make up a third of the firm's installed capacity—which it aims to increase substantially (IBERDROLA, 2013). In addition to wind generation, the company seeks to craft a portfolio with more predictable returns by focusing more on regulated generation activities in growth countries like Brazil and Mexico (IBERDROLA, 2013).

Enel is another case in point of a utility directing most of its growth CAPEX to markets outside of Europe. Enel is especially focused on Latin America,[27] although it also invests in eastern European growth markets (Enel, 2013). Nevertheless, the company also seeks to develop new activities in its western European home markets. There it seeks to build on its network activities by developing more "new" downstream activities like smart metering, smart cities, and energy efficiency services (Enel, 2013).

GDF SUEZ is the one firm perfectly placed to leverage the problems in the EU by investing in activities that are outside the European electricity sector, whether in other continents or in other sectors of the economy. The company invests in electricity generation, gas production and distribution, and LNG assets in a number of markets from Latin America, eastern Europe, and the Middle East to Southeast Asia (GDF SUEZ, 2013b). However, for its business in the EU, it focuses on increasing income from energy efficiency services and by investing in "selective and capital efficient development of renewables" (GDF SUEZ, 2013b).

Finally, it seems likely for the big 7 to become the big 6, as Vattenfall looks on its way to retreat to Scandinavia. In 2013, the company has split its portfolio into Scandinavian and mainland European operations. For the latter, it is looking for buyers to partly sell off previously acquired business. In addition, the company is looking for coinvestors in projects it wants to remain involved in (Vattenfall, 2013). By doing so, the company is likely to lose significant parts of its portfolio and some of its diversity. In light of these developments, it is presumable that Vattenfall will focus more on its home market and those in the surrounding Scandinavian countries.

5   CONCLUSION

While the outlook for power utilities in the EU has been affected by several developments, the impact of distributed renewable generation seems to have the most far-reaching consequences. By shrinking an already stagnant market and diluting growth projections, distributed renewables have proved to be fierce

---

27. In Latin America, Enel currently has 11 GW of capacity under construction, and in the announcement of its plans toward 2017, it has conveyed to increase its capital expenditures in Latin America by several hundreds of million EUR (Enel, 2013).

competitors to centralized generation. As the context has changed, the big 7 have had to revise their corporate strategies. In their initial response, the firms have aimed at cost reduction, through layoffs and asset disposal programs.

In an attempt to compensate for declining income from their European home markets in the long term, the major utilities partly shift their focus to developing new activities in their home markets. At the same time, the firms focus on securing a future for their existing conventional capacity through policy support. Firms that are capable of it will also direct more investments toward international growth markets, strengthening their diversified portfolios.

The development of new business in the mature markets of Europe will be challenging. Utilities already allude to a new approach to doing business in the electricity sector by taking a consumer-centered approach, as a facilitator of energy services and as an enabler of change. Finding new ways of generating income will require much pioneering, something proven to be hard for industry incumbents, used to predictable patterns of doing business for decades.

From a group of companies with relatively similar strategies and business models, major European utilities now seem to be diverging. As the firms take different approaches to the challenge of strategic restructuring, the nature of their business is changing—and the same can be said about the power utilities.

REFERENCES

Bloomberg, 27 July 2012. Germany's Largest Utilities Downgraded by S&P on Weak Profits.

Bloomberg, 1 July 2013. Iberdrola Boosts Divestments to $1.3 Billion Selling Chile Stake.

Clingendael Energy, 2013. European Power Utilities Under Pressure.

E.On, 2012. We Make Clean Energy Better: An Overview of Our Business Activities.

E.On, 2013. E.On Debt Investor Update Call.

EDF, February 2013. EDF Annual Results.

EnBW, 2013. EnBW 2020.

Enel, 2013. 2012 Results and 2013–2017 Plan.

Enerdata, 20 August 2013. Statkraft Idles Two Additional Gas-Fired Power Plants in Germany.

Enerdata, 10 July 2013. EnBW Plans to Shut Down 4 Power Units and to Mothball One.

Energy Post, 21 October 2013. Exclusive: RWE Sheds Old Business Model, Embraces Transition.

Energy Post, 3 December 2013. Interview with Johannes Teyssen, CEO of Eon: "Renewables can become biggest without subsidies".

EPIA, 2013. Global Market Outlook for Photovoltaics 2013-2017.

Euractiv, 11 October 2013. Energy CEOs Call for End to Renewable Subsidies.

Eurelectric, 2012. Powering Investments: Challenges for the Liberalized Electricity Sector.

Eurelectric, 2013b. Utilities: Powerhouses of Innovation.

European Commission, 2009. Directive 2009/28/EC on the Promotion of the Use of Energy from Renewable Sources.

EWEA, 2013a. Wind in Power.

EWEA, 2013b. Where's the Money Coming From? Financing Offshore Wind Farms.

ExxonMobil, 2013. The Outlook for Energy: A View to 2040.

Fraunhofer ISE, 2013a. Electricity Production from Solar and Wind in Germany in 2012.

GDF Suez, 6 December 2012. Investor Day Presentation.

GDF Suez, 2013a. H1 2013 Results.

GDF Suez, 2013b. At a Glance.

GSE, 2013. Rapporto Statistico 2012 Solare Fotovoltaico.

Iberdrola, 2012. 2012-2014 Outlook.

Iberdrola, 2013. Nine Months.

IHS, 2013. Keeping Europe's Lights On: Design and Impacts of Capacity Mechanisms.

Management Team, 17 July 2013. Peter Terium (RWE) over Transitie van de Energiemarkt.

OECD, 2013. Environmental Performance Review: Italy 2013.

Petroleum Intelligence Weekly (24 October 2013).

Platts, 5 August 2013. Spanish Reforms Hit H1 Profits, in Power in Europe issue 657.

Red Eléctrica de España, 2013. El Sistema eléctrico español 2012.

REN21, 2013. Renewables 2013 Global Status Report.

Renewable Energy Directive, 2009. Directive 2009/28/EC on the Promotion of the Use of Energy from Renewable Sources.

Reuters, 10 August 2012. RWE to cut 2,400 Jobs.

RWE, 2012. Annual Report.

RWE, 2013a. Value in Uncertain Times.

RWE, 2013b. 3 Steps to Long-Term Value.

SSE, 2013. Coalition of Progressive European Energy Companies Letter.

The Economist, 12 October 2013. How to Lose Half a Trillion Euros.

The Financial Times, 2013. Markets Data by Thomson Reuters. accessed 26 September 2013, http://markets.ft.com/research/Markets/Companies-Research.

The Financial Times, 23 July 2013. Write-downs Moves Vattenfall to Restructure.

The Financial Times, 6 August 2013. Renewables: A Rising Power.

The Wall Street Journal, 5 March 2013. RWE Warns Profit to Fall Amid Low European Power Prices.

The Wall Street Journal, 6 March 2013. Swedish Utility Vattenfall Cuts 2,500 Jobs.

Vattenfall, 2012. Our Strategy in Challenging Markets.

Vattenfall, 23 July 2013. Vattenfall Makes Substantial Impairments and Divides the Company.

Wood Mackenzie, 20 February 2013. The Role of Coal in European Power – One Last Push or a Long-term Player.

World Gas Intelligence (14 July 2012; 3 October 2012; 7 November 2012; 23 January 2013).

Chapter 7

Lessons from Other Industries Facing Disruptive Technology

Fereidoon P. Sioshansi and Carl Weinberg

ABSTRACT

Since the dawn of civilization, new and improved technology has replaced the old, less efficient, less convenient, or more expensive. The track record of how the incumbents dealt with the challenge of the new is decidedly mixed. In a few cases, the incumbents managed to find a way, sometimes by trial and error, while others missed the opportunity to see the telltale signs that eventually hollowed out their livelihood before they could devise an appropriate response. In many cases, the emergence of disruptive technology simply doomed the existing one since it was so much more superior, less expensive, or more convenient. In this context, do *decentralized energy resources* qualify as disruptive technologies, and if so, how will the electricity supply industry fare?

Keywords: Disruptive technology, Business strategy, Utility business model, Business survival, Distributed generation

1 INTRODUCTION

Disruptive technology refers to a new and fundamentally different way to meet customers' needs and/or deliver products and services that they desire faster, better, and less expensively than the traditional means.[1] Typically, disruptive technologies allow new players to enter a field or business dominated by the powerful and entrenched incumbents, allowing them to bypass the traditional ways and means of providing the service and/or the product. Generally, the new product or service can be delivered at a fraction of the costs offered by the incumbents, often by means or through channels that are faster, cheaper, and more convenient. The ease and convenience of the new service and new means of delivery plus lower costs and/or improved quality/level of service typically leaves the incumbents bewildered and unable to compete. In many cases,

---

1. For example, see Christensen, Clayton M. The Innovator's Dilemma: When New Technologies Cause Great Firms to Fail. Boston, MA: Harvard Business School Press, 1997.

Distributed Generation and its Implications for the Utility Industry. http://dx.doi.org/10.1016/B978-0-12-800240-7.00007-2

the disruptive technology is simply superior on so many dimensions as to doom the existing way of doing things.

The typical reaction of the beleaguered incumbents facing disruptive technologies is akin to the bad news given to a patient with a serious, perhaps incurable, disease. They go through the stages of denial, anger, and frustration, before they resign to the fact that they need to accept and cope with the inevitable. In many but not all cases, the incumbents are too slow and timid to react to the challenges posed by the disruptive technology and/or misinterpret its longer term devastating implications—which is to "hollow out" the customers and revenue source that traditionally kept them in business.

Initially, as some customers leave, the incumbent is usually left with fixed costs that must be spread among the remaining customers. This typically encourages more customers to leave—a vicious circle that leaves the incumbent with considerable stranded assets that can no longer be sustained, with a business model that is no longer viable, or with expensive-to-serve customers once the profitable ones have been skimmed off by newcomers.

Take the case of companies like Blockbuster Video facing the arrival of Netflix followed by technological advances that ushered in the option to download movies on demand, at any time and virtually from anywhere with added convenience and at much lower costs (Box 1).

The Blockbuster example illustrates the key features of disruptive technologies: rapid technological change coupled with a new business model to deliver a product or service faster, better, more conveniently, and cheaper—allowing a clever newcomer to bypass the entrenched incumbents' infrastructure and outdated business model. In the case of Netflix, it effectively turned Blockbuster's extensive and expensive infrastructure, 5500 retail stores in prime locations, into a massive stranded cost overnight. Moreover, Blockbuster's business model, which was based on customers walking into retail outlets, was made unnecessary and inconvenient.

Box 1 Blockbuster: Start to End in 25 Years

Blockbuster Video's rapid rise started in the 1980s with—what was at the time considered—a brilliant idea: renting DVDs. The company enjoyed phenomenal growth in the United States with 5500 outlets by 2005 before Netflix Inc. began to take its customers away with lower costs and increased convenience initially offering a mail-in service on disks rather than DVDs. Blockbuster was forced to close stores as revenues plunged; 1700 outlets remained by 2011 when the company was acquired by DISH Networks Corp. for $234 million. The new owner's attempts to revive the business did not succeed, leading to a decision to close the remaining stores by the end of 2013. Commenting on the closure of the remaining outlets, Joe Clayton, the company's CEO said, "It wasn't an easy decision, yet consumer demand is clearly moving to digital distribution of video entertainment."

Box 1 Blockbuster: Start to End in 25 Years—cont'd

Growth of Netflix's streaming US subscribers, 2011-2013, in millions

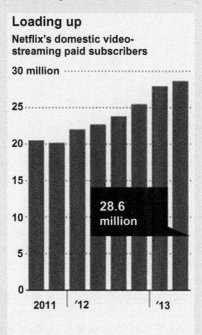

Source: *The Wall Street Journal, 23 July 2013 based on data from Netflix.*

Given the option, who wants to drive to an outlet to look for a decent movie, rent one, bring it home to watch, return it the next day to the same outlet when you can choose from thousands of titles on line, and get the DVD delivered to your mailbox, with the added convenience of mailing it back at *your* convenience without penalties or late fees? That, of course, was Netflix's original appeal before the arrival of broadband communication that allowed movies to be downloaded on demand, when and where consumers want it. By 2013, Netflix had acquired over 28 million paid subscribers for its video streaming service (accompanying graph). The convenience, improved service quality, and lower costs of the disruptive technology doomed the video rental business before the incumbent had time to realize or adopt.

What made Netflix a success, initially, was the substitution of video disks though mail for bulky DVDs. This was supplanted by streaming videos digitally via the Internet. The exponential growth of Internet traffic is often credited for rapid pace of change in a myriad of industries and businesses in ways that was hard to imagine only a decade ago (Figure 1).

Many believe that today's electricity supply industry (ESI) may be facing one or more types of disruptive technologies that could significantly erode

Political progress

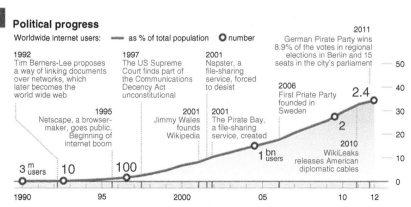

FIGURE 1   Rapid rise of the Internet as an enabling technology is being felt in numerous industries. *Source: The Economist, 5 January 2013.*

the industry's hold on its captive customers. Just as telephone companies have lost their once enviable position of having a hard-wired link to virtually every customer everywhere, electric utilities could conceivably end up with a future where a growing number of customers no longer rely on them exclusively—if at all—to meet their electricity service needs.

As a number of chapters in this book suggest, electric customers now have the option to meet a significant portion of their needs from on-site self-generation—rooftop solar photovoltaics (PVs), for example—reducing their dependence on grid-supplied electricity. This phenomenon has already allowed some customers to go "grid-parallel" or "grid-assisted," which refers to the fact that they rely on the grid for backup service when the self-generation option does not meet their needs—for example, at night when the PVs are not generating.

Further technological advances accompanied by falling costs could conceivably make it possible for some customers to go a step further by becoming virtually *grid-independent*, in the same way that many customers no longer depend on hard-wired landlines for telephony.

With advances in storage technologies and modifications in other systems within customers' premises, consumers can store excess electricity generated during periods when demand is low and generation is high for use in periods when the reverse is true. For example, as described in chapter by Platt et al., the batteries of an electric vehicle may serve as a convenient and distributed storage medium to "dump" excess generation for use at other times. Or a fuel cell could augment the customer's service needs. Or energy could be stored in variety of other forms, hot or chilled water, for example, to smooth out variations in local self-generation. Or perhaps, a group of customers, a community, or a cooperative with divergent consumption and generation patterns can pool their resources to ride out through daily and seasonal cycles of generation and consumption.

As described in chapter by Burger and Weinmann, *virtually* energy-independent communities are beginning to emerge in Germany, usually with the assistance of certain subsidies and/or support mechanisms. Some experts believe that with rapidly falling costs of decentralized energy resources (DERs) and advances in storage and microgrids, the day of reckoning may arrive sooner than many predict.

This chapter examines a few examples of how *other* industries have responded to disruptive technologies with the aim of drawing useful parallels that may apply to the ESI. The chapter also examines a few attempts already proposed in how to confront and/or prevent the spread of disruptive technologies.

The chapter is organized as follows. Section 2 provides anecdotal examples of how disruptive technologies have impacted and, in some cases, devastated existing businesses. Section 3 examines how incumbent stakeholders in the ESI have proposed to confront disruptive technology thus far and an assessment of what may be necessary to make the transition to a more sustainable future—the topic of a number of chapters in this volume—followed by the chapter's conclusions.

2 CONFRONTING DISRUPTIVE TECHNOLOGY

As illustrated in Table 1, history is full of new, superior technology displacing existing ones, sometimes with not so obvious outcomes.

Among the most often-mentioned examples of a giant, entrenched, well-endowed, and highly profitable company with global brand recognition that did not fare well when confronted with disruptive technology is Kodak. Here was a company that had perfected and dominated the manufacturing, distribution, and processing of photographic film for a very long time. Up until the time of its demise, it had virtually no rivals—Fujifilm was a latecomer to the game—but by then, the writing was already on the wall (Figure 2).

Much has been written about Kodak and how it missed multiple opportunities to reinvent and revitalize itself to face the mass migration to digital photography. Among the fundamental mistakes—there were many—was Kodak's adherence to the outdated notion that what people wanted were *copies of photographs* taken on cameras, and they wanted to preserve these on Kodak *film*. It is easy to sympathize with this definition of a product or service since it had served the company well for decades.

The rapid advent of digital photography, digital storage, screen displays plus the ability to edit/send/receive/share files, and, more recently, social networks such as Facebook, the basic definition of the product or service rapidly changed. What people increasingly wanted was to *share memories and experiences*, not necessarily print photographs on film, mail them to relatives, or stick them in a photo album as older generations had done. The younger generation would find photo albums as outdated as writing daily diaries, or God forbid, writing or reading letters sent/received via *snail mail*.

TABLE 1 Examples of Disruptive Technologies or New Technologies Replacing the Old

Old Technology	New Technology	Superior Features of New Technology
Inland waterways	Trains	Speed, vastly expanded geographic reach
Outcome: trains replaced inland waterways as the main means of transporting goods		
Gas lighting	Electric lighting	Safety, convenience, versatility
Outcome: electric lighting replaced gas over time		
Trains	Trucks	Customized, point-to-point delivery on demand
Outcome: trucks became the dominant mode of transport for all except heavy, bulky commodities on long haul		
Mechanical watches	Digital watches	Cheaper, more accurate, mass-produced
Outcome: surprisingly, expensive handmade watches have moved upscale and are thriving as pseudojewelry		
Kodak film	Digital photography	Images easily saved/edited/shared
Outcome: Kodak filed for bankruptcy protection having lost its main source of revenue, photographic film		
US Postal Service	E-mail	Speed, virtually free, instant confirmation/reply
Outcome: USPS has been a money-losing enterprise constrained by the Congress and labor laws, unable to revitalize		
Landline phone	Mobile phone	Convenience of instant wireless global connectivity
Outcome: millions of former consumers no longer use landlines as business has moved to wireless mobile devices		
DVD rental	Netflix or digital service	Ability to download movies on demand anywhere
Outcome: DVD rental business was displaced by CDs initially mailed subsequently downloading on demand		
Traditional education	Massive open online courses	Virtually free universal access to education anywhere
Outcome: elite schools OK; 2nd- and 3rd-tiered universities/colleges need to reinvent business/pricing model		
Trucking	Driverless trucking	Potentially major cost savings
Outcome: potential loss of millions of jobs as autonomous trucking may gradually displace truck driving obsolete		

Source: Compiled by the author.

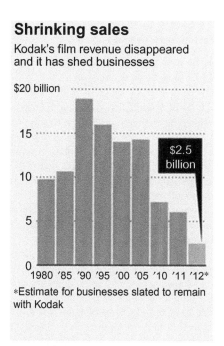

Shrinking sales

Kodak's film revenue disappeared and it has shed businesses

*Estimate for businesses slated to remain with Kodak

FIGURE 2   How do you go from $19 billion in revenues to bankruptcy? *Source: The Wall Street Journal, 29 September 2012 based on data from Kodak.*

It is, of course, easy to make such observations after the fact. Perhaps, Kodak could have done things differently and it might have saved the company from oblivion. But as the failed attempts of Blockbuster to compete with Netflix illustrate, once nimble newcomers, enabled by new technology and/or delivery channels appear, it is often too late for the incumbent with its culture stuck in the old way of doing things to change and compete fast enough. In the case of Kodak, the tectonic shift of digital photography rendered its single valuable product—photographic film—virtually irrelevant.

In Box 2, Carl Weinberg points out that the current challenges facing the ESI are not new: the industry has been undergoing change since mid-1970s. What is different is that the pace of change has accelerated in the recent past and may be even more rapid in the future.

The US Postal Service (USPS) offers another example of a series of developments gradually eroding profitable segments of the incumbent's business (Figure 3). The demise of the USPS started decades ago, first with the arrival of nimble rivals including FedEx and United Parcel Service (UPS), who relentlessly peeled away profitable niches away from the USPS, leaving mostly unprofitable or marginally profitable segments behind. While FedEx focused on fast, reliable, overnight delivery—light, high-value items that absolutely and positively had to reach their destinations overnight—UPS went after

Box 2 Disruptive Technologies: What Is New?

The unusual of today is the common of tomorrow

Chinese proverb

Despite the recent talk about new challenges facing the ESI, the reality is that the industry has been undergoing change over its long history, particularly since the oil shocks of the 1973 and 1979. Ever since, the industry's regulatory regime, central generation paradigm, and the traditional emphasis on more sales leading to more revenues and profits have been in slow retreat. What is different now is that the pace of change has dramatically accelerated, putting increased pressure on fundamental business and regulatory reforms.

Now, as always, changes in the cost of technologies and more stringent environmental design criteria are the drivers—with the difference that so many of the fundamentals have changed, are changing, or will change.

Looking back, it is clear that the boiler-based central station technologies had played out and peaked in the 1970s. For some time, utilities in the United States and elsewhere have been looking at emerging technologies and pondering how best to integrate them into the traditional utility value chain.[2]

Research at Pacific Gas and Electric Company, for example, examined the future of all aspects of the business from generation, transmission, and distribution to customer use of electricity as outlined in the following:

- Generation: examined clean technologies including a myriad of renewables such as wind, geothermal, biomass, photovoltaic, central solar, fuel cells, and gas turbines and their integration with the prediction that the costs of these technologies, over time, will decline, while they would get progressively cleaner.
- Transmission: studied real-time power flows on the network concluding that there was more capacity than assumed using standard practices; the precursor to smart grid, which has advanced substantially by real-time data collection, analysis techniques, and the Internet.
- Distribution: measured flows in the distribution wires and at substation level discovering that substation flows experienced peaks much sharper than the system peak. This peak was much shorter and existed only for a small percentage of the time. Reducing these local peaks increased the utilization of the distribution asset and diminished the system peak.
- Distributed generation: studies covered small gas generation, wind, photovoltaic, and fuel cells and measured their performance, costs, and their potential impact on the network. While at the time, PVs were far from grid parity, the expectation was that their costs would come down and their applications move from remote to central generation.
- Energy efficiency: examined customer use; measured demand, including standby use; monitored different types of buildings; and did extensive

---

2. The author was the head of R&D at Pacific Gas and Electric Company during its golden R&D years, 1983-1994.

Box 2 Disruptive Technologies: What Is New?—cont'd

measurement, concluding that through energy efficiency, consumption could drop 50-80% for all types of customers, devices, and buildings.

The research led to a few insights relevant to this volume's main theme. The most important was the realization that the utility business can be thought of as a balanced equation with the overall cost of service as a function of several key variables as indicated in the following:

$$\$ = f(\pm G, S, T, D \pm gs)$$

where $+G$ is large central generation and $-G$ is more efficient central generation, S is large-scale storage, T is transmission, D is distribution, $+g$ is distributed generation and $-g$ is energy efficiency, and s is small distributed storage.

Looking at the business from this perspective, it is easy to see that a change in cost of any parameter will change the cost of the system as a whole. For example, as $\pm gs$ approaches the retail cost of electricity or grid parity, it will displace $+$ or $-G$, S, or T. But grid parity at this point is retail cost not bus bar cost.

This is precisely what is now readily apparent with the falling costs of energy efficiency and/or PVs relative to the rising retail tariffs as described in Chapter 1. The reality facing the ESI is that the cost of customer-side generation and efficiency options, the DERs, has *declined* relative to the cost of grid-supplied electricity, which is on the *rise*.

The second insight was that relevant technologies were shifting from *economy of scale* to *economy of manufacturing*. This is because most of the important technologies, for example, solar PVs, and energy efficiency technology are modular and manufactured off-site. With organizational learning, this means that installation and market penetration could be rapid and operation sequential. The rapid rise of PVs in the recent past, for example, in Germany, Australia, or California, is a result of the economy of manufacturing and organizational learning.

The third important insight was that there are limited options to deal with the environment and sustainability issues. In the longer run, mankind must move closer to the flux of the sun, the blow of the wind, the heat of the earth, and the pull of the gravity—inexhaustible, plentiful, and nonpolluting renewable resources.

The fourth insight was the realization that the substation could become a flexible electricity management facility, managing both supply and demand that led to the concept of semi-independent microgrids. The final insight was that many of the emerging technologies are radical and disruptive in the sense that they are

- dispatched by nature, not the grid operator;
- efficient;
- localized;
- modular; and
- geographically distributed.

The bad news for the ESI is that organizations that are good at traditional technologies usually do not survive a shift to radical or disruptive technologies. For incumbents to adapt and embrace new modular, disruptive technologies, new organizational structures would have to emerge for both the utilities and their regulators.

Continued

Box 2 Disruptive Technologies: What Is New?—cont'd

Why have these developments caught the industry by surprise? Unfortunately, the industry was distracted by other priorities and not receptive to the message and insights that could have been embraced and internalized earlier. Consequently, much of the developments have taken place *outside* the industry, which is why the industry, broadly speaking, views decentralized generation as a threat to its survival rather than a different and potentially profitable way to meet customers' energy service needs.

At the same time, grid security has become a major concern due to the realization that the transmission grid is highly vulnerable. The military is aggressively establishing microgrids for security purposes. And reliability for a society addicted to electricity will become ever more important. For better or for worse, the electricity industry will have to change to survive these forces.

As others in this volume have proposed alternative future scenarios—e.g., Riesz et al., Felder, Kristov and Hou—I envision four possible future scenarios (accompanying graph):

- Going back to a tightly regulated structure—call it Triumph of the Good Old Boys
- Moving toward a competitive central station generation model—New Gladiators Old Weapons
- Migrating to a new regulatory future—Teaching Old Dogs New Tricks
- Evolving into everything deregulated—The Supermarket of Choices

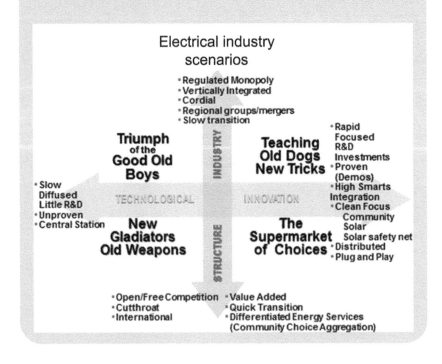

> **Box 2 Disruptive Technologies: What Is New?—cont'd**
>
> Utilities today are in *all* four quadrants, wondering where they may end up. If forced to pick one, my favorite scenario would be moving toward "Teaching Old Dogs New Tricks." Why? The utilities have the ability to aggregate, offer customer choices, and provide service options. They are also good in maintaining the complex transmission and distribution network, which by all measures will remain a vital natural monopoly. Moreover, today's industry has evolved over a century to meet many of the needs and the services that society demanded of it.
>
> In my view, the ESI *can* continue in that role but with a different regulatory structure, different types of services, and different pricing models, commonly referred to as a "new business model."
>
> *By Carl Weinberg*

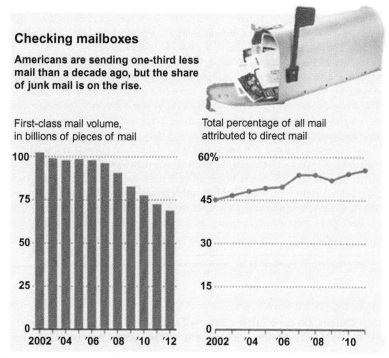

Checking mailboxes

Americans are sending one-third less mail than a decade ago, but the share of junk mail is on the rise.

First-class mail volume, in billions of pieces of mail

Total percentage of all mail attributed to direct mail

FIGURE 3   The demise of USPS: volume declining, price rising. *Source: USPS as reported in The Wall Street Journal, 24 January 2013.*

small- to medium-size package delivery service to, from, and within major metropolitan areas.

While this represented a major hemorrhage of revenues and client base, the USPS decided that first-class mail and third-class "junk" mail will allow it to survive. With the rapid penetration of the Internet and declining cost of

---

Box 3 Going Postal: Analogy of Fixed Costs of Maintaining a Massive Network

As technologies evolve, they delegate entire industries to history's trash can. Everyone's favorite is Kodak, which did not appreciate the impact of digital photography. Or American railroad tycoons in the 1900s, who saw their competition as other railroads rather than faster and more convenient *transportation* provided by cars, trucks, and planes. Today's publishing business is undergoing similar challenges by electronic readers and templates rather than printed matter. Likewise, the US Postal Service and its counterparts in other countries are facing the challenges of dropping revenues as more business gets done via the Internet and shrewd competitors like FedEx and UPS selectively strip virtually all the profitable traffic away from the incumbent monopoly.

The dilemma of falling volume—and revenues—afflicting the postal systems worldwide is analogous to that facing the utility industry with a future of flat or potentially falling volumetric kWh consumption and growing distributed generation. For the USPS, which delivers roughly 40% of global mail, it takes a massive network of local offices and distribution and sorting centers that serve 151 million addresses in the United States, 6 days a week, 52 weeks a year.

Using an analogy with the electricity industry, what if the USPS were to charge every postal address a fixed amount merely to pay for the costs of maintaining the vast network, the "postal grid"? After all, once in a while, we all look forward to getting the proverbial check in the mail. Using similar logic, utilities facing customer migration to distributed generation have proposed higher fixed charges to reduce their revenue loss.

Utility business has one advantage that may make them somewhat immune to getting entirely bypassed, at least until distributed generation plus reliable, low-cost storage becomes cost-effective. Customers need the reliability and the backup services provided by the hard wire that physically connects them to the grid, unlike the mobile phones, where multitudes of customers have cut the hard wire off. Until an equivalent way is found for electricity consumers to go entirely off-grid, they will rely on the grid.

Excerpted from EEnergy Informer, March 2013.

communications, however, first-class mail is gradually going away, as did the Kodak film (Box 3). Making matters worse, the USPS has to operate under many operational and budgeting constraints imposed by the US Congress—including 6-day/week delivery service and its postage-stamp pricing—flat rate regardless of the distance or the actual cost of delivering an envelope.

Despite a vast network and enviable reach, the USPS finds itself in a slowly dying business, with dwindling revenues, a declining customer base, and an expansive infrastructure that is expensive to maintain and operate. As the volume of first-class mail drops—because hardly anybody writes letters in the age of Facebook and instant mobile communications—and the lucrative business

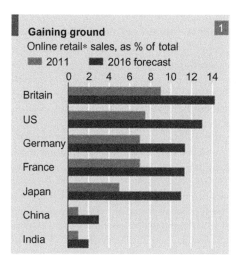

FIGURE 4   Online retail sales as percentage of total retail sales in selective markets, 2011 and 2016 projection. *Source: The Economist, 13 July 2013 based on sources listed.*

traffic is largely moving online[3], the USPS seems doomed as a patient diagnosed with slow but terminal disease.

Another example of disruptive technology is online retailing, already accounting for significant percentage of retailing sales in key global markets (Figure 4).

The battle between established *brick-and-mortar* retailing giant Walmart and the nimble online newcomer Amazon illustrates the powerful impact of technology and its disruptive effect on incumbent players. In this case, Walmart

---

Box 4 Can Walmart Beat Amazon at Its Own Game?

Amazon offers the best example of the *transformational* power of the Internet. In 2012, the company shipped $61 billion of merchandise from 40 distribution centers in the United States and elsewhere globally. Amazon, as everyone knows, does not have any retail stores. It has, more than anyone else, demonstrated the viability of a new business model based on orders received and processed online, with the goods shipped with stunning efficiency and speed.[4]

Continued

---

3. Utilities are among the last businesses gradually converting to electronic billing and/or automatic bank transfer services. At the current pace of conversion, very few businesses will be mailing, or receiving anything, to/from their customers within a decade.
4. On 26 November 2012, Amazon processed and shipped a stunning 26.5 million orders—306 per second— most delivered within a day.

> **Box 4 Can Walmart Beat Amazon at Its Own Game?—cont'd**
>
> Retailing is a tough business, with little growth and razor-thin margins. It is also a zero-sum game. Items bought on Amazon are losses on someone else's balance sheet. Traditional brick-and-mortar retailers are trying to emulate Amazon before it eats the rest of their lunch. Johnson & Johnson, for example, sees Amazon as a major competitor as it begins to sell many of its products online at prices that beat J&J's own distribution channels.
>
> The most noticeable effort to beat Amazon at its own game is coming from Walmart Stores Inc., the world's largest retailer with 4000 stores and 158 warehouses in the United States and annual sales of $469 billion—more than the GDP of many countries.
>
> The company, known for its mastery of logistic distribution, however, has a lot of catching up to do when it comes to online retailing. Speaking to The Wall Street Journal (19 June 2013), the company's former CEO, Mike Dude, said, "We're starting to gain traction," adding "I say starting because we know we still have a long ways to go." Walmart's e-commerce sales are a mere $7.7 billion, roughly a 10th of Amazon's.
>
> Why would a retailing behemoth with the size and clout of Walmart even want to bother with selling merchandise online? The answer is that it is where the growth is. In 2012, online sales in the United States rose 16% to $224.3 billion. That currently accounts for roughly 5% of overall US consumer goods sales, according to the WSJ. Forrester Research figures the 5% number will double by 2017. If Walmart wants to remain viable as a low-cost retailer, it must make a going of its online business.
>
> When asked how long it will take and how much it would cost to develop a competitive e-commerce business, Neil Ashe, president of Walmart's global e-commerce, said, "It will take the rest of our careers and as much as we've got. This isn't a project. It is about the future of the company."
>
> ---
>
> *Reprinted from EEnergy Informer, July 2013.*

appears to have belatedly recognized the seriousness of the threat it is facing and appears to have taken the initial steps to confront the enemy, gradual migration of sales online (Box 4), which Amazon virtually invented and continues to improve and refine.

Another interesting case is the experience of phone companies, who have gone through at least two phases of change, both brought about because of disruptive technology[5]:

- The first phase was the transformation of business from one of carrying voice over wires using the so-called landlines to wireless mobile phones. This entailed cannibalization of one type of service in favor of another. This

---

5. Refer to Chapter 12 by Brennan.

illustrates that incumbents, if they move fast enough, can retain large segments of their customers by rapidly embracing new technologies.

- The second phase, still in progress, is the gradual disappearance of voice communication—people actually talking to each other—in favor of written text messages, pictures, audio, video, files, data sharing, and other types of digital communication. With passage of time, talking on the phone is becoming passé and decidedly uncool, especially among the younger generation.

It is not that talking is disappearing, but other types of communications are growing at such stunning pace as to make voice communication a small and dwindling volume of the business, as described in Box 5.

Box 5 Talking Less, Paying More: How Is That Going to Work?

The current problems facing the ESI or the USPS are, by no means, unique. As technology dwindles demand for kWh brought from the grid due to self-generation or the need to send bills, receipts, correspondents via *snail mail*, utilities, and the USPS is forced to raise retail tariffs or the cost of postage stamps, respectively, to remain revenue neutral—as they would prefer to call it—or to spread their mostly fixed network costs among fewer paying customers. It is not a sustainable business strategy, but that is all they have been doing thus far.

Their dilemma has interesting parallels to the plight of telecom business, which continues to evolve with the rapid migration of traffic in digital format to the Internet and the equally rapid evaporation of voice communication, whether on mobile phones or landlines. An article titled "Talking less, paying more for voice," by Greg Bensinger, which appeared in the 6 June 2012 print issue of The Wall Street Journal, points out that "The largest U.S. wireless carriers are working on ways to keep their customers paying up for something they do less and less—making phone calls."

Excerpts from the article, which appear below, are particularly relevant to the urgency of developing a new business model that captures the rapidly changing nature of demand for traditional products or services brought about by technological advances in all these cases.

> In a sea change for consumer behavior, the amount of time spent making old-fashioned voice calls has fallen every year since Apple Inc. introduced the iPhone in 2007. The rub for carriers is that voice billings still account for about two-thirds of what they charge cellphone customers every month.
> To make sure monthly billings don't follow usage downhill, carriers expect to get rid of plans that let contract subscribers buy only the number of minutes they need and replace them with a flat rate covering unlimited calls.[6]

Continued

---

6. As explained in other chapters, one of the most favored solutions to address net energy metering laws in the United States is to raise fixed monthly fees regardless of the volume of kWh consumed.

Box 5 Talking Less, Paying More: How Is That Going to Work?—cont'd

Carriers say a move to unlimited-only calling plans would simplify what can be a confusing array of options. But it also would keep a cash cow healthy by depriving customers of the option to trade down to cheaper plans—even as their phone use drops as they spend more time texting and using Internet-based calling services such as Skype.

"The industry's definitely moving towards unlimited," AT&T Mobility Chief Executive Ralph de la Vega said in a recent interview. "Especially as more people adopt smartphones that have voice capabilities over the Internet, segmented voice plans will become less relevant."

Phone calls simply are no longer the primary reason people buy mobile phones. The shift is so pronounced that AT&T Inc. Chief Executive Randall Stephenson said at an investor conference ... "that he wouldn't be surprised if some carrier pops up in the next two years with cellphone plans that cover only data, no voice."

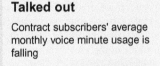

Talked out

Contract subscribers' average monthly voice minute usage is falling

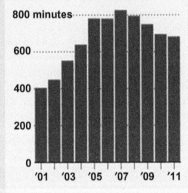

Carriers are trying to shift their billing practices to reflect the new data-driven reality, but they aren't keeping up with the pace at which consumer habits are changing. Meanwhile, subscribers are figuring out they could save money each month by scaling back their voice plans.

The problem is one carriers have largely brought on themselves. By opening up to the iPhone and then heavily promoting other smartphones, carriers gave their subscribers hundreds of new reasons to use their handsets, which in reality have become mobile computers.

Box 5 Talking Less, Paying More: How Is That Going to Work?—cont'd

People no longer need to call their loved ones for directions or restaurant reviews. They now have Google Maps and Yelp at their fingertips. And texting has become far easier, even with much-bemoaned touch screens, than using the old telephone keypad with multiple letters per button.

All of that has accelerated a shift driven by younger users who have adopted texting as their primary way of communicating while restricting calls to a small privileged circle of friends, parents and circumstances.

One customer said he typically used just 50 minutes of his 450-minute monthly allotment with AT&T, which includes free nighttime and mobile-to-mobile calling. "I am just getting more and more rollover minutes now and there's no way I can use them," he said about his $40/mo. plan.

According to CTIA, the trade group for the US wireless industry, the average phone call lasted 3.03 minutes in 2006, the year before the iPhone was rolled out by AT&T. At the end of 2011, it took 1.78 minutes, down by nearly half. Monthly use by the customers who buy contracts has followed suit. After peaking at an average of 826 minutes per month in 2007, the average fell to 681 minutes at the end of 2011.

Confident that data use is growing quickly, carriers have moved away from unlimited plans in favor of tiered offerings that charge subscribers more dollars when they use more megabytes. AT&T was first to shift to tiered data plans nearly two years ago. Verizon Wireless and Vodafone Group PLC, followed suit in 2011. Sprint and T-Mobile still offer unlimited data plans.

The moves have helped data revenue increase quickly, but it still is dwarfed by voice charges. Data accounted for 37% of carriers' $169.8 billion in wireless revenue in 2011, compared with 12% in 2006.

The material included in this box is excerpted from "Talking less, paying more for voice" by Greg Bensinger, 6 June 2012 print issue of The Wall Street Journal.

While still grappling with the first transition, telephone companies were blindsided by the second. It took them a while to realize that the very nature of service, *mobile telephony*, was changing and they needed to come up with a new business model that captured what consumers were increasingly doing with their wireless devices.

Virtually overnight, a generation of young and tech-savvy mobile phone customers were using their wireless devices—iPhones, iPads, laptops, tablets, and other mobile gadgets—to send text messages, photos, files, apps, videos, books, movies, tweets, and music—everything *but* spoken words—to each other.

Telephone companies with their outdated legacy systems and even more out of date business mentality entered the wireless world with service plans that mostly charged customers based on the number of minutes they used their devices with little or no fixed fees. This was a legacy of the landlines, when people were actually talking to one another. It is hard to believe, but only a few years ago, most mobile phone customers would select a plan, for example, with 500 min per month, and they were charged a penalty if they went over the limit.

With the rapid proliferation of high-speed broadband Internet and ubiquitous Wi-Fi, customers are talking less while exchanging more files, data, audio, and video and spending more time online searching for information, directions, reviews, news, games, and apps. What matters in this environment is not how many minutes you stay online but how many bits and bytes of data you are transmitting or downloading.

In time, mobile phone companies realized that charging customers by the minute did not make much sense. The nature of service had changed—what customers want above all are the following:

- *Universal coverage*, namely, a strong and reliable signal no matter where they are
- *Speed and bandwidth*, namely, the ability to download/upload huge files instantly, on the go

The first critical requirement is *connectivity*, and the second speed or *bandwidth*. These are what tech-savvy mobile device users increasingly want and are prepared to pay for, which explain why nearly all mobile companies have switched to service plans that charge customers for connectivity and bandwidth as described in Box 5. If the customer wants to be able to upload/download reams of data or be able to watch a live video while sipping coffee in a café in Timbuktu, he/she has to pay a premium for the service.

As described in a number of companion chapters in this volume, these two parameters, *connectivity* to the network and the *capacity to upload or download*, also happen to be critical to ESI. The utilities' future business model, like those of the mobile phone companies, must be changed to reflect what customers really want and need and how much they are willing to pay for it. In his chapter, Gellings points out that the critical functions provided by the grid will become even more valuable in a decentralized future, argument supported by numerous others.

Another example of disruptive technology, yet to play out, may be that of autonomous or driverless cars and trucks. Just as elevators no longer need an "operator," with advanced technologies, cars and trucks can be programmed to go to a designated destination by push of a few buttons, according to experts working in this field. While driverless cars have been in the press, driverless trucks may have more of an impact if the technology progresses as its proponents claim (Box 6).

> **Box 6 Will Truck Drivers Follow Meter Readers into Oblivion?**
>
> Few people remember milk delivery business, more recently followed by meter readers in places where smart meters have made them redundant. Now, there is talk of driverless cars. Bill Ford Jr., the CEO of Ford, predicts self-driving cars to be a reality by 2025.
>
> What may be more profound, however, is the inevitable rise of or autonomous trucks, in time putting millions of truck drivers around the world out of well-paying jobs. There are an estimated 250 million trucks of one size or another around the world and an equally big number of drivers, some 5.7 million in the United States alone.
>
> Driverless trucks have already emerged in a few mining operations replacing hard to find drivers operating heavy lift trucks 24/7, usually in 6 h shifts. Autonomous trucks can operate continuously and don't require expensive pension or medical coverage nor the need for breaks or holidays. Experts believe that driverless trucks running on exclusive roads are not far behind. Since most trucks follow designated routes, say from a warehouse, port, or distribution center to another, they will be less technically demanding than driverless cars, with significant savings.
>
> Does this pose a serious and near-term concern for today's trucking companies, and if so, what can be done? For today's truck drivers, there is nothing to worry about, especially if retirement is expected before the new technology.

3   ESI RESPONSE TO THE RISE OF DERs

Not everyone agrees that the traditional ESI business model is under serious or immediate threat. Among those who believe DERs are a serious threat, opinions vary on what may be the best solution or response, as reflected in various chapters of this volume and Chapter 1.

There is, however, reasonable agreement that the combined effect of consumers using less, due to energy efficiency investments/improvements, and generating more, due to the rise of self-generation options, is likely to contribute to flat or potentially declining revenues under the ESI's prevailing revenue collection model, which is mostly flat tariffs multiplied by volumetric consumption. If there is agreement on anything, it is that a larger portion of the cost of maintaining the network must be captured through fixed charges as described in Box 7.

Others offer alternatives, pricing schemes, and different ways of capturing the value of services delivered by the grid—investments upstream of the customer meter that are pertinent to customers and will remain critical for the foreseeable future. Among the many proposals in this volume is the concept of transactive energy, covered in chapters by King and Cazalet. The latter, in particular, proposes an elegant scheme that may, at least on paper, address many of the problems associated with DERs in the context of disruptive technology.

Box 7 How Much Should Self-Generators Pay for the Grid?

That is among the questions being asked not just in the United States but nearly in any country where self-generation, in one form or another, already is or is likely to become cost-effective.

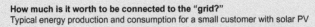

How much is it worth to be connected to the "grid?"
Typical energy production and consumption for a small customer with solar PV

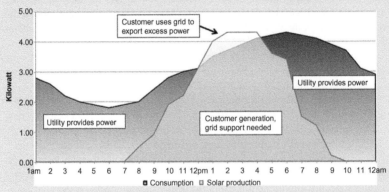

Source: Value of the grid to DG customers, IEE Issue Brief, Lisa Wood & Bob Borlick, September 2013.

Until the day arrives—and some say this may not be too far—when self-generators can cut the cord entirely and go off-grid, they will continue to rely on the grid for a range of services that are virtually priceless. In most cases, self-generators' production is variable and intermittent, as in the case of solar PVs, which means that they rely on the grid to balance their local consumption against variable generation. The grid, in other words, acts as a *free* battery, absorbing the surplus generation while making up the shortfalls.

In an IEE brief, Lisa Wood and Robert Borlick examined the value of the grid to a hypothetical distributed generation (DG) or microgrid consumer who may be producing sufficient amount of juice to be virtually self-sufficient in aggregate terms while relying on the grid for balancing and reliability services. Not surprisingly, they conclude what nearly everyone acknowledges to be true, namely, such customers are essentially free riding on critical services provided by the grid. Yet, if the prevailing tariffs are based on net volumetric consumption, they end up paying virtually nothing for grid's vital services.

Wood and Borlick argued, as have many others, that collecting revenues on a net volumetric basis, as is currently the case in many jurisdictions, is unfair and inequitable, assertions heard from many utilities these days. The accompanying table illustrates the nonenergy—i.e., fixed—costs of serving a typical residential US consumer, who on average uses ~1000 kWh/month.

Box 7 How Much Should Self-Generators Pay for the Grid?—cont'd

Costs are mostly fixed
Non-energy charges paid by a typical residential customer on a retail tariff

Average Residential Customer: Non-Energy Charges as Percent of Typical Monthly Bill	
Average Monthly Usage (kWh)*	1000
Average Monthly Bill ($)*	$110
Typical Monthly Fixed Charges	
Ancillary/Balancing Services	$1
Transmission Systems	$10
Distribution Services	$30
Generation Capacity ^	$19
Total Fixed Charges for Customer	$60
Fixed Charges as Percent of Monthly Bill	55%

*Based on Energy Information Administration (EIA) data, 2011
^The charge for capacity varies depending upon location.
  This is just an estimate.

Source: *Value of the grid to DG customers, IEE Issue Brief, Lisa Wood & Bob Borlick, September 2013.*

Using data from the Energy Information Administration (EIA), more than half of the avg. monthly residential US electricity bill of about $110 may be attributed to fixed, not energy-related, charges. It does not take rocket science to conclude that if such typical customers self-generate some or all of their 1000 kWh monthly allotment and if the prevailing tariffs are mostly or totally based on net volumetric usage, the resulting bill will be essentially zero. The $60 of fixed charges that are needed for the maintenance and upgrading of the grid is no longer recovered.

As reflected in a number of chapters in this volume, Borlick and Wood suggested three basic ways to address this missing money problem:

- Redesign retail tariffs to make them more cost-reflective—for example, by recovering a significant portion of the fixed costs through nonbypassable demand or fixed fees.
- Charge DG customers for their gross consumption under prevailing tariffs and separately compensate them for their gross generation.
- Impose standby transmission and distribution charges for DG customers.

The first approach is what many utilities currently favor. This may address some of the immediate issues, but will not fundamentally resolve the DG problem. So long as retail tariffs remain relatively high and/or are rising and as long as the cost of DG remains flat or continues to fall, consumers in high tariff jurisdictions will find it cost-effective to use less—through energy efficiency investments—and generate more.

Continued

Box 7 How Much Should Self-Generators Pay for the Grid?—cont'd

If fixed rates are raised substantially—Arizona Public Service Company (APS), for example, has proposed quadrupling current fixed charges for PV customers only—it may have the effect of pushing DG customers to go off-grid, cutting the cord, so to speak.

The second approach may be preferable to the first if the redesigned tariff fundamentally addresses the value of self-generation and consumption, both of which vary based on time of use/generation. With smart meters in place, it is in principle easy to reward self-generation for its true value to the network and charge usage according to its cost incidence on the same network.

The final approach may be the best, especially if it encourages consumers to self-select how much "grid" services they need/want and are willing to pay for. Giving the opportunity to rely on the existing grid for all the wonderful services it offers at a reasonable price should be a no brainer for most, if not all, DG customers. Getting the details right, however, will not be easy.

Adopted from EEnergy Informer, November 2013 based on IEE report by Wood and Borlick.

4   CONCLUSION

Not everyone is convinced that DERs pose as serious or immediate threat to the ESI's traditional business model. Those adhering to this point of view favor marginal or incremental adjustments to the existing revenue collection model, most notably charging all, especially self-generating customers, a fee that adequately captures the value of the services delivered by the grid and ESI's upstream investments.

Others believe that the rise of DERs, manifested by the expected continued fall of the price of alternatives to grid-supplied service, represents a growing challenge that cannot simply be fixed by incremental pricing adjustments. According to this line of thinking, the ESI must fundamentally rethink what it is that customers want and are willing to pay for followed by new pricing options.

Part II

Implications and Industry/ Regulatory Response

Chapter 8

Electricity Markets and Pricing for the Distributed Generation Era

Malcolm Keay, John Rhys and David Robinson

ABSTRACT

Wholesale and retail electricity markets have changed little over the past century, with flat kilowatt-hour pricing still the norm. This outdated model will have to change to meet the future challenges facing the industry.

Present market structures are ill-adapted to the cost and operating characteristics of a decarbonized supply side, especially in view of technological innovations in distributed energy resources. The prices that energy markets produce do not underpin investment in these sources or give useful operating signals for intermittent sources like wind and solar. Meanwhile, consumers are either sheltered from the impact of wholesale market volatility or faced with confusing and perverse price signals.

The chapter considers various options for a pricing structure linking supply and demand coherently, including a reconceptualization of electricity into two products: "as-available" and "on-demand" power. The implications of these developments for electricity business models and investment are examined.

Keywords: Electricity markets, Renewables, Distributed generation, Wholesale market design, Electricity prices

1 INTRODUCTION

This chapter explores pressures on the electricity sector arising from government policies and economic developments that support the rapid growth of distributed generation (DG) and intermittent renewable power. Other chapters in this book look at developments in specific countries (e.g., Burger and Weinman on Germany) or across Europe in general (Groot). This chapter looks at the longer term underlying problems that change in the generation mix will raise for liberalized markets and explains why most current reform proposals are inadequate.

There are of course often good policy reasons why governments wish to promote the new energy sources, particularly where decarbonization of electricity

Distributed Generation and its Implications for the Utility Industry. http://dx.doi.org/10.1016/B978-0-12-800240-7.00008-4

is a major policy goal. There are also good economic reasons why these forms of generation are becoming more attractive—for instance, solar photovoltaics have fallen rapidly in price in recent years and are particularly suited to DG applications. In some cases, there may also be less desirable economic drivers—in some countries, premises that generate their own power using DG can avoid paying their share of system and policy costs. But whatever the reasons, such sources are growing rapidly in most systems and are posing major new challenges.

The two broad categories of generation under discussion here—intermittent renewables and DG—are treated together. They raise essentially the same issues, being typically both policy-driven and nondispatchable; in any event, as noted earlier, there is a considerable overlap between the two sorts of generation. Conventional pricing and market models are not well adapted to systems with a significant share of these generation sources or to the incorporation of demand-side resources, including energy efficiency and demand response.

The chapter consists of five sections in addition to the introduction. Section 2 looks at traditional approaches to pricing and dispatch and the problems recent developments cause. Section 3 looks at current efforts to address these problems. Section 4 discusses the adequacy of these efforts and proposes new solutions. Section 5 looks at the implications for electricity business models. Section 6 sets out conclusions.

2 PRESSURES ON TRADITIONAL APPROACH TO ELECTRICITY PRICING

This section looks at traditional approaches to electricity pricing, the changes that have taken place since liberalization, and the sustainability of current approaches.

2.1 Traditional Approach to Dispatch and Electricity Pricing

Prior to liberalization, end-use price setting for vertically integrated monopolies was designed to ensure that prices should reflect the electric utility's costs. In some countries, average cost pass-through was—and still is—the norm, but in others, the goal was to reflect marginal costs; this was generally regarded as theoretically preferable (Della Valle, 1988; Turvey, 1968, 1969). The approach was designed to meet three policy objectives: allocative efficiency in production and consumption of electricity, revenue sufficiency to maintain financial viability of the power sector, and incentives for investment in the combination of plant types and vintages that would deliver the least-cost power sector over the long term.

Within the marginal cost approach, a fundamental distinction can be drawn between short-term and long-term approaches to pricing. Short-run marginal cost (SRMC) signals, used as the basis for dispatch, are important for operational efficiency. However, if prices are equal to SRMC, revenues will in general not be sufficient to reward investment and consumers will not pay the full costs of supply. In consequence, electricity tariffs in regulated monopoly systems have in principle often been designed to reflect long-run marginal costs (LRMCs). However, LRMC is difficult to define, even for a vertically integrated monopoly utility, and is almost impossible to identify in the market-driven environment that now characterizes EU countries.

This distinction between SRMC and LRMC lies at the heart of many of the current commercial and public policy issues associated with electricity tariffs.[1]

- *Resource adequacy*—Spot prices in energy-only markets tend toward SRMC-based outcomes, especially in periods of surplus capacity. Concerns over fixed-cost recovery and resource adequacy are increasingly leading governments or regulators to introduce capacity remuneration schemes—in the form of long-term contracts, administered capacity payments, or market-based capacity payments.
- *Uneconomic bypass*—Regulators may have to decide what should be paid to independent power producers outside a utility wishing to sell their surpluses to the grid. The economic value of this power is often defined by reference to the avoided cost for the utility, that is, its SRMC. Independent producers selling on this basis may believe they are getting well below the full market value of their output. However, if the utility pays more than SRMC, it will be increasing its total costs, leading to an increase in prices—in effect, a subsidy to the independent producers. Similar problems emerge in competitive systems when deciding how much should be paid to distributed generators selling energy to the system.
- *Tariff design*—Issues also arise in translating these costs into consumer tariffs, revolving around what are the optimal, and/or most equitable, methods for apportioning fixed costs and "capacity costs"—those costs that are not captured in SRMC estimates. For smaller consumers, the costs may be simply averaged over kilowatt-hour charges, with limited time of day (TOD) differentiation. For larger consumers, a wider range of options is available, including the metering of maximum demand or rated capacity of the premises. In the past, the sophistication of tariff structures has been constrained by metering limitations other than for the largest consumers.

2.2   New Developments

Recent developments discussed in other chapters in this book are having a profound effect on tariff and pricing issues.

---

1. For general discussion of electricity pricing issues see, for instance, Stoft (2002), Hunt (2002), and Harris (2006).

- *Unbundling of integrated utilities and the development of markets—* Liberalization in the EU has involved a separation (unbundling) of the competitive activities of generation and retail supply from the natural monopoly network activities of distribution and transmission, increasing the importance of wholesale electricity prices.
- *Technology changes in generation—*Generation technology is shifting away from fossil generation, with its relatively simple cost structures, to the more complex operational characteristics of low-carbon plant, including the intermittency of some renewable energy sources, and the growth of DG, as discussed extensively elsewhere in this volume.

  This change in the generation mix has three immediate consequences. First, the prevalence of occasions when system marginal cost is zero accentuates the gap between short- and long-term considerations and between SRMC and LRMC. The prospect of very low, zero, or even negative prices forces the issue of finding economically and commercially credible means of rewarding capacity. Zero or negative prices are already reasonably frequent in many European markets—negative prices arise because some generation receives income from outside the market or is expensive to shut down for short periods, so may continue to generate even when market prices fall below zero. In future, such situations could be almost the norm. For instance, in the GB system (with average demand of around 40 GW), there are likely to be over 60 GW of nuclear or renewable generation by 2020.[2]

  The second consequence is the need to incorporate DG within an efficient overall system. Utilities will have to learn how to deal with customers who are both producers and consumers of power and have much more in common with ordinary small-scale retail customers than with traditional autoproducers.

  The third is to call into question some of the basic assumptions about wholesale markets as having a single price, based on a conventional concept of merit order and short-term system marginal cost.

- *Metering and consumer control of own usage technology—*Developments in communication and control technology have created an explosion of possibilities in metering and service provision. They allow for the application of sophisticated TOD metering—even real-time pricing—previously seen as impractical or impossibly expensive. Given the interactive nature of the possibilities for controlling load, utilities need to consider how retail consumers and their loads should be incorporated into processes for the secure and efficient operation of the system.
- *Direct government interventions—*Increasingly governments are intervening in the choice of technology and using a variety of instruments to do

---

2. DECC (2009).

so, including subsidies, quotas, feed-in tariffs, and long-term contracts; this raises questions around the financing of the interventions and the maintenance of a level playing field in generation.

Particular problems arise for utilities and regulators when these interventions impose costs, which have to be recovered through the general body of consumers. For example, requiring utilities to purchase power and pay uneconomic amounts for the "capacity" credited to those purchases will automatically introduce the problem of uneconomic bypass. On the other hand, regulators may charge distributed generators for costs that they do not impose on the system and thereby discourage economically justified competition from DG.

Taken together, these developments will lead to fundamental changes in traditional business models.

3 REFORM PROPOSALS

The section summarizes efforts to reform market structures and regulation to cope with the problems addressed in the preceding text. There is a growing recognition of the need to deal at least with the following issues, further described in the succeeding text:

- Efficiency of dispatch, including renewable energy and demand response
- Efficient distribution network investment and operations
- Fixed-cost recovery for different energy resources

3.1 Efficient Dispatch Integrating Different Energy Resources

Existing wholesale markets are designed to incentivize the central dispatch of large-scale conventional generation in merit order, with market prices reflecting the cost of the marginal plant on the system or at a higher price when demand exceeds available generation. Absent market power, market-clearing prices provide incentives to run when a plant's costs are below the market price. This applies to the suite of energy wholesale markets that draw on existing supply sources, including day-ahead, intraday, balancing, and ancillary services (e.g., operating reserve) markets. Figure 1 shows the general principle in schematic form. The market price is determined by the generator with the highest accepted bid price—in this case generator 12. All generators (1-11) with lower SRMCs have an incentive to bid a price reflecting those short-run costs because they will receive the market price; these "inframarginal" generators will also receive a "rent" from the excess of the market price over their short-run costs, enabling them to make a contribution toward their capital costs. A generator that bids more than the market price will be out of merit and earn nothing.

Many aspects of this approach are now being reconsidered in the light of the fundamental changes outlined earlier. Some relevant issues include the following:

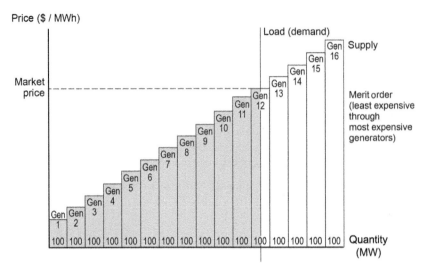

FIGURE 1   Merit order and market prices. *Source: Hunt (2002, p. 134).*

- *Reforming wholesale markets*—There is a debate about whether the increased share of plants with low or zero marginal costs justifies doing away with this wholesale market design as the basis for dispatch—the model illustrated earlier assumes a set of generators all with different but positive SRMCs. It is very different from the expected future UK market discussed earlier. There is an argument—discussed in Section 4—that the model is no longer sustainable and should be replaced. In practice, however, most current reform efforts involving remuneration for capital costs are aimed at ensuring that markets can deliver short-term operational efficiency, using existing mechanisms, without having to provide the sole basis for fixed-cost recovery. If wholesale prices are frequently low or zero, the argument goes, which is because short-term operating costs are indeed very low; other mechanisms, such as capacity markets, can be developed to recover fixed costs, without necessarily changing the logic of least-cost dispatch. An increasing number of EU countries either have introduced such markets or are in the process of doing so, as shown in Figure 2.

  However, there are also voices that oppose capacity markets; in some ways, such markets may run counter to the objectives of full liberalization and a level playing field for competition across Europe. A feature intrinsic to these markets is that they cannot normally be expected to arise through market interactions. They depend at some point on the intervention of some national authority, government, or regulator, to set an implicit or explicit standard of generation security and hence to define capacity requirements. In principle, this could be done by an authority at EU level, but it is currently unlikely that any national government would be willing to surrender such a fundamental responsibility to a supranational body.

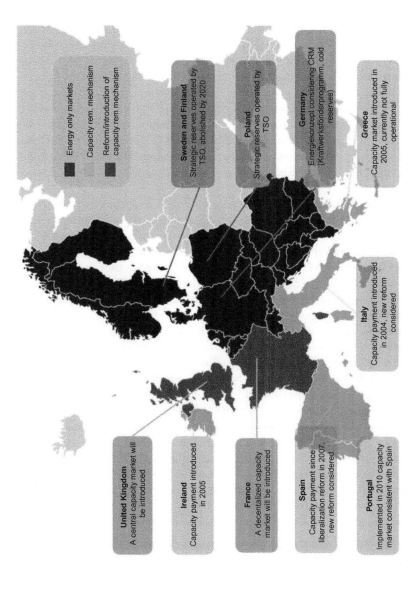

FIGURE 2 Energy and capacity markets in Europe. *Source: Based on IHS CERA March 2011, with update from Eurelectric (2012).*

In consequence, there is a real conflict between the aspirations of the European Commission (EC) in relation to competition policy and the creation of a single market in energy, on the one hand, and national policy interventions aimed at supply security, on the other. Such policy interventions necessarily impact very substantially on electricity market prices, not just in the country making the intervention, but across the market as a whole. National policies for capacity will therefore tend to undermine or "distort" an EU-wide energy-only market in much the same way as national carbon emissions policies are sometimes deemed to have undermined the EU carbon price.

The EC therefore tends to favor energy-only markets, arguably because it recognizes the difficulty in harmonizing capacity markets, and lacks the authority to impose EU-wide security standards.[3] The result is an impasse, to which it is currently hard to see a clear resolution. But the situation highlights some important economic and institutional aspects of the capacity markets' question.

- *Refining the algorithms in centralized dispatch and balancing markets*— Those who argue in favor of maintaining central dispatch and the merit order seek a refinement of the algorithms used to determine offer prices and set wholesale market prices. For instance, in the new era Combined Cycle Gas Turbine (CCGT) plants are frequently "flexed" to back up intermittent renewables, and this raises their variable operating and maintenance costs[4]; generators should reflect these additional costs in their bids and market operators should include the additional costs in their own algorithm where complex bids are accepted. Likewise, balancing charges and mechanisms can be reformed to provide sharper signals and encourage investment in flexible supply options from generation and demand sides of the market.
- *DER generation*—Reforms are now under way to allow for the active participation of new sources of energy in central dispatch and balancing markets, especially from DER, including both supply-side resources (DG) and demand-side response (DSR). For instance, in the United Kingdom, the regulator Ofgem has proposed changes in the balancing market which will, among other things, incentivize the participation of DER.[5]
- *New balancing markets and commitments*—Different forms of generation and DER have different characteristics, with implications for balancing regimes and operating timescales. Existing markets are designed around the characteristics of fossil generation but are not so well adapted to the newer entrants. For instance, it is more difficult for wind generators to make a firm

---

3. See the European Commission's "Delivering the internal electricity market and making the most of public intervention," November 5, 2013. http://ec.europa.eu/energy/gas_electricity/doc/com_2013_public_intervention_en.pdf.

4. Battle and Perez-Arriaga (2012).

5. Ofgem (2013).

commitment 24 h ahead of dispatch than it is for a fossil-fired plant, and it is much more difficult for a DER resource to make a firm commitment—say, to reduce demand for multiple hours—on its own. Reforms will need to reflect the different technologies that offer energy resources and enable them to manage their risks.[6] For example, an aggregator could combine multiple sources of DG, demand response, and batteries to provide a firm energy source of supply. Alternatively, the regulator could provide a coupling service whereby demand response was combined with spare capacity of fossil fuel plants so that together they could provide a steady supply of energy.

3.2   Efficient Distribution Network Investment and Operations

Currently, distribution system operators (DSOs) manage a network that assumes energy will be transferred from the high-voltage transmission system to the distribution system and then to the final customer. The DSO's role will need to be refined quite substantially in the new era, especially to deal with the reality of a growing amount of DER, whose impact will be felt primarily in the local distribution networks.[7] Changes in demand, DG, and storage decisions should and will influence DSO investment decisions and operations. For instance, customers may be offered interruptible access tariffs if they agree to the curtailment of service to assist the DSO in managing demand on the network. Faced with evidence of local congestion, the DSO could also organize auctions for services that would reduce congestion. In these auctions, the DSOs are likely to communicate with aggregators, not with individual customers.

DSO investment—DSOs will need to invest in new capacity and technologies to manage DER. In some cases already, for instance, Galicia in northwest Spain, DER already exceeds local demand for electricity by 20%.[8] In cases like these, the DSO needs to consider increasing its total capacity. More generally, DSOs will need to make a new set of investment decisions related to the provision of network capability to deal with multidirectional flows of energy, local storage especially from electric vehicles (see Chapter 17 by Platt et al.), and possibly the collection and management of customer information. This will require not only a change in regulation aimed at the efficient provision of new services (e.g., along the lines of the RIIO[9] in the United Kingdom) but also considering ways to include DER options in the investment decision, for instance, by offering interruptible connection tariffs as a way of reducing DSO investment to meet the peak.

---

6. Ofgem (2013).
7. See chapters by Kristov and Hou and Gellings.
8. The installed capacity of DG connected to the distribution networks of Union Fenosa Distribución (2203 MW) represents 120% of the area's total peak demand (1842 MW), reported in EurElectric (2013, p. 3).
9. RIIO stands for revenue = incentives + innovation + outputs. See https://www.ofgem.gov.uk/network-regulation-%E2%80%93-riio-model.

3.3   Fixed-Cost Recovery for Competing Sources of Energy

Hybrid systems—Today's wholesale markets allow for the recovery of fixed costs in different ways. Energy-only markets are supposed to allow energy prices to rise to levels that remunerate investments, but, as pointed out in Section 2, there are increasing doubts over whether such prices will provide sufficient certainty over revenues. Many wholesale markets therefore already include an energy payment and a separate payment for capacity. In the new era, energy-only markets will almost certainly be unable to provide the revenues necessary to induce investment in power stations or demand-side options.[10]

DER involvement in capacity markets—New capacity market mechanisms allow DER resources, as well as conventional generation and large-scale renewable generation, to participate in auctions or mechanisms to select new capacity. For instance, the UK government is keen to encourage investment in new and flexible capacity—from generation and from DER. This creates an opportunity for customers to participate in transitional arrangements to provide capacity and balancing services, before the capacity mechanism delivers new investment in 2018. There are a number of features of the proposed transitional arrangements that aim to facilitate DSR in the capacity market[11]:

- Auction of products that are time-banded (e.g., 9-11 a.m. winter weekends) or time-limited (e.g., maximum of 4 h in times of system scarcity) making it easier for DSR to comply
- More lenient penalty regime for DSR nonperformance than for nonperformance of generation
- Reservation of capacity solely for DSR
- Possible provision of a "matchmaking" service to marry DSR providers' peak products with spare generating capacity off-peak

Uneconomic bypass and the recovery of fixed costs from customers—The risk of uneconomic bypass increases as the new forms of generation take a greater share of the system, but, in many cases, do not pay an appropriate share of fixed infrastructure costs sometimes because of government policies to promote distributed renewable generation—for instance, some solar generators with "net metering" are effectively receiving free storage services from the utility.[12]

---

10. This view is, however, controversial, as noted earlier. Some experts in Australia (see chapter by Sood in this volume) also argue that energy-only markets will suffice. See Cramton and Ockenfels (2012) and Hesmondhalgh et al. (2010) for discussion.

11. See DECC (2013a).

12. Clearly, this depends on the characteristics of the system and the level of prices offered to local generators. But in some cases, certain generators (usually solar installations) can net off all exports to the grid from their consumption and pay only for the balance (e.g., if they export 1000 kWh and import 2000 kWh, they pay for only 1000 kWh, at the utility's normal retail rate). This is equivalent to free storage in the system (or a free option)—yet it is well known that the expense of electricity storage is one of the key aspects of electricity economics. See also chapters by Nelson et al. and by Mountain and Szuster.

The remedies for this are twofold:

- First, governments need to exercise much greater caution in using the electricity sector as a vehicle to finance wider aspects of policy (including decarbonization), and energy companies need to be alert to this danger.[13]
- Second, there needs to be a very clear identification and separation of infrastructure costs for the purpose of charging consumers and prosumers in order to produce genuinely cost reflective tariffs for all the relevant services, including network provision and backup. For instance, in its comments on the Spanish government's proposed introduction of a "backup" charge for prosumers (autoproducers), the regulator argued that this charge was high and included costs that they did not impose on the system, such as capacity payments made to other generators, and that the backup charge did not recognize that prosumers were reducing losses and other costs for the system.[14]

4 UNSOLVED PROBLEMS AND PROPOSED REFORM

This section advances the argument that the reform proposals outlined earlier do not resolve the fundamental problems associated with the growth of intermittent renewable power, including DG.

4.1 Why Current Reform Proposals are Inadequate

Although electricity "market reform" is much talked about, such proposals relate mainly to instruments for incentivizing investment in particular sorts of generation or refining algorithms and balancing arrangements as discussed earlier. Wholesale markets themselves remain in essence unreformed.

The problem is that markets as they exist were not designed to, and do not, reflect the cost structures and operating characteristics of the growing new energy sources. In traditional markets, basing short-run operating decisions on an energy market and the prices it generates leads to efficiency. However, for most "new" renewable and DG sources, such prices have less significance; they have no, or very low, marginal costs of generation and usually no opportunity costs—with present technology, the wind (and, in most cases, solar power) cannot be stored for some future date. In any event, in many systems, for instance, across the EU, renewable sources are not in practice subject to

---

13. Indeed, there is currently a review under way in the United Kingdom of whether certain policy costs should be borne in general taxation rather than by the electricity consumer. This argument has also frequently been made by the electricity companies in Spain, in the context of resolving the very large tariff deficit in that country.

14. CNE, *Informe 19/2013 de la CNE Sobre la Propuesta de Real Decreto Por el Que se Establece La Regulación de las Condiciones Administrativas, Técnicas Y Económicas de las Modalidades e Suministro de Energía Eléctrica con Autoconsumo Y de Producción Con Autoconsumo*, 4 September 2013. http://www.cne.es/cne/doc/publicaciones/cne85_13.pdf.

the same dispatch criteria as other sources.[15] Similarly, DG is generally used for self-supply and exports to the system when there is a surplus, rather than in response to particular price signals.

As regards investment recovery, this is mainly provided via government policy instruments such as feed-in tariffs and quotas rather than direct market signals. In other words, intermittent renewable and DG sources do not rely primarily on wholesale energy markets either to give operating signals or to justify investment decisions. Indeed, it is difficult to see how they could ever be remunerated via traditional market structures, at least if they are to be present at the levels desired by governments, *even if* they are competitive in the sense of producing power at a levelized cost below that of conventional power and *even if* there is a high carbon price. Where they form a significant component of the system, nondispatchable sources are less able to benefit from the peak-price periods that provide the main route for investment recovery; indeed, it is the very absence of nondispatchable sources on the system that usually leads to the high prices. On the other hand, their presence has the effect of depressing prices at times of high wind or solar activity, as illustrated in Figure 3 later. In energy markets, nondispatchable sources therefore tend to receive below average revenue per unit; and of course, they are in most cases unlikely to benefit significantly from capacity payments.

So, existing markets provide an inadequate basis either for operation or for remunerating investment in a system dominated by these new sources of generation, and current approaches offer no clear exit strategy from this problem. Furthermore, by their participation in the markets, these new sources have impacts on other participants and on overall market prices, creating a divergence

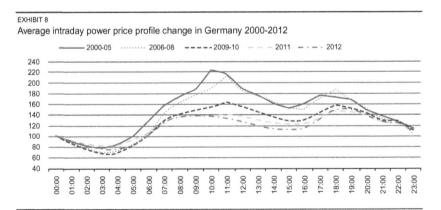

EXHIBIT 8
Average intraday power price profile change in Germany 2000-2012

FIGURE 3   Intraday power price development in Germany. *Bloomberg.*

15. Under Article 16(2) of the Renewables Directive of 2009, renewables are given priority in generation provided that security is maintained.

between overall system costs and market prices.[16] For instance, Burger and Weinman note that in Germany, while the cost of producing power is rising, wholesale market prices are falling, due to excess capacity and the impact of wind and solar power. Meanwhile, retail prices are rising, primarily because of the taxes and levies associated with the introduction of renewables. As prices become disconnected from costs, they can no longer achieve the objectives described in Section 2—allocative efficiency in operation, dynamic efficiency in investment, and revenue sufficiency. This is in some ways a more fundamental problem than that of uneconomic bypass—that can in principle be addressed by appropriate regulation, as noted in the previous section. But ensuring that wholesale markets give meaningful cost signals requires a reexamination of their basic operation.

4.2   Reform Proposal

In many discussions of the pricing structures needed at retail level to cope with increasing intermittency, the problem identified is that market signals are not coming through properly. So, for instance, responses include TOD pricing and "transactive energy" (discussed in more detail by King and Cazalet in this volume). But TOD pricing only produces efficient results if wholesale prices are a true reflection of underlying costs, which, this section argues, is ceasing to be the case. Transactive pricing has many attractions as an ultimate "steady-state" solution but may not be the best means of effecting the transition to that state. It does not directly address two central problems identified in this chapter, namely,

- the market distortions that result from mixing subsidized and unsubsidized sources in a single market
- the complexity and uncertainty (discussed later) that are likely to inhibit consumer investment.

It is also unclear whether such a fundamentally new approach would be compatible with existing regulatory mechanisms.

It is arguable that what is needed, at least during the transition, is a pricing system that can

- cope with a world of the "second best" while the new forms of generation are being introduced to the system via special support mechanisms and
- provide the transition to a self-sustaining low-carbon market where the support mechanisms can be removed.

---

16. The underlying problem is one of the so-called pecuniary externalities—that is, situations where the behavior of one market participant affects the prices received by another. Normally, this does not constitute a problem; it is just the operation of the market. Indeed, because of the pervasive system effects, pecuniary externalities are endemic in electricity markets. However, where one set of producers is subsidized, pecuniary externalities can have the effect of distorting the overall market, creating a gap between prices and costs. There is a good discussion of these issues in OECD (2012, p. 34-37).

A possible option would be to create separate markets—an energy pool for non-dispatchable plants and a flexibility market for load-following plants at wholesale level and "as-available" and "on-demand" products as their counterparts at retail level.[17]

The price signals created would allow utilities to test how far consumers really value reliability and to what extent they are prepared to adjust their consumption over time. *In addition to* the price distortions referred to earlier, there are reasons why existing approaches will not effectively reveal these preferences:

- *Complication*—In a low-carbon (or high DG) system, time of use pricing is not the relatively straightforward TOD concept we are familiar with from conventional systems, where it refers to predictable demand peaks (say, 4-8 on winter evenings). In a system dominated by intermittent generation, high prices will occur at essentially random times of system stress, when intermittent generation falls significantly below demand; direct consumer planning—that is, for each customer, without aggregation—will not be feasible.

- *Uncertainty*—It could be argued that consumers do not have to react consciously to these situations—automated devices could do it for them. But since neither the price of electricity at these times nor the frequency with which they will occur can be forecast with any confidence, consumers have no way of knowing if it is worth investing in such devices and are unlikely to do so. Indeed, recent experience in Germany has involved a flattening of the intraday price curve, because of the introduction of large volumes of renewable plants, as shown in Figure 3.

  As the price curve flattens, the value of demand response diminishes overall. Demand flexibility may, of course, still be needed for particular times of system stress, when prices may reach very high levels, but these are likely to be infrequent and difficult to predict. Governments have accepted that price volatility and uncertainty inhibit investment in nondispatchable plants and should be managed for producers via FiTs and similar arrangements. One could apply a similar logic to consumers.

- *Robustness of price signals*—"Pecuniary externalities" may be a recondite concept, but it will be apparent to consumers that electricity does not really cost less than nothing to produce at certain times (as negative prices imply). So they are not likely to have confidence in the operations of the wholesale market or the signals it produces; they will expect it to be put on a more rational basis at some point and this will create additional uncertainty for any demand-side investment.

---

17. As far as the authors know, no market is currently moving in this direction. However, the United Kingdom has recently started developing proposals for an offtaker of last resort mechanism that would provide independent renewable developers with a "backstop" route-to-market at a guaranteed minimum price—see DECC (2013b)—and could in principle form the basis for such a market.

A possible solution is to present consumers with a simple choice between different sorts of supply—"as-available" power at a low price and "on-demand" power at a significantly higher price. There is a rough precedent in the United Kingdom "Economy 7" and related tariffs that offer a lower nighttime rate, via separate metering, mainly intended to encourage electric storage heating. Prices under this tariff were for a long time significantly lower, at about 40% of daytime electricity; over time, they produced very significant changes in demand levels and patterns. Demand for consumers in this group was over twice that of the average consumer and the peak was shortly after midnight, rather than in the early evening, as shown in Figure 4. These impacts are orders of magnitude greater than what has been achieved with traditional TOD pricing; impacts on this scale would be needed to cope with the problems described earlier.

With the development of smart metering, separate meters and circuits should not be needed. Instead, consumers should have the option of using appliances fitted with microchips that could react to the presence of as-available supply and be designed to make best use of it. Price differentials could be based around the same principles as with Economy 7—unit prices could be set at around 40% of the normal unit price and the broad differential would be guaranteed over

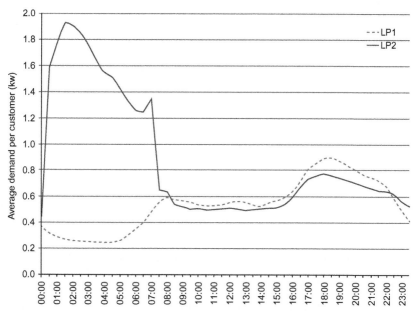

FIGURE 4   Profiles of customers with and without off-peak metering and billing in the United Kingdom. *Note:* In the chart, the dotted line LP1 represents the load profile of "unrestricted" customers (about 22 million) who pay a uniform rate at all times of day. The Solid line LP2 represents customers (about 5 million) who have separate metering for night use and are billed on a two- or three-part peak/off-peak tariff like Economy 7. *Source: Sustainability First 2012.*

a long period. It would therefore give consumers a secure basis for making the necessary investments in equipment.[18] Depending on the system, this sort of differential could be a product of regulation or simply of suppliers responding to the market signal from wholesale markets.

Consumers with the right equipment would receive the as-available price so long as that generation exceeded that class of demand and have their prices calculated pro rata between the two markets at other times (e.g., pay half the lower price, half the on-demand price when half of generation was from as-available sources), though they would have the choice, for nominated equipment, to have automatic cutoff at high-price periods. Consumers who did not have such equipment would pay the more expensive on-demand price, incorporating the wider system costs of reliability and flexibility. There would be a big incentive to have an as-available supply.

Wholesale markets could be constructed on the same lines. Generators would have the choice of either entering the on-demand market or being dispatched in the as-available pool. In the latter case, prices would initially incorporate support mechanisms as at present, in addition to the as-available pool price, though in the longer term, once a developed as-available market existed, such generators would be able to contract directly with suppliers or consumers. Dispatch in the as-available segment would be automatic—that is, the system operator would have to accept the power or use curtailment auctions in case of excess.

In the parallel flexibility market, the operator would call on additional generating plant (or demand response) as necessary to meet overall demand. This market would involve capacity and flexibility payments, as well as a kilowatt-hour market, in order to incentivize investment; the costs, including overall system costs, would be passed on to consumers via prices in this market. Over time, and assuming a high carbon price is also imposed, prices would move significantly above the cost of nondispatchable plant, allowing the support for such plant to be phased out and creating a viable exit strategy, once a low-carbon power system is established—an exit strategy that, as explained earlier, is not available in the present situation.

While the system involves intervention in pricing in the interim period, that is not new; the difference is that present interventions do not support the underlying policy goals—they simply "smear" the extra cost of policy-supported generation across the whole market, by adding a standard fixed uplift to the kilowatt-hour price or a fixed capacity contribution. This does nothing to mitigate the underlying distortions and gives no useful signals to suppliers or consumers. By contrast, under the arrangements proposed here, both wholesale and retail markets would be designed around the same long-term policy objective, and synergies between them would be realized. A schematic representation of the proposal is presented in Figure 5.

---

18. The 40% figure is an example; the actual differential could in practice be based on an assessment of long-term cost minimization for the whole system, taking both demand and supply resources into account, given the constraint of government objectives in terms of generation types.

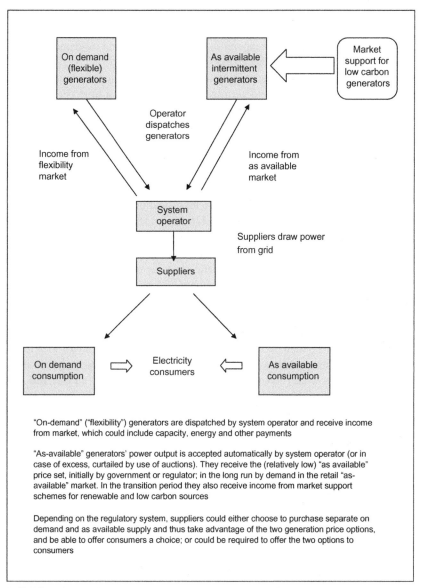

FIGURE 5   Schematic representation of on-demand and as-available markets.

4.3   Variation on the Reform Proposal: with One Wholesale Market

An alternative version of this proposal would involve the two separate retail offerings proposed earlier (among others), while retaining a single wholesale electricity market and central dispatch. The logic is that, in practice, one cannot distinguish among the different sources of electricity arriving at a customer's premises. Furthermore, it may be unnecessary to identify two physically separate wholesale markets to correspond to the as-available and on-demand retail offers. There are at least two ways that this might work.

One is that centralized dispatch would occur as it does now, and a single energy market price would be determined for each relevant time period. Customers with as-available retail contracts and corresponding appliances would consume the as-available energy when wholesale prices were below a defined threshold and pay the higher tariff when wholesale prices were above that threshold. In effect, customers would be buying an interruptible supply of low-cost energy to use in the appropriate appliances, with interruption being determined by a wholesale price threshold.

Another approach would also involve least-cost dispatch of all plant and a single market price but would distinguish between the volume of energy that was dispatched within the category of on-demand plant and the volume dispatched with the category of on-demand plant. These two categories would need to be defined, for instance, by reference to some price threshold or the type of plant that was being dispatched. Customers with as-available retail contracts could consume their share of dispatched as-available generation, as in the original proposal, but without any need to create separate wholesale markets.

One of the aims of this proposal is to provide incentives for the development of a supply chain to support a new approach to consumer behavior in a low-carbon world. Manufacturers of appliances, storage capability, in-house displays, and meters would have a target market, corresponding to demand for as-available (low-carbon) energy. Retailers would develop their service offerings to reflect customer preferences in terms of their willingness to pay for different kinds of energy service; they would be free to define other service offerings as well as the specific model described earlier.

5   NEW BUSINESS MODELS

If advantage is to be taken of the developments described earlier, then there are implications for future business and regulatory models.

5.1   Redefining the Consumer Offering: from Commodity to Service

The reform proposed in Section 4 was one step on the road to redefining the consumer offering. Consumer behavior needs to be brought into the picture as a much more active element in the system. Faruqui and Grueneich in this volume look at some strategies for regulated utilities; this chapter looks more generally at the sort of response that might be appropriate in competitive as well as regulated environments.

What is needed is to redefine the "consumer offering," treating electricity as a set of services, rather than a homogeneous commodity, always available "on tap." This requires starting with a clean sheet in defining the nature of the services that consumers need and the basis on which they will pay. So, for example, a consumer wanting to charge electric vehicle batteries might request X kilowatt-hour to be delivered within a specified period, which might be over several hours or several days, and the contract might specify that the consumption will only be for as-available energy. Similar arrangements could apply to the purchase of power for heat, a more complex definition of requirements for refrigeration, and so on. The requests would be put to energy service companies (ESCOs), one of which might or might not be part of the utility itself, which would be able to aggregate consumer requests and feed them in as part of the system optimization routines. The services offered might well, as with the Economy 7 example earlier, be packaged up with the provision of appropriate equipment. This approach might well develop along the lines suggested in the chapter by Cazalet but the assumption here is that consumers will at least initially need to operate via intermediaries.

This would allow consumers to purchase power for particular usages in the same way that they purchase other goods and services, as opposed to perpetuating the instantaneous commodity characteristics that have hitherto been a unique feature of the power sector; it would correspond to what consumers will actually want and need from the utility. At the same time, it would make many of these usages much more affordable. Consumers could still choose to take some power "on tap" and would normally pay a higher price for this, which could in principle be linked to TOD or "spot" prices where these were relevant.

This change is enabled by one set of technologies—those that surround metering, separate identification of different loads within each consumer household or business, and use remote control. But it also helps to resolve some of the problems posed by another set of technologies, those linked to intermittent or inflexible sources of nonfossil generation and DG.

5.2 Growing Role for DSR

Whereas traditional demand management involves shifting demand to periods with lower energy prices, new demand management lowers the costs of the system by offering flexibility—especially important to help in the management of intermittent renewable sources. Note that flexibility is needed both to increase and to reduce demand. Flexible demand increases are valuable because they avoid having to reduce production at significant opportunity cost from conventional sources (like coal or nuclear) when intermittent renewables increase their output. Flexible demand reduction is valuable because it reduces the need for spinning reserves to cope with a reduction in renewable generation.

Industrial customers may already have their own generation, storage, and demand response capability to offer in capacity and balancing markets. If not, this is a suitable time to be considering investments, especially in countries like Great Britain where reserve margins are tight and where flexible services will have high value due to the growth of intermittent renewable energy.

5.3   Distribution System Operators

DSOs have a new role in the electricity markets described here. Whereas the DSO currently manages the local distribution network to allow one-way flows from the transmission network to the customer, the new DSO will have a more active role in managing multidirectional flows, organizing auctions, and offering other incentives to involve DER in minimizing the DSO's investment and operating costs.

This new role raises questions about the independence of the DSO from the owner of the distribution network and from affiliated retail companies. It is illegal for distribution companies to discriminate in favor of their affiliated retail companies and against competing retailers. Regulators will need to be vigilant to ensure that the new DSO functions are not used in a discriminatory way to favor affiliated retail companies or the owner of the distribution wires. Indeed, the common ownership of the wires and the DSO function should be revisited, and regulation developed to ensure that the new DSO can carry out its functions independently.[19]

Another potential new role for the DSO has to do with the management of customer information. If the DSO is sufficiently independent of commercial retail service interests, it could be the repository of all data collected from smart metering that is made available to competing retailers, DER suppliers, software developers, and aggregators. This is one of the models currently being debated in the EU. Given the potential sensitivity of this information and the potential for it to promote competition and new services, this sort of management role will require regulatory oversight.

5.4   The Growing Role of Aggregators

For reasons of scale economies and to manage risk, aggregators are likely to play an increasingly important role at the interface between the suppliers of DER and the wholesale market. These demand-side aggregators already exist for large industrial customers (especially in the United States), who typically sell their demand-response services to the system and share the benefits with the aggregator. They also exist in the renewable energy sector on the supply

---

19. Chapters in this volume by Kristov and Hou and Brennan also look at the issues affecting distribution system operators.

side, for instance, in Spain, where multiple wind or solar generators use aggregators to sell their energy to the system. In the new era and with the help of smart technologies, regulators are likely to welcome aggregators who are able to offer their services to smaller customers, potentially including residential customers.

5.5   Reversion Toward Vertical Integration

The discussion earlier, like most of this chapter, has been mainly concerned with reconciling the new developments under discussion with the liberalized markets prevalent in Europe and many other parts of the world. However, one possibility is that some governments might find this task so complex that they prefer instead to intervene directly themselves. While this need not imply a return to the world of vertically integrated monopolies, it would be a step back from the trend of two decades towards increasing disaggregation, unbundling, and more liberalized markets. In the United Kingdom, the behavior of market participants has been consistent with this trend change, with a significant degree of business-driven vertical reintegration that has not in practice been inhibited by the dictates of competition policy.

The crucial factors pointing in this direction are the following:

- Perceptions of an inevitable increase in policy intervention by governments or national authorities, most obviously in order to promote an environmental or emissions reduction agenda; the security issue and creation of capacity markets are another manifestation of policy-driven intervention, with the driving force in this case being the interest of governments/regulators in ensuring supply security as a public good.
- The frequently repeated demand of infrastructure investors, in long-lived and nonmobile use-specific assets, for long-term regulatory commitment and/or contractual assurances.
- Unresolved coordination issues, highlighted at different points throughout this book, in relation to DER, new means of managing consumer demand, and so on.

A degree of central direction is one way of dealing with these considerations. Associated with a single buyer, for example, it provides a potential policy instrument for government, an entity that can provide long-term contractual assurances, and a mechanism for dealing with some of the coordination issues that arise in developing low-carbon strategies on both supply and demand.

The challenge is to achieve the benefits, in terms of policy, coordination, and a lower cost of capital, without losing important competitive disciplines in key areas of utility costs, particularly in the construction and operation of new generation plant. However, provided the single buyer is properly regulated and incentivized, this should in principle be possible while maintaining the disciplines associated with competitive tendering and optimum dispatch of plant.

6 CONCLUSION

The central argument of this chapter is that the growth of intermittent renewable and DG sources has rendered existing wholesale and retail electricity markets unsustainable. Changes will be needed in these markets and, at a fundamental level, in the way we conceive of electricity.

The markets of the 1990s were not designed with intermittent sources in mind, much less the incorporation of demand response. The growth of intermittent renewable and DG sources has created a significant and growing wedge between the total cost of providing electricity and the revenues being generated in wholesale markets. Retail price levels that reflect wholesale energy market prices do not therefore recover generation costs or promote efficient consumption or investment by consumers. So markets are no longer able to promote their central functions of promoting efficiency and providing sufficient revenue for producers.

New mechanisms and new ways of thinking about electricity are needed, to promote efficient markets and provide meaningful price signals. One option involves a new conceptual distinction between "on-demand" and "as-available" energy, to be traded and sold to consumers in separate markets, and a shift toward treating electricity as a service rather than a (kilowatt-hour) commodity.

Governments will also need to facilitate the transition to a low-carbon electricity system by encouraging the development of a supply chain, regulation, and markets that are consistent with this new approach. New technical possibilities vastly expand the range of services that can be offered to consumers, and these are likely to be associated with pricing and tariff mechanisms very different from those in conventional use. This should point to much greater consumer choice and a large and innovative role for new ESCOs. The electricity industry is entering an era of fundamental change—more fundamental than at any time in its past—and needs to show that it can rise to the challenge.

REFERENCES

Battle, C., Perez-Arriaga, J., 2012. Effects of variable energy resources on operating costs for different short term market rules. In: MIT CEEPR Spring 2012 Workshop, Cambridge, USA, May. http://web.mit.edu/ceepr/www/about/May%202012/may%20handouts/Batlle%20Perez-Arriaga.pdf.

Cramton, P., Ockenfels, A., 2012. Economics and design of capacity markets for the power sector. Z. Energiewirtsch 36, 113–134.

DECC, 2009. The UK Renewable Energy Strategy Cm 7686, July.

DECC, 2013a. Electricity Market Reform: Capacity Market – Detailed Design Proposals Cm 8637, June.

DECC, 2013b. The Backstop PPA Proposal, Report by Redpoint/Baringa for DECC, London.

Della Valle, A.P., 1988. Short-run versus long-run marginal cost pricing. Energ. Econ. 10, 283–286.

Eurelectric, 2012. The role of energy-only markets in the transition phase towards building a low carbon economy. In: Cailliau, M. (Ed.), Electricity Markets at the Crossroads: Which Market Design for the Future. Eurelectric (a trade association).

EurElectric, 2013. Active distribution system management: a key tool for the smooth integration of distributed generation. Full Discussion Paper.

Harris, C., 2006. Electricity Markets: Pricing, Structure and Economics. John Wiley and Sons, Oxford.

Hesmondhalgh, S., Pfeifenberger, H., Robinson D., 2010. Resource adequacy and renewable energy in competitive wholesale electricity markets. The Brattle Group Discussion Paper.

Hunt, S., 2002. Making Competition Work in Electricity. John Wiley and Sons, Oxford.

OECD/NEA, 2012. Nuclear Energy and Renewables.

Ofgem, 2013. Electricity Balancing Significant Code Review – Draft Policy Decision, July, Ofgem 120/13.

Stoft, S., 2002. Power System Economics: Designing Markets for Electricity. Wiley Interscience, Oxford.

Sustainability First Paper 2, 2012. GB Electricity Demand 2010 and 2025, London.

Turvey, R., 1968. Optimal Pricing and Investment in Electricity Supply. Allen and Unwin, London.

Turvey, R., 1969. Marginal Cost. Econ. J. 79 (314), 282–299.

Chapter 9

Transactive Energy: Linking Supply and Demand Through Price Signals

Chris King

ABSTRACT

Transactive energy is the concept of linking supply and demand in the electricity system primarily through the response of electric loads and generation on the consumer side of the meter to price signals. It provides a solution to growing challenges to system efficiency and reliability resulting from rapid adoption of distributed renewable energy. This chapter reviews both the technical and economic aspects of transactive energy. It explains the essential elements of a transactive energy system, including required technologies and how they interact. Price signals are discussed in detail, highlighting those prices with the greatest effect on the efficiency of electricity system resources—primarily generation and distribution. The chapter addresses practical approaches to implementation and provides several examples of transactive energy in action.

Keywords: Transactive energy, Demand response, Dynamic pricing, Market pricing, OpenADR

1 INTRODUCTION

Since electricity was first provided to the public in 1881 in the English town of Godalming—powered by a Siemens generator—the electric grid has largely been a centrally planned, one-way, outbound system.[1] Utilities traditionally developed forecasts of customer demand for electricity and then built and installed sufficient generation, transmission, and distribution resources to meet that demand. However, as described throughout this book, today's grid is characterized by increasing levels of distributed renewable resources, demand response programs, and energy efficiency initiatives. The utility business and physical model are being severely disrupted. On the physical side, utilities

---

1. "The Electric Light at Godalming, 1881" by Kenneth Gravett, Surrey History, Vol. II No.3, Surrey Archaeological Society, Guildford, 1981-82.

Distributed Generation and its Implications for the Utility Industry. http://dx.doi.org/10.1016/B978-0-12-800240-7.00009-6
189

and market operators must do their best to forecast short-term production by intermittent wind and solar resources, as described in Chapter 20 by Frank Felder. At the same time, forecasting long-term production by these resources is even more difficult, as renewable resource subsidies come under increasing political pressure. On the financial side, utilities are facing strong competition with their largest source of profitability—generation resources—on top of cost increases needed to operate the grid in a world of abundant renewable resources.

A partial solution to these challenges is a new concept known as "transactive energy," which is also discussed in Chapter 10. This chapter describes this new phenomenon, which, in brief, is the use of market forces and flexibility—as opposed to central planning—to manage and operate the grid. In contrast to Chapter 10, which presents a somewhat theoretical and long-term optimized approach, this chapter highlights a shorter-term, more practical approach.

According to the GridWise Architecture Council,[2] transactive energy is the "coordination of energy use and generation based on power price signals and grid conditions" (GWAC, 2013). This chapter considers the various elements of transactive energy. Section 2 explores the definition of transactive energy. In a sense, the definition sets the boundaries of what is and is not included in the concept. For example, transactive energy must be considered in light of distributed renewable resources, demand response, and energy efficiency.

The remainder of the chapter is structured as follows: Section 3 explores the concept of "coordination," describing how transactive energy plays a role in resource adequacy, maintenance of grid stability, and even utilization of specific resources such as wind, solar, storage, and others. Section 4 discusses a variety of issues related to price signals. If such signals are to elicit the proper responses of energy consumers, equipment, and even devices, then the signals must reflect resource constraints or other costs appropriately. At the same time, there are practical limitations on pricing from the perspective of being able to respond. Section 5 considers issues related to price signals in the context of the production side. Wind, solar photovoltaic, and renewable resources are usually "must take" for wholesale market operators. However, physical and financial constraints of other resources and the grid itself often results in curtailment, especially of wind power. Transactive energy provides the potential for reducing forced curtailment and otherwise increasing utilization of renewable resources. Section 6 addresses price signals related to distribution networks, which are characterized by high fixed costs—capital equipment and a fixed level of operations and maintenance labor—and low variable costs. Efficient use of these assets is best served by tariffs that reflect their largely fixed-cost nature. The amount of transmission and distribution capital investment—the sizing of these systems—is driven by each customer's highest instantaneous

---

2. The GridWise Architecture Council was created in 2004 by the Department of Energy as a group of industry experts working toward a common grid modernization vision.

consumption during a year (the peak demand, measured in kilowatts). Transactive energy is a way to reduce the peak demand and increase efficiency. Section 7 summarizes some of the technical infrastructure needed to implement transactive energy followed by the chapter's conclusions.

2   DEFINING TRANSACTIVE ENERGY

To intelligently discuss, and before implementing a new concept, one must understand it. Because "transactive energy" is a recent term of art, consensus and understanding continue to develop around its meaning. This section analyses this issue and concludes with a recommended approach.

The GridWise Architecture Council was formed by the US Department of Energy to promote and enable interoperability among the many entities that interact with the electric power system. This balanced team of industry representatives proposes principles for the development of interoperability concepts and standards. The Council provides industry guidance and tools that make it an available resource for smart grid implementations. GWAC defines transactive energy as

A set of economic and control mechanisms that allows the dynamic balance of supply and demand across the entire electrical infrastructure using value as a key operational parameter.

(GWAC, 2013)

The new element implied in this definition is control mechanisms on the demand side, as opposed to only on the supply side, which was the case in the past.

The key differentiator of the electricity system compared to other commodities is that supply and demand must be balanced across the entire grid and in real time. A temporary shortage cannot be tolerated, because it will bring the entire system down if it is not isolated to a smaller part of the grid. Grid operators—historically utilities, but more commonly regional transmission operators today—forecast loads annually, daily, hourly, and even in the next few seconds. They build or contract for and then dispatch generation resources sufficient to deliver the forecasted loads. Related issues are explored in Chapter 18 by Kristov and Hou.

In understanding transactive energy, it helps to consider where the industry is coming from. Historically, utilities have performed all of the balancing activity. They developed forecasts of future electricity demand by their customers and then built generation, transmission, and distribution resources to deliver the power. Customers were not asked whether they would like to pay for new investment. Regulators provided cost recovery, so long as the investments were actually used as planned. With relatively few exceptions, customers were expected to use electricity as and when desired, with no consideration for how their usage would affect the system. Customers were expected to turn lights on when they wanted them, and utilities were expected to ensure that enough power plants were available to deliver the electrons when called (RAP, 2011). Sioshansi discusses this "regulatory compact" in the Introduction: The monopoly utility is obligated to serve, and customers are obligated to buy from that utility.

As reflected in a number of chapters in this volume, the situation is changing today, with customers now involved in the decision to add generation on their side of the meter or for that matter invest in energy efficiency to reduce their demand for generation. Sometimes, customers add their own solar panels, and sometimes, they save energy rather than burden the grid. By making such decisions, customers are creating a *de facto* transactive energy market: they are conducting transactions—making decisions and investments—that affect the electricity grid.

The following elucidation of the key elements of the GWAC definition adds further clarity to the transactive energy concept:

- "Dynamic balance" means balancing the grid supply and demand in the context of operations (as opposed to planning and construction), with time periods from less than a day to as little as 5 min.
- "Using value as a key operational parameter" means (1) having prices that reflect both the short- and long-term cost of resources, including generation, transmission, and distribution; (2) making those prices transparent to all resource providers and users, be they large corporate power producers or individual customers; and (3) decision-making reliance on those prices by resource providers and users to increase or decrease either generation or load in order to balance the grid.
- "Economic mechanisms" refer to markets, both formal and informal, especially in the sense of collecting and exchanging detailed data on production, consumption, and prices at the individual customer or even device level, a concept known as "prices to devices," originally coined by EPRI (2010).
- "Entire electrical infrastructure" means traditional generation as well as customer-owned generation, transmission and distribution assets, and new infrastructure elements, such as energy storage and demand control equipment.

Based on this definition, "transactive energy" can be summarized as techniques to balance short-term supply and demand on the electricity grid by having energy producers and users making decisions to produce or use energy based largely on short-term prices.

3 TRANSACTIVE ENERGY TO BALANCE THE POWER SYSTEM

Transactive energy operationally balances supply and demand on the grid through the convergence and interaction of financial incentives, market participants, and fluctuating loads. Under federal law, the transmission system

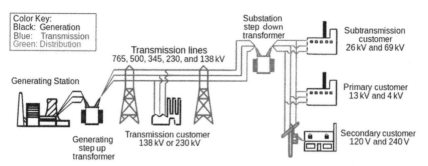

FIGURE 1 Elements of the power grid. *Source: Wikipedia.*

operator is responsible for managing and taking both the long- and short-term actions needed to balance supply and demand, as well as maintain power quality, including voltage and frequency (USA, 2005). These actions pertain to the high-voltage grid generally used for transporting power long distances and between cities and other large load centers as illustrated in Figure 1. The Distributor Grid Operator (distribution utility) is responsible for the long- and short-term actions needed to ensure that sufficient capacity is available to meet peak demand in the substations, lines and transformers. In addition, this extends to ensuring power quality, including constant voltage level, is properly maintained.

The TSO spends much of its operational effort on balancing supply and demand. This begins with forecasting total system loads on a day-ahead basis, hour by hour. In areas where there is competitive retail supply, the retailers provide forecasts of the total loads for their customers. The TSO incorporates these forecasts into its own. The TSO then identifies the source of generation to meet the forecasted load amounts and, in real time (every 4 s), dispatches specific generation sources to meet real-time load (PJM, 2013).

Historically, the TSO was a unit of a utility that owned its own generation, so the TSO was limited to dispatching the utility's own generation units. With the evolution of markets, TSOs dispatch resources owned by others and TSO-contracted resources. TSOs also must accept automatic dispatch of many renewable resources—which have priority—and resources dispatched by others. For example, in the Texas market, competitive retailers dispatch their own generation, based on plans submitted to the TSO. In such cases, the TSO dispatches generation units only at the margin, to meet unforecasted loads or to fill in for undispatched, but expected, generation (ERCOT, 2012).

Increasing levels of wind and solar resources, as well as customer-installed generation and storage operated autonomously, are causing problems of inter-mittent and overproduction that transactive energy can address. Intermittency is caused by weather changes, including changing cloud cover. Overproduction can mean the grid cannot physically handle the power injection, because the installed circuits and transformers are insufficient. It also can mean that generators of lower priority—for example, nuclear plants—would rather pay to keep operating than bear the expense of shutting down and restarting later. In markets such as Germany, abundant solar and wind generations have already led to negative wholesale prices as low as minus 100 EUR (about $135) per MWh (Economist, 2013).

As the proportion of renewables and self-generation increases, direct control of either of these situations by the TSO is becoming difficult or impossible. The TSO cannot control fluctuations caused by weather and generally must take all of the power produced by wind and solar resources. The only action the TSO can take is to curtail, or shut off, production of large wind or utility-scale solar plants when there is overproduction. For rooftop solar, distribution companies are increasingly requiring that solar inverters be equipped with remotely controllable shutoff switches to do the same. As the number of rooftop units proliferate, the control challenge becomes very difficult, growing by orders of magnitude. For example, in the 1990s, the California ISO monitored and controlled about 4000 grid intake and outtake points. As of this writing and as shown in Figure 2, California already has over 190,000 solar systems in place. For more on this, see Chapter 18 by Kristov and Hou.

At the same time, there is much less ability to plan centrally for the amount of generation, transmission, and distribution resources, decisions made by individuals or autonomous stakeholders not under the control of the TSOs. With customers producing their own electricity, TSOs and distributors must respond to a market that now both consumes power and produces it more or less as it pleases. The result is a dramatically different market, both operationally and fundamentally. Figure 3—the "Duck Curve"—shows the traditional bell-shaped demand curve of California and is representative of most markets being replaced with a curve dominated by the midday presence of solar production, not depicted, because this is on the "load" side, and a very narrow, sharp peak occurring just as the sun sets.

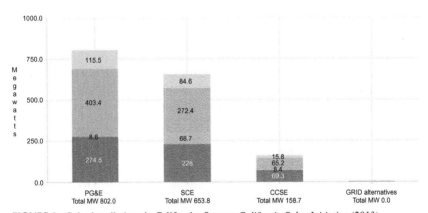

FIGURE 2 Solar installations in California. *Source: California Solar Initiative (2013).*

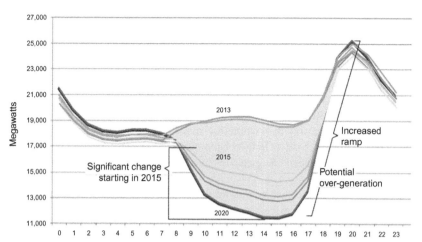

FIGURE 3 California's current and projected system load curve; the *x*-axis is hour of the day. *Source: CAISO (2013).*

In this context, transactive energy provides two types of solutions:

- First, it provides data that TSOs and distributors can use to plan the operation of the grid. Smart meters deliver hourly interval consumption data at least once a day that can be used in the next day's forecast. Some smart meters capture both inflows and outflows—as opposed to only a net total—which can further enhance forecasting ability. By having such detailed data, TSOs and distributors—through the use of analytics software that can process large amounts of data—could ultimately forecast production and consumption at the individual customer level. Another data element is detailed production data from solar panels that can be sent from inverters.

- Second, transactive energy can provide prices that can be used to adjust both production and consumption. On the production side, most distributed energy system owners will produce as much as possible to maximize their economic benefit. Maximizing production is especially true when fuel is free (e.g., solar). With respect to production, the primary use of a price signal would be for production curtailment. With transactive energy, customers—or, more likely, their equipment—would receive prices that tell them how much they would be paid to curtail production. At times of overproduction in the overall market, energy prices would drop—going negative—until they were low enough to cause enough customers to turn off their production to balance the system. Customers would preprogram their wind turbine controllers or solar inverters to undertake certain curtailment actions as different price thresholds are reached. This can happen on both a day-ahead and real-time (5-min ahead) basis.

Transactive energy price signals are already used somewhat to reduce peak demand. There is widespread access by competitive retailers and large commercial and industrial customers to wholesale markets in the US Northeast, Mid-Atlantic, Texas, and California; in Victoria, Australia; in Ontario, Canada; in Scandinavia; in the United Kingdom; and elsewhere. These markets have prices that change hourly based on supply and demand generally and even every 5 min for specific transmission grid locations (nodal pricing). In Texas, transactive energy price signals—wholesale market prices—already have a large effect on customers' loads. On the system peak day in 2011, customer response to price signals reduced the peak by an estimated 1700 MW, or about 4% (Brattle, 2012).

Transactive energy can provide simple control signals that result in the same load-reducing effect as price signals. There are a number of programs where TSOs or distributors send control signals—either directly or through intermediaries such as demand response aggregators—to customers or their devices; in California, over 200 MW of load are controlled by signals sent via the Open-ADR, "Open Automated Demand Response," standard (PowerIT, 2012). These signals may indicate that a demand response event will occur or may initiate a specific action such as adjust a thermostat's operation. The signals are a form of transactive energy, because the criteria for sending the signals are based on wholesale prices. An example was the PowerCentsDC residential pilot program in Washington, DC. For customers on the hourly pricing option, the utility sent control signals to their thermostats when prices exceeded a preestablished threshold in the PJM wholesale market. The result was peak demand reductions of up to 51% (ESC, 2010).

4 INFORMATION TECHNOLOGY UNDERLYING TRANSACTIVE ENERGY

The previous sections explained that the primary purpose of transactive energy is balancing supply and demand on the electricity grid and that the primary mechanism is price signals. To make such a system work requires what is often called a "smart grid."[3] In the context of transactive energy, a grid is "smart" when a digital layer is added.

This digital layer, illustrated in Figure 4, has five key elements:

- *Control* capability for equipment and devices that can be activated either automatically via a local computer responding to external data inputs or remotely via a communications network. Examples in the context of transactive energy would be thermostats, commercial building control systems,

---

3. For a detailed treatment, see *Smart Grid – Integrating Renewable, Distributed & Efficient Energy*, F. Sioshansi, editor, Academic Press, October 2011.

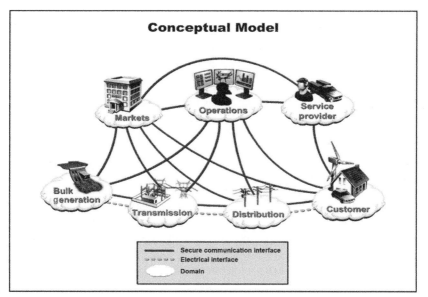

FIGURE 4   Smart grid conceptual model. *Source: NIST (2011).*

appliances, electric vehicle chargers, solar panel inverters, and grid storage systems.

- *Sensing* capability to measure power flows, voltage, and other grid characteristics at intervals from every few seconds to once an hour. Examples include line sensors, outage sensors, and smart meters—these typically sense current and voltage.
- *Communications* capability to send information between devices or systems. Four communications systems are commonly used: (1) the Internet; (2) utility-owned or controlled private networks, usually wireless, to send and receive data and control signals from devices mounted on distribution equipment (for distribution automation applications such as voltage control); (3) other networks, also usually wireless, to send and receive data from smart meters and, through smart meters, to devices that communicate wirelessly with smart meters; and (4) corporate networks, including virtual private networks, to send data between information technology systems. For example, outage alarms might come through the smart meter communications network and then be delivered from the computer head end of that network to the outage management system.
- *External data inputs* from systems outside of the transmission and distribution smart grid. The most important of these is price signals from the wholesale power market. Another very important input is weather, not only just temperature but also insolation, cloud cover, and wind speed and direction—all of which significantly affect solar and wind power generation.

- *Software* is the brain of transactive energy and is present in controllers, sensors, power consuming devices, power generating devices, communications networks, and, of course, all of the applications that process data and deliver outputs such as turning devices on and off.

These technologies/capabilities exist today and, in some cases, are already deployed. For example, the United States has approximately 50 million smart meters installed (IEE, 2013).

As mentioned earlier in the PowerCentsDC example, a transactive energy system can contain relatively few elements:

- *Control* via smart thermostats.
- *Sensing* was also via smart thermostats, namely, of indoor temperatures, which were allowed to drift up a few degrees during high-price events in the wholesale market.
- *Communications* of price signals started with the wholesale market at PJM, then via a VPN to the PowerCentsDC operations facility, then via another VPN from that facility to a paging transmitter, and then via a public wireless paging network to the smart thermostats.
- *External data inputs* to determine control events were next-day prices and historical prices, both from PJM. The latter were used to determine the frequency of high-price events and develop algorithms to determine when high prices should be converted to smart thermostat control signals and actually dispatched. They key constraint in this determination was how many hours per year the smart thermostats would be controlled. In a more interactive transactive energy system, customers would have set the price thresholds themselves or would have selected for how many hours per year they could be controlled (e.g., one customer might say up to 100, another up to 500.).
- *Software* in the program included billing (of hourly pricing rates), meter data management, demand response management, communications network management, an integration platform to connect the systems, and the internal software in the smart thermostats.

At the other extreme, the possibilities for extending transactive energy systems are nearly unlimited. For example, in its transactive energy framework, GWAC points out over 100 current and future electric power system control applications that are physically interconnected and would benefit from the coordination offered by transactive energy (GWAC, 2013).

5　AT WHAT PRICE GENERATION?

Transactive energy is fundamentally about providing the right price signals to the right customers or devices to elicit the right response. There is broad consensus that the essential goal of the electricity grid is to be stable, reliable, and efficient. This should be the goal of transactive energy price signals.

The difficulty is that price signals themselves have many more objectives to achieve. Looking at where the majority of electricity system costs occur—generation and distribution—allows us to conclude that the most important price signals relate to those costs.

It is widely agreed that the principles proposed by James Bonbright (1961) over five decades ago—in spite of changes in metering and other technology—remain valid for setting electricity prices today or "designing rates" as designated by regulated utilities:

- Rate attributes: simplicity, understandability, public acceptability, and feasibility of application and interpretation
- Effectiveness of yielding total revenue requirements
- Revenue (and cash flow) stability from year to year
- Stability of rates themselves, minimal unexpected changes that are seriously adverse to existing customers
- Fairness in apportioning cost of service among different consumers
- Avoidance of "undue discrimination"
- Efficiency, promoting efficient use of energy, and competing products and services

The problem is that these objectives have inherent inconsistencies. For example, rapidly changing hourly prices are highly effective in promoting efficient use of resources but are neither simple nor stable. In addition, too much simplicity—such as a flat rate—fails to promote economic efficiency and provides a barrier to the benefits of transactive energy. Increasingly, regulators are not attempting to satisfy all of these objectives with a single solution. Instead, policymakers are increasingly promoting customer choice, whether via retail competition (e.g., Texas and Pennsylvania) or rate options (e.g., Oklahoma and California). For regulated utilities, typical options are non-time-varying rates versus time-of-use or critical peak pricing rates. Consumers can decide which ratemaking principle is most important for themselves by choosing their rate from two or more options.

In this "rate options" context, one can now consider what is the "right" price for transactive energy. As discussed previously, transactive energy is focused on efficient balancing of supply and demand for generation, transmission, and distribution resources. Just as with Bonbright's principles (1961), it is difficult—in practice, it would be cost-prohibitive—to establish, communicate, and have customers and devices act on prices that, all the time, reflect the availability of each of these three resource types, all the time. Such an all-encompassing approach would include multiple wholesale generation markets (week-ahead, day-ahead, and hour-ahead), ancillary services such as frequency control and balancing energy, and nodal transmission prices that vary every 5 min. In Chapter 10, Cazalet discusses these concepts in detail. Here, the emphasis, instead, is a more pragmatic, balanced approach focused on the most important overall price signals.

Which resource is most important? Generation, transmission, and distribution are all essential; what's different is that generation is the lion's share of total electricity costs. According to the Energy Information Administration (2013), power costs in the United States break down for all consumers, on average, as follows:

- Generation: 57%
- Transmission: 11%
- Distribution: 32%

This breakdown suggests that the greatest efficiency gains from transactive energy can be had from focusing on generation and distribution costs. This section discusses the former; Section **6**, the latter.

Generation costs change hourly in most wholesale markets and vary widely. Figure 5 shows that in Texas—a typical case—prices went from below $0 to above $300 per MWh (below 0 cents to above 30 cents per kWh), with very high and very low prices occurring for only a few hundred hours per year (out of a total of 8760). The prices are very high when generation resources are tight and very low when there is overproduction, primarily from West Texas wind resources). It is these times when transactive energy can be most effective. Accordingly, hourly wholesale market prices are the most effective prices to be used for transactive energy.

As if there were not enough options, the final question remains: which hourly prices? There are the three options in most markets: day-ahead prices,

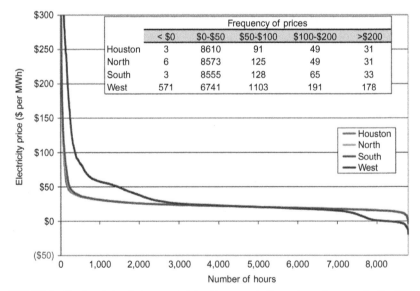

	Frequency of prices				
	< $0	$0-$50	$50-$100	$100-$200	>$200
Houston	3	8610	91	49	31
North	6	8573	125	49	31
South	3	8555	128	65	33
West	571	6741	1103	191	178

FIGURE 5 Hourly wholesale power price duration curve in Texas, by major zone, for the 8760 hours of the year. *Source: Potomac Economics (2013).*

hour-ahead prices, and real-time prices. Day-ahead and hour-ahead prices are most effective.

Day-ahead prices are produced by day-ahead bids submitted to market operators for electricity supply and demand. The market operator matches the bids and sets hourly prices at the point at which supply and demand are in balance. Market participants are then allowed to update their bids a couple of hours before actual dispatch (say, 10 a.m. for dispatch at noon); following a new set of balancing, the market operator publishes hour-ahead prices. As the transmission grid operator actually dispatches resources (every 4 s), additional resources are purchased for the final matching. These are the true "real-time" prices and known to the market only after the fact. Indeed, actual real-time prices may not be known for hours after the actual dispatch, after data have been delivered and contracts are finally settled.

Day-ahead and hour-ahead prices are much more effective than real-time prices, because end customers and devices can act on knowledge of prices only if the prices are known. Customers and devices can change their actions to respond to such prices by increasing or decreasing production or consumption. The only way to respond to real-time prices would be to predict them, a risky and challenging endeavor.

6   AT WHAT PRICE DISTRIBUTION?

After generation costs, distribution costs contribute the most to consumers' electric bills—except where taxes are very high, such as the Scandinavian countries. In considering the relative importance of different price signals, it is essential to note that distribution costs (substations, circuits, transformers, etc.) have very limited variability; they are essentially fixed costs best recovered through tariffs that reflect that fact and vary by peak load, not by time of day; for example, the distribution fee would be assessed based on monthly peak demand of the individual customer[4] GEODE (2013).[5] Here, too, transactive energy plays a role, by actively minimizing a customer's peak demand.

First, why emphasize demand charges for distribution? The distribution tariff structure should contribute to the overall efficiency of the system by giving the right incentives with regard to distributors' costs. It should also encourage customers to optimize energy behavior in the long term while providing incentives to encourage energy savings in the short term.

A peak demand based tariff structure is cost reflective for the distributor because the pricing principle is the same as the key cost basis of the electricity distribution. For example, distribution grid operators have found it to be most cost-effective to estimate long-term peak load on an asset, such as a transformer, and install a device that meets that requirement. Going out and changing

---

4. For billing purposes, demand is typically measured over 5- or 15-min periods.
5. GEODE is an association of local electricity distribution companies across Europe.

transformers frequently would be very expensive. Demand charges reflect the size of the investment and, thus, directly correlate to the distributors' costs. Importantly, even with distribution demand charges, energy efficiency is still promoted. The majority of the bill—generation charges—is collected via per kilowatt hour charges that provide a continuing incentive to reduce total consumption. Second, how does transactive energy play a role with demand charges? The goal in a situation where a consumer is paying demand charges is to orchestrate consumption to minimize that peak. In a commercial context, a building management system—such as those provided by Siemens—can ensure that, when there are multiple chiller units, only one at a time may operate. Thousands of buildings are already operated this way, responding to peak demand charges that are already common for large commercial customers.

In the residential context, a consumer would not want to use their air conditioner, electric oven, and electric clothes dryer at the same time. With automated appliance controls, a home energy management system could ensure the three do not operate at the precisely the same time. By actively managing the devices—for example, alternating the air conditioner and over every 5 min and then drying the clothes after the food is done—comfort is maintained, food is cooked, and clothes are dried.

7   CONCLUSION

Transactive energy has recently emerged as a concept that not only improves efficiency and reliability of the electricity grid but also helps transmission and distribution grid operators to manage the increasing complexity of the grid. Transactive energy helps grid operators coordinate energy use and generation based on power price signals and grid conditions. Implementing transactive energy requires a clear mental model that is implementable and actionable. This chapter offers the following definition in that regard: "transactive energy" can be summarized as techniques to balance short-term supply and demand on the electricity grid by having energy producers and users making decisions to produce or use energy based largely on short-term prices.

An examination of the options suggests that the most important and effective prices for transactive energy are day-ahead and hour-ahead wholesale energy prices. In this day of increasing levels of distributed renewable resources, demand response programs, and energy efficiency initiatives, transactive energy appears to have rare and promising potential as a solution to grid management and operational challenges.

This vision is already a reality in some places and promises to continue growing—subject to regulators empowering energy consumers with pricing options via either regulated utilities or competitive markets. From the United Kingdom to Texas to Victoria, Australia, large commercial customers are responding to fluctuating wholesale prices passed through by competitive retailers, and they are increasingly responding automatically via building control systems receiving signals utilizing the OpenADR standard. In places such as Oklahoma (Oklahoma Gas & Electric) and Florida (Gulf Power), tens of thousands of residential consumers have smart thermostats that receive price signals from their utility and automatically reduce load at high cost times. In Illinois (regulated), Finland (competitive), and Sweden (competitive), residential consumers have direct access to wholesale market prices, with some having smart devices that automatically respond as prices rise above threshold levels. In spite of such progress, we are seeing only the very beginning of the long-term potential, which is promising indeed.

REFERENCES

Bonbright, James C., et al., 1961. Principles of Public Utility Rates. Public Utilities Reports, Inc.
Brattle, 2012. ERCOT Investment Incentives and Generation Resource Adequacy. The Brattle Group. June 1, 2012 (accessed November 11, 2013), http://www.ercot.com/content/news/pre sentations/2013/Brattle%20ERCOT%20Resource%20Adequacy%20Review%20-%202012-06-01.pdf.
CAISO, 2013. In: Long Term Resource Adequacy Summit. Presentation on February 26, 2013 by Mark Rothleder of the California Independent System Operator. (accessed November 11, 2013), http://www.caiso.com/Documents/Presentation-Mark_Rothleder_CaliforniaISO.pdf.
California Solar Initiative, 2013. California Solar Statistics. The California Solar Initiative. updated November 6, 2013 (accessed November 10, 2013), http://www.californiasolarstatistics.ca.gov/.
Economist, 2013. European Utilities – How to Lose Half a Trillion Euros. October 12, 2013.
EIA, 2013. Annual Energy Outlook 2013, Energy Prices, Table Electricity Supply, Disposition, Prices, and Emissions. The Energy Information Administration of the Department of Energy (accessed November 11, 2013), http://www.eia.gov/oiaf/aeo/tablebrowser/#release=AEO2013&sub ject=0-AEO2013&table=8-AEO2013&region=0-0&cases=ref2013-d102312a.
EPRI, 2006. Advancing the Efficiency of Electricity Utilization: Prices to Devices. The Electric Power Research Institute (accessed November 10, 2013), http://www.nema.org/Products/Doc uments/Advancing-Efficiency.pdf.
ERCOT, 2012. ERCOT Operating Procedure Manual. Electric Reliability Council, Texas. May 1, 2012 (accessed November 10, 2012), http://www.ercot.com/content/mktrules/guides/proce dures/Real%20Time%20Operating%20Procedure%20V1Rev26.doc.
ESC, 2010. PowerCentsDC Final Report. eMeter Strategic Consulting. September 2010 (accessed November 11, 2013), http://energy.gov/oe/downloads/powercentsdc-program-final-report.
GEODE, 2013. GEODE Position Paper on the Development of the Distribution System Operator's Tariff Structure. September 2013 (accessed December 16, 2013), http://www.geode-eu.org/ uploads/GEODE%20Position%20Paper%20Tariff%20Structure.pdf.
GWAC, 2013. GridWise Transactive Energy Framework. Draft Version, The GridWise Architecture Council. October 2013 (accessed November 9, 2013), http://www.gridwiseac.org/pdfs/te_ framework_report_pnnl-22946.pdf.
IEE, 2013. Utility-Scale Smart Meter Deployments: A Foundation for Expanded Grid Benefits. August 2013 (accessed December 16, 2013), http://assets.fiercemarkets.com/public/sites/ energy/reports/IEE_SmartMeterUpdate_0813.pdf.
NIST, 2011. Draft NIST Framework and Roadmap for Smart Grid Interoperability Standards, Release 2.0. National Institute of Standards and Technology. Special Publication for public comment, October 2011 (accessed November 2011), http://www.nist.gov/smartgrid/.

PJM, 2013. PJM Manual 12: Balancing Operations, Revision: 29, November 1 2013. PJM Intercon-
    nection Inc. (accessed November 11, 2013), http://www.pjm.com//media/documents/manuals/
    m12.ashx.

Potomac Economics, 2013. 2012 State of the Market Report for the ERCOT Wholesale Electricity
    Markets. Potomac Economics, Ltd. June 2013 (accessed November 11, 2013), http://www.
    potomaceconomics.com/uploads/ercot_reports/2012_ERCOT_SOM_REPORT.pdf.

PowerIT, 2012. Powerit Solutions Now Controls an Estimated 30 Percent of California's Automated
    Demand Response Capacity. (accessed December 16, 2013), http://www.poweritsolutions.
    com/index.php?id=709.

RAP, 2011. Electricity Regulation in the U.S.: A Guide. Regulatory Assistance Project (accessed
    November 10, 2013), www.raponline.org/document/download/id/645.

USA, 2005. The Energy Policy Act of 2005. signed into law August 8, 2005, (accessed November
    10, 2013), www.gpo.gov/fdsys/pkg/PLAW-109publ58/pdf/PLAW-109publ58.pdf.

Chapter 10

Transactive Energy: Interoperable Transactive Retail Tariffs

Edward G. Cazalet

ABSTRACT

It is increasingly apparent that existing retail tariffs will be inadequate in a rapidly evolving environment where massive amounts of intermittent generation plus potentially equally large amounts of decentralized generation, storage, semi-independent microgrids, and smart energy use devices are joining the existing network, changing both the quantity and the direction of power flows. This chapter examines the ramifications of these developments and proposes an elegant framework for addressing the many operational, investment, and economic complexities resulting from the decentralized revolution, now sweeping the industry. The fundamentals of the proposed tariff scheme, coined interoperable transactive retail tariffs, are its simplicity in addressing the myriad of transactions taking place as consumers become prosumers, as decentralized generation, storage, and microgrids offer new opportunities to coexist and transact in new ways with the existing network.

Keywords: Transactive energy, Retail tariffs, Electricity markets, Decentralized control, Forward transactions

1 INTRODUCTION

The retail electricity tariff is central to the economic and reliable operation of the electric grid especially in a world with increasing decentralized resources, variable renewables, and automation of customer energy usage. An *interoperable transactive retail tariff* (ITRT) enables both forward and spot retail transactions for unbundled energy and distribution services as a way to deal with the challenges of decentralized resources and energy usage automation that are the subject of this volume.

There are many retail tariffs currently used in the electricity industry including flat price, increasing block price, time-of-use price, critical peak price, and real-time price (RTP) tariffs. In some cases, both energy ($/kWh) and demand

Distributed Generation and its Implications for the Utility Industry. http://dx.doi.org/10.1016/B978-0-12-800240-7.00010-2

($/kW) prices are specified. Block-and-index tariffs are used in some jurisdictions. These tariffs, however, are generally unable to address the vexing problems of decentralized generation, variable renewables, smart devices, two-way flows, and investment recovery on the distribution grids, topics examined in companion chapters of the book.

This chapter proposes an ITRT for all retail sectors including residential, commercial, industrial, governmental, agricultural, and electric vehicle sectors. An ITRT supports

- buy and sell transactions and two-way flow of electricity between customers, decentralized energy resources (DERs), retail energy providers (REPs), and distribution operators (DOs[1]),
- forward and spot transactions for recovery of both grid or network investment and operating costs,
- unbundled energy commodity and distribution products, and
- interoperation among all types of usage, supply, storage, distribution services and parties, and devices.

An ITRT is an enabling agreement among two parties—i.e., a REP and a retail customer or DER operator—the agreement enables buy and sell transactions between the retailer and another party at specific prices, quantities, time intervals, and delivery locations.

The ITRT employs transactive energy (TE) technology and standards that have been developed within the OASIS standards development organization under the direction of the Smart Grid Interoperability Panel.[2] Chapter 9 by King focuses on the communication of spot dynamic prices to devices; this chapter focuses on forward and spot transactions enabled by priced forward and spot, buy and sell tenders[3] among parties on an electric grid or network.

TE is designed to support all end-to-end, wholesale and retail transactions on an electric grid. This chapter focuses on retail transactions because of the opportunity to make substantial improvements over current retail tariffs and retail grid operation that were designed in an era without advanced interval metering, DER, and two-way flows on the distribution grid. The practical application of TE to the end-to-end grid is also discussed by the Transactive Energy Association.[4]

---

1. The functions of DO as defined here include only the operation of distribution facilities and do not include the broader functions of a DSO as defined by Keay et al. in Chapter 8. See also Section 7.

2. OASIS Energy Interoperation Version 1.0, 18 February 2012, OASIS Committee Specification 02. http://docs.oasis-open.org/energyinterop/ei/v1.0/energyinterop-v1.0.html.

OASIS Energy Market Information Exchange [EMIX] Version 1.0, 11 January 2012, OASIS Committee Specification 02. http://docs.oasis-open.org/emix/emix/v1.0/emix-v1.0.pdf.

3. A tender is defined as a binding offer to sell or a binding bid to buy a product or service. A tender is made by a party to a counterparty. If the counterparty accepts all or part of a tender, a binding transaction among the two parties is created.

4. Transactive Energy Association, www.tea-web.org.

The ITRT supports four major innovations, as explained in this chapter:

- First, the ITRT avoids the need for large fixed charges for revenue stability that are recommended by many other authors in this volume; instead, forward transactions provide more stable cost recovery than fixed charges.
- Second, the ITRT can help overcome the customer and regulatory resistance to spot dynamic pricing; the ITRT will accelerate dynamic pricing implementation and benefits by combining it with forward transactions that provide stable customer bills.
- Third, the ITRT avoids the complexity and cost of specialized tariffs for specific groups of parties or types of usage and DER. An interoperable ITRT is understandable to people, organizations, and automated devices and energy management systems in any jurisdiction or geography. A standard, ITRT will accelerate the development and benefits of the smart grid.
- Fourth, the ITRT supports a decentralized business and market model where the DO buys and sells distribution transport services and manages the monitoring, switching, and control of the wires, transformer, and similar devices, but does not take part in dispatch of any generation, storage, or customer usage. This is very different from the proposal for an ISO[5]-like, centralized DO that combines distribution operation and generation, storage, and demand response dispatch as presented by Hanser and van Horn and Kristov and Hou in Chapters 11 and 18.

The chapter consists of seven sections in addition to "Introduction." Section 2 defines TE and describes the basic TE technology, standards, business process, and vision. Section 3 summarizes current challenges in retail markets that create the need for ITRTs. Section 4 presents an example of an interoperable transactive tariff. Questions about whether such an ITRT will work are addressed in Section 5. Section 6 discusses the benefits of transactive tariffs from a number of perspectives. Section 7 suggests a future transactive business and regulatory model for the electricity industry followed by the chapter's conclusions.

1.1   Forward Transactions

TE transactions are for energy delivered over intervals of time such as a year, month, day, hour, 5 min, or 4 s—whatever is appropriate in the context of the transactions taking place among the parties. Tenders resulting in transactions may be executed years, months, days, hours, minutes, or seconds ahead of delivery or at the time of delivery.

---

5. An Independent System Operator (ISO) in the United States is essentially the same as a Regional Transmission Organization (RTO). The chapter uses the term ISO to refer to both an ISO and an RTO.

Forward transactions among parties are necessary for four basic reasons:

- Device, system, and grid operation generally must be planned and must reflect the physical limits of devices and systems to consume, produce, or store energy; turn on or off; ramp up and down; and provide the services.
- Devices, systems, and grids must be manufactured, constructed, installed, and maintained, and fuel must be purchased and scheduled for delivery ahead of actual operation. Customers may want to plan operations and investments and purchase electricity in advance at known prices.
- Parties prefer stability in costs and revenues, which can in part be accomplished with forward transactions for energy. Suppliers need stable long-term revenues to support investment financing.
- Forward transactions reduce the volume of spot market transactions and thereby reduce the leverage of large suppliers over spot market prices.

The threat of a utility death spiral as decentralized generation displaces centralized generation has been raised in several other chapters. The use of forward transactions provides a way for customers and decentralized resources to enter into longer-term commercial transactions with DOs and energy suppliers to recover fixed operating and investment costs and variable operating costs. Tariffs based on fixed and variable charges without such an enforceable commercial contract can present risks to both customers and DOs. In fact, simply increasing fixed charges in a retail tariff may encourage more customers to leave the grid entirely as the only way to avoid the fixed charges!

2 WHAT IS TE?

TE engages customers and suppliers as participants in decentralized markets for energy transactions that strive toward the three goals of economic efficiency, reliability, and environmental enhancement.

TE provides convenient and powerful means to address all aspects of coordinating both operations and investments associated with the production, storage, and use of energy including DER that are the theme of the present volume.

The TE technology described in this chapter is based on a formal profile of the OASIS TE standards called Transactive Energy Market Information Exchange (TeMix). TeMix is designed for automation of transactions and massive, decentralized implementation across the entire grid.[6] TE in this chapter uses only the TeMix profile, which is an open and free standard. Practical implementations of TE using TeMix will interface with existing transaction systems that do not use the TeMix profile of TE.

TE is based on buy and sell transactions of energy among parties that consume, produce, store, and transport electric energy as illustrated by Figure 1.

---

6. Cazalet, Edward G., "Automated Transactive Energy (TeMix)," Grid-Interop Forum 2011, http://temix.com/images/GI11-Paper-Cazalet.pdf.

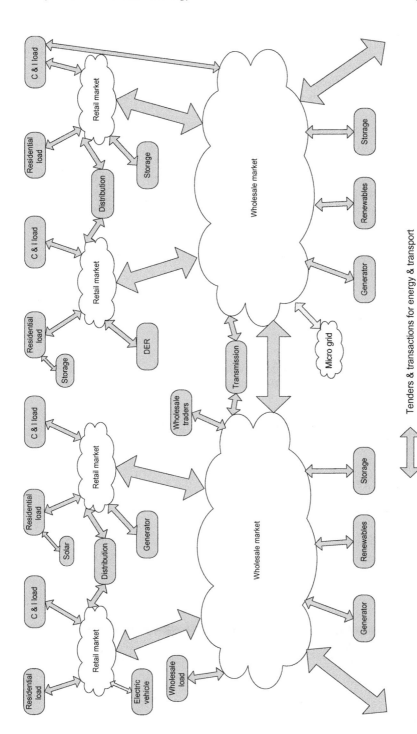

FIGURE 1   Transactive energy market structure.

Transacting parties can include end users owning energy-consuming devices and storage and generation devices, central generation owners, and distribution and transmission grid operators. The two-way arrows in the figure illustrate the two-way interactions between the parties using priced buy and sell tenders for energy and the transport of energy that may result in transactions between the parties.

TE as illustrated by Figure 1 can allow *any* party to transact with *any* other party under virtually any imaginable future scenario, including those mentioned in Chapters 18, 20, and 23 by Kristov and Hou, Felder, and Riesz et al. For example, a party such as an industrial customer can transact with a neighboring industrial customer, a commercial or residential customer, a retailer, a local DER, a wholesale generator, etc. Regulatory policy may limit the transactions that are permitted, but TE is applicable to all permitted interparty transactions.

In this environment, conceptually, a party takes one of two sides in a TE interaction: buy or sell. TE parties interact with tenders that may result in transactions as illustrated in Figure 2.

A party may asynchronously interact with several parties taking a different side in each interaction. All tenders and transactions are pairwise; if a buy tender by party B to party A is accepted by A, then A takes the sell side and B the buy side of the transaction. The framework is comprehensive, allowing virtually any physically feasible transaction among two consenting parties to take place: across time and location and for any products or services. The framework also supports purely financial transactions among the parties that may be used for financial risk management.

The key elements of TE including forward and spot transactions, products, interfaces, subscriptions, coordinated decentralized control, and business model and vision are outlined in the succeeding text.

2.1 Spot Transactions

In addition to forward transactions, transactions at the time of delivery—e.g., spot or real-time transactions—are necessary to instantaneously balance energy production and usage and to assure that the grid operation is stable. A balancing operator may take responsibility for posting spot tenders to parties to balance supply and demand in each interval.

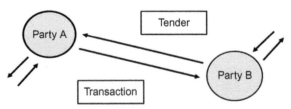

FIGURE 2   Parties interact with tenders and transactions.

The price of spot tenders is not just short-run marginal cost; see Chapter 8 for a contrary view. The price of spot transactions in TE is set by supply and demand. Generally, variable costs will set a floor on the spot price, and when demand approaches supply, the price will rise above the variable costs to also contribute to the recovery of fixed costs and profits. Keay et al. in Chapter 8 also made the observation that the existing wholesale spot prices can be extremely volatile and uncertain. TE uses forward tenders and transactions to allow customers to manage price uncertainty while participating in price discovery processes near the time of delivery. Volatile spot prices are often necessary for reliability and to balance supply and demand without overinvestment.

Contrary to often expressed misconceptions, transactive tenders are not just prices-to-devices broadcasting of price signals. Such price broadcasting has many problems including risks of grid instability, market abuse, and volatile costs and revenues.

2.2   Products and Interfaces

TE has two primary products: *energy* and *transport*.[7] TE energy transactions occur at *interfaces*, for example, areas, hubs, locations, or nodes on the grid. TE transport transactions allow a party to transact and pay for transport between an injection interface, where energy for a transaction enters the grid, and a take-out interface, where energy for a transaction leaves the grid. This point-to-point transport product applies to both high-voltage transmission and lower-voltage distribution services. In this chapter, the focus is on transport services between the high-voltage grid and retail customers and decentralized resources.

Critically, TE energy and transport transactions can be used to balance supply and demand and account for losses and grid constraints. For example, a retail party can purchase energy at its delivery interface, such as a business, from a supplier party at a tender price that includes transport; or a retail party can purchase energy at another interface, such as a substation, and also purchase transport from that location to the delivery interface.

The tender price of transport is an all-in, point-to-point price that covers marginal losses, congestion costs, and other fixed and variable costs between two grid locations.

2.3   Subscriptions

A subscription, as the term implies, allows a series of transactions over time. For example, a subscription for a year might be tailored to match the typical patterns of hour-of-day, day-of-week, and month-of-year usage of a particular customer. The cost of the subscription will be based on the price and quantity of each

---

7. The TeMix profile also defines call and put options on energy and transport that act like capacity products. These options are not discussed in this chapter.

transaction that makes up the subscription. Customers will generally pay similar transaction prices for the same interval and interface, but the costs of their subscriptions can differ because of their different usage patterns. A subscription can be a useful construct in communicating the concept of forward transactions to customers.

As mentioned earlier, subscription transactions over longer periods of time are critical to solving the difficult issue of fixed investment and operating cost recovery and avoidance of stranded costs described in other chapters. In contrast to retail tariffs with fixed per kilowatt charges and per kilowatt hour energy charges in a regulated monopoly jurisdiction, subscriptions support longer-term commercial contracts between energy and transport providers.

Ultimately, most consumers and prosumers, whose usage and/or self-generation pattern are approximately known and may be repetitive, are likely to enter into long-term subscription service with counterparties. For example, an energy-intensive industrial user with a predictable load and/or some self-generation may opt for a 10-year subscription for a specified amount and pattern of service. Similarly, a residential customer with solar PVs and an EV may prefer the stability of long-term subscriptions for energy and transport. Such long-term subscription contracts provide a measure of revenue/cost stability to all parties.

2.4   Coordinated Decentralized Control

TE supports real-time, coordinated decentralized control of electrical devices by the users and owners of these devices. Such coordination is accomplished using explicit priced forward tenders and transactions among parties to pay for electric energy consumed or produced by devices.

Coordinated decentralized control is an alternative to both uncoordinated, decentralized control and centralized control of devices. In electric grids with fixed prices for retail customers, for example, there is little coordination of retail decentralized device operation and grid conditions. Centralized control of retail devices conflicts with the desires of retail customers and is very complex and expensive because of the amount of information on device physics and customer preferences that needs to be collected. In fact, centralized control is generally not feasible, and coordinated decentralized control using priced tenders and transactions is the only practical alternative if coordination is desired.

As illustrated in Figure 3, TE supports self-control or self-dispatch, usually automated, of generation, load, and storage devices in facilities owned or controlled by parties that receive the benefits and pay the costs of their operation.[8] Based on the prices of forward tenders, a device management system

---

8. Cazalet, Edward G., and Sastry, Chellury R., "Transactive Device Architecture and Opportunities," Grid-Interop Forum 2012, http://www.temix.net/images/GI12-Paper-12032012-Final_Cazalet_Sastry.pdf.

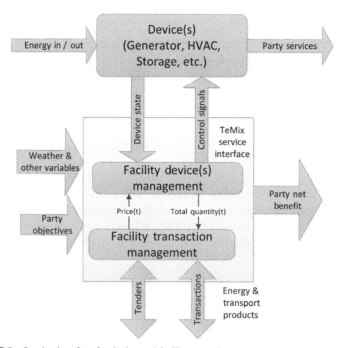

FIGURE 3   Service interface for device and facility control.

automatically manages, optimizes, and controls devices that use, make, or store energy for the party's net benefit. The total quantity of energy used, the net of any production, in each interval is the basis for the facility transaction manager to accept forward buy and sell tenders at levels indicated by their proposed operation as illustrated. Parties can also post buy and sell tenders at prices and in amounts that will be to their benefit.

2.5   Decentralized Transport System Control

DOs can similarly plan, self-control, and optimize the operation of their systems based on forward tenders they receive for distribution transport products while satisfying reliability and physical constraints. They can also post buy and sell tenders for transport products at prices and in amounts that will implement their optimal plan of operation. Ideally, DOs are not permitted to buy and sell energy, except for the energy needed by their systems to cover losses.

2.6   TE Business and Market Processes

TE business processes employ the most basic concepts of ordinary business as illustrated by Figure 4.

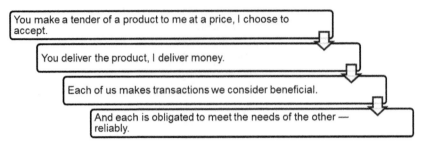

FIGURE 4　Transactive energy business process.

TE supports decentralized decisions and coordination using near-continuous, asynchronous communication[9] of tenders among parties and acceptance of tenders to create transactions. Many different market processes to reach transactions may use the TE model. Different parts of the energy market may employ different market processes.

The complete market process is illustrated in Figure 5 where indications of interest lead to tenders, which may be accepted as transactions; buy and sell transactions for an interval accumulate to a position for the interval; then, actual delivery occurs, which is measured by an interval meter. Any difference between the position quantity at delivery time and the metered quantity results in another transaction with a party that provides balancing service tenders.

Indication	(1) A request for a tender, (2) a forecast of usage or supply, or (3) a forecast of price for an interval
Tender	A price and quantity for an offered transaction with an expiration time
Transaction	Formed by accepting all or part of a tender
Position	Net quantity and net cost for a sequence of buy and sell transactions for an interval
Delivery	The metered quantity delivered

TE may employ a wide range of algorithms and specific business processes including e-commerce for electricity. Formal specifications of the TE business process as well as messages and protocols are critical to enable automation of the transaction.

FIGURE 5　Transactive energy business process stages.

---

9. Most commerce is carried out in markets where transactions can be carried out at any time. This is in contrast to centralized auctions, such as used by ISOs wherein many tenders are transacted at prescribed times.

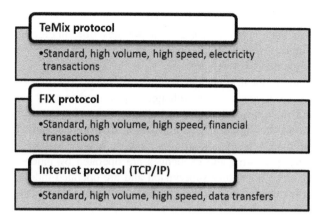

FIGURE 6   The TeMix Protocol and related protocols.

The foundation for the TE business process is the TeMix Protocol.[10] As illustrated in Figure 6, the TeMix Protocol is designed for massive volumes of small electricity transactions, just as the FIX Protocol[11] is designed for massive volumes of financial transactions and the Internet Protocol[12] is designed for massive volumes of data transfer in small data packets.

The TeMix Protocol provides the messaging interface to the TeMix Platform as shown in Figure 7.[13] The two-way, broad arrows in the figure indicate the two-way messaging of tenders and transactions using the TeMix Protocol,

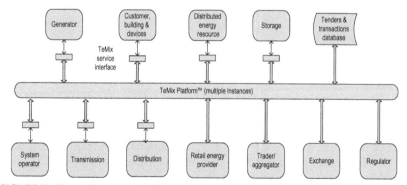

FIGURE 7   Transactive energy platform and interfaces.

10. Cazalet, Edward G., "Automated Transactive Energy (TeMix)," Grid-Interop Forum 2011, http://temix.com/images/GI11-Paper-Cazalet.pdf.

11. http://www.fixtradingcommunity.org/.

12. http://www.ietf.org/rfc/rfc791.txt.

13. http://www.temix.com.

which is an open and free standard. For facilities and devices that require control, the TeMix Service Interface is used, as was described in Figure 3. The control signals and device state feedback for parties that control devices are indicated by the thin two-way arrows.

The platform is not an exchange; it supports bilateral, peer-to-peer, and continuous exchange processes; the exchange is just another party hosted on the platform, and multiple exchanges, including existing futures exchanges and energy auctions, are supported. A transaction platform, including payments and collateral management supporting all parties, may be operated by various parties. Ideally, platforms and exchanges are operated by independent third parties under strict regulatory supervision and are operated independently by all other parties transacting on the platform as proposed in Section 7 of this chapter.

Typically, many instances of the TeMix Platform will be deployed. For example, separate platform instances may be deployed for a building, campus, local feeder, microgrid, substation, neighborhood, city, control area, state, region, and continent. The local platforms will service local parties and devices. Separate instances can be dedicated to transport energy products or to forward transactions with long-duration intervals and spot transactions with 5-min or 4-s duration intervals. Instances may be hosted on secure cloud, utility, substation, pole top, building, and microgrid servers.

Parties with authorization can transact on any platform instance and view their transactions and positions as if all their transactions were made on a single virtual platform. This use of platform instances supports scaling to massive transaction volumes hosted on multiple instances to provide resiliency and decentralized operation for hundreds of millions to billions of smart end point devices!

2.7 TE is a Vision for the Electricity Industry

The TE vision is an evolutionary process from today's grid to a mature TE grid. Different regions and elements of the grid within each region will evolve at different paces. A multistage, evolutionary road map to achieve this TE vision has been developed as illustrated in Figure 8,[14] and a business and regulatory model for this vision is suggested in Section 7 of this chapter.

There are many incremental ways to work toward this vision; the introduction of ITRT is an excellent first step and can be accomplished without a full implementation of TE. Microgrids can implement TE at any time and then interoperate with each other using TE.

---

14. Cazalet, Edward G., "Draft Transactive Energy US Roadmap" October 31, 2012, http://www. temix.net/images/Transactive_Energy_Roadmap_Cazalet_GWAC_103112.pdf.

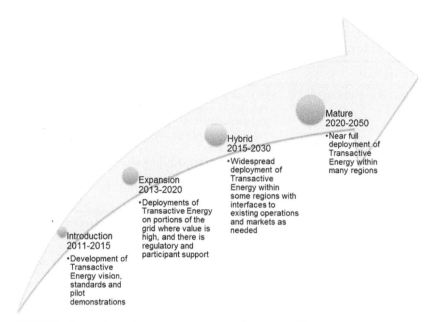

FIGURE 8 Stages of a road map to a mature transactive energy grid.

3 CURRENT RETAIL ELECTRICITY CHALLENGES

As has been described in other chapters of this volume, both cost-of-service (regulated monopoly franchise) and competitive jurisdictions are facing vexing challenges with the rapid rise of decentralized generation, storage, and other options on the customer side of the meter. These challenges can be viewed from the perspective of the different parties as further summarized in Table 1.

TABLE 1 Current Retail Electricity Challenges from Different Perspectives

Perspective	Challenges
Customer	• Need to satisfy different customer preferences and needs
	• Some customers are not interested in energy management, but others are
	• Need to support automated retail energy management
	• Some consumers are becoming prosumers
	• Bidding into markets is not natural for most electricity users
	• Most customers desire lower, stable bills

Continued

TABLE 1 Current Retail Electricity Challenges from Different Perspectives—cont'd

Perspective	Challenges
Retail energy provider (REP)	• Increasing need to manage load to follow rapid changes in renewables supply • Continuing need to manage peak loads • Fixed energy costs are increasing relative to variable costs • Increasing price and usage volatility increases financial risks to REPs • Need automated interfaces to both ISO and non-ISO wholesale markets
Distribution operator (DO)	• Distribution costs are mostly fixed, but revenues vary • Decentralized generation and zero net energy buildings may shift costs among customers • Net metering may increase cost shifts among customers
Decentralized generator and storage	• Many devices have unique characteristics and current capabilities and other uses • Many operators are resistant to centralized dispatch • Resilience, backup, local ownership can be important • Owners do not want to pay for services they do not need
Wholesale supplier and transmission owner	• Increasing fixed costs • Long-term contracts are desired but not always available • Increasing risks of stranded assets in a shift to more decentralized generation and storage
Wholesale market operator (ISO)	• ISOs have little visibility and control over decentralized resources and customer usage • Wholesale market operators' dispatch of decentralized resources may conflict with DOs' dispatch
Regulatory policy	• Continuing need for oversight of both cost-of-service and competitive markets • Proliferation and complexity of tariffs • Need to accommodate policy mandates and subsidies
TE (transactive energy) technology	• Need for solutions that scale to many millions of end points • Need to automate transactions, billing, payment, and customer information • Need for interoperable solutions and business processes

4   HOW WILL AN INTEROPERABLE TRANSACTIVE TARIFF WORK?

This section illustrates an example of an ITRT for REPs and DOs that addresses the retail electricity challenges highlighted in Section 3.

4.1   Customer Interactions with ITRTs

We first present the example from the perspective of retail residential, commercial, or industrial customer with or without decentralized generation or storage. The customer buys or sells commodity energy from or to an REP. In competitive jurisdictions, the customer can select one or more REPs; in cost-of-service jurisdictions, there is one REP. As a departure from most current REP business models, we assume that the REP is both a buyer and a seller of energy with its retail customers.

The customer is also served by a cost-of-service DO that owns and operates the wires and other equipment that transport electric energy between a retail facility (home or business) and a location on the transmission grid such as a substation.

The customer or the customer's automated agent may interact with the REP and DO in a "passive" or "automated" mode or somewhere between. Passive interactions may appeal to customers with small bills and little interest in management of their electricity usage and with little or no self-generation, storage, or EVs. Automated interactions may appeal to larger customers who desire more self-control over their energy usage or those with decentralized generation, storage, and EVs who want to lower the cost of usage and increase payments from any net production. Both passive and active automated customers are therefore served by the same ITRT.

Figure 9 illustrates the interactions with passive customers, and Figure 10 illustrates the interactions with automated customers.

Next, the automation of customer actions in the highlighted, fourth step in Figure 9 is illustrated in Figure 10.

The afore-mentioned example is explained from the perspective of customers, decentralized generation, storage, electric vehicles, and prosumer parties. Next, we will explain the perspective of the REP and the DO.

4.2   REP Interactions with Transactive Retail Tariffs

The process for determining the tender prices posted by REPs will depend on the regulatory model:

- An REP that is a cost-of-service retailer that also owns generation with fixed costs will want to set forward energy subscription tender prices to recover all fixed and variable costs for the energy subscription. The forward tender

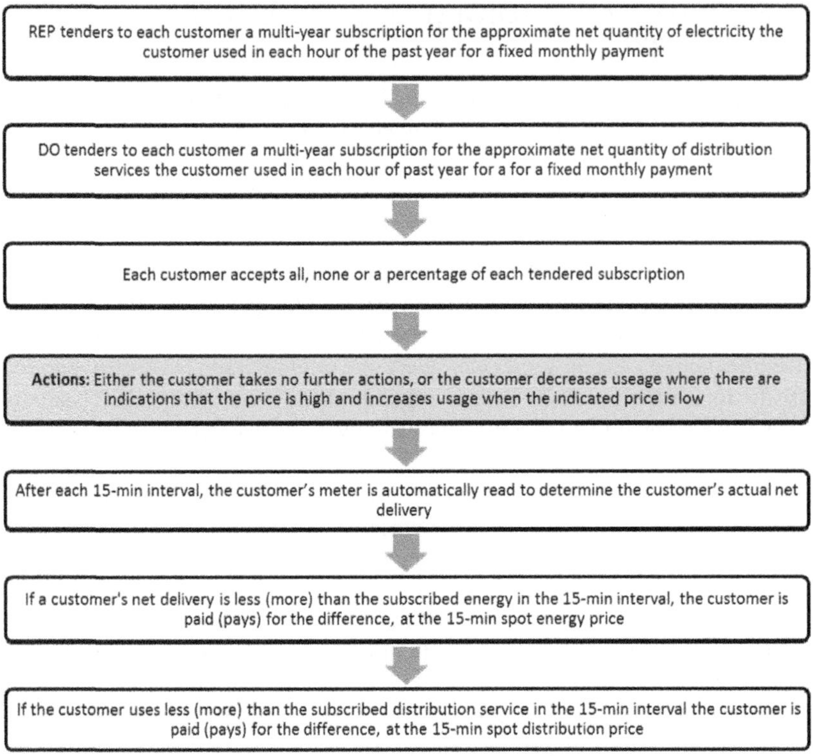

REP tenders to each customer a multi-year subscription for the approximate net quantity of electricity the customer used in each hour of the past year for a fixed monthly payment

DO tenders to each customer a multi-year subscription for the approximate net quantity of distribution services the customer used in each hour of past year for a for a fixed monthly payment

Each customer accepts all, none or a percentage of each tendered subscription

Actions: Either the customer takes no further actions, or the customer decreases useage where there are indications that the price is high and increases usage when the indicated price is low

After each 15-min interval, the customer's meter is automatically read to determine the customer's actual net delivery

If a customer's net delivery is less (more) than the subscribed energy in the 15-min interval, the customer is paid (pays) for the difference, at the 15-min spot energy price

If the customer uses less (more) than the subscribed distribution service in the 15-min interval the customer is paid (pays) for the difference, at the 15-min spot distribution price

FIGURE 9   Passive customer transactive tariff interactions.

prices will be influenced by wholesale forward and future wholesale market transactions by the REP. The spot energy tender prices will likely depend on internal marginal dispatch prices or ISO locational marginal prices plus a regulatory approved dynamic adder that recovers more fixed costs per kilowatt hour when energy usage is high and less per kilowatt hour when energy usage is low. Such pricing will discourage high levels of usage that may increase the fixed costs of energy investment.

• An REP in a competitive jurisdiction will develop a portfolio of wholesale positions that closely matches its forward retail positions in each interval. The REP will post forward and spot tenders to retail customers that reflect current actual tenders or forecasted tender prices in the wholesale markets.

An REP, cost-of-service or competitive, that is a participant in an ISO spot market will post spot tenders to retail customers, based on forecasted ISO prices. The ISO can improve this process by posting hourly forecasted RTPs for the next 24 h and 5-min prices for the next hour. Additionally, the ISO may enhance its systems to post buy and sell tenders at these forward prices that may result in forward transactions with the retailer and by the retailer with

At any time a customer's automated energy management system may decide to reduce or increase energy use, by shifting or reducing usage, or discharging or charging storage depending on the prevailing forward tender prices or changes in personal or business schedules. Accepting tenders adjusts the customer's subscription quantity in the forward interval and locks in the price of the tender

The customer might also buy more if they have recently installed new energy-consuming devices such as an EV, or sell if they plan reduced usage, a vacation, or their own generation such as PV

On the day of energy use, the REP and DO continuously post forward buy and sell tenders for all 15-min intervals in the next 24 h. The customer may have an intelligent energy management system that monitors weather and business or personal schedules

A customer's smart appliance displays the cost of running the appliance now or later. The customer selects when to run the appliance, and their automated energy management agent locks in the forward price of energy and distribution with a forward purchase or sale transaction to align the customer's usage and subscription amounts

Smart energy management systems determine when to run appliances, HVAC, refrigeration, pumping, decentralized generation or other devices based on cost, comfort, other parameters that a customer determines and the prices of the forward tenders. Once the operation of the devices is planned, the energy management system locks in the forward prices of energy and distribution with a forward purchase or sale transaction

FIGURE 10   Optional active customer transactive tariff interactions.

the customers and DER. This will improve the integration of the wholesale ISO and retail markets.

4.3   DO Interactions with Transactive Retail Tariffs

A DO may post tender subscriptions for a fixed share of the distribution grid capacity on a feeder or substation. For example, each of five customers on a feeder may subscribe for a 10-kW share of a 50-kW feeder in each direction although different amounts may be transacted by each customer. The monthly charge for these subscriptions across all customers will recover the total fixed and variable costs of the distribution service.

The spot price for a distribution transport tender will have two elements, a congestion price and a fixed cost recovery price:

- *Variable transport spot tender price.* If a customer uses more than the subscribed distribution transport in an interval, the customer will pay a spot congestion price for the excess transport use in the interval; if the customer uses less, the customer is paid a spot price for the unused transport use in the interval. In this case, the congestion payments from customers will be paid to other customers; the distribution provider is fully compensated by the subscriptions. Typically, the distribution grid will be uncongested and the congestion price will be zero. However, the marginal cost of losses on the distribution grid is always included in the price to customers.

- *Fixed cost transport spot tender price.* A U-shaped function will set the fixed cost recovery spot price of distribution transport to a high price when the total usage by all customers is high and near zero prices when the net usage of all customers is low and a high negative price when the total usage is high in the reverse direction from the normal usage. The U-shaped function will be scaled to recover annual fixed costs over a period of a year, for example.

5   WILL AN ITRT WORK?

An ITRT based on TE addresses many of the deficiencies of current tariffs including the lack of dynamic pricing and forward transactions to recover fixed costs. The typical questions and associated answers in Table 2 provide a way to communicate some of the potential objections and highlight the solutions provided by ITRT.

TABLE 2 Typical Questions About Transactive Retail Tariffs

What if a customer's usage this year is different from last year?	The customer may either receive larger payments if oversubscribed or pay more if undersubscribed. At any time, the customer or the customer's automated agent can increase or decrease the customer's subscribed amounts and pay more or be paid for these changes
What if the spot prices are volatile?	In general, this should n't affect a customer's total bill much, but if the customer adjusts usage automatically to take advantage of high or low prices, then the customer may save more money
Isn't it too complicated to bill all of these transactions?	No, think of this like a phone billing system where each call or text message is a transaction. The customer only looks at the billing detail if they want to verify the correctness of a bill
Why not eliminate the forward subscriptions and just use 15-min spot prices?	Just 15-min prices can be used, but then, the customer's total bill may be much more variable. And REPs and DOs need stable forward subscription revenues to reduce their cost of financing investments, a cost saving that may be passed to consumers who buy forward subscriptions
Can't a customer stay on a flat, full-requirement tariff?	Sure, but the customer may pay more than with the transactive tariff because the REP may pay more for expensive and seldom-used resources to supply the customer when both prices and the customer's usage and other's usage are very high at the same time

Continued

TABLE 2 Typical Questions About Transactive Retail Tariffs—cont'd

Can this be done with hourly or 5-min spot transactions, if a customer asks?	Yes, it will work exactly the same way as for 15-min transactions. Customers with more automation will save more with 5-min transactions; smaller customers with little automation can work with hourly transactions with little increase in cost
How do I know that the energy prices charged or paid are fair or correct?	If an REP is a cost-of-service, regulated REP, then it is the role of a regulator to assure that the energy prices are proper. If an REP is competitive, then such REPs still have regulatory oversight, but if a customer does not like an REP's energy prices, the customer can change to another REP. Or if the regulations allow, the customer can combine subscriptions and transactions from multiple REPs
How does the system guarantee that there will be power available?	A customer can guarantee that they will get power (subject to transport failures) by purchasing forward energy and transport subscriptions for physical delivery to their facility and/or by self-supply, perhaps within a customer's own islandable microgrid. Regulators may also require certain customer classes to subscribe to minimum levels of service. However, for those who overrely on spot purchases, they can expect to pay very high costs for electricity service when demand is high relative to supply. And those with adequate forward purchases may have the opportunity to sell some of their purchases at very high spot prices
Can someone "corner the market" or "manipulate the market?"	Continued oversight of transactive tariffs is necessary and can be supported by data from a transactive platform. Parties may be subject to position limits, and all parties may be required to provide collateral appropriate to the risk they place on counterparties. Forward subscriptions held by millions of customers and many suppliers can limit the ability of parties to corner and manipulate the market. Also, a tariff obligation requiring REPs and large sellers and buyers to continuously offer both forward buy and sell tenders with a relatively small price spread can help mitigate market manipulation
How are customers protected against high charges because of a mistake or a system glitch?	Customers can protect themselves against high charges for any reason by sufficient forward subscriptions. If there is a high price, their automated agents and devices may be able to sell at those high prices. Default prices, default device control rules, and settlement rules can help deal with system glitches and failures. Transaction platforms and communications need to be designed to be simple, decentralized, redundant, and secure
What system gathers prices from various suppliers to determine the lowest-cost supplier?	In the case of a single cost-of-service REP, the REP determines its supply portfolio subject to regulatory oversight. In the case of a competitive REP, each REP determines its lowest-cost supply portfolio, and the customer selects an REP. In the case where subscriptions and transactions can be combined from multiple competitive REPs, each customer determines their lowest-cost portfolio of suppliers

Continued

TABLE 2 Typical Questions About Transactive Retail Tariffs—cont'd

What system assesses potential delivery problems because of feeder capacity limits?	For each delivery interval (e.g., 15 min, 5 min, or an hour), the DO is responsible for monitoring the loading of feeders and other equipment in relation to capacity limits in both directions. The DO supports an algorithm that issues forward buy and sell tenders to all customers. For intervals where forward subscribed usage approaches feeder capacity limits, the algorithm increases buy and sell prices for tenders, which induces some customers to sell distribution services to those customers who are willing to pay more. If there is a surplus of distribution capacity, the algorithm sets buy and sell prices very low to allow customers to take advantage of excess distribution capacity
How does this transactive tariff work for net metered customers with rooftop PV or self-generation?	Because this tariff unbundles distribution and energy commodity services and uses both forward and spot transactions, there is more flexibility than with current tariffs to create tariffs that recover distribution fixed costs and still provide incentives for self-generation and efficient use of the distribution grid. For example, with transactive tariffs, the spot prices of commodity energy in each interval may vary significantly, thus rewarding customers able to sell at high prices and buy at low prices—perhaps also using their own storage and on intervals as short as 15 or 5 min or s. The forward subscriptions for both energy and distribution services can be designed to increase the assurance of the recovery of fixed costs without the inefficiency of lowering spot prices by imposing large fixed charges to recover increasing fixed costs

The proposed ITRT, as indicated by this example, addresses the vexing problems that arise with the rise of decentralized generation such as solar PVs, net energy metering, EVs, and decentralized storage. The benefits of this tariff are further developed in the next section.

6 BENEFITS OF ITRTS

Section 3 outlined some of the current retail electricity challenges. This section describes how ITRTs address these challenges from the perspectives of several classes of parties as presented in Table 3.

7 A FUTURE TRANSACTIVE ELECTRICITY BUSINESS AND REGULATORY MODEL

The ITRT and the TE business model suggest a future business and regulatory model for the electricity sector as outlined in the succeeding text.

TABLE 3 Benefits of Interoperable Transactive Retail Tariffs
from Several Perspectives

Perspective	Benefits
Customer	• Customers may pay lower prices for subscribed electricity because their REP saves money knowing how much is subscribed • In the long run, customers pay less for distribution services because all customers have an incentive to shift usage away from periods when the distribution grid is congested in either direction; this reduces future needs for the DOs to invest money that they must recover from customers • Customers, who shift usage from times when the price is high to times when the price is low, get bill credits to reduce their overall bill. New smart automated thermostats, appliances, and energy management systems help them do this • Customers who buy automated energy management systems and smart appliances can use the forward tender prices to save money • Prosumers with PV, other self-generation, and storage devices can use the forward tender prices to maximize their profits from energy production and storage
Retail energy provider (REP)	• REP revenues will be relatively stable if they sell forward subscriptions with fixed payments to most of their customers to recover the increasing fixed costs of energy • The prices of spot retail tenders can follow the volatility of spot wholesale markets (including ISO markets) while reducing risk to REPs • Customers' net energy metering can be settled on 15-min, 5-min, or shorter intervals. They are offered buy and sell tenders that reflect the actual value to REPs of energy in each interval. If expensive generation is needed in an interval, then the tender prices for energy in that interval can be high. If there is surplus solar PV in a given interval, then the tender prices may be very low or may reflect the costs of storing energy to another interval with a higher price
Distribution operator (DO)	• Regulated cost-of-service DOs with large fixed costs can lock in recovery of their costs with forward subscriptions with their customers where each customer subscribes to a slice of the distribution grid for a stream of fixed payments • Decentralized generation such as rooftop PV by customers and highly efficient buildings may change the pattern of the distribution grid usage toward more two-way flows. Net metering of distribution services on a monthly basis may not properly compensate DOs for the usage by such customers. However, subscriptions for distribution service with a fixed monthly subscription payment can assure fair recovery of costs. And spot transactions at variable prices help reduce the need for new grid investments by paying customers who do not use their subscribed service level when the local distribution grid is congested and charging customers who use more than their subscribed service level

Continued

TABLE 3 Benefits of Interoperable Transactive Retail Tariffs from Several Perspectives—cont'd

Perspective	Benefits
Decentralized generator and storage	• Providers and owners of rooftop PV and other decentralized generator systems can sell their generation and storage services to other parties using both long-term and spot transactions that recognize their location and time of generation • Spot tender prices on short intervals for energy and distribution will help value the flexibility of storage behind the customer meter or on the distribution grid
Wholesale supplier and transmission owner	• Wholesale energy and transport suppliers can use TE forward subscription transactions to stabilize the recovery of increasing fixed costs, thereby reducing the risks of stranded wholesale assets in a shift to more decentralized generation and storage
Wholesale market operator (ISO)	• ISOs can post spot tenders close to delivery that result in a response by decentralized generation, storage, and usage equipped to receive and act on tenders. Feedback of transactions to the ISO will increase the operating certainty for the ISO • DOs will post distribution transport tenders that when combined with retail energy tenders that account for the ISO tenders will allow wholesale retail coordination without conflicting instructions
Regulatory policy	• ITRTs will continue to require legislative and regulatory oversight • ITRTs apply equally well to various customer sectors, thus possibly reducing the number of tariff types. This increases transparency with customers and regulators while reducing information technology costs • ITRTs based on straight forward concepts of forward and spot transactions simplify tariff administration • ITRTs can be implemented on an opt-in or opt-out basis by customers, so customers can learn over time the benefits of transactive tariffs to them • Transactive tariffs provide the benefits of real-time pricing that many regulators desire while also providing customer bill and supplier revenue stability • Legislation can provide subscription subsidies for low-income customers
TE technology	• TE platforms can support hundreds of millions of device end points and customers using standard communications and messages • Providers of intelligent devices and energy management systems can use ITRTs to design and sell products and systems in many jurisdictions with more capability at a lower cost

7.1   Structure of the Transactive Business and Regulatory Model

The suggested structure of the transactive business and regulatory model for the electricity sector is illustrated in Figure 11.

In this future business model, all parties transact with each other using tenders and transactions that are communicated and recorded on transaction platforms. Transactions are for energy and transport products. The four types of entities in this model are the following:

- Energy service parties (ESPs) own and control all facilities for the usage, production, and storage of energy. A customer, prosumer, DER, and generator are ESPs.
- Transport services parties (TSPs) own and control all facilities for transport of energy, DOs, and transmission operators.
- Intermediaries provide exchange, market-making, matching, arbitrage, hedging, and financing services.
- Transaction platform providers (TPPs) furnish transaction platform information processing technology and are not a party to any tenders and transactions or ownership and control of energy and transport facilities.

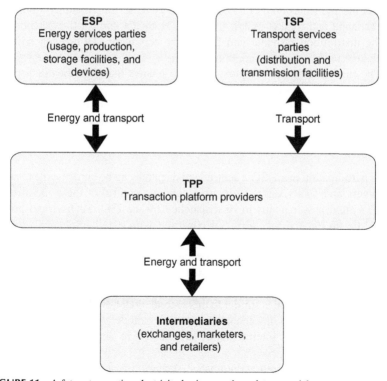

FIGURE 11   A future transactive electricity business and regulatory model.

Oversight of all parties, systems, processes, and tenders and transactions is provided by legislative, regulatory, reliability, and industry associations as in most current regulatory models.

7.2   Differences with the Current Business and Regulatory Models

The principal differences between this transactive business model and the current model are the following:

- Vertically integrated utilities combine all of the functions in Figure 11, except for customer and DER functions not owned by the utility. Such utilities can benefit by deploying ITRTs to better interface with consumers, prosumers, and DERs.
- Bundled energy and distribution utilities also combine the functions in Figure 11, with the exception of the transmission-level function. Such utilities also can benefit by deploying ITRTs even with no change in their business model.
- In restructured jurisdictions, such as in Texas with an ISO, cost-of-service DOs, and competitive energy retailers, the retailers are intermediaries, and the DO is the TSP for distribution transport. The ISO is the transmission TSP and the transmission TPP. The ISO also provides intermediary wholesale exchange and market-making services. Both the competitive retailer and the distribution provider can benefit by deploying ITRTs even with no change in their business model.
- The current ISOs use complex, multipart tenders that describe the dispatch characteristics of various generation, storage, and usage facilities. This complexity is required to carry out a centralized dispatch of these facilities. TE avoids this complexity using only single-party tenders and self-dispatch of generation, storage, and usage. Nevertheless, TE can interface with existing ISOs.

Since participation in an ISO's centralized dispatch can be burdensome to retail customers and DERs, aggregators may combine control of these facilities to create virtual power plants to be dispatched by the ISO. Such virtual power plants are an unnecessary artificial construct with a full implementation of TE and the conceptual industry structure in Figure 11.

Kristov and Hou in Chapter 18 describe two future models: One is where the ISO expands its scope to all DER at 50-100 kW. The second model creates an ISO-like entity to operate the distribution grid and distributed resources. Hanser and van Horn's proposal is similar to this second model. Both models, however, are very different from this chapter's transactive model where the functions of TSP, ESP, TPP, and intermediaries are hosted in separate, decentralized entities.

Felder's "decentralization dominates" scenario is consistent with this transactive business model, although he has DOs that manage distribution grid generation, which is self-managed in this transactive model. Keay et al. (Chapter 8) also have the DOs that manage distribution grid generation, which they call a DSO.

It is clear that achievement of the full benefits of the ongoing transition to decentralized resources that is the theme of this volume will be impeded by any further centralization of grid management. The decentralized revolution sweeping the industry will be better supported by a decentralized business and regulatory model as illustrated in Figure 11. Hosting each of the functions in the model independently assures the independence of decisions in the decentralized industry. Major parties in the current industry structure who adopt a decentralized, transactive business model and those regulatory agencies and legislatures who support or mandate decentralized business and regulatory models will be in a much better position going forward.

8   CONCLUSIONS

This chapter proposes an ITRT that is an elegant, practical solution to the vexing challenges facing the electricity sector with the rise of DERs. The tariff is built on the TE foundation, which also offers a decentralized future business and regulatory model for the electricity sector.

In summary, the proposed ITRT, together with TE, will

- avoid the need for large retail tariff fixed charges and instead use forward transactions and subscriptions with fixed payments for fixed services to provide more stable cost recovery for utilities than large fixed charges,
- overcome much of the customer and regulatory resistance to retail spot dynamic pricing by combining it with forward subscriptions that provide stable customer bills,
- avoid the complexity and cost of specialized retail tariffs for specific groups of parties or types of usage, generation, and storage and accelerate the demand for smart devices and systems and the development of the smart grid and support a decentralized business and regulatory model that will accelerate innovations in decentralized resources and smart energy usage.

The ITRT and TE do not require massive new investment—it is a regulatory and tariff solution that involves public decision making and information technology investment. In most cases, the new tariffs will provide the correct incentives for private innovation and investment. The trend toward decentralization in the electric industry is established. ITRTs will support and accelerate the inevitable transition to a decentralized electric grid.

Chapter 11

The Next Evolution of the Distribution Utility

Philip Hanser and Kai Van Horn

ABSTRACT

The electricity supply industry has undergone some of the most significant structural changes over the last two decades, and for electricity distribution utilities (EDUs), the greatest apparently are yet to come. Chief among the drivers is the widespread implementation of distributed generation (DG), a development partly driven by customers' reliability concerns and partly by economics, but which will radically redefine the core mission of EDUs. Despite attempts by EDUs to stem the financial bleeding resulting from decreased sales volumes, EDUs continue to be faced with bleak sales revenue projections. The potential impacts of DG on their livelihood cannot be addressed by mere tariff redesign, but rather requires a redefinition of their fundamental purpose and their structure. This chapter lays out the challenges facing EDUs as the distribution system's functions and customer's use modes significantly transform and envision alternative EDU business models to address those challenges.

Keywords: Utility business models, Distributed generation, Distribution systems, Retail rates, Net energy metering

1 INTRODUCTION

The distribution system is undergoing a transformation that is driving a fundamental shift in the role of the electricity distribution utility (EDU)[1] and the expectations of electricity consumers. On the one hand, decreasing demand growth rates; deepening penetrations of distributed generation (DG), energy efficiency, and demand response programs; and increasing retail rates have coalesced into a growing threat to EDU bottom lines and long-term business viability through their impact on EDU's volume-based tariffs. On the other hand,

---

1. In this work, EDUs are defined as those companies tasked with the delivery of electricity on the distribution network from the high-voltage transmission network to the end-use customer. This task is typically assigned as an exclusive franchise to a government-owned or cooperative enterprise or to an investor-owned utility subject to some form of price regulation (Kassakian et al., 2011).

Distributed Generation and its Implications for the Utility Industry. http://dx.doi.org/10.1016/B978-0-12-800240-7.00011-4
231

as the installation of DG, in the form of technologies such as solar photovoltaic (PV) and microturbines (MTs) or fuel cells with combined heat and power (CHP), continues to gain traction and the adoption of electric vehicles and community-level battery storage becomes more widespread, the conventional pattern of power flow *from* the transmission system *to* the distribution system is giving way to bidirectional, more volatile distribution system electricity flows.

As described extensively throughout this volume, the increasing number of customers who self-supply or participate in energy efficiency or demand response programs and the increasing number of transactions that take place among customers themselves on utility distribution systems will slash EDUs' sales volumes. To maintain their revenue requirements[2] and avoid stranded distribution investments,[3] EDUs will be forced to raise their incremental rates, resulting in further pressure on customers to reduce volumes, a positive feedback process shown in Figure 1 that has ominously been referred to as the "death spiral." Furthermore, the upward price pressure put on retail rates by the DG supplanting load on the system will fall largely on those customers who do not self-supply, raising equity issues. The interaction between DG implementation, retail rates, and declining demand and the implications for utility revenues cannot be adequately addressed with existing business models.

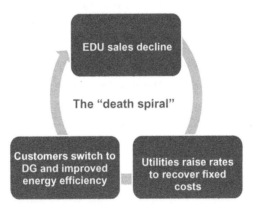

FIGURE 1   The EDU revenue death spiral.

---

2. Most of today's EDUs profit through a business structure based on a regulated rate of return on capital investments recovered through kilowatt-hour charges to their customers. The costs of doing business plus a reasonable return on investment—collectively referred to as the EDU's "revenue requirement"—are subject to review and approval by state-level public utility commissions and used as the basis for the determination of the level of the per kilowatt-hour charge.
3. Asset stranding is also likely in wholesale markets as a result of these trends, but that market is not the focus of this chapter.

To avoid being drawn into a death spiral, EDUs must reimagine their conventional role as solely the "local streets" for kilowatt-hours and explore new business models for meeting their financial needs outside those of today. Further, EDUs must proactively face and embrace challenges to their conventional role in the distribution system. Other challenges include the rapid adoption of new technology, such as advanced metering infrastructure, which is already taking place, and other communications and control technologies.[4] Enabled by such technologies, EDUs must broaden the scope of their interactions with customers offering new services, such as reliability-differentiated service, heat and cooling, and backup power, and begin to share the regional transmission organization (RTO)[5] task of balancing an increasingly complex power system.[6]

This chapter delineates the visions of the distribution system of the future and outlines the challenges EDUs will face in that future. Moreover, it puts forth possible paths forward for the EDU. The work provides context for the transition taking place in the electricity system at the distribution level and, hopefully, provides insights into what EDUs require to remain a relevant posttransition electricity system player. While this work focuses on EDUs operating in regions balanced by an ISO/RTO, many of the same challenges—and opportunities— are present for EDUs in vertically integrated utility structures.

In the remaining four sections of this chapter, the role of the EDU today and in the future and four alternative business models for EDUs to remain relevant in their new role are described. In Section 2, attention is drawn to the drivers of the changing role of the EDU and the incentives behind those drivers. In Section 3, a vision for EDUs' future roles is laid out. In Section 4, four EDU business models in this future role are discussed. In the final section, the key challenges faced by EDUs and opportunities for new business models are reiterated and major conclusions drawn.

2   THE CURRENT STATE OF THE EDU

The conventional role of EDU is to provide kilowatt-hours of electricity on demand to the customers it serves for which customers pay a volumetric rate. The magnitude of these per kilowatt-hour payments has historically been determined by EDUs and the state public utilities commissions (PUCs) through a regulated rate-setting process.[7] EDUs have reliably provided electricity

---

4. The impacts of new technology on the role of the EDU are laid out in Chapter 5 by Gellings.
5. The same will hold true of EDU's in areas balanced by an independent system operator (ISO). However, for the sake of brevity, only RTOs are referred to in this chapter.
6. For an in-depth treatment of the RTO-EDU interface, see Chapter 18 by Kristov and Hou.
7. In the rate case process, EDUs propose volumetric rates that allow the EDU to meet its revenue requirement including a reasonable rate of return on their investments. EDU rate cases are subject to PUC review and approval. The rate case process typically takes place with multiyear periodicity. The rate determined through this process is largely fixed from one rate case to the next save adjustments indexed to fuel prices or large deviations from expected consumption.

service and met their revenue requirements through volumetric charges for more than half a century. The success of the volumetric rate model can be attributed in large part to consistent decreases in the cost of service over time driven by consistent and significant year-upon-year demand growth. EDUs made capital investments to maintain reliability and meet their obligation to serve existing and new customers, recovering the cost of those investments primarily through the volumetric charges. Because electricity demand was consistently growing, investment and business costs were spread over an increasing volume of kilowatt-hours, thus keeping rates relatively stable.

The days of reliable and meaningful growth in electricity consumption, however, seem to have come to an end (Faruqui, 2013). The forecast for future growth in annual electricity consumption is projected to be in the neighborhood of 0.7% annually (EIA, 2012a, 2012b). As discussed in detail in Chapter 15 by Faruqui and Grueneich, the projected low demand growth is attributable to changes in consumer psychology and energy efficiency programs and changes in energy-use-related building codes and appliance standards, DG, and fuel switching. Figure 2 shows the electricity industry retail sales revenues—in real 2012 dollars—for the past 20 years and projected forward to 2020.[8] It is clear that the retail sales revenue outlook is grim at best. Furthermore, just as EDUs are facing the challenges of declining sales volumes, a shrinking customer base and potentially stranded assets and the needs for investments in communications, control, and electric infrastructure in the distribution network are on the rise.

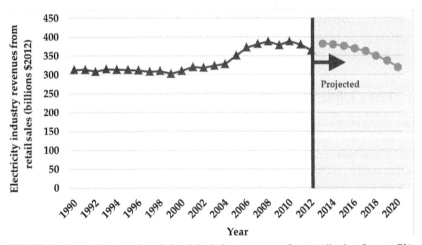

FIGURE 2   Historical and projected electricity industry revenues from retail sales. *Source: EIA, Electricity Data (2012a).*

---

8. The projected revenues were obtained from a cubic regression—for which the coefficient of determination is 0.92—of the historical revenues over the years from 1990 to the present.

The challenges EDUs face today are propelled by a number of drivers that have been evolving at an accelerating pace over recent decades. The primary drivers have been significant cost reductions in installed solar PV and other forms of DG now paired with regulatory support for jump-starting such technologies' industries, through programs such as net metering. Complementary developments—driven by deepening penetrations of intermittent and variable energy sources—are reductions in the installed cost and improvements in the lifetime of batteries and other energy storage devices. A third major driver of the transition has been relatively low natural gas prices, which have increased the economic attractiveness of MTs and fuel cells, especially for CHP applications. In the following paragraphs, a brief overview of these developments is provided. A comprehensive overview of the developments is given in Chapter 7 by Sioshansi.

The costs of installed solar PV have been cut in half over the past decade as a result of improved cell and system component efficiency accompanied by decreases in manufacturing and installation costs (Barbose et al., 2012). Further, policy makers at the national, state, and local level have enacted generous subsidy programs, including tax credits, production-based incentives and net metering, or up-front cash rebates, to provide further incentives for electricity customers to invest in solar PV to support existing renewable portfolio standards or other policies aimed at reducing the carbon intensity of electricity supply (DSIRE, 2013). Such programs have jump-started the solar PV industry and put it on a path to become a major source of electricity.

At the time of this writing, there is over 9370 MW of installed solar PV capacity and about half of the newly installed capacity is distributed solar PV—solar PV is one of the fastest-growing sources of new capacity (GTM and SEIA, 2013). In some EDU jurisdictions, the cost of solar PV—on a levelized cost of energy basis—is already competitive, in some cases without subsidy, with the retail rate for electricity, thus providing a strong financial incentive for customers to invest in solar PV systems and, at least in part, go "off the grid" (Lazard, 2013). Large companies such as Google and Walmart have sounded the solar battle cry and have committed to generating a portion of their energy with DG. Walmart, for example, which already self-produces 4% of its electricity with solar PV, has a target of self-producing 20% of its electricity from solar PV by 2020 (Smith and Sweet, 2013). Figure 3 shows the total US installed PV capacity and its prices over the past decade and the projected capacity and its prices in the next half-decade illustrating the expected decline in price and corresponding steep increase in installed solar PV capacity.

Complementary drivers to the cost reductions in solar PV are the technological and cost improvements that have been made in battery and other energy storage technology. Driven in part by the resurgence of interest in the deployment of electric vehicles, battery costs are coming down while lifetimes and energy densities are increasing. In addition, there are a number of innovative storage solutions, such as sodium sulfur batteries, in the pilot testing phase that

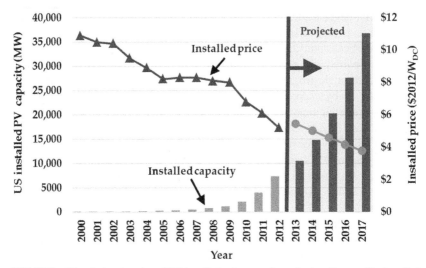

FIGURE 3   Historical and projected US installed solar capacity and price. *Source: The installed price is the average price for residential installations in the United States. The installed price of commercial- and utility-scale installations are typically $1-2 less per watt (Barbose et al., 2012). Further, the projected installed prices are estimated from a linear regression of the historical installed prices; the regression coefficient of determination is 0.94. EIA (2012a, 2012b), GTM and SEIA (2013), and Barbose et al. (2012).*

may prove to be transformative technologies for grid-scale storage (Akhil et al., 2013; Sandia, 2013).

A recent EPRI study found that the benefits of grid-scale batteries exceeded the costs in several applications—most notably to provide ancillary services, such as regulation and spinning reserve, or to provide bulk energy storage at the distribution level (EPRI, 2013). Furthermore, of the applications studied, the distribution-level applications had the highest value proposition. Distribution-level battery storage, whether through stand-alone battery storage installations or through aggregations of electric vehicles, facilitates deeper penetrations of DG contributing to the challenges EDUs face. While battery technology ramps up to meet the time-dependent power demands of consumers, MTs are making a comeback as a form small-scale, distributed, controllable generation.

Both MT and fuel cell technology improvements and persistent low natural gas prices, driven by the discovery of substantial domestic formations of shale gas, have considerably increased the economic attractiveness of MTs and fuel cells for medium-scale (100 kW-1 MW) commercial- or community-level applications. Localized thermal generation has an inherently attractive feature: it brings with it the opportunity to utilize the electricity-generating process' "waste" heat and thereby increase the overall cycle efficiency. The overall efficiencies of such CHP applications can reach as high as 80%; such efficiencies have important implications for the economics and environmental impacts of energy needs satisfied through CHP.

In New York City, for example, 28 MTs, with an average capacity of 250 kW, have been installed since the year 2000 by commercial and residential buildings seeking to reduce annual electricity and energy payments and increase electricity supply reliability (DOE and ICF International, 2013). Further MT and fuel cell cost reductions and the persistence of low natural gas prices may drive a rapid expansion of these resources at the community level. The widespread adoption of MTs and fuel cells—which are largely behind the meter—has the same impact of widespread distributed PV implementation on eroding the EDU bottom line.

Falling installation costs for solar PV, MTs, fuel cells, and battery technology are driving an explosion in the proliferation of such technologies and fueling a distribution system paradigm shift. This shift is also driven by a number of public policies implemented over the past two decades that provide strong incentives for the expansion of DG and increasing consumer demand for self-supply for reliability purposes. A prominent example of such a policy is net metering.

Net metering[9]—the implementation of which was made mandatory for all electric utilities in the Energy Policy Act of 2005 (EPAct, 2005)—enshrines a right for consumers to reduce their payments to the EDU by self-generating and, in some jurisdictions, to receive payment at the retail rate if those customers generate more than they consume. Since the passage of EPAct, 43 states and the District of Columbia have implemented net metering policies. Eligibility for net metering credits has been restricted to systems no greater than a few MW. Despite these limits, net metering continues to be a major force driving consumer investment in DG.

However, from the EDU perspective, the incentives provided by net metering distort the cost of energy supply and introduce cross-subsidies that result in inequities between those customers who do and those who do not install DG. Further, such cross-subsidies distort the incentives for customers to invest in DG leading to overinvestment in DG to the detriment of the EDUs bottom line (Price et al., 2013; Wood and Borlick, 2013).[10]

An increasingly important component of the incentives for customers to adopt DG is their desire to ensure reliability and achieve a measure of self-reliance.[11] Major blackouts due to weather-related events, such as hurricane

---

9. Under EPAct, net metering "means service to an electric consumer under which electric energy generated by that electric consumer from an eligible on-site generating facility and delivered to the local distribution facilities may be used to offset electric energy provided by the electric utility to the electric consumer during the applicable billing period" (109th Congress of the United States, 2005).
10. An extensive discussion of the issues raised by Wood and Borlick (2013) is provided in Box 7 of Chapter 7 by Sioshansi.
11. Further discussion of the increased importance and value customers place on self-reliance is given in Chapter 3 by Burger and Weinmann.

Sandy in 2012, have increased the value customers place on self-generation via DG to reduce their risk of losing access to electricity. Such events also galvanize regulators to pursue bold policies towards grid development (President's Council of Economic Advisers, 2013). With the apparent increasing frequency of extreme weather-related events in the past decades, the incentive to self-generate for reliability purposes may become a significant component of the erosion of the EDU revenue base, and EDUs may find regulatory bodies amenable to policies that promote the growth of DG to build robustness into distribution system as a hedge against future outages.

The widespread implementation of DG and the expansion of energy efficiency programs are likely to continue for the foreseeable future. The incentives driving these developments, such as low natural gas prices and falling solar PV installation costs, provide persistent pressure for customers to invest in such technologies and participate in such programs. These pressures and their outcomes are transforming the distribution system and will fundamentally redefine the role of the EDU in that system. In the following section, a vision for the electricity system of the future is described.

3   A VISION FOR THE FUTURE ELECTRICITY SYSTEM

Today's distribution systems were designed to provide for the unidirectional flow of power from large-scale, centralized generation facilities on the transmission system to loads on the distribution system. In restructured areas of the electricity system, RTOs maintain the supply-demand balance around the clock by scheduling generation resources to meet the loads, the majority of which have up to now been passive, price-insensitive consumers. The EDU participates in maintaining the supply-demand balance insofar as it procures electricity in the RTO-run markets to meet its obligation to serve the load within its jurisdiction. There is little participation from the EDU in controlling resources to meet the supply-demand balance. Distribution systems built to serve the load are radial in nature; there are few interconnections between customers or groups of customers on the distribution system and few customer-to-customer transactions on the distribution system. Up-to-now, due to the limited penetration of distribution-level generation, such generation has largely had a negligible impact on reliability and on the flow of electricity across distribution-level substations. Thus, the need to explicitly consider and control such resources in operations and planning has not yet been a priority for EDUs and system operators.

The paradigm of large-scale, centralized generation serving the load and unidirectional power flows across the substation is on the cusp of a tectonic shift. The pressure put on existing distribution infrastructure by the widespread implementation of DG, namely, the emergence of bidirectional, and more volatile power flows on the distribution system is driving a revision of the structure of that system. In the new paradigm, each distribution system resembles the transmission system of today: a highly meshed system—one in which there are a large number of customer and customer group interconnections—with thousands of load, storage, and controllable generation resources, such as MTs and fuel cells, and many noncontrollable generation resources, such as distributed rooftop solar PV.[12]

Further, the loads become active participants in maintaining the supply-demand balance adding to the complexity of the grid-balancing problem. The adoption of electric vehicles could make available vast quantities of grid-connected energy storage that will be used to help balance the output of ubiquitous household- and small commercial-level DG. Moreover, community-level MTs and fuel cells will make available local sources of electricity and heat for the needs of local processes. The distribution system will become more independent from the transmission system for its electricity needs, and the transmission system will become largely a conduit to deliver large-scale renewable generation from remote, high-quality renewable resource locations and provide geographic diversity between renewables on the distribution systems. Figure 4 is a pictorial representation of a vision for the future electric distribution landscape.

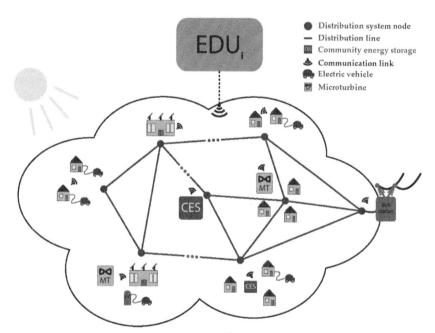

FIGURE 4   The distribution system of the future.[13]

---

12. In Chapter 5 by Gellings, a similar vision for the future distribution grid, termed the "local energy network," is explored.
13. For clarity in Figure 4, only a subset of the distribution nodes, as indicated by the purple ellipses, are shown. Further, elements of the distribution grid, MTs, CES, EVs, customers, etc., are assumed to be connected to the node to which they are closest and to have a communication link with the EDU.

With a large proportion of electricity demands met via DG resources that output DC power, such as solar PV and battery storage, there will also be pressure to overhaul the internal structure of customer loads. To eliminate the losses in converting from DC to AC and back to DC, a typical distribution customer may have two electricity systems: one system operating on low-voltage DC to meet the increasing proportion of customer loads that are electronic, such as LED lighting, and a second system operating on 240 V AC to efficiently meet the demands of motor loads, such as washers and dryers and refrigerator compressors. Such a hybrid AC/DC system, while more complex than today's single AC systems, offers loss reductions and reduced complexity at the end-use level as the converters in each device can be eliminated.[14] There will also be pressure on automobile manufacturers to standardize their electric vehicle batteries and design them to work as household storage systems.[15]

In the emerging paradigm, the balancing of the large number of controllable resources at the distribution level and reliability monitoring will take place both on the level of each distribution system and at the level of the transmission system. Distribution-level control will be facilitated by the widespread implementation of communications infrastructure in the electricity system. The prospect of tens of thousands of DG resources, which have a significant impact on the net load, will require a fundamental rethinking of the division of control between RTO and EDUs in order to manage the scale of the balancing and reliability problem. Figure 5 shows a possible division of control between the RTO and the EDUs in the system of the future.

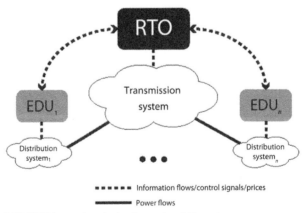

FIGURE 5   RTO-EDU interactions in the future electricity system.

14. See the recent article in *The Economist* regarding the USB (http://www.economist.com/news/international/21588104-humble-usb-cable-part-electrical-revolution-it-will-make-power-supplies).
15. Currently, automobile manufacturers have expressed reluctance to enable the ready use of EV batteries for local grid storage.

The RTO will maintain its duty to balance resources on the bulk transmission system. In line with the second scenario outlined in Chapter 18 by Kristov and Hou, the major difference between the control approach today and the future control approach described here will be the prominent role played by EDUs in the future in balancing supply and demand on their local distribution systems. The EDU, taking on the role of a mini-RTO, will coordinate with the RTO to balance supply and demand and maintain adequate reliability for customers in its jurisdiction. The EDU focus will shift from meeting the load to meeting the net load, which in some cases may even be negative.

In this future, each repurposed distribution system can be a load or a source of generation from the perspective of the RTO, and the EDU will be an important bridge to economically and reliably coordinate distribution-level resources with those on the transmission system. A key challenge will be to develop effective hierarchical control strategies and means of sharing information between the RTOs and EDUs.[16] Data management of the vast quantities of data from the thousands of sensors and distributed resources will also present a significant challenge in the future. Table 1 summarizes the differences between the electricity system of today and that of the future.

The future system described here is not without precedent. Distribution companies in countries such as Denmark and the Netherlands are already experimenting with alternative models of balancing distribution systems with deep DG penetrations in ongoing pilot projects (eLab, 2013). Germany, which has a very deep penetration of distributed solar PV, is developing a body of experience in local balancing from which US-based EDUs can draw.[17]

The question faced by EDUs is not whether the electricity system will undergo a significant transformation, but rather what it will look like after that transformation and how EDUs can position themselves with respect to the coming changes. In the next section, several EDU business models are proposed as a means by which the EDUs can adapt to the vision put forth in this section.

4   THE FUTURE BUSINESS OF THE EDU

The decline of the EDU is by no means inevitable and there are numerous opportunities for EDUs to repurpose themselves to maintain relevancy and profitability in the future electricity system. For examples of companies that have successfully weathered the storm of disruptive technological change, EDUs need only look to the success of telecommunications companies such as AT&T and

---

16. Chapter 10 by Cazalet describes an alternative, completely decentralized approach, referred to as the "transactive retail tariff" to meeting the challenges of the future system.
17. The developments in the German system are covered extensively in Chapter 3 by Burger and Weinmann. A broader assessment of the impacts of widespread DG implementation in the EU is provided in Chapter 6 by Groot.

TABLE 1 The Electricity System of Today and that of the Future

Key Electricity System Features	Today	In the Future
System structure	Radial distribution systems; strong transmission system	Highly meshed distribution systems; modest transmission system
Responsibility control/ balancing/ reliability	Primarily RTO duty	Duty shared by RTO and EDU
Source of generation	Large-scale, centralized facilities	DG, large-scale remote renewables
Load participation	Passive, consumers of electricity	Active participants in maintaining the supply-demand balance
Role of energy storage	Limited role, mostly pumped hydro	Major role in balancing output from variable and intermittent DG and providing ancillary services
Flow of power across the substation	Unidirectional, transmission system to distribution system, low level of uncertainty in magnitude	Bidirectional, lower in magnitude, higher volatility/ uncertainty in magnitude
Customer internal structure	AC system, many converters	Hybrid AC/DC system
Customer expectations	Electricity available when called upon	Electricity available when called upon, open access to distribution system

Verizon in the past two decades as that industry underwent a transformative shift driven by mobile phone technology.[18]

The emergence of affordable mobile phone technology catalyzed the decline in demand for wired-phone service. Telecommunications companies faced a shrinking base of customers from which to recover their investments in the infrastructure to support wired-phone service and the risks associated with prospective investments to support the development of mobile networks. By embracing

---

18. For additional examples of disruptive technological change and companies that have successfully and unsuccessfully managed the transition to a new industry paradigm, see Chapter 7 by Sioshansi.

mobile phone technology, such companies have maintained their relevance and continue to be leaders in their industry, to grow, and to be profitable (Kind, 2013). While the wired-phone network exists and is still utilized today, many people rely on their mobile phones as their primary method of communication and maintain their wired service as a backup against loss of service in the wireless network—analogous to the way DG customers could potentially use the distribution grid.

The changes taking place in the utility industry are akin to those that have taken place in telecommunications: long-standing incumbent service providers with significant existing capital investments are under pressure from transformative technology and competitive pressure to redefine their role. For EDUs to repeat the transitional successes of the telecom industry—for example, to follow the path of companies such as Verizon instead of that of companies such as Kodak—they must embrace the new reality of DG, jettison the aspects of past business models that are no longer applicable, and adopt new business models and, with them, new tariff structures.[19] EDUs must rethink and diversify the range of services they offer to customers and recognize changes in the distribution of value customers place on their various forms of energy use and be adept at developing innovative ways to capitalize on the highest valued among them.

For example, the grid provides DG owners with services, such as regulation and voltage control, a means to inject energy during the hours of excess generation and withdraw energy during the hours of deficit generation, and backup power when on-site generation is inoperable due to equipment maintenance, unexpected physical failure, or prolonged overcast conditions, for which they do not yet pay (Wood and Borlick, 2013). These services must be monetized and integrated into the EDU business model to fully capture the costs imposed by DG on the distribution system in their connection charges and provide correct incentives for DG investment. The monetization of such services, however, does not fit well into the current volumetric model.[20] Recovery of the costs of the range of new investments in electric and control and communications infrastructure at the distribution level to realize the future grid will require a fundamental change in the way consumers pay for their connection to the distribution system.

With declining electricity sales volumes, it is clear that volumetric charges alone, which has been an industry standard for decades, are no longer a tenable revenue recovery mechanism. There is some experience in the United States and abroad with tariffs that rely less heavily on incremental charges. One such approach is the so-called straight-fixed variable (SFV) rate. Under such a rate, the fixed investment costs of the EDU are recovered through a fixed charge and

---

19. An interesting tariff structure, termed the "transactive retail tariff," that is a radical departure from the status quo is discussed in Chapter 10 by Cazalet.

20. Chapter 13 by Nelson et al. provides an in-depth treatment of the challenges and limits of tariff design in the future system.

the variable costs, such as energy market payments or fuel costs, are recovered through a variable rate. SFV rates have typically been restricted to the realm of industrial customers. EDUs in some states in the past 5 years, however, have begun to experiment with SFV rates for other customer classes (Newton-Lowry et al., 2013). Experience with such rates is also accruing to utilities abroad. In the United Kingdom, the "use of system charges"—based on an administratively determined cost of maintaining the distribution system—is socialized to all electricity users (eLab, 2013). While fixed charges address the EDUs investment and revenue adequacy issue and are an important component of the future EDU business models, they do not capture the full range of income opportunities for EDUs in the future and address all of the challenges of the changing distribution grid.

EDUs must transition out of the role of being primarily a conduit for electricity purchases and step into the new role of a mini-RTO, balancing the thousands of new resources on the distribution system up to the point of injection with the transmission system. As consumers begin to use the grid mainly as a backstop against generation shortfalls from DG, EDUs will become primarily wire companies responsible for the maintenance, control, and reliability of a robust and interconnected distribution network. The distribution system will take on a "common carrier" aspect as those thousands of new resources interconnect and interact. EDUs or their affiliates will be able to offer new services, such as reliability insurance, heat for thermal processes, and DG leasing. The new role of the EDU will require a new business models.

There are a large number of possible future business models that would mitigate the financial impact of DG on EDUs' bottom line.[21] In the succeeding text, four models are described.

4.1   The "Local Reliability Provider" Model

Under the local reliability provider model, EDUs become the "reliability provider of last resort." Borrowing from established actuarial practices in the insurance industry, EDUs will determine the reliability needs of customers depending on their equipment and assess fixed charges accordingly to ensure a customer's desired level of reliability.[22] In other words, instead of paying an incremental fee for energy consumed, customers will pay a fixed monthly fee for their desired level of reliability. Under this business model, unlike current models, the EDU will offer reliability-differentiated service.[23]

---

21. The results of a survey of utility executive perspectives on the future of EDU—showing that most utility executives believe strongly that DG will play a large role in the future electricity supply—are given in Chapter 2 by Nillesen et al.

22. See, for example, http://www.actuarialstandardsboard.org/ for information on insurance companies' current approaches.

23. The success of reliability-differentiated service offerings depends on the implementation of customer-level communication and control infrastructure.

A key strength of this model is its explicit recognition of the changes in the way consumers use the distribution network and of a new EDU value stream: the provision of reliability as opposed to the provision of kilowatt-hours. Furthermore, the model can be extended to the provision of other distribution-level services, such as heating and cooling. This model breaks from the current practice by transitioning from revenue recovery primarily through incremental consumption charges to revenue recovery primarily through fixed charges. Fixed charges will allow EDUs to recover their investments in the system despite changes in the ways customers use the system. Furthermore, the fixed customer charges, by explicitly integrating customer-specific information about their equipment, will reduce the cross-subsidy, which arises under current tariff structures in systems with DG and net metering. The local reliability provider model makes a direct connection between customer reliability demands and the investments required to ensure those demands are met.

4.2   The Membership Model[24]

The membership model is based on the idea that all users of electricity have a stake in the existence and maintenance of a robust distribution system. Under this model, the distribution system is thought of as a "club" and users of the system pay fees to be a "member." The fees customers pay would be proportional to connection voltage, peak demand, size of DG system, and/or some similar metric(s) to account for differences between customer use modes and equipment. The fees could also be differentiated to account for customers who do not have DG or those who have backup batteries and customers charged or credited accordingly. By differentiating fees by customer equipment, this model allows the EDU to reflect more closely in the rates paid by consumers the costs of the system services they utilize—including those services, such as ancillary services and backup power—that are not accounted for under current rate structures. As is true with many clubs and social networks, the strength of this model will likely grow as the number of members increases and electricity users feel increasing social pressure to participate.

A major benefit of such a model is the simplicity with which it could be implemented. Unlike the local reliability provider model, this model requires no risk assessment. Instead, the revenue needs of the EDU would be determined, and the charges for each class of customers chosen on the basis of a predetermined metric to ensure the EDU's revenue requirements are met. These membership charges would be reevaluated periodically to ensure the EDU's revenue requirements are met but not exceeded. Like the local reliability provider model, the membership model relies primarily on fixed charges to meet the EDU's revenue requirements and thus addresses the same issues with declining sales volumes addressed by the local reliability provider model.

---

24. This might also be viewed as a "Network Model."

4.3   The DG Ownership Model

Solar PV development via third-party ownership of distributed solar PV is a rap-idly growing industry in the United States. The third-party solar PV development model has so far taken primarily two approaches. Under both approaches, a third-party developer pays the up-front system installation cost and ongoing mainte-nance costs. The difference between the approaches is in the way the customer pays the developer. Under the first approach, the customer leases the system from the developer for a fixed monthly payment and collects benefits from the electricity the system generates through net metering. Under the second approach, the customer enters into a power purchase agreement (PPA) with the third-party developer and pays for electricity generated by the system at a fixed price, which should be no greater than the customer's retail price.

The third-party ownership business model has been capturing US rooftop solar market share over the past decade at an astounding rate. In 2007, just 10% of new systems installed on California rooftops were owned by third parties and leased by homeowners. By 2012, that proportion had risen to 75% (Tweed, 2013). Under the DG ownership model, EDU affiliates would follow in the foot-steps of third-party solar providers and become an installer and owner of DG sys-tems.[25] This model could also be applied on a larger scale through collective access to a large PV array by a group of consumers, such as a neighborhood.[26]

In exchange, customers would share in the benefits of DG through the industry-established arrangements described in the preceding text or through new arrangements that may emerge. Under this model, the primary sources of EDU revenue, in addition to the existing incremental payments for electric-ity, would be incremental payments from PPAs and lease payments. These addi-tional DG payments would offset some of the EDU revenues lost to decreasing electricity sales volumes. The contracts the EDU enters into with its DG cus-tomers also provide an opportunity for the EDU to exercise greater control over those DG resources, for example, by including a contractual clause allowing the EDU to curtail some amount of energy from a resource per year. DG ownership would provide EDUs with additional revenue to bolster their bottom line and additional flexibility that will be crucial for the reliable management of the distribution system of the future.

4.4   The Mixed Model

Under the mixed model, the EDU combines the components of the existing model with those of the other models described earlier in the text. For example, the EDU may retain the existing model for customers without DG at their

---

25. The third-party ownership model, particularly through a leasing arrangement, would also lend itself well to MTs, fuel cells, and CES units.
26. Described in Chapter 3 by Burger, an example of the community/group ownership model is being implemented in Germany's "bio-villages."

current rate with some automatic adjustments for changes in fuel prices and apply the insurance model to customers with DG to explicitly value and charge for the value of the system as a source of backup power and services to those customers.

Under this model, the EDU would meet its revenue requirements through a combination of fixed and incremental charges and alternative revenue sources. Furthermore, the EDU can integrate new business model components slowly so as to minimize the disruptive impacts of tariff changes, new sources of revenue, and changes in the regulatory structure on their customers while still addressing the challenge of meeting revenue requirements in the future system described in the previous section. The mixed model will provide a stepping stone for EDUs to transition from the old to the new distribution system paradigm. Table 2 provides a comparison of the key components of the four EDU future business models.

TABLE 2 A Comparison of the Key Components, Revenue Sources, and Tradeoffs of the Four EDU Future Business Models

EDU Business Model	Key Model Characteristics	Primary Revenue Source	Main Tradeoffs
Local reliability provider model	Access charges are based on customer reliability requirements and system impacts of customer equipment	Fixed charge based on the customer's desired level of reliability	Requires potentially complex risk assessment, reduced use efficiency incentives
Membership model	Access charge to customers based upon the size of connection or a similar metric	Fixed charges to different customer classes	Reduced use efficiency incentives, fewer opportunities for real-time control
DG ownership model	EDU owns and controls DG, customers pay leases or via PPA, EDU collects revenue for DG	Fixed and incremental charges to customers based on contractual arrangements	May raise conflict of interest issues if EDU also distribution-level mini-RTO
Mixed model	Combination of current and emerging models	Fixed and incremental charges depend on the combination of models	Those of the included models, "fairness" issues if large difference between models applied to different customer groups

Needless to say, the radical changes to the EDU business model will require buy-in from regulators at all levels. A key component of a successful EDU transition will be coordination with regulators to develop the regulatory structures that will allow EDUs to undertake new revenue-generating opportunities, develop new rate structures, and redefine the role of the EDU in the future system.

5 CONCLUSIONS

The role of the EDU is on the threshold of a fundamental change. Low demand growth and the accelerating DG implementation have the potential to erode significantly EDU bottom lines. Redefining tariff structures to bring them in line with the way consumers will use the distribution system in the future in order to allow EDUs to continue to meet their revenue requirements is part of the solution to the challenges EDUs face; new tariff structures, however, are not a complete solution.

To adapt to the electricity system of the future, EDUs must proactively seek to harness new technology, reassess consumer demands of the electricity system, and reenvision the suite of services they provide. The service-diversified EDU of the future will bear little resemblance to the energy-only focused EDUs of today. EDUs must draw on the lessons of the other industries that have successfully weathered the storm of disruptive technological change and look to the structure of the transmission system and other energy service providers as they restructure to meet the demands of the future system. By incorporating a diversity of energy delivery, electricity, communications and control infrastructure, data management, and reliability services, EDUs can emerge as a dominant force in the future electricity industry. Harnessing the strength of diverse service offerings will also provide EDUs a hedge against the uncertainty in the structure of the future electricity system. While the future of the electricity system is by no means certain, it is very likely that the future electricity system will be vastly different than that of today. It is clear that without fundamental changes in the way EDUs do business, their financial and corporate viability is in jeopardy in the next era of the electricity service industry.

REFERENCES

109th Congress of the United States, 2005. "Energy Policy Act of 2005." Washington, DC, January 4.

Akhil, A.A., Huff, G., Currier, A.B., Kaun, B.C., Rastler, D.M., Binquing-Chen, S., Cotter, A.L., Bradshaw, D.T., Gauntlett, W.D., 2013. DOE/EPRI 2013 Electricity Storage Handboook in Collaboration with NRECA. SAND2013-5131, Sandia National Laboratories, Albuquerque.

Barbose, G., Darghouth, N., Wiser, R., 2012. Tracking the Sun V: An Historical Summary of the Installed Price of Photovoltaics in the United States from 1998 to 2011. LBNL-5919E, Lawrence Berkeley National Laboratory, Berkeley.

DOE, ICF International, 2013. Combined Heat and Power Units Located in New York. Accessed October 8, 2013, http://www.eea-inc.com/chpdata/States/NY.html.

DSIRE, 2013. Database of State Incentives for Renewable & Efficiency. Accessed September 16, 2013, http://www.dsireusa.org/.

EIA, U.S., 2012a. Electricity Data. Accessed November 25, 2013, http://www.eia.gov/electricity/data.cfm#gencapacity.

EIA, U.S., 2012b. Market Trends—Electricity. Accessed October 8, 2013, http://www.eia.gov/forecasts/aeo/MT_electric.cfm.

eLab, 2013. New Business Models for the Distribution Edge: The Transition from Value Chain to Value Constellation. Rocky Mountain Institute, Boulder.

EPRI, 2013. Cost-Effectiveness of Energy Storage in California. 3002001162, Electric Power Research Institutes, Palo Alto, CA.

Faruqui, A., 2013. Surviving sub-one-percent growth. In: Electricity Policy, pp. 1–12.

GTM, SEIA, 2013. U.S. Solar Market Insight. Accessed October 1, 2013, http://www.seia.org/research-resources/us-solar-market-insight.

Kassakian, J.G., Hogan, W.W., Schmalensee, R., Jacoby, H.D. (Eds.), 2011. Utility Regulation. In: MIT Study on the Future of the Electricity GridMIT Energy Initiative, Cambridge, pp. 175–196.

Kind, P., 2013. Disruptive Challenges: Financial Implications and Strategic Responses to a Changing Retail Electricity Business. Edison Electric Institute, Washington, DC.

Lazard, 2013. Lazard's Levelized Cost of Energy Analysis–Version 7. Accessed September 5, 2013, http://gallery.mailchimp.com/ce17780900c3d223633ecfa59/files/Lazard_Levelized_Cost_of_Energy_v7.0.1.pdf.

Newton-Lowry, M., Makos, M., Waschbusch, G., 2013. Alternative Regulation for Evolving Utility Challenges. Edison Electric Institute, Washington, DC.

President's Council of Economic Advisers, 2013. Economic Benefits of Increasing Electric Grid Resilience to Weather Outages. Executive Office of the President, Washington, DC.

Price, S., Horii, B., King, M., DeBenedictis, A., Kahn-Lang, J., Pickrell, K., Haley, B., et al., 2013. California Net Energy Metering Ratepayer Impacts Evaluation.

Sandia, 2013. ARRA Energy Storage Demonstrations. Sandia National Laboratory, Albuquerque, NM. http://www.sandia.gov/ess/docs/ARRA_StorDemos_4-22-11.pdf (accessed 01.10.13).

Smith, R., Sweet, C., 2013. Companies unplug from the electric grid, delivering a jolt to utilities. Wall Street J. Published September 17, 2013. http://online.wsj.com/news/articles/SB10001424127887324906304579036721930972500 (accessed online on 01.10.13).

Tweed, K., 2013. Solar leasing gains momentum, costs to federal taxpayers fall. Greentech Solar. Greentech Media, July.

Wood, L., Borlick, R., 2013. Value of the Grid to DG Customers. Innovation Electricity Efficiency, Washington, DC.

Chapter 12

An Expanded Distribution Utility Business Model: Win-Win or Win-Maybe?

Timothy Brennan[1]

ABSTRACT

Challenges to the sustainability of the traditional distribution business model have brought calls for diversification into energy services, distributed generation, and other markets. As policy makers consider responses, prior justifications for quarantining regulated monopolies from competitive markets risk being forgotten. The rationale is neither a lack of entrepreneurship nor difficulties in changing to a business model predicated on reducing electricity use. Rather, utility expansion creates incentives to discriminate against rivals, primarily in access to regulated services. It may also lead to cross-subsidization, charging competitive services costs to the regulated sector. These can distort competition in unregulated markets and raise prices of utility services. The chapter assesses reasons for permitting diversification despite these risks, noting that a distribution utility could retain market power even as its market shrinks. It also notes that utilities may be unduly vulnerable from pricing regimes that allow consumers to essentially get backup electricity capacity for free.

Keywords: Diversification, Discrimination, Cross-subsidization, Market power, Pricing

1 INTRODUCTION

As many of the authors in this volume have observed, the electric utility industry, particularly on the distribution side, is facing potential challenge to its traditional dominance in the delivery of energy to end users. The challenge is competition from distributed generation (DG)—the ability of end users to generate electricity on or close to their homes or businesses. Industry observers have characterized these challenges as "disruptive" and "game changers," particularly threatening the ability of utilities to maintain the high credit ratings to

---

1. The views expressed here are those of the author and not necessarily those of Resources for the Future (which does not take institutional positions) or of anyone on its staff.

Distributed Generation and its Implications for the Utility Industry. http://dx.doi.org/10.1016/B978-0-12-800240-7.00012-6

which investor had become accustomed (Kind, 2013). Utilities around the world face these challenges (Chapters 2–4, 6). Other industries, such as newspapers and mail, face severe challenges because of the advent of new technologies (Chapter 1). When utilities are threatened, however, it becomes a policy problem because of the need for their regulators to respond (Chapter 16).

As a consequence of this threat, DG is leading distribution utilities to consider new business models. In addition, many states charge utilities with implementing "energy efficiency resource standards" that set percentage or quantity targets for reducing electricity use (Palmer et al., 2013). Advocates of energy efficiency believe that utilities should change their business model from supplying electricity to supplying electricity services such as lighting, heating, and cooling (Fox-Penner, 2010). As a number of chapters in this volume also point out, charging a price for services provided rather than electricity used would give utilities incentives to conserve electricity.

With technological change, policies to reduce electricity use, and falling demand overall, utilities are considering diversification, e.g., "business models where utilities can add value to customers and investors by providing new services" (Kind, 2013, p. 19). Some observers suggest that utilities enter the DG market directly, particularly on the solar side (Hanelt, 2013). The impetus to diversify is understandable, but profitability is not given. Competition may reduce the ability to do more than cover one's costs. In addition, a regulated entity long responsible for electricity delivery may lack the entrepreneurial spirit to do well in new markets, especially for services that reduce electricity sales.

These issues, while potentially important, are not the chief concern of this chapter. The central point here is that regulators need to consider whether utility diversification, specifically expanding into otherwise competitive service or product businesses, assists or impedes their control of market power.[2] The thesis here is that diversification is such an impediment. Such concerns have been at the core of electricity policy since the FERC's Order 888 in 1996.[3] Because their lessons may be forgotten in the rush to diversify, it is timely to review them.

The remainder of this chapter is organized as follows. Section 2 presents the rationale for concern when regulated monopolies diversify into other businesses. Section 3 describes historical applications of this concern regarding diversification in telecommunications and the electricity sectors. Because

---

2. Hanser and van Horn, Chapter 11, present an interesting proposal to have utilities expand their role in regulated distribution to include delivery of from (DG) customers, not just too them, as if it were more akin to a common carrier. Cazalet, Chapter 10, embellishes their proposal by describing the multiplicity of "transactive retail tariffs" that the future distribution utility might adopt in its dealings with DG users, storage providers, and energy service managers.

3. Federal Energy Regulatory Commission, Promoting Wholesale Competition Through Open Access Non-discriminatory Transmission Services by Public Utilities; Recovery of Stranded Costs by Public Utilities and Transmitting Utilities, Order No. 888 (24 April, 1996).

regulation and market power are both necessary conditions for this concern with diversification—it largely disappears if the very survival of distribution utilities were at stake—Section 4 explains why falling demand as a result of a disruptive technology need not mean that market power will disappear. Section 5 examines factors that could mitigate concerns associated with diversification. Utilities, however, are needlessly vulnerable because of usage-based pricing, which essentially subsidizes DG, as Section 6 shows. Section 7 concludes.

2   RATIONALES FOR LIMITING DIVERSIFICATION

There are a number of reasons motivating concerns with distribution utility diversification into competitive markets in products or services, for example, solar PV financing, installation, and maintenance. After reviewing concerns regarding such diversification that ought not apply, two primary concerns based specifically on the regulated nature of distribution are discussed in the succeeding text, followed by a summary table.

2.1   What the Rationales Are Not

Much of the concern with monopoly diversification is that it may preempt competition in related markets through vertical integration, tying, or exclusive dealing. For example, if a monopoly brings in-house a certain service that used to be available from independent suppliers, those independent suppliers may lose the opportunity to compete for that business. Such foreclosure is thought to allow the firm to leverage its monopoly from one market to another. A famous recent example involved Microsoft's bundling its browsers with its operating system.[4]

Over the last few decades, economists have developed abstract theories that explain how these practices might be anticompetitive. They may preempt competition among complement providers through exclusive dealing (Brennan, 2007) or preventing entrants from acquiring scale economies (Rasmusen et al., 1991). More strategic theories involve tying a competitive good to a monopoly good to convince rivals that the firm will refuse to cede share in the competitive market (Whinston, 1990). Vertical integration may make credible an exclusive arrangement to raise input prices (Ordover et al., 1990).

None of these theories justify a general presumption that diversification is undesirable. Coase (1937) explained that whether to diversify or vertically integrate depends on the trade-off between (i) the gains from going outside to take advantage of market competition and (ii) the costs of going outside associated with searching for and verifying the quality of goods and services. This explains, for example, why utilities undertake a great deal of maintenance in-house rather than entirely rely on procuring services from outside contractors.

---

4. United States v. Microsoft, 253 F.3d 34 (D.C. Circ. 2001).

The "Chicago" school of antitrust economics pointed out that without special circumstances, an unregulated monopoly would forego the benefits of competition at other stages only if other efficiencies, e.g., from improved promotion, quality control, or coordination, exceed losses from foregone competition in the supply chain (Farrell and Weiser, 2003).

However, if a firm is regulated, vertical integration and diversification could have an additional motive—to evade the regulatory constraint on profits. The two primary tactics (Brennan, 1987) and their consequences for utility diversification are described below.

2.2   Discrimination

Discrimination involves tying access to a regulated service to an unregulated service that the regulated monopoly also provides. Because of regulation, this is accomplished by discriminating against rivals in that unregulated market in nonprice terms, such as the quality or timeliness of access. This creates an artificial competitive advantage, allowing the regulated firm to evade the regulatory constraint on profits through this discriminatory access.

Suppose that distribution utility diversifies into providing a particular energy management service, such as controlling cooling in a home. This service generally costs $100 over and above the cost of the electricity used to provide the relevant services. Suppose further that the distribution utility can reduce the reliability or timeliness to competitors in that energy management service. Another possibility is that it may have information on load variability and reliability that it could provide to its energy management affiliate but not to its rivals. These impediments mean that clients of energy management would pay a $30 premium above the $100 for the management service provided by the utility's affiliate. In effect, the utility diversifies and then uses nonprice discrimination against rivals in the management market to, in effect, raise the price of distribution by $30 over the regulated rate.

Moreover, this discrimination could discourage the entrepreneurial development in energy services, harming its customers and perhaps reducing environmental and grid benefits that might be reaped from increased innovation. For example, suppose that utility entry into solar DG discouraged other firms from participating in the market, because they would not be able to compete effectively against a distribution utility that could favor its solar DG affiliate in terms of better connections for backup power or delivery of surplus power back into the grid. Fewer solar DG participants are likely to reduce the rate of innovation both because there are fewer independent potential innovators out there and because reduced competitive pressure in the solar DG market will reduce incentives to come up with better and less expensive ways to meet the demand.

It is important to note that the incentive to discriminate begins with regulation—the inability to exploit market power directly by raising the price of a monopoly service. It also requires a vertical or complementary relationship between the monopoly service and the unregulated market, e.g., that the latter needs to work with the former or procure service from the former, as with energy management services and electricity distribution in this example.

2.3 Cross-Subsidization

The second concern goes by the name of cross-subsidization, but the problem at heart is cost misallocation. The basic idea is that costs of a competitive service (e.g., energy management, as in the preceding text) could be charged to the regulated service (e.g., the distribution utility). If regulators set distribution rates based on reported costs, this will then lead to higher rates for distribution. The profits from those higher rates show up on the books of the competitive service, because it has nominally lower costs in selling at the prevailing market price.

Cross-subsidization has not played a large role in electricity policy making. However, it could be important as distribution utilities diversify. A specific illustration may be useful. Suppose that some of the IT equipment used to provide energy management service is also used to provide distribution. Suppose that a utility provides energy management service and is then able to charge this IT equipment to the regulated distribution service. To apply numbers, suppose that the average reported deliver cost per hour increases from $30 to $34 because of these artificially added IT costs. Under cost-based regulation, this will increase the price charged to distribution ratepayers by $4. The regulated utility looks like it's just covering costs. The profits from this cross subsidy show up on the books of the unregulated management services affiliate, which appears to have lower costs of supplying services, which it can still sell at the prevailing market price.[5] The direct harm is that the ratepayers pay higher prices, as the regulatory price constraint is effectively evaded. Moreover, rivals in the competitive markets, such as energy management, may limit operations or stay out altogether rather than face a utility-owned competitor that may be able to have captive regulated ratepayers cover the costs of its operations.

The relationship between the distribution utility and the cross-subsidized service need not be direct. While the lynchpin of discrimination was a vertical or complementary relationship between the regulated and unregulated services, the lynchpin of cross-subsidization is having a common input. A common input a regulated service has with any business is financial capital. If a regulated distribution service backs bonds issued to finance diversified operations of

---

5. A variant on cross-subsidization occurs when the regulated firm chooses a technology for producing both the unregulated and regulated services using a common plant rather than separate plants. This could be inefficient yet profitable if it facilitates cross-subsidization, for example, when the entire cost of the common plant is charged to the firm's regulated operations (Baseman, 1981; Brennan, 1990).

any sort, the risks of those enterprises are shifted to the utility's customers. The capital expenses of the distribution utility's efforts in these markets are cross-subsidized. The harm may be especially consequential if the local utility can enter the DG market and, through cross-subsidization, create a credible predatory threat against those threatening the utility's monopoly over electricity delivery.

The incentive to cross-subsidize is created not just by the fact of regulation, but, as noted earlier, that regulated rates are based on costs. To the extent that one adopts a regulatory mechanism where the link between costs and rates is kept tight, as under revenue decoupling (Brennan, 2010), the potential for cross-subsidization will be larger. If the connection between costs and rates is loose because of regulatory "lag" between incurring costs and setting rates, the incentive to misallocate falls. Under "price caps," in which price adjustments over time are based on inflation and forecasts of productivity but not actual costs, the link between costs and rates would be broken altogether, and the incentive to cross-subsidize disappears (Brennan, 1989).

Table 1 summarizes these two potential tactics for exercising market power through diversification, their requirements, and their potential harms.

TABLE 1 Potential Harms from Diversification

	Discrimination	Cross-Subsidization
Essence of practice	Tying competitive energy product to regulated distribution through nonprice service delay or degradation, e.g., provision of distribution lines	Misallocating costs of providing competitive energy product to regulated distribution utility revenue requirement, directly or by artificial "common costs"
Required regulation	Price ceiling below monopoly distribution price	Price below monopoly level set based on reported average cost of distribution
Related markets	Complementary markets that use electricity distribution, such as energy services or distributed generation	Business that uses the same types of inputs as the regulated service, especially borrowed capital, including but not limited to energy efficiency and distributed generation
Harms	Higher prices in these related markets, displacement of more efficient competitors, reduced innovation in the energy sector	Higher prices paid by all ratepayers, including non-DG ratepayers, displacement of efficient entrants, and reduced innovation in related markets, perhaps a credible predatory threat

3   NOT A NEW STORY

Academic journals are full of theories of harm that look good as paper exercises but have had no practical implications. The contrary holds here. These theories of discrimination and cross-subsidization sit at the core of perhaps the most significant industrial policies of the last three decades.

Although some hints of these ideas can be found in some US policy developments regarding oil pipelines in the 1970s (Mitchell, 1979),[6] their main manifestation arose in regulatory and antitrust cases against AT&T, leading up to AT&T's divestiture of its local telephone monopolies in 1984. During the middle of the last century, policy makers became concerned with efforts by AT&T to impede entry in customer telephone equipment and long-distance calling and potential interference with new information services.[7] The analogy here would be concerns that regulated distribution utilities would impede competition in related energy services and energy efficiency markets, with potential interference with new DG technologies. Evidence supporting these concerns led to the US Department of Justice's antitrust case against AT&T and the divestiture of its local operating monopolies.[8]

More directly relevant here is that concerns regarding diversification by regulated transmission utilities into competitive generation markets were at the core of policies to open the electricity sector. Federal Energy Regulatory Commission Orders 888 and 889 were designed to prevent transmission access discrimination against independent generators. FERC did not opt for full divestiture, as in the AT&T setting, but instead defined rules for "independent system operators"—later called "regional transmission operators" in FERC's Order 2000 issued in 1999[9]—to limit the degree to which generation owners can control transmission operations.

The central point is that concerns with diversification are not merely the stuff of economics textbooks. They have been central to the development of competition in two major utility sectors—telecommunications and electricity—that had been regulated from top to bottom. The need to impose separation

---

6. The discrimination-like concern was that if an oil company, refinery, or group of them owned a regulated pipeline, they would have an incentive to limit the capacity of the pipeline to the monopoly size, driving up the price of oil or oil products at the pipeline's terminal. They would profit from this market power through higher oil or oil product prices. The discrimination sustaining this pipeline would be a refusal to expand the pipeline at the request of new shippers.
7. For excellent histories of the telecommunications network and policy during this time, see Brock (1981, 1998).
8. The "information services" restriction on the divested local companies was lifted in 1990, as both parties to the case, AT&T and DOJ, no longer believed it necessary. The long-distance ban was lifted through regulatory orders following passage of the 1996 Telecommunications Act and decisions that the divested local companies were no longer acting to thwart entry into their markets.
9. Federal Energy Regulatory Commission, Regional Transmission Organizations, Order No. 2000 (December 20, 1999).

TABLE 2 Diversification Parallels

	Telecommunications, 1984	Wholesale Electricity, 1996, 1999	Local Utilities, 2014
Regulated monopoly	Local telephone service	Transmission	Distribution
Unregulated competitive markets	Customer premises equipment, central office equipment, long-distance service, information services	Generation	Retailing, energy services, energy efficiency, distributed generation
Separation	AT&T divestiture of local telephone operating companies	Orders 888, 2000: Functional unbundling through ISOs, RTOs	?

to prevent discrimination is so consequential that it defines the term used to denote the opening of electricity markets to competition: "restructuring."

Table 2 describes how these concerns affected past policy in telecommunications and wholesale electricity markets and what the parallels may be if distribution utilities diversify in response to the DG threat.

4   LOSS OF MARKET DOES NOT MEAN MARKET POWER DISAPPEARS

These concerns would disappear if, after DG entry, distribution utilities retain no market power and can be deregulated. In the extreme, if the distribution utilities' very survival is at stake because of the game-changing nature of DG, one could treat them as just another competitor in a broader electricity supply market and let them diversify as they see fit.[10] However, even if DG takes a large amount of business away from a distribution utility, that utility need not lose market power. Reducing demand and constraining price are not the same thing.

---

10. One could ask whether the disappearance of the regulator would cause the regulated firm's prices to rise by a small but significant amount (Brennan, 2008) as competition authorities ask when two firms merge (Department of Justice and Federal Trade Commission, 2010). One needs to exercise this test with some care. To the extent that regulatory prices included internal cross-subsidies, e.g., keeping prices equal across geographic areas when the costs of serving them differ, some prices may go up because they were below costs. In addition, because the costs of regulation may be greater than the costs (if any) of preventing an otherwise desirable merger, one might want to allow a higher increase in price for deregulation than one might tolerate from a merger.

For example, while many people have dropped landline phones for mobile services, mobile phones limit the ability of landline service providers to increase price only when consumer choices between mobile and landline are highly sensitive to price.

Perhaps surprisingly, an incumbent utility have a higher monopoly price after entry by DG than without it (Brennan and Crew, 2014). The criterion for when a loss of demand for utility distribution from growth in DG enhances the utility's market power is that the non-DG users are less sensitive to price than DG adopters. If the remaining customers are highly averse to adopting DG, the profit-maximizing price for utility distribution to those customers could increase above what it was when the utility was the only electricity game in town, despite having lost some of its market to DG.

Whether this would happen depends on who leaves and who stays when DG appears. If the primary adopters are those with relatively limited electricity needs that can be satisfied with DG, those left behind will have with greater energy demand and a greater willingness to pay for utility service. On the other hand, those who adopt DG may be those who can come up with the necessary funds to make the investment, leaving the utility with less wealthy customers unwilling to pay much higher prices for distribution.

5   OTHER FACTORS MITIGATING CONCERN

Even if the utility distribution market is not fully competitive, other considerations can justify diversification despite concerns over cross-subsidization and discrimination.

5.1   Economies of Scope

There may be benefits of having a distribution utility engage in energy services, energy efficiency management, or DG itself. These benefits come from "economies of scope," the savings in cost when a firm supplies two goods and services relative to the costs of having two firms separately providing different goods. If these savings are sufficiently large, they can outweigh the potential costs from diversification and cross-subsidization (the latter discussed in Brennan and Palmer, 1994). Where economies of scope are present, regulators should consider permitting diversification even if discrimination or cross-subsidization cannot be ruled out.

Rhetoric of "value generation" aside, economies of scope are not automatically present. Experiences with scope economies differ across sectors that have seen separation efforts. In telecommunications, the development of independent equipment markets suggests that providers of local telephone service had no cost advantage in supplying telephones and other customer devices or in producing large-scale switches and network equipment. Long distance was more complex but proved feasible, until the growth of packet-switched transmission (the Internet) eliminated the market altogether.

In electricity, confidence that no economies of scale were lost by unbundling of generation from transmission has been a tougher call. Unlike the telephone sector, electricity generation and transmission both require major indivisible and simultaneous investments. Coordinating investment across sectors might be inconsistent with independent entrepreneurial competition on the generation side of the sector. Michaels (2004) reviewed a number of studies showing cost savings and benefits from continued integration of generation and transmission. Taylor and Van Doren (2004) argued that the coordination benefits of integration are sufficiently great that it may have been a mistake to open wholesale markets to competition and end-to-end regulation should be reimposed, a notable argument as both authors are researchers of the Cato Institute, not known for its tolerance of economic regulation. Inability to coordinate could be a factor supporting claims of a decline in transmission investment (Brennan, 2006).

Consequently, whether there are economies of scope between utility distribution and other enterprises into which they might diversify is an empirical matter. Claims of economies of scope merit close examination. There may be reasonable methods for coordinating across sectors, sharing information, and minimizing joint costs that do not require the degree of integration that could lead to discrimination and cross-subsidization. Regulators will likely have to make difficult judgments regarding the relative magnitudes of economies of scope and anticompetitive harm when considering whether to permit utilities to move forward with plans to diversify into competitive services.

5.2 Ex Post Regulation

Another reason regulators should not worry about diversification up front is that they can detect and potentially mitigate harmful conduct after the fact. As observers have noted (Faulhaber, 1987), concerns with cross-subsidization and discrimination involve not just regulation but also particular and almost paradoxical kinds of regulatory failure. For cross-subsidization, the harms depend on having a regulator that diligently connects prices to reported costs—otherwise, neither the regulated rate rises nor are competitive operations subsidized when costs are misallocated to the regulated side of the enterprise. However, this diligent regulator must be unable (or unwilling) to detect that costs of competitive services have been incorrectly attributed to the regulator. Detecting cost misallocation can be difficult, especially if it is done by choosing technologies that increase joint fixed costs and reduce the incremental costs that could be easily attributed by the regulator to the competitive enterprise (Baseman, 1981).

Discrimination may be even more paradoxical. On the one hand, it depends upon the regulator being unable to detect that the utility is favoring its affiliate over its rivals in related energy services or DG markets. On the other hand, for discrimination to create a competitive advantage, customers in those related markets must be able to detect that the utility's offerings in that market are superior to those of its competitors. Although those customers may neither be able nor need to understand that this superiority stems from discrimination, this suggests that customers would be detecting something that the presumably expert regulator is not. However, the regulator's ability to detect discrimination in part relies on its ability to determine that the favorable treatment of the affiliate is not the result of a genuine economy of scope between the utility and its affiliated competitive products.

5.3   Ex Post Antitrust Enforcement

A third reason regulators need not be concerned with diversification into other lines of business is that if the problems discussed here arise, they could be subject to antitrust enforcement. Antitrust was how diversification was ultimately resolved in telecommunications, while regulatory solutions remain the method for electricity. The telecommunication experience provides a cautionary tale, in that, while it did succeed, the antitrust case took over 7 years from filing to settlement, with an additional 2 years before the divestiture was official and continuing quasi-regulatory oversight from the court until the 1996 Telecommunications Act.

The United States faces a perhaps idiosyncratic impediment in using antitrust enforcement to justify allowing utilities to diversify. In a series of cases in the last 10 years,[11] the US Supreme Court has ruled that if a regulator has the authority to set the rules for competition for the sectors it regulates, antitrust enforcement is not desirable.[12] In addition, US courts could decide under existing doctrines that state regulation regarding diversification trumps antitrust law. However, this may be desirable as any detrimental effects of diversification are likely to fall within state boundaries (Brennan, 1984; Easterbrook, 1983). In addition, as different state regulators approach this and other issues raised in this volume differently, we will have a larger body of experience from which to learn which policies work best. Having different states adopt different policies regarding retail electric competition meant that the flaws found in the California approach in 2001-2002 were not felt nationwide.

---

11. Verizon v. Trinko, 540 U.S. 398 (2004); U.S. Postal Service v. Flamingo Industries, 540 U.S. 736 (2004); Credit Suisse v. Billing, 551 U.S. 264 (2007). For discussion of the competition policy implications of *Flamingo* and *Trinko*, see Brennan (2005a,b).

12. Some (Federal Trade Commission, 2010) suggest that government (rather than private) antitrust cases may still be viable.

6   SUBSIDIZING THE COMPETITION?

It needs to be noted that the competition from DG, which motivates utility diversification, is in some degree due to a subsidy to those competitors. Weather conditions may disrupt the output of on-site generation, such as clouds for solar or still air for wind. Moreover, on-site generation will predictably fail occasionally.

Hence, backup power is a necessary complement to DG. The expense of on-site generation may leave little in the budget for on-site backup power. But most DG users need not worry about installing just-in-case generation because they can get that capability for low to no cost from the distribution utility.[13] The capacity to get power from centralized generation through the distribution utility is typically paid for on a per kilowatt-hour basis. If so, maintaining a backup connection to the grid is essentially free until used. When energy is needed, these users simply pay the going rate for electricity. As noted by numerous industry observers in this volume and elsewhere (Borenstein, 2013), some of that will contribute to the cost of maintaining a line to that user, but not as much as that paid by customers who continue to get all of their electricity from the utility.

In essence, the distribution utility, through its ratepayers, is forced to subsidize its DG competitors.[14] If DG users connected to the grid do not cover the cost of their connections because of little or no use, the remainder of those costs are folded into the per kilowatt-hour prices for electricity paid by everyone. Were the grid priced efficiently, everyone connected to it would pay (at least) the same monthly charge for connection. The per kilowatt-hour component would be (closer to) the actual marginal cost of the energy, and not include a component to cover the cost of the distribution network.[15]

Decoupling distribution utility revenues from aggregate usage might appear to address this problem, but does not. Decoupling fundamentally keeps distribution utilities from opposing policy initiatives to reduce electricity use, such as subsidizing energy-efficient appliances. Moving away from usage-based pricing would be a step in the right direction, as added usage would not make as large a marginal contribution to utility coffers and thus reduce objections to energy efficiency policies. However, decoupling as implemented does not

---

13. Sioshansi points out to me that California has no fixed fee for grid connection.

14. Other policies, such as feed-in tariffs for solar DG covered through higher electricity prices overall, also subsidize competition to standard utility services (Borenstein, 2013). Fereidoon P. Sioshansi has noted that this can transfer wealth from those in cloudy parts of a country to those in sunny parts.

15. I say "at least" and "closer to" because if the revenues from charging a monthly connection cost equal to the incremental cost of connection fail to cover the total cost of the grid, the remainder will need to be covered through some combination of surcharges over marginal cost. However, this will still have a higher fixed monthly charge and lower usage charge than under a regime where all distribution costs are covered on a usage basis.

involve that switch. Recovering distribution costs through fixed monthly fees would reduce the marginal price of electricity and thus encourage consumption, against the wishes of decoupling's advocates.

As with so much else here, this issue has historical precedent. Prior to the breakup of AT&T in 1984 and extending past it, much of the cost of providing telephone lines to residences was subsidized through surcharges on use of the telephone, particularly for calling beyond the local exchange area. This led some bulk users to come up with ways to bypass the local exchange and connect directly to other locations for long-distance calling, just to escape paying the subsidy (Gerald, 1998, pp. 182-200). Some argue that the entry by long-distance competitors was not so much a matter of providing better service at lower cost, but the result of keeping long-distance prices above costs, resulting in what was called "cream skimming" (Temin and Galambos, 1989, p. 49). Eventually, technological change in the telecommunications sector was sufficiently great to render this dispute moot, but it may not be good to count on similar changes to deal with this problem in the electricity sector.

7   CONCLUSION

Utilities are responding to the challenge of DGs and the demands of energy conservation policies by proposing to diversify into other lines of business. Assessments of this response pay too little attention to concerns regarding the potential for access discrimination and cross-subsidization by regulated firms that enter related competitive markets. These concerns exist not just between the covers of a textbook; they have been at the core of telecommunications and bulk electricity policy for decades. One cannot conclude that utilities lack market power and will be deregulated if DG takes a substantial bite out of their market; utilities may retain the ability to charge monopoly prices to those customers that choose not to adopt. However, while market power may not fall or even rise, the shrinking size of the market certainly reduces the potential profits available to a distribution utility and, eventually, its ability to cover its costs even at a monopoly price for distribution.

However, for a number of reasons, regulators might want to permit diversification despite these concerns. Economies of scope could provide benefits of diversification that counter the risk of incurring these costs, especially when the benefits of coordinated planning between utility distribution and related competitive services are substantial, but they should be shown, not just presumed. Regulation could be invoked after the fact. For cross-subsidization to occur, regulators must be diligent about tying rates to costs but are unable or unwilling to verify the legitimacy of those costs. Discrimination requires that regulators be unwilling or unable to detect it but that the market can observe its results. These seeming inconsistencies can be overcome, but they do justify some pause in presuming that diversification is harmful.

Antitrust enforcement could be a remedy. Not only is antitrust enforcement long, expensive, and somewhat clumsy, however. Recent Supreme Court decisions indicate that if a regulator has authority over the competitive circumstances in a market, antitrust has no role to play. As the harms of diversification are likely to be intrastate as well, it may be best to let different states experiment with different diversification policies rather than impose a nationwide approach.

Finally, it is important to remember that the distribution utilities are threatened by DG in part because of inefficient pricing of its capacity to provide backup power. Covering the cost of distribution lines through usage-based rates effectively subsidizes the provision to DG adopters of backup power delivered by the utility. The same problem created controversies in telecommunications for decades. Before coping with the potential results of utility diversification, it might be useful to change the pricing structure creating or exaggerating the DG challenge that in large measure is engendering this diversification.

REFERENCES

Baseman, K., 1981. Open entry and cross-subsidization in regulated markets. In: Fromm, G. (Ed.), Studies in Public Regulation. MIT Press, Cambridge, MA, pp. 329–370.

Borenstein, S., 2013. Rate design wars are the sound of utilities taking residential PV seriously. Energy Economics Exchange (November 12).

Brennan, T., 1984. Local government action and antitrust policy: an economic analysis. Fordham Urban L. J. 12, 405–436.

Brennan, T., 1987. Why regulated firms should be kept out of unregulated markets: understanding the divestiture in U.S. v. AT&T. Antitrust Bull. 32, 741–793.

Brennan, T., 1989. Regulating by 'capping' prices. J. Regul. Econ. 1, 133–147.

Brennan, T., 1990. Cross-subsidization and cost misallocation by regulated monopolists. J. Regul. Econ. 2, 37–51.

Brennan, T., 2005a. Trinko v. Baxter: the demise of U.S. v. AT&T. Antitrust Bull.. 50, 635–664.

Brennan, T., 2005b. Should the flamingo fly? Using competition law to limit the scope of postal monopolies. Antitrust Bull. 50, 197–221.

Brennan, T., 2006. Alleged transmission inadequacy: is restructuring the cure or the cause? Electricity J. 19 (4), 42–51.

Brennan, T., 2007. Saving section 2: reframing U.S. Monopolization Law. In: Ghosal, V., Stennek, J. (Eds.), The Political Economy of Antitrust. Emerald Group Publishing, Amsterdam, North-Holland, pp. 417–451.

Brennan, T., 2008. Applying 'merger guidelines' market definition to (de)regulatory policy: pros and cons. Telecommun. Policy. 32, 388–398.

Brennan, T., 2010. Decoupling in electric utilities. J. Regul. Econ. 38, 49–69.

Brennan, T., Crew, M., 2014. Gross substitutes vs. marginal substitutes: implications for market definition in the postal sector. In: Crew, M., Brennan, T. (Eds.), The Role of the Postal and Delivery Sector in a Digital Age. Edward Elgar, Cheltenham, UK.

Brennan, T., Palmer, K., 1994. Comparing the costs and benefits of diversification by regulated firms. J. Regul. Econ. 6, 115–136.

Brock, G., 1981. The Telecommunications Industry: The Dynamics of Market Structure. Harvard University Press, Cambridge, MA.

Coase, R., 1937. The nature of the firm. Economica 4, 386–405.

Department of Justice and Federal Trade Commission, 2010. Horizontal Merger Guidelines. U.S. Department of Justice, Washington.

Easterbrook, F., 1983. Antitrust and the economics of federalism. J. Law Econ. 26, 23–50.

Farrell, J., Weiser, P., 2003. Modularity, vertical integration, and open access policies: towards a convergence of antitrust and regulation in the internet age. Harv. J. Law Technol. 17, 85–134.

Faulhaber, G., 1987. Telecommunications in Turmoil: Technology and Public Policy. Ballinger, Cambridge, MA.

Federal Trade Commission, 2010. Is there life after Trinko and Credit Suisse? The role of antitrust in regulated industries. In: Before the United State House of Representatives Committee on the Judiciary, Subcommittee on Courts and Competition Policy (Washington, DC, 15 June 2010).

Fox-Penner, P., 2010. Smart Power: Climate Change, the Smart Grid, and the Future of Electric Utilities. Island Press, Washington.

Gerald, B., 1998. Telecommunication Policy for the Information Age: From Monopoly to Competition. Harvard University Press, Cambridge, MA.

Hanelt, K., 2013. Making Friends with Solar DG. Public Utilities Fortnightly. 10–13, September.

Kind, P., 2013. Disruptive Challenges: Financial Implications and Strategic Responses to a Changing Retail Electric Business. Edison Electric Institute, Washington, DC.

Michaels, R., 2004. Vertical integration: the economics that electricity forgot. Electricity J. 17 (10), 11–23.

Mitchell, E.J., 1979. Oil Pipelines & Public Policy Analysis of Proposals for Industry Reform & Reorganization. American Enterprise Institute, Washington.

Ordover, J., Saloner, G., Salop, S.C., 1990. Equilibrium vertical foreclosure. Am. Econ. Rev. 80, 127–142.

Palmer, K., Grausz, S., Beasley, B., Brennan, T., 2013. Putting a floor on energy savings: comparing state energy efficiency resource standards. Util. Policy 25, 43–57.

Rasmusen, E., Ramseyer, M., Wiley Jr., J., 1991. Naked exclusion. Am. Econ. Rev. 81, 1137–1145.

Taylor, J., Van Doren, P., 2004. Rethinking Electricity Restructuring. Policy Analysis No. 530, Cato Institute, Washington.

Temin, P., Galambos, L., 1989. The Fall of the Bell System: A Study in Prices and Politics. Cambridge University Press, Cambridge.

Whinston, M., 1990. Tying, foreclosure, and exclusion. Am. Econ. Rev.. 80, 837–859.

Chapter 13

From Throughput to Access Fees: The Future of Network and Retail Tariffs

Tim Nelson, Judith McNeill and Paul Simshauser

ABSTRACT

This chapter critically evaluates how utilities will need to adjust their tariff structures to compete with new forms of energy production. The structure of these tariffs will be a critical determinant of the success or failure of utilities in the future. The authors assess existing and emerging tariff designs to address the advent of grid-connected substitutes such as distributed generation. The economic efficiency and equity implications of changing tariff designs are assessed based on existing and emerging technologies. The chapter concludes that utilities must be cautious in rapidly redesigning their business models or rebalancing their tariffs, as they attempt to recover revenue previously obtained through rising volumetric consumption. Importantly, adjusting tariffs to recover revenues in the short term may hasten the adoption of energy storage technologies, further undermining the financial stability of utilities in the long term.

Keywords: Australian electricity markets, Electricity tariffs, Network regulation, ToU pricing, Microeconomic reform

1 INTRODUCTION

The rise of distributed generation has fundamentally changed the way in which economists consider the electricity supply industry. Historically, electricity supply has been a monopoly. While microeconomic reform has introduced retail competition, consumers have still been reliant upon a monopolistic network for their electricity supply. As distributed generation has become more economic, utilities have been adjusting to a world in which grid-based electricity supply now has a genuine substitute product.

Australia is one of the most relevant markets in which to consider this fundamental shift. This is because the convergence of a number of critical factors has placed enormous strain on the revenues of existing network businesses: significant uptake of distributed generation driven by generous feed-in tariff

Distributed Generation and its Implications for the Utility Industry. http://dx.doi.org/10.1016/B978-0-12-800240-7.00013-8
267

policies and capital subsidies, large capital expenditure to upgrade and replace aging electricity networks, significant increases in retail electricity prices, and declining electricity demand.

Simshauser and Nelson (2013) provided an overview of the reasons why Australian electricity prices have increased by almost 100% since 2008. In many ways, Australia is experiencing an "energy market death spiral" because flat "average cost" tariffs provide no incentive for customers to reduce their peak load consumption (see Section 3 and Simshauser and Nelson, 2014). Air-conditioning penetration has resulted in peak demand growing at almost double the rate of underlying energy demand (Quezada et al., 2013). As prices have increased to recover the costs of significant capital expenditure to meet peak demand, consumers have reduced their consumption but at a time most convenient to them, not on the hottest days of the year when system peak demand occurs—an observation predicted by Houthakker (1951) more than 60 years ago. Such a scenario has resulted in poorer capacity utilization and a need to further increase electricity tariffs. In this context, consumers are increasingly purchasing embedded solar PV as a substitute for at least part of the electricity they purchase from the grid.

As a result, tariff design is critical for encouraging consumers to use electricity efficiently. Although there are 6 National Electricity Market (NEM) regions in Australia,[1] there are 13 distribution network "patches," and in each patch, one network operator has a different tariff structure. However, while each network tariff is different, at the residential level, they are conventional two-part tariff structures comprising a modest daily fixed charge with flat "average cost" usage charges for each unit of throughput.

Redesigning electricity tariffs has been a topic of debate for some time in Australia. With peak demand growth being driven by air-conditioning uptake, many energy economists have been calling for time-of-use (ToU) pricing to be introduced to unwind current cross subsidies between households with and without significant contributions to system peak demand (see, e.g., Borenstein, 2005; Faruqui and Sergici, 2010; Simshauser and Downer, 2012). However, actual reform has been slow. Only Victoria has seen a significant rollout of smart metering technologies, and ToU pricing has only recently been permitted.

However, the emergence of a genuine "partial" grid substitute, namely, embedded solar PV, has reignited the debate about tariff design. A recent study commissioned by the Energy Supply Association of Australia found that the average household with solar PV is paying substantially less in network charges than other users (see Section 3 and Chapter 4, for a different estimation of cross

---

1. The NEM is a wholesale market covering much of Australia's east and southern geography including the states of Queensland, New South Wales, South Australia, Victoria, the Australian Capital Territory, and Tasmania. Only Western Australia and the Northern Territory are not part of the NEM.

subsidy). This is despite household peak demand occurring after solar PV systems are producing material amounts of electricity. But at the same time, households that have installed solar PV have become politically active through the creation of the Solar Citizens[2] organization. The challenge for Australia's utility industry is to undertake the necessary reform to tariff design while addressing the concerns of households that have already installed embedded generation, legitimate or otherwise. Such circumstances are not unique to Australia. California is experiencing similar issues with a recent report concluding that "Non-solar customers are said to be subsidizing solar customers, who tend to be more affluent" (Sioshansi, 2013a, p. 6).

This chapter is structured as follows: Section 2 outlines the economic inefficiency of existing electricity tariffs; Section 3 provides an overview of the convergence of events that have led to utilities in Australia, prioritizing tariff reform; and the implications of moving toward capacity-based demand tariffs are explored in Section 4; while other options for utilities are considered in Section 5. The chapter's concluding remarks are in Section 6.

2   ECONOMIC INEFFICIENCY OF EXISTING ELECTRICITY TARIFFS

Before considering the unique convergence of events that have placed significant strain on the revenues of Australian utilities, it is necessary to examine existing tariff structures. It may well be that the existing poorly designed residential tariff structures have made adverse circumstances even worse. There are generally four types of end-user pricing for electricity further described as follows:

- *Average cost flat tariffs*: Flat tariffs are based on the average cost incurred to supply each unit of energy throughput, typically structured as a two-part tariff. Flat tariffs have traditionally been used by utilities in electricity markets because interval metering was not economically viable and as such pricing as a function of time was therefore impracticable. The limitations of simple accumulation meters have somewhat been overcome with the installation of two meters to allow general and controlled consumption to be recorded.
- *Inclining/declining block tariffs*: An inclining block tariff system is based on "blocks" of energy consumption having different pricing structures. For example, a consumer might pay $0.20/kWh for the first 7000 kWh of consumption and then $0.30/kWh for any kilowatt-hour consumed over and above 7000 kWh. The purpose of these tariffs is to discourage (inclining) or encourage (declining) electricity use by utilizing a form of second-degree price discrimination.

---

2. See http://www.solarcitizens.org.au for further information on the group that represents Australia's "1 million solar households." Accessed online on 30 September 2013.

- *Time-of-use (ToU) tariffs*: ToU tariffs are based on pricing, which varies between individual time periods. The advent of smart metering technologies where electricity consumption can be recorded digitally in very small discrete time periods has been a necessary development for the adoption of ToU pricing. Apart from commercial and industrial customers, ToU pricing is still not widespread despite smart metering becoming increasingly available. According to Faruqui and Palmer (2011, p. 17), "only 4 of 1755 respondents to a 2009 Federal Energy Regulatory Commission survey of utilities indicated they had non-experimental dynamic pricing programs in place for residential customers." Only 3% of US electricity consumers are billed based on ToU tariffs (Sioshansi, 2013a, p. 15).
- *Capacity and energy blended tariffs*: Where smart metering technologies are available, utilities are able to charge customers for their use of both energy and capacity. Energy usage is priced using either average cost or ToU tariffs, while capacity charges are based upon a customer's maximum demand measured in kilovolt-ampere or kilowatt.

Most residential electricity tariffs in Australia, as in many other parts of the world, are largely based on conventional two-part tariffs, with energy throughput measured in kilowatt-hour, utilizing either average cost or inclining block structures. As an example, current Queensland residential tariffs are 20.1 cent/kWh with a fixed charge, irrespective of individual household connection peak demand requirements, of 79 cent/day (Simshauser, 2014). For a household consuming an average 7000 kWh per annum, 83% of the total annual bill of $1695 is based on energy throughput with the remaining 17% effectively a smeared average cost connection charge. Tariffs in New South Wales and South Australia are a mix of average cost and inclining block pricing structures. Only Victoria has permitted the widespread use of ToU pricing due to a mandated rollout of smart metering technologies.

"Average cost" flat tariff pricing is used despite some regions of Australia's NEM being among the peakiest in the world.[3] Figure 1 shows the average and peak day load for a sample of 145,854 AGL customers that have smart meters installed.

In Figure 1, the peak demand of these households is approximately double that of underlying average maximum demand. The consequence of such a situation is extremely poor capacity utilization, further described in Section 3. Australia's Productivity Commission recently found that some 25% of retail electricity bills are required to meet around 40 hours of critical peak demand each year (Simshauser and Nelson, 2013).

The limitations of average cost and inclining block tariffs within the context of the peak load problem have been known for over 60 years. Boiteux (1949),

---

3. One of the most "peaky" regions in the NEM is South Australia. Only 0.5% of customers in South Australia have a smart digital meter installed.

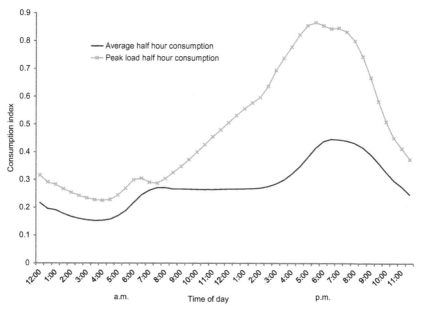

FIGURE 1   Average and peak load residential electricity consumption profiles for a sample of AGL customers. *Source: Data provided by AGL Energy.*

Houthakker (1951), Williamson (1966), and Turvey (1967) provided useful explanations of how average cost or inclining block tariffs misallocate resources in an electricity system characterized by high peak demand relative to underlying energy demand. Houthakker (1951, p. 5) summed up the consumer response to higher average cost or inclining block tariffs succinctly: "Consumers will naturally economize at the times least inconvenient to them, and these are likely to be off-peak periods when no reduction of demand is required."

The consequence of continuing to persist with average cost or inclining block tariffs in a market with a significant peak load problem (in a nonenergy-limited power system) is an "energy market death spiral." This can be explained as the effect of declining underlying energy demand accompanied by rises in critical peak demand, causing poorer capacity utilization. Prices must increase and this, in turn, results in further declining energy demand, further per unit cost increases, and so on, in an ever downward spiral (Severence, 2011, p. 13; Simshauser and Nelson, 2014).

ToU or capacity-based demand tariffs are likely to provide a more efficient economic outcome. Households that consume energy at times of peak demand, where significant infrastructure is required, would pay more for their electricity than those households that consume energy at other times.[4] This may present

---

4. Emerging tariffs are being proposed that attempt to "blend" capacity and ToU pricing. Nelson and Orton (2013) provided an example of this type of approach.

equity considerations, however. Households that have installed a solar PV system may be entitled to object to any rebalancing of tariffs in just the same way that utility-scale investors in electricity generation consistently cite the need for investment certainty (see Nelson et al., 2010, 2012b, 2013). In addition, the essential service nature of electricity supply requires utilities to consider the affordability impacts of redesigning electricity tariffs on low-income and hardship consumers.

3   OTHER REASONS WHY UTILITIES ARE CONSIDERING ALTERNATIVE TARIFF STRUCTURES

Apart from the economic inefficiency of existing tariff structures and the prospect of a death spiral, there are new factors emerging that are leading utilities to reconsider tariff design. Quezada et al. (2013) had argued that the Queensland electricity system is facing a "perfect storm" of socioeconomic, demographic, and technological changes largely driven by a need to address climate change adaptation and mitigation policies. Specifically, large capital expenditure is required to upgrade and replace aging electricity networks; electricity demand is declining; and as long as pricing continues to be largely based on "average cost" tariffs, significant increases in retail electricity prices will result.

Australian residential electricity prices have indeed risen substantially in recent years—doubling or close to doubling in most jurisdictions (Simshauser and Nelson, 2013). The Australian Energy Market Commission (2011, p. 2) stated that higher tariffs are largely due to "peak demand, higher commodity prices, replacing aging assets and higher costs of capital due to the global financial crisis." This view is reinforced in Figure 2, which shows the increase in Sydney (New South Wales) residential electricity prices between FY08 and FY13. Prices in Sydney over this time period have more than doubled overall. Notable among the increases are the tripling of network distribution costs and the more than doubling of fuel costs for generation. The significant rise in "average cost" electricity tariffs has had two primary consequences: a reduction in energy demand and an uptake in substitute products, not only primarily embedded solar PV generation but also energy efficiency.

In relation to reductions in electricity demand, Sood and Blanckenberg in Chapter 19 provide interesting insights into historical demand and renewable energy projections in Australia. Figure 3 highlights the gap between industry expectations and actual whole of system demand growth. The revised projections for the future are also shown. These projections are developed by the Australian Energy Market Operator (AEMO).

Underlying energy demand is at the same level today as it was back in the middle of last decade. Furthermore, under the low scenario demand projection of Figure 3, it is forecast that demand may be no higher in 2020 than it has been at any time in the last decade. IES (2013) established that reduced energy

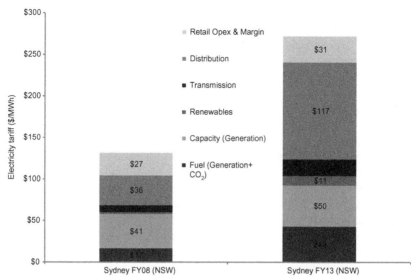

FIGURE 2   Increase in Sydney electricity prices between FY08 and FY13. *Source: Simshauser and Nelson (2013).*

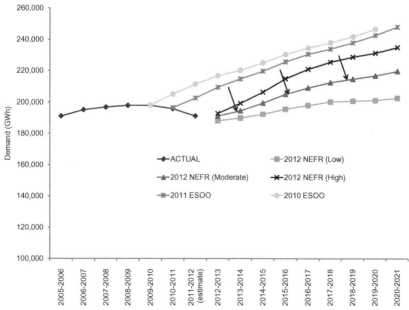

FIGURE 3   Reductions in actual and underlying National Electricity Market (NEM) energy demand. *Source: AEMO forecasts reproduced by AGL Energy (2013).*

demand over the past few years may be attributed to a number of causes: variations in weather, increased uptake of solar PV, solar hot-water installations,
closures of large industrial facilities such as aluminum smelters, and greater
energy efficiency.

However, while underlying energy demand has been declining for several
years, peak demand has not (Figure 4). Peak demand across the NEM has grown
by between 20% and 37% over the period 2001-2012—approximately double
the rate of underlying energy demand growth (Simshauser and Nelson, 2013,
p. 69).

There has been a material reduction in peak demand in 2012 (Figure 4).
However, this is because the weather has been significantly more temperate.
High-temperature days (over 35 °C) during FY12 were 9 and 6, respectively—down from 25 and 16 in previous summers. Furthermore, much of
the distribution-related capital expenditure is related to localized network peak
demand. As material reductions in demand have occurred outside many of these
network areas, peak demand is still driving capital expenditure. Indeed, Energex, which serves metropolitan Brisbane in Queensland, continues to forecast
higher peak demand growth compared to underlying energy demand. Accordingly, Australian utilities are right to be concerned about continued peak
demand growth, reduced underlying demand growth, and reduced capital
utilization.

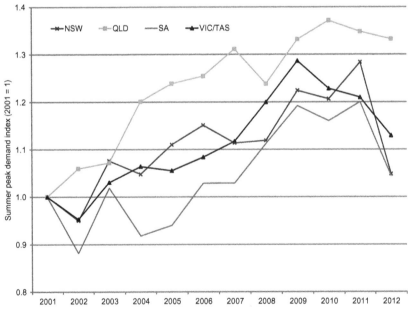

FIGURE 4 Non-weather-corrected whole of system peak demand in NEM jurisdictions. *Source:
Data provided by AEMO.*

The emergence of embedded solar PV and its widespread adoption have compounded the problem of capacity utilization. Mountain and Zsuster, in Chapter 4, explain the factors contributing to the success of solar PV adoption in Australia—where it has now reached one in eight Australian households. After several decades of increasing capacity utilization, and falling long-run average costs, there has been a sharp decline in capacity utilization since 2008, coinciding with the emergence of embedded solar PV (Figure 5). This uptake of solar PV as a genuine partial substitute to expensive grid-supplied electricity has become the new focus of policy makers.

Analysis by AGL Energy shows that at the end of 2012, embedded solar PV output now comprises an estimated 1.9% of total aggregate NEM demand and 5.5% of aggregate residential demand. In some states, solar output as a share of residential demand is as high as 12%, while in Queensland, almost one in five customers now has a PV unit (AGL Energy, 2013). But while there have been significant reductions in grid-supplied energy demand from households with solar PV installed, it has not occurred at the time of peak demand as illustrated in Figure 6.

For New South Wales and Queensland, for example, average peak household consumption and peak day household consumption both occur significantly after the time period in which solar PV is producing material amounts of energy (Figure 6). This has significant consequences for the efficiency and equity of tariffs charged by utilities. "Average cost" tariffs that are based

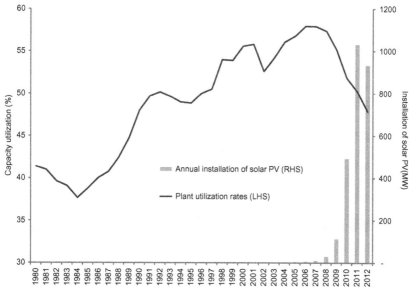

FIGURE 5   Capacity utilization and installation of solar PV. *Source: ESAA (2012) and Clean Energy Council (2013).*

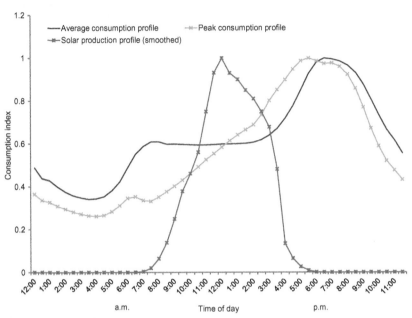

FIGURE 6   Solar PV production and household consumption indexes. *Source: Data adapted from data provided by AGL Energy and Energex (2010).*

solely on energy throughput rather than capacity requirements allow households with solar PV to significantly reduce their electricity bills, while peak demand infrastructure requirements remain the same. That is, households that have reduced consumption from the network have avoided (throughput) charges associated with using the network. Net metering and net feed-in tariffs make this problem worse. The problem is by no means unique to Australia. In Chapters 3 and 6, Burger and Weinmann and Groot discuss the significant impacts of renewable energy on European utilities, while in Chapter 8, Keay et al. conclude that consumers are faced with "confusing and perverse pricing signals."

In 2013, the Energy Supply Association of Australia commissioned an economic consultant to review current tariff designs, with a particular focus on the equity impact of embedded generation. In its report, ACIL Tasman (2013) found that the average household with solar PV is paying $241 per year less in network charges than other users. This is illustrated in Figure 7. Based on this analysis, the reduction in distribution network revenue alone for the 1 million households with solar PV installed is likely to be in the order of $1.25 billion over a 5-year period. This is a nontrivial problem for utilities given the infrastructure required to meet demand has not diminished. It is also an issue of equity between those households with and without solar PV, something that has been raised in Australian policy debates regularly (see Mountain and Zsuster in Chapter 4; Nelson et al., 2011, 2012a).

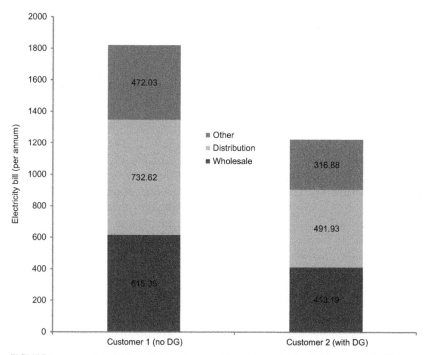

FIGURE 7   Annual electricity bill for customers with and without solar PV. *Source: ACIL Tasman (2013, p. v).*

Not surprisingly, utilities are considering alternatives to current "average cost" tariff structures for a number of reasons: addressing the worsening capital utilization rates caused by the "peak load problem," climate change adaptation and mitigation requirements as identified by Quezada et al. (2013), and the growth in embedded generation reducing revenues as consumers have avoided costs without lessening peak infrastructure loads. Sioshansi (2013a, p. 8) summarized the problem with average tariffs succinctly: "Under rate of return regulations, to keep the local utility whole, retail tariffs must be raised. That simply encourages more customers to bypass the grid-supplied electricity." In the European context, it has been estimated that tariffs would need to increase by around two-thirds over the coming decade to recover lost revenue due to volumetric declines from both embedded generation and reductions in electricity demand (Citi Research, 2013, p. 35).[5] This is not surprising given Germany has

5. It is unclear whether wholesale market redesign will be required to facilitate changes in technology. Keay et al. and Sood and Blanckenberg in Chapters 8 and 19, respectively, provide insights in relation to wholesale market reform. Importantly, the latter conclude that "there is no fundamental reason to believe these developments would undermine the efficient functioning of the market."

more than 1 million independent producers of power now connected to the grid (Chapter 3), and Groot estimated that installed distributed capacity had increased from 10 to 70 GW in 2010-2012 (Chapter 6).

4   THE INDUSTRY RESPONSE—NEW TARIFF DESIGNS

Australian utilities are considering a number of options for addressing the inefficiency and inequity of existing "average cost tariffs." These options include significantly increasing the "fixed" component of the "average cost" tariff; introducing a peak demand or kilovolt-ampere charge; and ToU pricing, including dynamic pricing. The advantages of considering these tariff options are obvious. Recall from Figure 6 that household electricity load factors in Australia are very poor. By moving toward ToU or kilovolt-ampere peak demand charges, utilities can deploy price signals that attempt to reduce critical peak demand, thereby reducing the need for future investment and improving capital utilization rates. Simshauser and Downer (2012) provided a good overview of why ToU pricing leads to a more economically efficient outcome in the Australian context, and Nelson and Orton (2013) illustrated one example. In Chapters 8, 9, and 10, tariff design is discussed by Keay et al., King, and Cazalet, respectively, with the latter's discussion particularly important given its focus on longer-term pricing and consumer contracts.

Australian electricity retailers are the most prominent advocates of the introduction of ToU-based electricity tariffs in Australia (see Stewart and Nelson, 2012). This is most likely due to the significant costs associated with wholesale price risk management in the NEM, where prices can increase from an average of around $30/MWh (ex carbon price) to $13,100/MWh in a 30-min trading interval. Networks may not prefer ToU electricity tariffs because individual localized network peak demand infrastructure requirements at the substation level may well be correlated with an alternative time relative to system-wide peak demand. Individual components of the network may peak at different times and capacity cannot be transferred from one isolated part of the network to another.

Accordingly, much of the debate in Australia has focused on increasing the daily fixed charge component of existing two-part tariffs (i.e., irrespective of the level of the household's individual peak demand). Done in complete isolation, it would seem to us to be somewhat inappropriate to introduce higher fixed charges unless it formed part of broader reform toward cost-reflective pricing. Increasing daily fixed charges across all customers in isolation would be arbitrary in nature and penalize many customers without regard to their contribution to industry peak load infrastructure requirements. On the other hand, the introduction of kilovolt-ampere peak demand or ToU pricing in conjunction with changes to daily fixed charges would be a more efficient and equitable policy for pursuing cost reflectivity for individual customers.

However, the application of a peak demand kilovolt-ampere pricing methodology to customers with embedded solar PV is likely to require greater consideration by network companies. At face value, it would appear that introducing kilovolt-ampere demand-based fixed charges would be ideal for ensuring allocative efficiency. From Figure 6, the average customer with a solar PV system installed would be required to pay a fixed charge based on their peak demand, which occurs around 5.30 p.m. Given the profile of solar production in Australia, solar output appears to do little to reduce this peak demand given that "peak" solar output and peak demand do not coincide. The household would therefore pay network charges to reflect their use of the network at 5.30 p.m. irrespective of the *output* produced by their solar PV system.

To demonstrate the impacts of networks rebalancing their tariffs along these lines, two scenarios have been modeled that calculate the payback period for the installation of a 1.6 kW system with a current capital cost of $2490.[6] Based on current tariff structures, a consumer installing a 1.6-kW system that produces approximately 2500 kWh/year would achieve an economic payback of between 6 and 7 years. Much of the benefit that accrues to the customer is due to net metering arrangements whereby households consume energy behind the meter to the extent they can with energy that is surplus to requirements exported to the grid. However, if the customer is levied $241 per year in additional network charges due to the introduction of new network tariff structures, the payback period would increase to between 11 and 12 years. Therefore, cost-reflective pricing for the *average* Australian customer would significantly reduce the economic attractiveness of solar PV.

Australian utilities should be cautious though in assuming that such a scenario would simply restore revenues to levels experienced prior to the significant uptake in solar PV that began in 2009. There are now over 1 million Australian households with solar PV on their roofs. Each of these households is likely to view their investment in solar as a "sunk cost." In other words, the investment case for any technology that allows the consumer to shift the output of their solar PV system to the time of their peak demand would not include the cost of the existing solar PV system.

It is therefore quite possible that if utilities reduce the economic attractiveness of existing solar PV systems by introducing peak demand or kilovolt-ampere-related network tariffs, they may bring forward investments in batteries that allow consumers to shift their PV output from the middle of the day to the time in which their peak demand occurs. This is important given current battery prices and estimates of how they may trend. At present, a 5-kWh battery

---

6. Capital cost obtained from Origin Energy Web site (http://www.originenergy.com.au/174/Solar-power). Accessed online on 22 September 2013. Origin is one of the largest utilities and installers of solar PV in Australia.

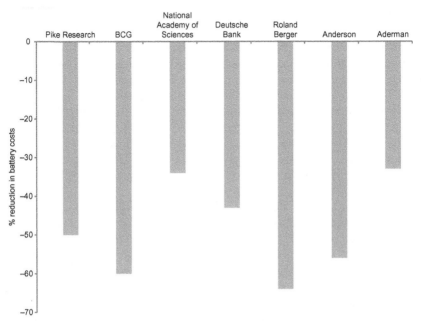

FIGURE 8   Predicted battery cost reductions to 2020. *Source: Adapted from ESAA (2013, p. 35).*

requires a capital investment of approximately \$16,250.[7] But most commentators expect that battery prices will decrease significantly over the coming decade (Figure 8). This would of course be a game-changing event for utilities. Beyond batteries, future advances in fuel cells and other devices including EVs to store extra generation may become reality.

Analysis completed within the Australian context on the economics of a solar PV system combined with battery storage provides insights that utilities will increasingly focus on. SunWiz (2013), one of Australia's leading solar consultancies, had estimated the payback period on solar PV with storage. Storage in a ToU pricing environment effectively lengthens the payback period for an embedded generation system from around 6 years to approximately 13 years. At a total cost of \$21,900, the current 3 kW solar PV system with storage may well be beyond the reach of most Australian households as average Australian household disposable annual income is only \$47,736 (ABS, 2012). However, if battery prices were to halve, the system installation price could fall as low as \$14,000—effectively providing a household with a payback period of less than 10 years. This is not too dissimilar to the economics of current

---

7. Capital cost sourced from *RenewEconomy* (http://reneweconomy.com.au/2013/battery-storage-the-numbers-dont-add-up-yet-97438) reporting of the International Battery and Energy Storage Alliance and adjusted using an exchange rate of \$AUD 1 = €0.70. Accessed online on 23 September 2013.

stand-alone solar PV investments in Australia—which are still occurring in significant numbers.

This analysis is based on households that *have not* installed solar PV to date and would be making a decision on the combined economics of solar PV and storage. For those 1 million households that have installed solar PV, which would be considered a "sunk cost," battery costs declining by half may raise the interest of consumers in such a proposition. This proposition would be accelerated where utilities provide an arbitrage opportunity by shifting to peak demand or ToU tariffs. Ironically, if "average cost" pricing remained in place, no arbitrage opportunity would exist, and it would be uneconomic for households to install a battery even when they have already installed solar PV.

The consequences of this straightforward analysis are important for utilities to consider. If battery technology did not exist, shifting away from flat "average cost" tariffs to peak demand or ToU tariffs would be an ideal solution for utilities to "compete" with solar PV—the new partial substitute to grid-based electricity supply. However, if utilities were to shift to peak demand tariffs today and battery costs did decline, one would expect a nontrivial number of the approximately 1 million Australian electricity households with solar PV to install a battery. The 1 million are also a significant political voice: it may not be so easy or popular to change tariffs overnight.

If households are inclined to install batteries, there are two implications for utilities. Firstly, nonfixed charge utility revenue would continue to decline. The problem being experienced today (declining revenue from reduced grid throughput) would be replayed once again in the future. Households would pay less network charges as batteries were progressively deployed following the introduction of demand tariffs. Secondly, significant components of the grid would be effectively stranded as households no longer utilize them to the extent they did in the past. This would strain the relationship utilities have with regulators and policy makers and may lead to asset write-downs.

Electric utilities are presented with an interesting dilemma. Moving away from "average cost" tariffs to more ornate designs such as ToU or demand tariffs will reduce the incentive for the installation of new solar PV systems in the short run—thus reducing pressure on the utility businesses. However, in the long run, such tariff designs may well accelerate the deployment of energy storage technologies. This may be a more damaging prospect for utilities than solar PV alone. Wright (2013) succinctly summed up the overarching issue in the Australian context:

Australian power companies (and state government owners) are seeking to maintain their old world revenue via increases in per kWh usage charges, or they are pushing for higher fixed daily network service charges that you can't do anything to minimise. By doing either of these two things they will encourage energy efficiency and the installation of ever more rooftop solar with the addition of storage to eliminate as much household demand as possible.

5   OPTIMIZING THE UTILITY RESPONSE—EMBRACING CHANGE

A recent Boston Consulting Group (BCG) report argued that "driven by its increasingly favorable economics in some locations, solar PV paired with battery storage is now on the cusp of viability" (Pieper et al., 2013, p. 1). The report continues that "for many consumers, it is now less expensive to consume self-generated electricity from solar PV with battery storage than it is to use electricity from the grid" (Pieper et al., 2013, p. 3). The study found that the technology combination is likely to be cost-effective for German residential consumers from 2014. It is not surprising then that Navigant Research has forecast that annual installation of embedded solar PV will increase from around 30 today to around 75 GW in 2020 (Sioshansi, 2013b).

Firms are rarely successful in "holding back the tide." If solar PV technology is economically competitive relative to grid-based supply, consumers will vote with their feet. Even with existing and emerging technological trends, there are at least two complications that will make utilities necessary in the future: the physical limitations associated with going "off-grid" and the essential service nature of electricity supply. It would appear that solar PV supply and grid-based electricity supply are not perfect substitutes, hence our use of the term "partial substitutes." In other words, there is not a constant marginal rate of substitution.

First, there are some rather obvious limitations to the adoption of solar PV by households as a genuine substitute to grid-based electricity supply. Average Australian electricity usage in 2013 is around 6.5 MWh/year (ESAA, 2012). To generate sufficient supply to meet this consumption, households would be required to install a large 5 kW system. Such a system requires 50 m^2 of rooftop space.[8] This is problematic given that one-quarter of Australia's population does not live in a free-standing house (ABS, 2013). Furthermore, many households do not have rooftops that would allow 50 m^2 of solar PV panels to be deployed in a way that would enable optimum output and esthetics to be maintained. In fact, the CSIRO has estimated that because of these types of limitations, the medium cost of household disconnection is $0.47/kWh with a low and high range of $0.23 and $0.93/kWh, respectively (Graham et al., 2013). These costs are significantly higher than current tariffs and ignore the additional costs that would be required to produce another 2500 kWh if the household requires electric vehicle charging capabilities at some point in the future.[9]

---

8. Solar PV output, system size, and rooftop space requirements obtained from Origin Energy Web site (http://www.originenergy.com.au/174/Solar-power). Accessed online on 22 September 2013.
9. Some of these limitations can be overcome by shared ownership of large units collectively, and such models are beginning to emerge. Energy "cooperatives" increased almost tenfold from 66 in 2001 to 656 in 2012 in Germany (Chapter 3), and there are likely to be few barriers to cooperatives evolving in Australia.

Secondly, one of the most critically overlooked aspects of modern electricity systems relates to their provision of an essential service. Quezada et al. (2013) made the salient observation that utilities must be cognizant of "fuel poverty" in how they respond to technological and demographic changes. At present, these same social obligations are not placed on the providers of solar PV. For solar PV to be a true "grid substitute" technology, providers would be required to have systems and processes in place to respond when consumers lose supply. Such arrangements require material economies of scale, which is one of the very reasons large centralized grids emerged as the industrial organization of choice for electricity supply in the twentieth century.

Electric utilities will need to consider how to design their tariffs and business models to "coexist" with the rapid deployment of embedded solar PV. Tariff redesign to ensure more cost-reflective pricing signals is an important reform that utilities and regulators should pursue. Rather than attempting to slow down the uptake of these new technologies, utilities should embrace these trends and consider how they can add value for consumers in adopting them. It is also important that utilities remind customers why a centralized grid has been so successful to date, promoting the benefits of reliable electricity supply and immediate responsiveness to outages, which restores interrupted supply quickly.

Interestingly enough, on the matter of asset stranding, it could be argued that there is an "implied" social contract between society and regulated electricity network businesses. Network businesses have been constrained from earning monopoly profits through economic regulation. Any reduced utilization of the network in terms of energy throughput may require society to reassess how it values the network. Higher fixed charges may in fact be acceptable to society if the alternative is reduced reliability. In fact, the European utility industry association, Eurelectric, has stated that "future business models could be based on revenue per customer rather than volumetric supply" (Citi Research, 2013, p. 34). It is important that utilities begin conversations with policy makers and civil society today to address these issues of future energy supply.

Electric utilities do not have a particularly innovative track record. NRG Energy's CEO, David Crane, recently commented that the traditional US utilities are "an industry of Neanderthals" and that it has been "the least innovative industry in America, maybe the world, in history" (Sioshansi, 2013a, p. 7). Utilities will need to change the image society has of their operations to be successful in the future, particularly in relation to discussion of tariff reform.

Riesz et al. in Chapter 23 provide an interesting summary of the potential futures facing Australian utilities. Their Big Green, Small Green, and Off Green scenarios are worthy of further consideration. Decarbonization, energy security, affordability, reliability, electrification, and other drivers will impact the speed and direction of change within the energy industry. An important conclusion that all utilities will need to address is that pricing and tariff design are no longer important just to achieve "allocative efficiency." Tariff design is a critical component of the industry's ability to adapt given the entrance of new partial substitutes to grid-connected electricity supplies.

6   CONCLUSION

In considering how to respond to a new substitute product, any firm in any industry must consider a number of aspects of its own products and the substitute. Most important is the degree of homogeneity between the two products. In the case of embedded solar PV and grid-based electricity, a key consideration relates to security of supply. Even with a battery, most households will not want to be able to disconnect themselves from the grid. As such, embedded solar PV is at best a partial substitute for grid-based electricity as the grid provides consumers with utility that solar PV cannot—effectively unlimited energy supply. However, utilities need to adjust their business models to address the rise of distributed generation. The headwinds being experienced by utilities the world over are only likely to intensify. Analysts are questioning the value of existing business models and infrastructure. Importantly, utilities can draw upon the experiences of other industries where partial or complete substitute products have emerged and new business strategies have been developed by necessity. Sioshansi considers some of these examples in Chapter 7.

This chapter has assessed different types of tariffs that utilities may consider introducing in response to the range of challenges associated with the rise of distributed generation. At one level, the conclusions appear obvious. In the short run, moving away from "average cost" tariffs to demand tariffs or ToU pricing may lead to a reduction in the demand for new solar PV units. But ironically, such structural changes to tariff design may also accelerate the deployment of batteries for existing solar PV customers who consider their investment sunk. This may create even greater financial duress for utilities. In a practical sense, the introduction of ToU pricing at the retail level and demand charges at the network level is an important way of providing the right pricing signals for the efficient consumption of energy network use. Ultimately though, utilities will need to think carefully and creatively as to how they charge customers for the implicit benefits they provide, the most important one being reliability. Such a decision should not be taken lightly and should be made only after detailed discussion with policy makers and civil society.

REFERENCES

ACIL Tasman, 2013. Distributed Generation: Implications for Australian Electricity Markets. ACIL Tasman Publication, Melbourne.

AGL Energy, 2013. Market Update—Falls Creek. Investor Presentation. Available at: http://www.agl.com.au/about-agl/media-centre/article-list/2013/may/merchant-energy-presentation-may-2013 (Accessed on 27 August 2013).

Australian Bureau of Statistics (ABS), 2012. Household Income and Income Distribution. Available at: http://www.abs.gov.au/AUSSTATS%5Cabs@.nsf/mediareleasesbyCatalogue/1C05E178211SD933CA25732A0021E90E?Opendocument (Accessed on 23 September 2013).

Australian Bureau of Statistics (ABS), 2013. Australian Social Trends. Available at: http://www.abs.gov.au/AUSSTATS/abs@.nsf/Lookup/4102.0Main+Features30April+2013#back7 (Accessed on 22 September 2013).

Australian Energy Market Commission, 2011. Retail electricity price estimates—final report for 2010-11 to 2013-14. AEMC Publication, Sydney.

Boiteux, M., 1949. La tarification des demanded en pointe: application de la theorie de la vente au cout marginal. Revue Generale de l'Electricite. Translated by H. Izzard to Boiteux, M., 1960. Peak load pricing. J. Bus. 33 (2), 157–180.

Borenstein, S., 2005. The long-run efficiency of real-time electricity pricing. Energy J. 26 (3), 93–116.

Citi Research, 2013. Pan-European utilities: the lost decade, where next. Citi Research Equities Note, 17 June.

Clean Energy Council, 2013. Clean Energy Australia 2012. Clean Energy Council Publication, Melbourne.

Energex, 2010. Networks of the future. In: Presentation to QLD Power and Gas Conference, October 2010.

Energy Supply Association of Australia (ESAA), 2012. Electricity Gas Australia. ESAA Publication, Melbourne.

Energy Supply Association of Australia (ESAA), 2013. Outlook for Electric Vehicles in Australia. ESAA Publication, Melbourne.

Faruqui, A., Palmer, J., 2011. Dynamic pricing of electricity and its discontents. Regulation Magazine. (October), 16–22.

Faruqui, A., Sergici, S., 2010. Household response to dynamic pricing of electricity: a survey of 15 experiments. J. Regul. Econ. 38, 193–225.

Graham, P., Dunstall, S., Ward, J., Reedman, L., Elgindy, T., Gilmore, J., Cutler, N., James, G., 2013. Modelling the Future Grid Forum Scenarios. CSIRO, Clayton South, Victoria.

Houthakker, H., 1951. Electricity tariffs in theory and practice. Econ. J. 61, 1–25.

Intelligent Energy Systems (IES), 2013. What is Driving the Decline in Electricity Demand. IES Publication.

Nelson, T., Orton, F., 2013. A new approach to congestion pricing in electricity markets: improving user pays pricing incentives. Energy Econ. 40, 1–7.

Nelson, T., Kelley, S., Orton, F., Simshauser, P., 2010. Delayed carbon policy certainty and electricity prices in Australia. Econ. Pap. 29 (4), 446–465.

Nelson, T., Simshauser, P., Kelly, S., 2011. Australian residential solar feed-in tariffs: industry stimulus or regressive form of taxation? Econ. Anal. Policy 41 (2), 113–129.

Nelson, T., Simshauser, P., Nelson, J., 2012a. Queensland solar feed-in tariffs and the merit-order effect: economic benefit, or regressive taxation and wealth transfers? Econ. Anal. Policy. 42 (3), 277–301.

Nelson, T., Simshauser, P., Orton, F., Kelley, S., 2012b. Delayed carbon policy certainty and electricity prices in Australia: a concise summary of subsequent research. Econ. Pap. 31 (1), 132–135.

Nelson, T., Nelson, J., Ariyaratnam, J., Camroux, S., 2013. An analysis of Australia's large scale renewable energy target: restoring market confidence. Energy Policy. 62, 386–400.

Pieper, C., Ostermeyer, H., Konecny, P., Hering, G., Rubel, H., 2013. Solar PV Plus Battery Storage: Poised for Takeoff. bcg.perspectives, Germany.

Quezada, G., Grozev, G., Seo, S., Wang, C., 2013. The challenge of adapting centralised electricity systems: peak demand and maladaptation in South East Queensland. Reg. Environ. Chang. (May), 1–11.

Severence, C., 2011. A practical, affordable (and least business risk) plan to achieve 80% clean electricity by 2035. Electr. J. 24 (6), 8–26.

Simshauser, P., 2014. When does retail price regulation become distortionary. Aust. Econ. Rev, Forthcoming.

Simshauser, P., Downer, D., 2012. Dynamic pricing and the peak load problem. Aust. Econ. Rev. 45 (3), 305–324.

Simshauser, P., Nelson, T., 2013. The outlook for residential electricity prices in Australia's national electricity market in 2020. Electr. J. 26 (4), 66–83.

Simshauser, P., Nelson, T., 2014. The consequences of retail electricity price rises: rethinking customer hardship. Aust. Econ. Rev, 47 (1), 13–43.

Sioshansi, F., 2013a. EEnergy Informer: The International Energy Newsletter. 23 (11).

Sioshansi, F., 2013b. EEnergy Informer: The International Energy Newsletter. 23 (8).

Stewart, A., Nelson, T., 2012. Leveling the high cost of peak power. The Australian Financial Review. (January), 47.

Sunwiz, 2013. Storage: The business case. In: Presentation to Clean Energy Week—Clean Energy Council Annual Conference, Brisbane.

Turvey, R., 1967. Peak load pricing. J. Polit. Econ. 76 (1), 101–113.

Williamson, O.E., 1966. Peak-load pricing and optimal capacity under indivisibility constraints. Am. Econ. Rev. 56 (4), 810–827.

Wright, M., 2013. Utilities creating their own battery-powered demise. Climate Spectators. Available at http://www.businessspectator.com.au/article/2013/8/18/smart-energy/utilities-creating-their-own-battery-powered-demise (Accessed on 19 August 2013).

Chapter 14

Industry Response to Revenue Erosion from Solar PVs

Jonathan C. Blansfield and Kevin B. Jones

ABSTRACT

Residential solar photovoltaic installation is the fastest-growing source of distributed energy generation in the United States and globally. This chapter presents three case studies examining the relationship of increased distributed energy resource (DER) adoption and the evolving electric utility. Arizona Public Service is struggling to curtail the growth of solar photovoltaics in its service territory to limit cost shifting and maintain profitability. San Diego Gas and Electric offers a comparison between the two poles of acceptance, as it works with policy makers for rate reform and integration of new technologies. In Vermont, Green Mountain Power is aggressively pursuing a wide range of cooperative solar projects in the City of Rutland and leveraging public benefit to achieve policy success in other areas. These three examples demonstrate that achieving regulatory and public support for DER and ensuring utility profitability will require a more sophisticated and strategic long-term approach beyond rate reform.

Keywords: Solar PV, Distributed generation, Net metering, Electric utility, Customer generation

1 INTRODUCTION

Electricity derived from distributed energy resources (DERs) is rising rapidly, with solar photovoltaics (PV) accounting for much of the recent growth. By the end of 2013, the amount of cumulative solar PV capacity installed in the United States will surpass 10 GW (Tech Media Research, 2013). Additionally, there will be over 100,000 individual solar systems deployed and the United States' global share of solar installations will reach 13%, up from 5% in 2008. Further, the real cost of solar is rapidly approaching grid parity in many parts of the United States and in western Europe. Recently, former FERC Chairman Jon Wellinghoff stated, "solar is growing so fast it is going to overtake everything" (Trabish, 2013a). These sentiments echo the progress that solar is making around the globe and underscore the challenges presented to utilities domestically and internationally.

Distributed Generation and its Implications for the Utility Industry. http://dx.doi.org/10.1016/B978-0-12-800240-7.00014-X

While these gains are impressive, controversy has arisen as solar PV capacity grows and threatens to displace traditional electric utility revenue, which funds the existing electric grid. This tension between growing solar PV and the traditional electric grid is driving a need for electric utilities, legislators, regulators, and other stakeholders to cooperatively address the technologies' benefits and challenges, to ensure that both DER and public utility goals are achieved.

This chapter is organized into three sections, each analyzing an industry response to a growing role of solar PV and the attendant revenue erosion issue. Section 2 will introduce foundational information to contextualize the specific threats that utilities perceive, including the politically contentious regulatory mechanism of net energy metering (NEM). Section 3 presents three case studies covering the challenge that DERs pose to utilities and explores the three utilities' varied approaches to the problem. Section 4 describes lessons learned from these varied experiences that offer insight into future models for reform, followed by the chapter's conclusions.

2   UNDERSTANDING THE THREAT POSED BY DERs

The foundation of the traditional utility business model in many jurisdictions remains volumetric sales,[1] meaning the more electricity the customer uses, the more profitable the utility. The energy delivery system put in place decades ago, largely still in place today, served the needs of a growing economy well. The distribution utility is an integral component of the traditional electricity generation and delivery paradigm. In exchange for guaranteed access to electricity for everyone, everywhere, state governments have guaranteed distribution utilities a fair return on investment. In fact, the system has worked so well that it has rendered innovation in certain key respects—for example, customer engagement, delivery technology, decentralized generation, alternative energy, and storage to name a few—unnecessary. This system has cemented in place the status quo, and created a consumer that is largely unaware, uninterested or ambivalent at best about the nature and cost of energy delivery (Tomain, 2008). The desire for the proverbial cold beer and hot shower for a reasonable price remains dominant. Sioshansi and Weinberg discussed how the status quo in the electricity service industry is further threatened by various factors.

The totality of circumstances surrounding DER, including improvements in technology, declining costs for solar PV, and increased concern over greenhouse gas emissions from the electric sector, has changed the DER dynamic. DERs allow host customers to generate their own electricity to reduce and, in some cases, eliminate the amount of power they draw from the traditional

---

1. Some jurisdictions, such as Arizona, rely on this cost recovery methodology. However, some jurisdictions, such as California and Vermont, have instituted revenue decoupling, essentially breaking the link between utility profit and the amount of power produced and sold.

electric grid. DERs, largely in the form of distributed solar PV, have grown dramatically in recent years largely because of the expansion of utility NEM programs in combination with the rapidly declining cost of solar panels. Net metering laws generally allow a customer to connect DER to the grid on the customer side of the electricity meter to offset customer demand. Some states, such as Vermont, allow the customer to not only reduce customer kilowatt-hour demand but also require that the utility pay DER generation a per kilowatt-hour premium for all kilowatt-hour's generated (usually requiring a second utility meter).

Generally, net metering programs allow a customer—on a monthly basis—to at best reduce their monthly bill to zero, though often bill credits can be carried over toward future months' usage. Some states, including California and Vermont, allow what is also called group or virtual net metering, which allows bill credits from DER to be allocated to customer accounts beyond the physically net-metered account. As a result of the success of net metering programs, solar distributed generation (DG) directly challenges the traditional paradigm of who pays for electricity delivery in the United States. These challenges must be met to allow utilities the financial incentive to maintain and improve the electric grid.

Expanding customer access to DER can both be economically beneficial and important in reducing GHG emissions from electricity production. Local distributed generation can reduce utility fuel cost and defer or eliminate future investment in T&D system upgrades. Current solar PV growth is aided by a number of market mechanisms including state renewable incentives, federal tax incentives, growing use of lease financing models, and expanding net metering incentives.

Some industry analysts and electric utilities, given this growth, characterize solar PV as a disruptive technology (Kind, 2013a; Kind, 2013b; Nelder, 2013). Sioshansi details the fundamentals of this argument in Chapter 7. Through the traditional industry lens, the decreasing cost of solar arrays and increased incentives leading to record levels of installation constitute a threat to the traditional utility service model (Kind, 2013a). DER raises utility revenue adequacy issues since net metering customers often have the ability to significantly lower or eliminate their payment toward utility fixed costs. Under the current utility rate paradigm, when one DG customer avoids their share of fixed costs, the rates for customers not participating in net metering can rise. As this trend continues, rising electric rates could lead to further revenue erosion as customers conserve more or seek cheaper energy sources. In the extreme, this is referred to the "utility death spiral" as the distribution utility continues to lose revenue and rates spiral upward. Hanser and Van Horn described the utility death spiral concept. Additionally, while this challenge is formidable, Nelson, McNeill, and Simshauser pointed out that issues such as customer rooftop space inhibit using DERs to replace utility service.

From the utility perspective, while a DER customer may significantly reduce or even eliminate utility charges, the customer still relies on the utility transmission and distribution system for services. At the heart of this debate often lies the issue of solar subsidies and NEM. While net metering programs are popular in some jurisdictions, the possibility for these incentives to allow DER customers to bypass a substantial portion of utilities' fixed cost-based revenue has raised significant debate in many others.

3　THREE DIFFERENT UTILITIES, THREE DISTINCT APPROACHES TO SOLAR PV

A business model that includes the prosumer as a meaningful component of a utility service territory can be sustainable.[2] If anticipated and planned for, the rise of the prosumer does not mean the death of the traditional utility.[3] It is under this backdrop that the authors discuss the approaches to solar PV and net metering at Arizona Public Service (APS) in Arizona, San Diego Gas and Electric (SDG&E) in California, and Green Mountain Power (GMP) in Vermont. These utilities offer different but potentially complimentary approaches to the same issue. For each case study, the authors begin by presenting some historical background on the utility and then present the current state of play of solar DG in each utility's service territory from both an institutional and regulatory perspective. The authors then offer some insights regarding how each utility is attempting to integrate solar PV while addressing the fundamental issue of revenue erosion. While this chapter focuses on three utilities in the United States, as discussed in Chapters 4 and 6 by Mountain and Szuster and Groot, utilities in Australia and the EU are facing similar challenges.

3.1　APS: A Contentious Approach to Solar PV and Net Metering

By the end of 2013, APS will have roughly 296 MW of distributed solar PV cumulative nameplate capacity installed on its system (SAIC, 2013). This includes both residential and commercial installations. By 2025, APS projects that it could have between 2.7 and 5.4 million megawatt-hour of distributed solar PV on its system (Puttre, 2013). APS has deployed the second most solar capacity in the United States, behind California (Green Tech Media Research, 2013). APS—the state's largest investor-owned utility—is fighting for a reformation of the Arizona Corporation Commissions (ACC) net metering policies. Solar PV development and the net metering policy that supports its growth is a contentious issue for APS.

---

2. The prosumer is an electric customer who generates electricity and sells excess back to the utility. This customer relies on the distribution grid to deliver electricity to the utility in addition to back up services when on-site generation is insufficient to meet host demand.

3. "Traditional utilities" may also include a distribution utility. As such, a utility for distribution purposes could remain vertically integrated and is rate-regulated, or it could be one that has fully divested its generation portfolio in a state that requires retail choice. Under either model, the revenue adequacy issues raised by DER growth are similar.

As an investor-owned utility, APS sees solar PV as eroding its ability to earn a fair return on investment and has taken the position that the state should reduce the NEM subsidy for solar. APS is not alone in its view that direct subsidies for solar—such as net metering—are bad policy. In fact, the battle in Arizona is indicative of a wider battle within the electricity industry as to the best way to accommodate solar DER, while continuing to promote cleaner yet stable sources of electricity.

Net metering policies in Arizona have been very successful in achieving significant levels of solar PV penetration. As of June 2013, Arizona had roughly 18,000 rooftop systems, up from 900 systems in 2009 (Puttre, 2013). There is no cap on capacity or percentage of load within a utility's service territory served by solar. APS would like to increase daily demand charges or decrease the amount it pays to generating customers under a standard fixed fee arrangement. APS views the ACC's net metering policies as unfairly burdening the utility with stranded fixed costs and allocating disproportionately higher rates to customers without solar.

Recently, the Edison Electric Institute (EEI), the investor-owned utilities' trade association, captured the debate on the future of the utility in a paper entitled "Disruptive Challenges: Financial Implications and Strategic Responses to a Changing Retail Electric Business" (Kind, 2013a). This report takes the position that while utilities and states should be in the business of exploring and developing renewable and DERs and technologies, this must be accomplished in a way that promotes fairness and reliability and contends that current net metering policy does not (Wilson, 2013). To the contrary, Arizona solar industry and consumer advocates asserted that the existing net metering policy is necessary to ensure that solar continues to grow and offset fossil fuel generation. Solar advocates also point to the importance of a solar serviced load in offsetting or eliminating new infrastructure costs such as construction of new generation and transmission assets.

The apex of the net metering controversy in Arizona was a July 2013 APS submittal to the ACC requesting "modification to the Arizona net metering rules." APS estimated that each of its solar customers was avoiding roughly $1000 annually for traditional grid services they still use, or about $18 million in total. According to APS, this increases costs by $6 to $8 million a year for every year the cost shift remains. To compensate for this cost shift, APS proposed two alternatives for new net metering customers while grandfathering existing customers. As one option, APS proposed a bill credit approach that would allow customers to choose their rate of choice. They would be credited for all of their solar generation at a market rate significantly lower than the retail rate. The second alternative proposed by APS was to allow solar PV customers to continue to net meter, but would be required to take service under an existing demand rate that ranged from $13.50 per on-peak kilowatt in the summer to $9.30 per on-peak kilowatt in the winter (Arizona Corporation Commission Utilities Division, 2013).

In September 2013, ACC staff recommended the commission postpone consideration of the APS request until their next rate case in 2016. The staff noted evident potential for cost shift. However, given that the parties involved dispute the value of distributed generation to the utility versus customers, and given that the issues involved are primarily cost-based questions, the issues raised by APS should be dealt with in a rate case. According to ACC staff analysis, both APS proposals would reduce the average customers' savings by as much as 50% (Arizona Corporation Commission Utilities Division, 2013).

In November 2013, the ACC voted 3-2 to institute a $0.70 per kilowatt charge for net metering customers. Solar installations complete by the end of 2013 would be grandfathered under the current rules. While the change altered the net metering framework by instituting a kilowatt demand charge, the charge was significantly lower than APS proposed. Four of the five commissioners did not want to delay acting on the charge due to cost-shifting concerns with one commissioner supporting the APS proposal. Neither side seemed pleased with the resolution. Solar proponents pointed out that there will be future installations that will no longer be cost-effective, while net metering opponents believed that the ACC did not go far enough in eliminating the cost shift (Trabish, 2013c). Given that the charge is significantly lower than APS proposed and likely to be only a modest monthly surcharge for residential customers, it is difficult to perceive this as more than a token victory for APS.

3.2   SDG&E: Customer-Driven Policies and DG Penetration

At the beginning of 2013, SDG&E customers owned more than 37 MWs of rooftop solar capacity. San Diego has more solar than any other city in the United States, with over 26,000 customers having installed PV systems—a number that grows by the month (San Diego Gas and Electric, 2013; SDG&E, 2013). The state of California is dedicated to achieving a clean energy economy. California has legislatively mandated its utilities achieve 33% renewable energy production by 2030 (California Legislative Service, 2013). Aside from Hawaii, this is the most aggressive state renewable portfolio standard program in the United States (United States Department of Energy and North Carolina Solar Center, 2013). As a part of this larger trend, SDG&E's customer base is largely supportive of solar, and SDG&E customers are well suited to take advantage of solar PV electric generation because of the southern Californian climate. While the number of SDG&E customers that have installed solar (a total 17,969 systems connected to the grid) represents just over 1% of SDG&E's customer base (Puttre, 2013), the issue of fixed cost recovery for DER customers is very much a concern because of the projected growth of residential solar generation. Importantly, the utility is concerned about the issue of revenue erosion yet recognizes the value of DER. SDG&E is taking steps to align their operations with growing residential solar penetration while working with state officials to ensure that the California regulatory paradigm reflects the economic needs of the utility without chilling or halting the growth of solar.

In early 2012, SDG&E petitioned the PUC to allow consideration of a network usage charge (NUC). The NUC would be allocated to solar customers based on the amount of electricity the customer consumed from or delivered to the grid. Under the proposed NUC, the average net-metered customer was estimated to pay $20-$30 dollars more per month. Absent the NUC revenues, any avoided costs would be recovered from all residential customers through future rate increases. In rejecting the NUC, the CPUC viewed an individualized recovery mechanism as discriminatory toward solar PV customers (The Public Utilities Commission of the State of California, 2012).

While the CPUC rejected the fixed charge approach, the debate about solar cost shifts in California continued before the legislature. In October 2013, the bill AB 327 was signed into law by California Governor Jerry Brown changing net metering requirements for the state's three large investor-owned utilities: SDG&E, SCE, and PG&E. The bill clarifies the requirement that these three utilities make available a standard offer contract for eligible customers who apply for inclusion in the NEM program, representing up to 5% of the utilities' aggregate peak load through July 01, 2017. The bill also allows the CPUC to institute a maximum ten dollar per month fee applicable to all customers (Western Electric Industry Leaders Group, 2013). This agreement can be seen as a compromise between the utility and solar advocates. California made a public policy choice to continue to incentivize DER installations while providing utilities the time necessary to counteract or plan for possible revenue erosion from increasing levels of DER penetration (Western Electric Industry Leaders Group, 2013).

In terms of SDG&E's institutional response, in 2012, the utility launched a community solar demonstration project. The "Share the Sun" program will eventually provide up to 10 MW of community solar power. The utility will own the project but the leaseholders will own the Renewable Energy Credits/Certificates (RECs). Additionally, SDG&E rolled out the "SunRate" program, in which utility's customers can elect to have their power supplied from solar projects the utility owns directly. SDG&E is committing up to 10 MW under this project, and customers can elect to have 50%, 75%, or 100% of their usage covered by solar. In this way, the utility is still securing revenue and providing an outlet for customers who wish to invest in green energy and offset their carbon contribution resulting from energy use.[4]

In addition to deploying community solar, working with state legislators to implement cooperative regulatory reform and encouraging private industry to innovate, SDG&E invests heavily in smart grid infrastructure that will support

---

4. More information on these SDG&E initiatives is available at the SDG&E Web site: http://www.sdge.com/clean-energy-options/solar-pilot-programs.

a growing reliance on DG. SDG&E estimates that the projected benefits of deploying smart grid initiatives to increase the capability of accommodating large-scale residential solar PV onto the grid and the related operational efficiencies would outweigh the costs. SDG&E includes in these estimates the quantification of significant environmental and social benefits, amounting from $760 million to $1.9 billion.

Lastly, San Diego, in addition to being a national leader in solar PV penetration, is also a national leader in the penetration of plug-in electric vehicles (EVs). Transportation accounts for almost 40% of California's greenhouse gas emissions, and California has thus been highly successful in promoting vehicle electrification. Among other initiatives, SDG&E has partnered with ECOtality to facilitate the implementation of EV charging infrastructure in San Diego. SDG&E has also been a leader in promoting time of use pricing for EVs in order to encourage charging during the most cost-effective time period. EV charging sales may provide a significant offset to revenue erosion from DER in regions of the country such as San Diego. With similar progressive policies, the growth of PV systems in conjunction with EVs could offset revenue loss from customer DERs as well as more efficient utilization of the electric grid.

3.3　Green Mountain Power: VT and the Solar Capital of New England

The Vermont-based utility Green Mountain Power (GMP), which in 2012 acquired and absorbed Central Vermont Public Service (CVPS), pledged to install at least 10 MW of solar capacity by the end of 2015, giving this Vermont city the most solar capacity per capita than any city in the New England (GMP Solar Power, 2012). In fact, largely as a result of its initiative in Rutland, GMP has been named by the Solar Electric Power Association as the investor-owned utility of 2013 (SEPA, 2013). As part of the 2012 merger between CVPS and GMP, they agreed to maintain a vibrant presence in Rutland Vermont.[5] GMP is committed to the economic redevelopment of Rutland, and the Solar Capital program is central to this effort. While most utilities will not have the same commitment to the cities in which they operate, GMP's investment offers insight into ways in which a utility might engage customers to maximize revenue opportunities.

To achieve the goals of GMP's Solar Capital program, it has and continues to deploy a wide range of solar options, well beyond customer rooftop PV. GMP benefits from Vermont's historical commitment to renewable energy and operates in a more progressive political environment than the average regulated utility. Nevertheless, GMP does not see the growth of DER within its service

---

5. While the corporate headquarters are in Colchester, both GMP's operations headquarters and Energy Innovation Center are located in Rutland, VT.

territory as a threat (Telephone Conversation with Green Mountain Power Executive David Dunn, 2013).

Direct utility involvement in DERs is not only environmentally advantageous; according to GMP, it makes good business sense as well. GMP is developing two 150 kW community solar projects in partnership with NRG Solar Solutions. NRG Solar owns and operates the community arrays, and up to 50 customers per installation sign leases to receive a credit on their GMP electricity bill for the value of the power produced, which participants will compare against the lease payment for the community array.[6] The value of the solar power is divided virtually, with longer leases enjoying a higher credit. Additionally, in February of 2013, the Vermont Public Service Board (PSB) authorized GMP to begin work on a solar streetlight pilot program. This program will be rolled out on a limited basis with the hope of converting each light pole in the city to a generating facility. The entire 9.8 kW system will be net-metered (Email Conversation with Green Mountain Power executive John Tedesco, 2013).

Under Vermont's progressive group net metering rules, GMP will use its distribution network and smart meters to administer the delivery and virtual billing, allowing the third-party generation owner, NRG, to subscribe the asset's electricity and extract its value. GMP sees private ownership of distributed generation assets increasing within its service territory in the near and long term. GMP sees the net benefits of these independent power projects as offsetting, and even eclipsing, the lost revenue from decreased electricity consumption (including load management, less reliance on-peak load plants and O&M cost savings).

To encourage customer adoption and continuing increase in the solar-served load in Rutland, GMP created community solar grants, which are awarded to IRS Code 501(c)(3) legal nonprofit organizations. The grants match investment by the nonprofit organization dollar for dollar, up to $20,000, for proposed solar projects built on-site at the nonprofit organization. The grants are funded through revenue sources that would otherwise accrue to customers rather than utility net income. As of October 2013, GMP awarded four community nonprofit solar grants, resulting in 45 kW of installed solar capacity at a cost of $80,000 to the utility. Three of the grants are roughly 10 kW systems, at PEG-TV (Rutland Regional Community Television), Good Shepherd Lutheran Church, and Vermont Farmers Food Center. In October 2013, GMP awarded Rutland Regional Ambulance Services a grant to help the organization install a 15 kW system. In addition to helping these organizations lower their electricity bills, GMP sees these investments as part of larger load management program to help lower their operating costs particularly by employing peak

---

6. It should be noted that since Vermont does not have an RPS, it is generally the policy of Vermont utilities including GMP to separate the Renewable Energy Certificates (RECs) from these projects and sell the RECs into neighboring state's RPS programs. The utility then uses the REC revenue to offset customer rates. While this negates the benefit from a Vermont GHG perspective and raises policy concerns, it does not affect the utility revenue adequacy issues.

power generation and reducing the need to expand the distribution system (Telephone Conversation with Green Mountain Power Executive David Dunn, 2013).

GMP sees investing in utility scale solar as a vital component to the overall mix of generation within its service territory. As of October 2013, the Vermont PSB approved a certificate of public good for a 2.3 MW utility-owned solar installation on a former landfill in Rutland. Brownfield development poses a viable model for utility partnership with local and regional municipalities. Similarly, GMP also received approval for a utility-owned 150 kW solar array on the grounds of Rutland Regional Medical Center. The utility and medical center entered a 25-year lease, under which the utility will own, operate, and maintain the installation in exchange for a credit to the Medical Center of 10% of the system's output. GMP has also recently completed construction of the Creek Path Solar Array, a 150 kW installation on a brownfield site owned by the utility in Rutland, which formerly was a manufactured gas plant site. Mutually beneficial property use is a viable mechanism to achieve a positive flow of revenue from decentralized renewable generation.

Vermont net metering policies continue to encourage customer solar generation. Vermont requires electric utilities to credit customers with $0.20/ kWh produced from small renewable systems.[7] They are required to do so for all requests up to 4% of peak demand of the previous year, or 1996, whichever is greater, within their service territory. Past 4%, they are not required to pay the additional fee. As of October 2014, three Vermont electric utilities have hit this cap. GMP is not concerned with exceeding an aggregate 4% legislative cap on net-metered installations in the utility's service area.[8] Given the possible advantages GMP perceives as flowing from solar, they are willing to push the envelope past the 4% mark and are currently unconcerned with a threat to their financial stability at least partly as a result of their direct investment in solar facilities.

GMP is also balancing its potential loss of revenue from net-metered customers by encouraging novel and efficient uses of electricity. GMP is using various grants and other sources of revenue to help incent early development of EV charging infrastructure. Transportation is Vermont's largest source of greenhouse gas emissions. GMP is a state leader in promoting the infrastructure for this new source of electricity revenue. Additionally, while Vermont has historically discouraged the use of electricity for space heating through a variety of state policies, GMP has embraced the emergence of new highly efficient airsource heat pumps. GMP has gained both approval for a heat pump leasing program and direct incentives to support customer ownership of the technology.

---

7. In practice, that works out to a 6 cent/kWh solar adder on top of the net-metered energy with the dollar credit "carried over" to reduce a customer's bill for up to 12 months.

8. The 4% cap is determined as a percent of peak kW demand and thus represents substantially <4% of kilowatt-hour sales, given the relatively low capacity factor of solar PV.

GMP is confident that the increased revenue derived from these new electro-technology initiatives will counterbalance lost revenue resulting from DER. Given previous state policy discouraging the use of electric space heating, the current predominant role of expensive fuel oil for space heating, the relative lack of availability of cheaper natural gas in much of the state, and the extremely high efficiency of new electric heat pump technology, it is not unreasonable to assume that GMP's strategy supporting electrotechnologies will have significant benefit to revenue generation.

The utility does not foresee any negative implications from investing in solar as a result of careful corporate planning, community partnerships, and strategic promotion of efficient electrotechnologies. It remains to be seen what the grid or economic threshold for solar saturation might be in Vermont. What is clear is that GMP has embraced continued solar growth and is working to ensure it is managed as part of a larger corporate strategy, so as to be better prepared once this threshold is appreciable on the horizon.

4   LESSONS LEARNED AND POLICY SUGGESTIONS FOR INCREASED SOLAR ACCEPTANCE

In its seminal report, the EEI stated that "if we can identify actionable disruptive forces to a business or industry, then history tells us that management and investors need to take these threats seriously and not wait until the decline in sales and revenues has commenced to develop a new strategy or, in the case of investors, realize their loss" (Kind, 2013a). Unfortunately, identifying the disruptive force is often an easier task than developing a strategic response to counter or take competitive advantage of that force. For regulated electric utilities, the task at hand can be even more challenging because implementing a successful strategy often requires the support of the state regulator, customers, and other stakeholders (Aanesen et al., 2012). These three case studies have identified three varied utility strategies to the current and forecasted growth of DER, particularly solar PV.

In Arizona, APS has led the opposition to increased solar penetration. Revenue erosion, cost shifting among customers, and potentially higher rates, together threaten to dramatically increase the cost of utility provided electric service. APS would either reduce the net-metered rate or increase fixed charges to solar PV customers, which would have a significant chilling effect on solar installations. The solar PV industry, customers, and ACC staff viewed the APS proposals as a direct attack on a popular customer program. While the ACC went further than staff's recommendation to delay action and instead instituted a fixed charge for net-metered customers at $0.70/kW, the new charge can hardly be seen as a victory for APS. It is an order of magnitude lower than the fixed demand charged it proposed and is unlikely to cause a meaningful reduction in DER installations. It is likely that if APS had taken a more collaborative approach toward discussing the challenges presented by the cost shift issue, it may have both achieved a better outcome and not expended so much political capital attacking a popular customer program.

In California, SDG&E is on the frontlines of renewable energy development and faces increased pressure from potential revenue erosion given the geographic and political demographics of Southern California. As a result, SDG&E has taken a two-pronged approach. SDG&E is promoting growth in its leading edge smart grid and clean energy initiatives while exploring new rate mechanisms that are more equitable for all customers. SDG&E has found more support for its proposals when they are framed as encouraging rather than resisting the introduction of DER. While the CPUC rejected SDG&E's proposed NUC, finding the cost allocation method is unfairly discriminatory; it had more success in the legislative process. The legislature laid the groundwork for a comprehensive redesign of net metering policy to be completed by the CPUC by 2017 and gave the CPUC the authority to implement up to a $10 per month fixed charge for all customers. Furthermore, SDG&E has been able to maintain the goodwill of its customers, state regulators, and other stakeholders, which translates to political capital. SDG&E (and APS) has found that proposals to institute substantial fixed charges are currently untenable. Furthermore, Sioshansi and Nelson, McNeill, and Simshauser described situations where instituting high fixed charges could result in further incenting solar customers to exit the grid. As a result, alternatives such as modest fixed charges as one component, seem prudent to pursue.

In stark contrast to APS, GMP has embraced DER and used its support for solar PV development to enhance its corporate image in Vermont. GMP effectively used its support for solar, especially its Rutland solar capital program, to win public support for a potentially controversial merger with Vermont's largest IOU, Central Vermont Public Service. Rather than resisting the growth in solar development, GMP has embraced it and become an investor in a number of innovative projects. Additionally, GMP has used some of the good will it has generated to promote new uses of electricity that, on balance, could offset some or all of the lost revenues from increased DER penetration. While GMP operates in a unique political climate, successful modern utilities cannot ignore the political environment in which they operate and must engage customers with attractive energy options. Nelson, McNeill, and Simshauser similarly noted that utilities should embrace these trends and find ways to add value for customers in adopting this technology. Similar to the SDG&E and GMP examples, Hanser and Van Horn argue that the DG ownership model can provide both additional revenue and load-serving flexibility to the utility of the future.

Utilities like GMP and SDG&E have achieved some progress in focusing on new investment in clean energy technologies as well as the promotion of new uses for electricity through innovative and efficient technologies. Further, 19 states legally preclude utility ownership of generation assets. Policy makers and regulators might consider allowing utilities to own distributed energy generation to allow them to invest in DER in a manner that both benefits their customers and their investors (Caperton and Hernandez, 2013). Additionally, it makes for good public policy to promote clean DER alongside the increased electrification of our transportation sector. Faruqui and Grueniech (Chapter 5) identify increased electrification as a viable strategy for utilities, and Groot (Chapter 6) recognizes that EV infrastructure and services can be a significant business for European power utilities as well. Joint promotion of solar PV and EVs can have the effect of significantly reducing state carbon emissions while also better managing the grid by smoothing peak demand on hot summer days and filling the valleys of off-peak periods through smart charging of EVs. Increased electricity use from EV charging may offset decreased traditional residential use. As new technologies become more cost-effective, electricity storage and microgrid options could also become part of an "a la carte" menu of available services that utilities might offer in conjunction with a growing proportion of residential solar PV.

5 CONCLUSION

The modern utility is grappling with increased customer demand for solar and the economic and environmental realities that make increasing amounts of solar more likely in the foreseeable future. Customer generation of electricity changes the traditional paradigm and constitutes a threat to the utility's traditional revenue source. Answers to these challenges are still being developed domestically and internationally. These three early examples of utility response suggest that rather than opposing net metering policies and installation of solar PV, utilities should embrace the dynamic change and experiment with methods to generate revenue by fostering new electrotechnologies, finance clean generation, and work toward collaborative rate solutions. The structural changes to the long-standing paradigm of the utility and electric industry require strategic leadership. For the industry to grow and remain relevant in an evolving and decentralizing electricity market, utilities must develop cooperative policy and regulatory processes to ensure that their institutional needs align with the realities of a modern and engaged prosumer.

REFERENCES

Aanesen, K., Heck, S., Pinner, D., 2012. Solar Power: Darkest Before Dawn. McKinsey & Company, New York.

Arizona Corporation Commission Utilities Division, 2013. Staff Report to The Commission Regarding Arizona Public Service Company—Application for Approval of Net Metering Cost Shift Solution (Docket No. E-01345A-132048) (September 30, 2013).

California Legislative Service, 2013. Chapter 611 (A.B. 327) (West).

Caperton, R., Hernandez, M., 2013. The Electrical Divide: New Energy Technologies and Avoiding an Electric Service Gap 4. Center for American Progress (July 15, 2013).

Email Conversation with Green Mountain Power executive John Tedesco, 2013. October 19, 2013 (on file with the author).

GMP Solar Power, 2012. Rutland: The Solar Capital of Vermont Project Plan. Available at, http://www.greenmountainpower.com/upload/photos/369Rutland_Solar_City_Implementation_Plan.pdf (August 21, 2012).

Greenmountainpower, n.d., http://news.greenmountainpower.com/manual-releases/2013/Ambulance-service-wins-GMP-solar-grant.

Green Tech Media Research, 2013. U.S. solar market insight report, Q2 2013, Executive Summary 2.

Kind, P., 2013a. Disruptive Challenges: Financial Implications and Strategic Responses to a Changing Retail Electric Business. Edison Electric Institute, January, p. 3.

Kind, P., 2013b. The utility Industry Can Survive the Energy Transition—it's leading it, A response to a recent GTM article on disruption in the utility industry—by the author of the cited report. Green Tech Media. http://www.greentechmedia.com/articles/read/The-Utility-Industry-Can-Survive-the-Energy-Transition-Its-Leading-It (April 16, 2013).

Nelder, C., 2013. Can the Utility Industry Survive the Energy Transition? A new paper from the Edison Electric Institute raises numerous doubts. Green Tech Media. http://www.greentechmedia.com/articles/read/can-the-utility-industry-survive-the-energy-transition (April 09, 2013).

Puttre, M., 2013. Arizona Utility, Installers Clash Over Solar. Solar Industry. 6 (7), August; see also, Trabish, H.K., June 21, 2013. The Continuing Saga of Solar Policy in Arizona. Green Tech Media. Available at, http://www.greentechmedia.com/articles/read/The-Continuing-Saga-of-Solar-Policy-in-Arizona (last accessed November 30, 2013).

SAIC, 2013. Updated Solar PV Value Report for Arizona Public Service, at 2-3–2.5, May 10, 2013.

San Diego Gas & Electric, Press Release, Media Statement in Response to Protest of AB 327. Available at http://www.sdge.com/newsroom/2013-08-28/media-statement-response-protest-ab327 (August 28, 2013).

SDG&E, 2013. Smart Gird Development Plan 2011–2020, at 15 (June 6, 2011). Available at, https://www.sdge.com/sites/default/files/regulatory/deploymentplan.pdf (last accessed November 30, 2013).

SEPA, 2013. SEPA Awards Honor Leadership in Customer Programs and Innovative Business Models for Solar Electric Power. SEPA Press Release. Available at, http://www.solarelectricpower.org/about-sepa/sepa-news/press-releases/sepa-announces-2013-utility-of-the-year-award-winners.aspx (October 22, 2013).

Telephone Conversation with Green Mountain Power Executive David Dunn, 2013. October 07, 2013.

The Public Utilities Commission of the State of California, 2012. Application of the San Diego Gas and Electric Company (U902E) for Authority to Update Marginal Costs, Cost Allocation, and Electric rate Design, January 18, 2012, Available at http://docs.cpuc.ca.gov/efile/RULC/157634.pdf.

Trabish, H.K., 2013a. Wellinghoff: solar is going to overtake everything. Green Tech Media (August 21).

Trabish, H.K., 2013c. Arizona preserves net metering by charging a small fee to solar owners. Green Tech Media (November 15, 2013).

United States Department of Energy and North Carolina Solar Center, 2013. Database for State Incentives for Renewables and Efficiency (DSIRE). RPS Data by State available at, http://www.dsireusa.org/rpsdata/index.cfm (last accessed October 25, 2013).

Western Electric Industry Leaders Group, 2013. Letter to Governors, Commissioners and Legislators. Available at, https://www.sdge.com/sites/default/files/documents/1843346665/WEIL%20Smart%20Inverters%20Letter%20Final%20Aug%207%202013.pdf?nid=8436 (August 07, 2013).

Wilson, R., 2013. Sunny Arizona, a battle over solar power. The Washington Post. Available at, http://www.washingtonpost.com/blogs/govbeat/wp/2013/10/16/in-sunny-arizona-a-battle-over-solar-power//?print=1 (updated October 16, 2013 at 6:00 am).

Chapter 15

Making the Most of the No Load Growth Business Environment

Ahmad Faruqui and Dian Grueneich

ABSTRACT

The slowdown in electricity sales growth has persisted, prompting a debate about whether it is temporary or permanent, the significant impact on electric utilities, and future options for utilities. This chapter begins by analyzing five major reasons for the slowdown in addition to the weak economy: changes in consumer psychology, ratepayer-funded energy efficiency programs, governmental codes and standards, the rise of distributed generation, and fuel switching to natural gas. The slowdown is expected to persist in the indefinite future. The second half of the chapter presents four strategies and three tactics that utilities can pursue to deal with this new reality, including a transition to simply becoming a wires company or a provider of distributed generation. The tactics include redesigning electric rates, revamping load forecasting efforts, and reinventing load and market research. The chapter also lays out a decision-making process that utilities can use to pick the winning strategy.

Keywords: Slowdown in sales, Strategies, Tactics, Demand forecasting, Rate design

1 INTRODUCTION

As shown in Figure 1, declining growth has been the norm and not the exception since 1950. However, the slowdown in electricity demand growth is now dramatic, will continue, and is likely to expand. Because current utility business models are almost universally built on sales growth, decreases in growth impede the recovery of utility fixed costs and threaten to erode earnings over the long term. This slowdown poses a serious threat to the financial viability of utilities and calls for changes in the current business model of most utilities.

According to John Caldwell of the Edison Electric Institute (Caldwell, 2012), based on the experience of the past five recessions, normal load growth usually resumes within 5 months after the recession ends. The longest it has ever taken has been 12 months.

Distributed Generation and its Implications for the Utility Industry. http://dx.doi.org/10.1016/B978-0-12-800240-7.00015-1

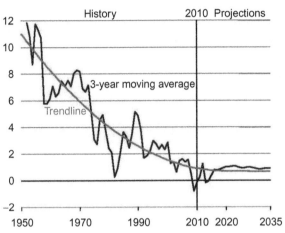

FIGURE 1   US electricity demand growth, 1950-2035 (%, 3-year moving average). *Source: EIA, 2012 Annual Energy Outlook.*

The Great Recession ended in 2009. Yet normal growth in the sales of electricity has not resumed. There is definitely something different about this recovery.

Traditional US electricity sales growth has dropped by half. At the national level, the US Energy Information Administration (EIA) is predicting growth in the sub-one percent range, down from the prerecession average of 2%. The EIA's March 2013 *Short-Term Energy Outlook* projects that aggregate electric retail sales will grow by about 0.7% between 2013 and 2014. In the residential sector, sales will grow at about 0.3% in 2013 and at about 0.8% in 2014. The EIA also projects that total electricity sales will not return to adjusted prerecession levels until 2014.

Load forecasters in a cross section of two dozen utilities around North America report they are seeing similar drops. The consensus projection of these forecasters is 0.7-0.9% growth in the years to come. While some parts of the country may grow faster than that, other parts will grow even slower. Unsurprisingly, similar trends are being observed in areas as far away as Australia and Hong Kong and as close as the province of Ontario in Canada.

This chapter consists of seven sections in addition to the Introduction. Section 2 analyzes five major causes of the electric sales growth slowdown, beyond the economic recession. Section 3 discusses looming electric industry investment needs, since these are critical in thinking about the future of electric utilities and changes in the business model. Section 4 offers "The Survival Toolkit" of four different utility strategies going forward, and Section 5 presents a suggested analytic approach for selecting among these strategies. Section 6 identifies three essential tactics for utilities to consider, regardless of the strategy they pursue. Section 7 summarizes the chapter's conclusions.

2   FIVE MAJOR FACTORS CAUSING THE ELECTRIC SALES GROWTH SLOWDOWN

Five forces, beyond the recent economic recession, are the primary cause of the slowdown. Most importantly, these forces are likely to continue and become even stronger going forward. The five forces are the following:

- First, *consumer psychology* has shifted in both younger and older populations.
- Second, *ratepayer-funded energy efficiency programs* have been increasing significantly, often prompted by state-imposed energy efficiency standards and legislation.
- Third, state and federal governments are also continuing to push ahead with aggressive efficiency enhancements to *building codes and appliance standards*.
- Fourth, *distributed generation* has shrunk utilities' customer base, as more and more customers turn to rooftop solar to produce their own power generation and are even exporting generation to the grid.
- Fifth, *fuel switching* in North America has changed over the past few years due to the revolution in shale oil and gas, which is pushing fuel prices downward, making natural gas more competitive in some customer end-use applications than electricity.

We address each of these five factors in the succeeding text.

The *first factor* is shifting *consumer psychology*. A new generation of consumers has arrived with new values and concerns. At the same time, the older generation of consumers, the baby boomers, are engaging in an unprecedented level of belt-tightening, to cope with continued economic uncertainty and anxiety. This becomes evident when one looks at the drop in the consumer confidence index, which is shown in Figure 2.

And, with the continuing drag in the US economy and those of many other countries, young adults have moved back in with their families, resulting in large-scale delays in starting new, independent households. This delay filters into the utility world through decreases in construction of new residential buildings, lower appliance sales, and the like. Likewise, much of younger generation desires to live in cities where the average utility bill in an apartment or small house is lower than that of larger, suburban houses of prior generations.

The *second and third factors, ratepayer-funded energy efficiency programs and building codes and standards*, are both the result of an increasing focus on energy efficiency and demand response. What does the future portend in these areas? To get an answer, after the recession ended, 50 energy experts in North America were contacted by Brattle and asked: *How much will energy efficiency programs reduce energy sales and how much will demand response programs reduce peak demand by the year 2020?* Unsurprisingly, the experts differed in their projections with some seeing big impacts and other seeing small impacts. But even the small impacts were significant. The impact of energy efficiency on

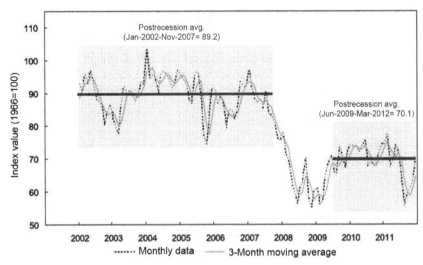

FIGURE 2   Index of consumer sentiment: recent changes. Note: Sentiment index for March 2013 increased to 78.6 from 77.6 a month earlier but fell again in April 2013 by several points. *Source: University of Michigan Survey of Consumers.*

the sales forecast was projected to range from 5% to 15% compared to what it would have been in the absence of these programs. The corresponding reduction in peak demand was projected to range from 7.5% to 15%, compared to what it would have been in the absence of these programs.

The impacts of energy efficiency and demand response will be at the higher end of those surveyed. Public support and funding for these programs, as will be explained later, continue to grow. Very importantly, new technologies, such as advanced analytics and automated demand response, are now possible and are enhanced with the extensive deployment of advanced metering infrastructure (so-called smart meters). Consumer-friendly products, Internet apps, and more engaged customers are also driving forces. In fact, the impact of energy efficiency and demand response is likely to be as great or even a greater cause than distributed generation, on utility sales. While decoupling has been used to offset erosion of utility revenues for energy efficiency to date, it is unclear if it can be sufficiency effective in a world of widespread distributed energy resources (DER).[1]

There has been a significant increase in ratepayer-funded energy efficiency programs since 2000, with a large portion of the increase taking place in states that historically have been minor players (Arizona, Arkansas, Indiana, Michigan, Ohio, and Pennsylvania). US ratepayer-funded electric efficiency budgets totaled over $6.8 billion in 2011—*a 25% increase from*

---

1. Distributed energy resources (DER) typically refer to energy efficiency, demand response, and distributed generation. DER is now expanding to include customer-level energy storage.

2010 levels.[2] Electric utilities are by far the largest providers of energy efficiency in the United States and their programs comprise 84% of the total ratepayer-funded energy efficiency budget nationwide.[3] Currently, 20 states have Energy Efficiency Resource Standards (EERS) in place, while 7 states have goals.[4] Lawrence Berkeley National Laboratory (LBNL) forecasts that:

Energy efficiency programs funded by utility customers are poised for dramatic growth over the course of the next 10 to 15 years, especially in the Midwest and South – with a contingent of populous Midwest states ramping up to meet statutory EERS targets, and in the South, the expectation that a collection of relatively modest EERS policies and nascent IRP/DSM planning processes in states with a large base of energy consumption will push spending upward from currently low levels. As a result, program spending is expected to become more evenly distributed nationwide by 2025.[5]

As seen in Figure 3, LBNL forecasts that program spending is projected to roughly double to $9.5 billion in 2025 and could reach $15.6 billion under aggressive assumptions about the policy support, implementation, and effectiveness of the current policies. Program administrators in many states are projected to achieve annual electricity savings of between 1.5% and 2%, surpassing the achievements of most leading states today.[6] According to LBNL's

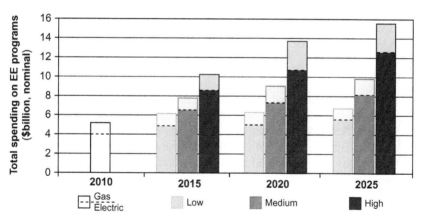

FIGURE 3   Projected electric and gas energy efficiency program spending.[7] *Source: 2010 spending based on CEE (2012).*

2. Institute for Electric Efficiency, 2012. Summary of Ratepayer-Funded Electric Efficiency Impacts, Budgets, and Expenditures (2010-2011). Prepared by Adam Cooper, Lisa Wood and IEE, January 2012. p. 4.
3. Ibid.
4. Database of State Incentives for Renewables & Efficiency (DSIRE): http://www.dsireusa.org/documents/summarymaps/EERS_map.pdf
5. Barbose, Galen, et al., Lawrence Berkeley National Laboratory, "The Future of Utility Customer-Funded Energy Efficiency Programs in the United States: Projected Spending and Savings to 2025," January 2013. p. 30.
6. Ibid.
7. Ibid.

projections, the projected growth in electric program spending and savings does not occur smoothly over the forecast period, but rather is "front-loaded," with much faster growth projected through 2015. LBNL states, "this dynamic is partly due to the fact that, in many states, recent multiyear DSM plans entail significant spending increases over the next several years, but no longer-term targets or resource planning process currently exists to guide program activity beyond the time horizon of the DSM plan."[8]

The third factor for the utility sales decrease, building codes and standards, is growing in impact. According to LBNL:

In recent years, state adoptions of building energy codes have increased, and federal minimum efficiency standards for appliances and end-use equipment have been tightened. These policies affect utility customer-funded programs by essentially raising the baseline against which savings are measured, thereby influencing both the size of the remaining potential that can be harvested through those programs and the mix of technologies targeted.[9]

Energy codes and standards set minimum efficiency requirements for new and renovated buildings, assuring reductions in energy use and emissions over the life of the building. The US Department of Energy (DOE), through the Building Energy Codes Program (BECP), supports energy efficiency in buildings through the development and implementation of model codes and standards. DOE provides technical assistance to states and localities as they adopt and enforce energy codes. As stated on DOE's BECP website, current program goals include:

- model energy codes and standards targeted at 50% energy savings over the 2006 International Energy Conservation Code (IECC) for residential buildings and ASHRAE Standard 90.1—2004 for commercial buildings,[10]
- adoption of the 2009 IECC for residential buildings and Standard 90.1—2007 for commercial buildings by 70% of US states and territories by 2015,
- a 90% compliance rate with the 2009 IECC for residential buildings and Standard 90.1—2007 for commercial buildings by 2017.[11]

DOE's Building Technologies Office also implements minimum energy conservation standards for more than 50 categories of appliances and equipment.[12]

---

8. Ibid., pp. 18-22.

9. Ibid., p. 26.

10. The US model standard that provides minimum requirements for energy-efficient designs for buildings except for low-rise residential buildings.

11. US Department of Energy's Building Energy Codes Program website: http://www.energycodes.gov/about

12. US Department of Energy's Building Technology Office website: https://www1.eere.energy.gov/buildings/appliance_standards/

As a result of these standards, DOE estimates that energy users saved about $40 billion on their utility bills in 2010. Since 2009, DOE has issued 18 new or updated standards, which are expected to help increase annual savings by more than 50% over the next decade. By 2030, cumulative operating cost savings from all standards will reach $1.7 trillion, with a reduction of 6.5 billion tons of carbon dioxide emissions, equivalent to the annual greenhouse gas emissions of 1.4 billion automobiles. Products covered by standards represent about 90% of home energy use, 60% of commercial building use, and 29% of industrial energy use.[13]

The future is likely to be shaped not only by the three forces discussed earlier but also by two new forces that are currently in an embryonic stage—distributed generation and fuel switching.

The *fourth force, distributed generation*, is being led by the revolution in rooftop solar and supplemented by microturbines. As discussed elsewhere in this book, experts agree that rooftop solar is approaching grid parity, spurred on by tremendous drops in cost due to technological advancements and large-scale demand. The buildup in demand has been brought on by upfront taxpayer-funded cash subsidies and propelled by net energy metering (NEM) tariffs (see Box 1) that compensate solar customers by paying them not only for the energy they feed into the grid but also for the transmission and distribution services.

Competitive leasing and financing options have also become a major factor in driving solar photovoltaic demand. The leasing model pioneered by Solar-City, in which the customer does not make any cash outlay, has changed the paradigm. Others are copying it rapidly. This new business model for solar is making it accessible, for the first time, to middle- and low-income residents, which are now the fastest-growing sector using leasing. According to some experts, the growth in distributed generation alone—with its attendant loss in

---

Box 1 Net Energy Metering

NEM is an electricity tariff where customers receive a bill credit for energy that they generate and export to the grid. Bill credits match the same retail rate (including generation, distribution, and transmission components) that a customer would pay for consuming energy. Currently, 43 states plus Washington, DC have adopted NEM policies. In many states, the establishment of NEM programs has been met with opposition. Investor-owned utilities are concerned that NEM programs in their territory result in increased rates for nonparticipating customers and an overall loss of profit for investors. Arizona, California, Texas, and other states are seeking to quantify these rate impacts and solar benefits in order to direct future policy decisions.

---

13. Ibid.

Box 2 California Assembly Bill (AB) 327

In 2013, California Governor Jerry Brown signed AB 327 into a law that restructures California's electricity rates and expands net metering in the state.

AB 327 repeals California's previous restriction on residential rate increases, including the rates of low-income customers in the California Alternate Rate for Energy (CARE) program. The bill largely eliminates rate restrictions enacted during the energy crisis of 2001 and authorizes the California Public Utilities Commission (CPUC) to redesign that rate structure, preserving at least two tiers of usage. The bill also authorizes the CPUC to approve new, or expand existing, fixed monthly charges of up to $10 a month for all utility customers beginning in 2015 ($5 a month for low-income customers).

AB 327 also gives the CPUC authority to raise California's 33% RPS without an act of the Legislature.

---

end-use sales and imports into the utility distribution grid—will offset all load growth.

NEM has been a major factor enabling the expansion of distributed generation. In 2003, there were less than 7000 US customers on net metering. In 2010, there were 156,000, of which roughly half were in California and producing what amounted to 0.1% of total US electricity sales. In California, newly signed AB 327 (see Box 2 later in chapter) expands the net metering caps of the investor-owned utilities beyond a 5% cap previously in statute and specifies new maximum capacities for each investor-owned utility service territory. The law requires the California Public Utilities Commission to develop a new net metering program with standard contracts or tariffs by 2015 to be used beginning in 2017. This new program eliminates any restriction on new eligible NEM customers entitled to receive service pursuant to the new standard contract or tariff.

Another new distributed generation option is community solar. California's new 2013 law, SB 43,[14] creates the Green Tariff Shared Renewables Program, the largest community solar program in the country at 600 MW. Under the program, customers—including local governments, businesses, schools, homeowners, municipal customers, and renters—of California's three largest IOUs can purchase up to 100% of their electricity from small- to medium-sized solar and other renewables projects. Participants can then receive a credit on their utility bill for the clean energy produced. Currently, there are a total of nine states with enacted shared renewables policies and many more with active campaigns.[15]

---

14. California SB 43: http://leginfo.legislature.ca.gov/faces/billNavClient.xhtml?bill_id=201320140SB43
15. http://sharedrenewables.org/

It is important to point out that while energy efficiency's impact on utility revenues can be addressed by decoupling, distributed generation schemes, especially when melded with some sort of NEM rate structure, not only impact sales revenues but also can be a loss in the split between wholesale/retail rates.

This book's other chapters provide a more in-depth discussion regarding distributed generation. Chapter 6 by Groot analyzes the impact of distributed generation on European utilities, whereas Burger and Weinmann in Chapter 3 provide a description of distributed generation's developments in Germany. Additionally, Chapter 8 by Keay et al. presents a "description of the longer term underlying problems changes in the generation mix will raise for liberalized markets and explains why current reform proposals are inadequate."

With distributed generation and expanded energy efficiency, zero net energy buildings are also becoming a reality. In Austin, Texas, the Zero Energy Capable Homes program requires that new single-family homes be zero net energy capable by 2015. The largest community of net-zero homes in the United States is rising in West Village on the grounds of the University of California at Davis. California policymakers have embraced the future of zero net energy buildings and are planning to standardize all new residential construction at zero net energy by 2020 and all new commercial construction at zero net energy by 2030.[16] They are also developing "whole building retrofit" programs that lower usage in existing buildings throughout California by 20-40% over the same period. In a world of zero net energy new and retrofitted buildings, net load growth not only disappears but sharply declining sales can be the new reality.

The *fifth force is fuel switching*. For years, fuel switching in North America favored electricity. Now it is poised to switch sides and work in reverse. The revolution in shale oil and gas is pushing fuel prices downward, making natural gas more competitive in some customer end-use applications than electricity. The use of gas for commercial air-conditioning and in industrial process, long a theoretical possibility and proven in demonstration sites by the Gas Research Institute, may become economic, leading to significant interfuel substitution away from electricity in the commercial and industrial sector. Even the residential sector will not be immune to such trends. Gas-fired residential heat pumps will begin making inroads into the home HVAC market.

3 ELECTRIC UTILITY INVESTMENT NEEDS ARE ALSO UNPRECEDENTED

Thunderstorms, ice storms, and tornadoes during the recent past have exposed the vulnerability of the power grid. Unable to face these extreme weather conditions, the grid has broken down in many places, plunging large numbers of

---

16. California Long-Term Energy Efficiency Strategic Plan. September 2008. http://www.cpuc.ca. gov/NR/rdonlyres/D4321448-208C-48F9-9F62-1BBB14A8D717/0/EEStrategicPlan.pdf

people in darkness and forcing them to unlive the life they have been conditioned to living not just for hours but for days.

The President's Council of Economic Advisers and the US DOE released a report in August 2013 that estimates the annual cost of power outages caused by severe weather between 2003 and 2012. Over this period, weather-related outages are estimated to have cost the US economy an inflation-adjusted annual average of $18 billion to $33 billion.[17] The costs of these outages include lost output and wages, spoiled inventory, delayed production, inconvenience, and damage to the electric grid.

Major new investments in the grid, all the way from the meter to the distribution lines, circuits, feeders and transformers, to the transmission network, are critical. In a report that was completed a few years ago for the Edison Electric Institute, The Brattle Group estimated that nationally, investments to the tune of $1.3 trillion will need to be made to modernize the grid and to make it resilient not only to adverse conditions arising from nature (bad weather) but also to newer threats arising from human intervention such as cyber-attacks.

Additional investments are needed to connect remote renewable energy resources to load centers. Furthermore, there is a long-standing problem of having poor load factors. Nationally, the number is under 60%. For some utilities, it is even lower. This forces utilities to invest in peaking power plants that run for only a few hundred hours a year and are idle during eight thousand plus hours a year.

The upshot is higher-average costs for all customers, while the number of customers and their utility-load requirements are diminishing.

4   THE SURVIVAL TOOLKIT: FOUR STRATEGIES

These realities put electric utilities in a bind. They need to invest at a time of a significant slowdown in sales growth that is reflected almost proportionately in a slowdown in revenue growth, putting an enormous pressure on their bottom line. In 2012, according to the financial report from the Edison Electric Institute, both sales and revenues of electricity fell last year, which was the third full year after the recession ended.

How should utilities respond to the challenge? Few doubt that the old "tried and tested" strategies will work in a business environment that bears no resemblance to the past. Companies that start with a clean slate, by dismissing the old strategies as passé, will have a greater chance to survive what is obviously a grim transition period and redefine their future.

---

17. "Economic Benefits of Increasing Electric Grid Resilience to Weather Outages", President's Council of Economic Advisers and the US Department of Energy's Office of Electricity Delivery and Energy Reliability, August 2013, p. 3.

What should be the new strategies and tactics? While the possibilities are endless, this section focuses on four strategies and the next section presents three tactics.

Utilities need to focus on both strategies and tactics. In business as in war, there is a temptation to put more weight on strategy than on tactics. In board-rooms as in war rooms, strategies get more time and attention than tactics because the latter are viewed as glamorous and tactics as dull. Utility executives ignore tactics at their own peril. Slightly modifying a military maxim, one can say: "Amateurs discuss strategy; experts discuss tactics." Thus, there is a need to go beyond strategies and embrace changes in tactics as well.

4.1   First Strategy: Stay the Course

This strategy assumes that the slowdown in sales growth is short-lived and not "the new normal." This strategy draws its inspiration from the fall in oil and gas prices. This price fall is presaged to bring about an industrial revival, possibly in the form of organic new growth and the return of manufacturing from offshore locations. This revival is expected to boost the sale of electricity to the industrial sector. CERA's Larry Makovitch has put forward a provocative argument along these lines.[18] However, it is unclear how widespread this revival is going to be and ignoring the slowdown is a high-risk strategy.

The electricity business will never be the same again. To pretend otherwise is dangerous. Take the example of the car industry. It is beset with change aris-ing from technological innovation and the emergence of substitutes. Plug-in electric vehicles represent a big change. Self-driving cars represent an even big-ger change. Bill Ford, the Executive Chairman of Ford Motor Company, noted recently: "The car as we know it, and how it's used in people's lives, is going to change really dramatically and it's going to change fast. If we don't start imag-ining this future, and then start trying to help shape this future, we're going to be left behind, because this future is going to happen with or without us."[19]

The stay-the-course strategy is predicated on the validity of the old normal. It is possible, but likely, that the old normal theory is valid. Betting the company on a nostalgic and wistful yearning for the past is a high-risk strategy. But some companies are likely to make that bet anyway. There are enough examples from other industries in the pages of the *Harvard Business Review* to prove that companies do not want to venture out from what they regard as their comfort zone. By the time it becomes clear that the comfort zone has actually become the irrelevant zone, it is too late to pull out of the slide. They slip quietly into

---

18. http://www.powermag.com/issues/features/Expect-U-S-Electricity-Consumption-to-Increase_5634.html
19. http://www.chicagotribune.com/business/la-fi-hy-autos-bill-ford-milken-talk-20130501,0,3825298.story

oblivion, not missed by their customers or their suppliers, barely getting a mention in the press.

4.2   Second Strategy: Electrification

This strategy concedes that the old normal is gone. Unlike the first strategy, which is essentially a do-nothing strategy, it actively seeks to create conditions that will restore the old normal of significant load growth.

An obvious example is electric transportation, driven by economic as well as environmental concerns. Until recently, electricity regulation occurred in relative isolation, focusing on development and delivery of power at reasonable rates. Likewise, transportation policies developed separately, since gasoline-, diesel-, and other nonelectric fuel-powered vehicles. Climate change and various states' commitment to reducing GHG emissions, however, have dramatically changed this historical approach, as have major private, nonutility investments in electric vehicle technologies, including batteries.

A recent paper by Stanford University's Shultz-Stephenson Task Force on Energy Policy, "California's Electricity Framework," proposes the integration of electricity planning with climate and transportation goals. It states:

...looking past 2020, pursuit of increasing GHG emissions will produce major impacts on the electricity sector – through huge demands for energy efficiency and substantial new supplies of decarbonized electricity. In turn, the transportation sector will be deeply challenged to move from direct fuels to a system largely powered by electricity. These transformations and the private investment needed to support them cannot be done through disconnected decisions by the state's climate, energy, and transportation policymakers.[20]

The emergence of plug-in electric vehicles, anticipated as far back as 1979, is finally beginning to bear fruit. A number of new models are on the road, but mostly in states such as California and Michigan. President Obama's goal of putting a million electric cars on the road by 2015 is far from being accomplished. At this point, only 5% of that goal has been realized and there are only 3 years to go to achieve the other 95%. *The Economist* in a recent editorial listed a number of electric vehicle companies that have gone out of business.[21] They are more expensive and have short-range problems that have deviled the concept since its very beginning. Even under the most optimistic scenarios, electric cars will not make much of an impact on electric sales a decade out and perhaps even longer. Platt et al. in Chapter 17 presented an analysis of the challenges and

20. Dian Grueneich and Jeremy Carl, "California's Electricity Policy Future," Shultz-Stephenson Task Force on Energy Policy, Hoover Institution, 2013. See also Grueneich, Carl "Renewable and Distributed Power in California: Simplifying the Regulatory Maze—Making the Path for the Future," 2013.
21. The Economist, "Flat Batteries," 1 June 2013.

benefits electric vehicles may pose to the electricity system and explored their potential to act as a DER.

Some of the other electrification efforts being undertaken by organizations such as EPRI on new industrial processes that are electricity-intensive may bear fruit sooner than electric vehicles. But the tilt in the price equation in favor of natural gas will stymie their adoption. And there are tremendous barriers to electrification of transportation on a large scale. Thus, it is unlikely, at least in the near-term, that electric utilities will realize much load growth from electrification. However, in the longer term, electrification could become an important revenue source for electric utilities.

4.3   Third Strategy: The Safe Haven

This third strategy epitomizes a strategic retreat. The electric utility concedes the business of selling electricity to the competition, whether it originates from new entrants such as SolarCity or the buyers themselves, in the form of energy efficiency. The utility withdraws into a safe haven and becomes simply a wires company. That preserves its status as a natural monopoly.

Of course, many electric utilities in the northeast, eastern Midwest, and Texas are already wires companies. So are electric utilities in Australia and the European Union. But that was not because they made a strategic choice. It was because their regulators decided to restructure the power industry. In contrast, with this strategy, utilities find ways voluntarily to become wires companies. This strategy is less risky than the two prior strategies. But it is not risk-free. All wires companies face the risk of collecting insufficient revenue since the bulk of distribution charges are tied to sales, and as sales growth slows down or disappears, they will not be able to cover their fixed costs.

4.4   The Fourth Strategy: Go On the Offense

This strategy is premised on being able to out-sun solar companies by creating a nonregulated affiliate that operates in other service areas. It requires the creation of a new enterprise culture that is nimble and customer-centric without which competition with mainstream solar companies will remain a fatal conceit. The payoff from this strategy might be significant. But this strategy does not fit well with the core competency of most electric utilities and therefore it is high risk. The last time utilities ventured into the world off diversification, they failed spectacularly. This strategy assumes that this time, it will be different.

Another risk is worth nothing. If regulators and legislators realign rates to change much of the retail customer benefit from net metering, then the value proposition for distributed generation will decrease, and at least in the short

term, utilities that venture into the solar leasing business will not generate substantial shareholder returns.

5   AN ANALYTIC APPROACH TO STRATEGY SELECTION

When all is said and done, each utility has to pick a single strategy, knowing that
it will have the opportunity to change strategies later on. The choice of strategy
will vary by utility and depend on a number of variables:

- First and foremost, what are its earnings goals?
- Second, how much risk is it willing to tolerate?
- Third, what will the future look like? It will be delusional to define a single
  future.

To quote Churchill, "The future, though imminent, is obscure." So it will be
prudent to lay out a few alternative futures. In each, the nature of the business
environment will have to be characterized: What will tomorrow's customers
look like, who will be the competitors, what will be the size of the business,
and so on. And fourth, how will each of the four strategies described earlier fare
in each of the business environments?

Once the utility has answered these questions, it will be able to construct a
decision matrix in which the rows are strategies and the columns are alternative
futures. Each cell of the matrix should be populated with an estimate of earnings. These can be obtained by gathering intelligence about customers and competitors and estimates of how much it would cost to pursue each strategy. The
utility should be able run its corporate financial models to estimate how much
each strategy would yield by way of earning. The utility will then populate each
cell with its estimated earnings.

At this point, the objective part of the analysis has been completed. Now comes
the subjective part having to do with management preferences for risk versus
reward and its perceptions of cultural and organizational fit. Two bookends can
be imagined. In one, management would circle the worst case for each strategy
and then pick the strategy with the "best" worst case. This is called *conservative
gamesmanship*. In the other, the best case is circled for each strategy and the strategy with the best "best" case is picked. This is called *aggressive gamesmanship*
and also sometimes called the "bet-your-company" strategy. Various in-between
combinations can be defined. And so, a strategic choice will be arrived at that
reflects the analysis of alternative futures and alternative strategies.

Some utilities will be inclined to take this approach one step further. They
would want to interject probabilities into the strategic calculus. The estimate of
earnings in each cell of the decision matrix will not be regarded as a definite
number. Uncertainty in that estimate will be recognized and it will be regarded
as the mean of a probabilistic distribution. These would be estimated either
objectively if data exist to make such a determination or subjectively if pertinent
data can only be gathered through expert opinion. Management will then be

interviewed and its attitudes toward risk quantified in the form of risk-reward trade-off curves. These curves will then be used with the decision matrix to zero-in on an optimal strategy.

In all but the simplest examples of future uncertainties, it is impossible to find a final strategy that will yield the best outcome under each future. The electric utility environment is marked by "deep uncertainty," to use a term coined by the RAND Corporation. But the optimal strategy will represent the best outcome under a range of possible futures, consistent with management preferences toward risk.

6 THREE TACTICS REGARDLESS OF THE SPECIFIC UTILITY STRATEGY

Regardless of which strategy is chosen, success will require the selection and deployment of complementary tactics. Three essential tactics are discussed below. In the opinion of the authors, it is critical for utilities (and their regulators) to focus on these three tactical areas as soon as possible, so that they have the tools needed to implement whichever strategic vision they select.

6.1 The First Tactic: Rethink Rate Design

For most utilities, revenues from residential and small business customers are collected through volumetric charges. However, a good share of utility costs is fixed. So there is a fundamental discrepancy between costs and revenues. This does not pose a problem when sales are growing as in the old normal: "A rising tide lifts all ships." But this model falls apart when the new normal kicks in. So the way forward will require a fundamental change in rate design toward one that relies on straight fixed and variable designs. The fixed charge will have to cover the costs of investing, operating, and maintaining the grid. Anecdotal evidence suggests that the national average is around $8 per customer per month. But at the low end are states like California where the fixed cost is almost nonexistent. Distribution costing studies suggest that the cost is somewhere between $25 and $45 per customer per month. Raising the fixed charge has become an imperative but making the transition will not be easy (see Box 2). Regulators and customers will have to be convinced that this is fair and equitable. And some temporary buffers may have to be created to protect small consumers who may see their bills rise. Additionally, some of the fixed charge could be recovered through demand charges that vary by size of customer.

The second change that should be made is to move the volumetric charges to a time-based character. This change will follow the time-based character of costs. It will also encourage customers to reduce their peak loads and to shift their peak usage to off-peak periods. That movement will improve system load factors and lower average costs. With the worldwide movement on installing advanced metering infrastructure, movement toward time-based rates with feedback to consumers is possible. Much research has been done showing the effectiveness of such rates. Forward-thinking utilities and their regulators need to begin movement in this direction sooner rather than later.

6.2 The Second Tactic: Reimagine Forecasting

Many utilities' sales forecasting models have been overforecasting sales for the past 3-5 years, creating doubt in their management's minds about the credibility of the models. No one seems to know what is causing the models to misforecast but it seems likely there is a missing link in the math that underpins the models. The models have no way of capturing changing customer tastes and behavior or changes in their purchases from the grid that arise due to the installation and use of customer-owned generation. A way has to be found to incorporate those insights.

A new generation of models has to be built, possibly as adjuncts to the existing models and ultimately as replacements. The new models can build on the techniques used by firms in competitive industries to do their sales forecasts. In those industries, the forecasters do not just rely on historical data to forecast the future, which is essentially an exercise in trend projection. These models incorporate insights from observational market research that involves frequent and ongoing interactions with consumers. This update will require that new data be collected and added to the econometric equations.

Access to better models is not, in and of itself, a solution, obviously. However, models and their ability to run differing scenarios of the future are an essential analytic tool of all complex industries. Unless utilities change the capabilities of their industry models to make them appropriate for analyzing a rapidly changing, competitive landscape, utility forecasters will be planning to a false future.

6.3 The Third Tactic: Reinvent Load and Market Research Functions

Both load and market research functions have existed in electric utilities for decades. But most of the studies are embarrassingly outdated. Budgets have been cut and it shows in the poor quality of the findings. Another problem is that the customers who are surveyed in market research studies are not the same customers whose hourly load profiles are tracked in the load research studies. The latter were designed primarily to support cost-of-service studies for rate making; the former used to gain insights in how many customers owned which appliances. This unfortunate bifurcation of samples prevents the drawing of deep insights, which are necessary to execute future strategies.

A new approach is required, which is based on systematically tracking customer behavior over time using a common sample of customers. Economists call this a panel data set. But the measurements need to focus not just on the "hard" variables such as loads and demographics but also on the "soft" variables that track tastes and perceptions—often these are called psychographic variables because they have something to do with the mind of the customer. Such data allow the gleaning of insights about changing tastes and about customer perceptions of competitor offerings. The end result is insights not only about the past and current patterns of use but also about likely future changes.

As with the two other tactics, this third tactic of developing better electric utility research tools is not a solution to the revenue erosion caused by the five factors we identified at the beginning of this chapter and particularly that due to distributed generation, as discussed in detail in other chapters. However, in identifying a future in which utilities can develop a new business model for revenues, they must understand the markets in which they operate and can succeed. In order to obtain this understanding, utilities must embrace a new approach to market research as proposed in this third tactic.

7   CONCLUSION

The slowdown in sales growth is not an aberration but very much in line with the downward trend being seen over the past six decades. However, this slowdown is unique in its own way. Five forces, three of which have already manifested themselves and two of which are beginning to emerge on the horizon, are driving it.

To use a term coined by Andrew Grove, the former CEO of Intel, the electric utility industry is at a strategic point of inflection, "a time in the life of business when its fundamentals are about to change. That change can mean an opportunity to rise to new heights. But it may just as likely signal the beginning of the end." To survive, companies will need executives with the ability to recognize that the winds have shifted and "take appropriate action before they wreck the boat." The worst thing they can do is to "fritter away their valuable resources while attempting to make a decision. The greatest danger is in standing still."

REFERENCES

Axford, Mark, 2013. Expect U.S. Electricity Consumption to Increase. POWER Magazine (June 01). http://www.powermag.com/issues/features/Expect-U-S-Electricity-Consumption-to-Increase_5634.html

Barbose, Galen, et al., 2013. The Future of Utility Customer-Funded Energy Efficiency Programs in the United States: Projected Spending and Savings to 2025. Lawrence Berkeley National Laboratory, January 2013.

Caldwell, John, 2012. Demand fallacy. Electric Perspectives (May).

California Long-Term Energy Efficiency Strategic Plan. September 2008. http://www.cpuc.ca.gov/NR/rdonlyres/D4321448-208C-48F9-9F62-1BBB14A8D717/0/EEStrategicPlan.pdf

California SB 43. Website: http://leginfo.legislature.ca.gov/faces/billNavClient.xhtml?bill_id201320140SB43

Database of State Incentives for Renewables & Efficiency (DSIRE): http://www.dsireusa. org/documents/summarymaps/EERS_map.pdf

Grove, Andrew S., 1998. Only the paranoid survive. Profile Business.

Grueneich, Dian, Carl, Jeremy, 2013. California's electricity policy future. Shultz-Stephenson Task Force on Energy Policy. Hoover Institution.

Institute for Electric Efficiency, 2012. Summary of Ratepayer-Funded Electric Efficiency Impacts, Budgets, and Expenditures (2010–2011). Prepared by Adam Cooper, Lisa Wood and IEE, January. 2012, p. 4.

President's Council of Economic Advisers, U.S. Department of Energy's Office of Electricity Delivery and Energy Reliability, 2013. Economic Benefits of Increasing Electric Grid Resilience to Weather Outages, August 2013.

Shared Renewables HQ. Website: http://sharedrenewables.org/

The Economist, 2013. Flat batteries: the electric car stalls in the race to be the green wheels of the future. That is not a tragedy. The Economist (June). http://www.economist.com/news/leaders/21578679-electric-car-stalls-race-be-green-wheels-future-not.

U.S. Department of Energy's Building Energy Codes Program. Website: http://www.energycodes. gov/about

U.S. Department of Energy's Building Technology Office. Website: https://www1.eere.energy.gov/buildings/appliance_standards/

U.S. Energy Information Administration, 2013. Short-Term Energy Outlook (STEO), March.

Undercoffler, David, 2013. Bill Ford: The Future of Self-driving Cars Is Closer than You Think. Chicagotribune.com (April 30). http://www.chicagotribune.com/business/la-fi-hy-autos-bill-ford-milken-talk-20130501,0,3825298.story

Chapter 16

Regulatory Policies for the Transition to the New Business Paradigm

William C. Miller[1], Roland J. Risser and Steven Kline

ABSTRACT

The challenges that current and advancing technology present in the utility industry are well documented in this volume. Yet—due to the unique nature of utility regulation—the near-term evolution of the industry is significantly in the hands of organizations that already regulate or oversee utilities. These organizations determine the terms under which the utility can charge its customers: which costs can be charged to customer and the structure of prices (rates or tariffs) the utility can use for each type of customer to collect those allowed costs. This chapter explores some of the rate and regulatory options those regulators typically have available in charting the near-term course of the transition to a new utility business paradigm, "Utility 2.0." It also highlights several of the technology developments and policy issues that may shape the planning of the early stages of this transition.

Keywords: Utility 2.0, Utility policy, Future utility, Transactive control, Utility rates

1 INTRODUCTION

The US electricity industry is undergoing major changes, yet most consumers are unaware or even resistant. Electric energy, the service which electricity provides, has unique characteristics: some of which are intrinsic, some a consequence of the technology that provides it, and some dependent on the role electricity plays in modern technological society. Historically, it has required large capital investment, lacked storability at any scale except for the smallest devices, and exhibited numerous and complex network effects; has ubiquitous

---

1. The views and opinions expressed in this chapter are wholly those of the authors and do not reflect the policy or position of the Lawrence Berkeley National Laboratory or the US Department of Energy.

Distributed Generation and its Implications for the Utility Industry. http://dx.doi.org/10.1016/B978-0-12-800240-7.00016-3
321

application; is regarded as essential to modern life; and is critical to support the twenty-first-century transformation into information technology-based societies. For the last century, the technology that provided the lowest cost and most reliable access to electricity was large central station power plants and distribution to consumers for immediate use through a network of wires and other infrastructures with power designed to flow "one direction," from generation to the customer.

Significantly, the electric system has never had any cost-effective means of storing significant amounts of electricity, so power must be generated and delivered exactly as it is being consumed at any point in time. Accordingly, grid infrastructure (e.g., generators, wires, and transformers) is sized to handle peak electric demand, which means that maximum capacity is only used intermittently, yet the costs of procuring and maintaining this infrastructure need to be recovered through the rates charged to customers.

The power grid over the last century has grown to regional scale, with power generated at one point supporting delivery of power at any point of grid interconnection. A complex system of organizations and markets developed with utilities providing vertically monopoly service from generating station/bulk generation to final customers. At its core, the regulated utility model included the following:

- Creation *by statute* of a regulator with rate or rate-setting authority (RSA)[2] and a mandate to ensure the fulfillment of specific social obligations by electricity providers.
- The regulator's ability to grant a monopoly franchise within a fixed territory in which one company had the exclusive right and *obligation* to provide electric energy to anyone who wanted it (universal service) under terms shaped by social concerns under statutory authority.
- Recovery of costs to the utility by the regulator providing oversight, as deemed in the interests of those receiving electric service.
- Costs recovered by the monopoly utility included appropriate return on capital invested to provide service, and regulation included oversight of how that capital was raised.
- The monopoly had exclusive rights to make new investments within the designated territory as customer service needs grew or equipment needed replacement.

From a business perspective, beginning in the 1990s, the electricity supply industry restructured from a regulated, vertically integrated monopoly model to one of mixed regulation, with supply at least partly privatized and parts still under regulatory control. This was a consequence of advances in technology

---

2. For investor-owned utilities, this is typically a state public utilities commission; for publically owned utilities, this is often an independent board or branch of city or county government.

(more efficient and smaller, natural gas-fueled combined-cycle units based on turbine technology) compared to much larger consumer demand. In every state that undertook restructuring, distribution utilities remain the dominant organization in the US electric system for delivering electricity to customers. Most importantly, distribution utilities are at "ground zero" where the latest technology is having an unprecedented impact: where the customer is increasingly adding new generation and interconnecting to the (local) electric grid. This means the local distribution utility's facilities designed to meet one load pattern are doubly disrupted: Some loads are being met by nonutility, even on-site customer-owned production, and the usage pattern of those loads is changing too.

As described in the previous chapters, another far-reaching development is the introduction of modern information technology along the electric system (the so-called smart grid). This certainly includes not only "smart meters" but also much more advanced sensing and control technology all along the system of both actual energy flows and business transactions that enable those energy flows.

The customer's expectation of more reliable electric service is growing at a time when utilities are incurring significantly higher costs to deliver it and, in a growing number of cases, lower revenue growth. Most US utility infrastructure is old, maintenance costs are increasing, and, in some areas, demands for reliability are rising. In fact, demands for reliability may even be accelerating, as new technology on the customer side of the meter requires higher quality and more reliable power. Other areas too are experiencing a growing challenge of extreme weather (e.g., hurricanes) and its impact on utility infrastructure.

This means utilities need to undertake capital investment to increase system reliability and resilience and add renewable forms of generation, taking advantage of advances in technology at a time when traditional mechanisms to fund that investment are challenged. Load and associated revenue growth is low, and in some areas, increasing customer-owned generation is reducing the revenues needed to maintain existing utility infrastructure.

A modern, more reliable, and resilient grid has the potential to better integrate customer-owned assets (including intermittent generation and "smart" loads) into grid operations at a lower cost than traditional utility assets[3] and thus has the potential to keep future rates down. This occurs because the "smart" grid can sense or even predict local usage patterns and "transact" with the customer to keep the system in balance. In the near term, this could include (perhaps limited) control of customer's generation, load, and storage assets. Or it could a means by which grid and customer assets can coordinate to optimize grid management at the lowest incremental cost (King's and Cazalet's chapters

---

3. Historically, this has occurred through "demand response" programs.

describe in detail the transactive energy approach).[4] The combination of these events impacts the viability of the utility business model. The forces in play are as follows:

- Flat sales undermine the industry's historical pattern of spreading increased costs over a growing sales base, which in the past kept rates flat or falling.
- Uncertainty about revenue growth/stability is directly linked to the cost of financing (capital) for maintaining and modernizing infrastructure and service. The higher the risk of cost recovery, the higher the financing costs, which in turn drives up rates.
- Customer expectations of more reliable service from an aging infrastructure will drive investment in new facilities and technology, raising costs and rates.
- Market roles of utilities and nonutilities allowed or prevented the opportunity of new investments, along with who will control how those investments come onto the electric grid and who will pay for them.

This chapter is organized as follows. Section 2 describes the rate design tools regulators have to address the near-term issues, particularly revenue issues, confronted by utilities. Section 3 describes some of the technological and policy challenges regulators face as the electric utility industry continues to absorb ongoing technical innovation. Section 4 concludes.

2   NEAR-TERM ISSUE: TRANSITION GRID MODELS FOR COST RECOVERY AND RATES

The rates set by each utility's rate-setting authority to recover the costs for generating and delivering electricity generally include traditional elements, the implementation and magnitude of which depend upon the objectives of the regulator or utility administrator. The regulator or administrator must balance various needs as they establish rates to address service area needs. Rate setting often balances such goals as trying to keep overall rates low, maintaining reliable service, driving social or environmental policies (e.g., support for low-income customers, energy efficiency program funding, renewable generation standards, and net metering[5]), balancing customer cost causation with revenue generation, and other factors. The tools include forecasts of usage by customer groups and various (and sometimes complicated) rate designs. These latter include rates with fixed charges and demand charges and rates that vary based

---

4. Also see the GridWise Architecture Council at http://www.gridwiseac.org/about/transactive_energy.aspx which uses "transactive energy" to refer to techniques for managing the generation, consumption, or flow of electric power within an electric power system through the use of economic or market-based constructs while considering grid reliability constraints.

5. "Net metering" covers a number of forms of payment from the distribution utility to the owner of the PV system based on the components of the cost of electricity to the PV system's site for the electricity provided to the distribution systems beyond the amount used at the site.

on the level of usage, time of usage, or historical usage pattern. The actual outcomes from rate design and implementation are often not fully realized until considerable time after the rates go into effect, which makes ongoing fine-tuning of this delicate balance difficult.

This already complicated set of issues is becoming more challenging. Some customers are concerned with the reliability, cost, or generation sources accessed by their utility and the speed with which improvements are occurring. The cost of customer-owned renewable generation has steadily decreased, and with ongoing cash flow from subsidies, net metering,[6] and the presence of innovative "pay as you save" lease structures for PV systems, the financial decision is becoming easier for more and more building owners. For a purchased system, combined with some energy efficiency measures, the payback of installing on-site generation can be only a few years, depending on utility rates and rate design.

This is also happening in whole communities (neighborhoods, campuses, and affiliated buildings), where a microgrid is developed for a local area and the utility is left supplying a limited amount of energy when the microgrid's generation is unable to supply all the customer's energy needs. Since microgrid generation is often closer to the customer, there are reduced losses as electric energy traverses the transmission and distribution system. Further, microgrids can have higher reliability and resilience because events far away will not affect the local delivery of energy, and recovery from local events can be quicker. Microgrid customers may also be able to choose their own preferred generation type (e.g., more renewable power) different from the blend provided by the local utility.

As the cost for renewable energy goes down and as regulators utilize rate options such as net metering to attract more of this renewable energy, the utility may be unable to recover their full costs to serve these customers since net-metered customers' charges are typically fully offset for every kilowatt-hour of energy delivered back to the grid. With most rate structures, this means that the fixed costs to supply those customers are also offset. As described in the chapter by Sioshansi, on average, about one-half of the utility bill is associated with fixed utility costs. To accomplish other objectives, many regulators have approved rate structures that depend on energy deliveries to recover a larger share of costs. This exacerbates the potential under-recovery issue described in the preceding text since a self-generation customer is being compensated for much more than the cost of the supplied energy. Customers who do not self-generate are subsequently paying a higher share of the fixed system costs. Over time, this effect on "nonparticipating" ratepayers not installing on-site

6. Some utilities have been concerned about net metering that pays at the full retail rate (i.e., electricity and nonelectricity costs). A recent situation in Arizona is described at http://www.washingtonpost.com/blogs/gov/beat/wp/2013/10/16/in-sunny-arizona-a-battle-over-solar-power/.

renewables under net metering can shift the burden of paying for the "system," an aspect net metering regulators should monitor to ensure costs are collected from ratepayers as regulators intend.

To the extent that the microgrid or net-metered self-generating customer is still dependent upon the utility for power (when the local generation is insufficient or intermittent), this exacerbates the issue of equitable utility cost distribution, since the energy sold under most rate structures covers only variable costs and does not provide sufficient revenue to maintain or improve the reliability of the utility infrastructure to serve these customers. Not only does this scenario not allow the utility to cover system costs or modernization expenses, but also it often requires the utility to incur new expenses to interconnect the new generation, again without the ability to adequately recover those costs in ongoing rates.

Whether for customers in general or for microgrids, this can create a downward spiral where more customers self-generate or join microgrids to improve energy reliability or access to renewable power, and utilities have decreasing revenues yet need to maintain or improve their infrastructure to keep more customers from taking this option, and a greater revenue burden falls to remaining customers (or their usage) putting upward pressure on rates. The immediate solution may be to change the rate structure moving toward greater "fixed cost" rate recovery mechanisms. The chapter by Borlick and Wood describes several options to address this financial challenge.

Many utilities are working to improve their reliability and customer service, often with cost recovery through special rate components on customer bills. For example, there are numerous utilities who have installed modern meters that can bill for energy used at different time intervals so that charges can more accurately represents the utility's actual costs and can communicate with the utility to support billing functions and power outage restoration. This along with utility system improvements will provide customers with enhanced service for years to come. However, the added costs for upgraded systems increase customer rates, and this has a tendency to drive some customers to add their own generation, lowering their contribution to utility costs, which exacerbates the challenge to the rate-setting authority.

Rate setting strives to achieve multiple and sometimes contradictory goals. The final result is often a workable balance to achieve a number of sometimes contradictory goals. These include the following:

- Rate options should be as simple as possible for the customer to understand and still achieve the desired objectives.
- Rates should be designed to provide allowed revenues to the utility with reasonably predictable consumer bills.
- Rates should reflect the prevailing notion of fairness in terms of allocating costs across customer groups (who pays) with how the costs are incurred (who benefits).

- Rates should accommodate regulatory (or statutorily) required social and environmental activities.
- Rates should promote efficient economic choices or use of resources. This means that the cost (rate) for the last energy unit consumed should reflect the costs with providing that unit.

Historically, social needs have included a number of characteristics. These have included lower rates for lower-income customers or customers with energy needs driven by medical necessity. Environmental policies have included energy efficiency and preference for preferred resources (e.g., rooftop solar supported through net metering). However, as other customers absorb these additional costs, higher rates can cause customer dissatisfaction and desire to defect to other service options (microgrids, self-generation, and municipalization).

These considerations must be balanced by the rate-setting authority in a process usually as much political as economic in nature. Fortunately, the new technologies that are forcing consideration of change also allow more effective solutions. In particular, "smart meters" permit rate structures that better meet the goals suggested in the preceding text, and "big data" analysis allows for more complete and effective examination of the effects of the advent of new technology and solution development.

From a practical perspective, this has meant that costs are generally divided into several categories: fixed (often monthly customer charges) that economically reflect the cost of access to the network, demand (often set by considering the customer history of maximum usage) that economically reflect the maximum service a customer would require (e.g., the voltage at which the customer is served), and commodity rates that reflect the cost of energy served.

Rate design is extremely complex. In the succeeding text are a few simple options and their implications. In most cases, the "actual" rate is a combination of several of these components. Given the challenges facing the rate-setting authority, the authors suggest the following components be considered in the process of setting new rates:

(1) Fixed and variable charges: Fixed charges ($/period) can align with true fixed system costs and lower the risk for a utility to recover the fixed costs of serving customers (e.g., metering, billing, and energy delivery) no matter what their energy usage in a given time period. Variable charges can be set to achieve revenue recovery and reflect various costs that are dependent on the forms of customer usage. A simple form of fixed charges is one-time connection fees, but then, fixed cost recovery follows new customer connections. But new connections only partly track when fixed costs are incurred and are of little relevance.

(2) Demand charges: Demand charges ($/kW/period) provide a way for utilities to recover their costs to serve loads based on the maximum load during specific periods. Customers needing grid power only periodically are averse to these charges but to the extent that they represent actual utility costs to serve, this provides a recovery mechanism. These can be complex to design and can take advantage of loads that can be moved in time (classic load shifting) or in space, such as an electric vehicle.

(3) Energy charges: Energy charges ($/kWh/period) provide a way to recover the costs of the energy itself and any associated costs. Like demand charges, they vary by time period and by type of customer.

(4) Time-of-use rates: Time-of-use rates are energy charges differentiated by time to allow the alignment of energy use with the actual costs to generate. These can set a rate for a fixed time period (seasonal, daily, etc.) or dynamic (e.g., responding to a wholesale market price), which will depend upon system changes (e.g., customer loads and availability of lower cost generation). With modern metering, data management, and billing systems, these can be fixed for the minimum duration which the metering and billing system can accommodate, which is likely to be no longer than an hour and perhaps shorter.

(5) Inclining block rates: Inclining block rates are a type of energy charge where the rate increases by the amount of energy used. The first block of energy has a lower rate and subsequent blocks at higher rates, providing incentives for customers to reduce energy use or self-generate. These also magnify the effect on revenues of changes in sales, both at the level of the individual customer and for total utility revenues. This latter effect increases the volatility of an investor-owned utilities' equity earning, influencing its market risk premium.

(6) Combinations of fixed and variable rates: Fixed and variable rate components can be defined for different time units and categories of customers based on costs or other considerations important to the regulator. An example is the straight fixed and variable rate combination—fixed monthly charge covers all costs except those that vary in the short run (e.g., short-term generation costs). The variable component can be based on energy ($/kWh) or an indicator of maximum demand over varying periods ($/kW). A simple straight fixed-variable combination would have a fixed monthly charge and a low-energy charge, while a more complex form could combine a monthly charge, a demand charge, and rising charges across an inclining (or declining) block structure. See Box 1 for a moderately complex example.

Box 1 PG&E's Residential Fixed and Inclining Block Structure

In December 2012, PG&E's average residential rate was $0.186/kWh, but it was calculated with a *daily* meter charge of $0.148 and five blocks of usage, with the least costly at $03.128 and the highest two at $0.336.

Source: http://www/pge.com/tariffs

> **Box 2 PG&E's Residential Rate and Incentives**
> With the highest marginal charge at $0.336/kWh, customers had a strong incentive
> to use less through efficiency, conservation, or on-site generation.

Rates in different combinations incent different responses by customers. Higher variable charges ($/kW h) provide an incentive for customers to reduce their energy bill by reducing energy use (energy efficiency and customer-owned generation). This reduces the incentive for customer self-generation. These charges can be balanced with variable charges based on indicators of usage or cost to achieve preferred outcomes. The straight fixed-variable structure described earlier in the text reduces the incentive and reward for customers to self-generate (or to conserve) by the low-energy charge. In its simplest form, it may not work well to achieve the multiple objectives presented in the preceding text. For an example, see Box 2. Some utilities have proposed monthly "grid access" charges to reflect some or all of the costs of maintaining the utility grid. Such a fixed charge could be calibrated with variable usage charges to better reflect cost causality while maintaining incentives for efficiency.

In addition to using these types of rates, rates and revenues can be connected in innovative ways. Rate designs that separate the immediate link between energy deliveries/sales and revenues allow regulators to set revenue recovery (rates) based upon historic data, and if the utility sells more or less energy than projected to recover their approved fixed costs, rates are reset to keep the fixed revenue recovery neutral to energy sales, reducing the short-term impact of reduced sales. Such "decoupling" mechanisms exist in many states for energy efficiency programs and could be instituted for local self-generation.[7] While this alleviates the utilities' revenue concerns, it shifts any fixed cost recovery shortfall onto the remaining ratepayers.

Modern data analysis permits customer groups to be defined at finer resolution than the historic Residential, Commercial, Industrial, and Agricultural, allowing better alignment between drivers of costs and cost recovery. Of course, this capability also expands the debate on rate equity, potentially making rate-setting processes more difficult to manage.

In the past, customers had received credit for energy-related grid services that they provided. One category of such activity occurred through "demand response" programs, but with current and emerging technologies, this could include load interactive or transactive energy and other ancillary services. Modern rates could provide customer incentives to add more capability into their building loads (sensors, communications, and controls) and even invest in

---

7. Decoupling is discussed in detail in "Revenue Regulation and Decoupling: A Guide to Theory and Application," by the Regulatory Assistance Project, August 2011, available at www.raponline. org/document/download/id/861.

electric or thermal storage on select circuits where the utility needs to invest in the local distribution system to assure reliable service. This approach may require a more transparent and systematic approach to local distribution planning, as stakeholders will have incentives to understand and interact with the process in a much more robust manner than has historically been the case.

Historically, detailed analysis of cost drivers on distribution systems has not been available, as monopoly provision of electric services allowed rates to be set in its absence. With multiple services (generation and storage) becoming available on the customer's premise, methodologies may need to be developed that allow for more precise calculation of the value of grid access to these different resources and the cost to the grid of managing diverse, customer-owned resources. Key to this determination of value is how reliable these customer-sited resources are in positively impacting the grid. Today, maximum value requires controllability by those responsible for grid reliability: locally, the distribution utility and, globally, the operator of the transmission system.

For large, well-defined, significant new infrastructure additions (e.g., disaster recovery and grid modernization), regulators have used special rate recovery mechanisms to clarify the process of approving the investment and tracking the utility's cost recovery. This can often be done as part of a normal rate proceeding or between general rate cycles. In some cases, the rate design recovers these costs on a per customer basis, and in others, the costs follow the benefits. For some investments (e.g., advanced metering infrastructure), benefits can be initially hard to quantify, which complicates this approach to rates design.

Utility incentives could be put into place to encourage utilities and facilitate the expansion of the full range of distributed resources (e.g., solar PV, customer-facility storage, and demand response). While decoupling removes the short-term financial "pain" resulting from reduced sales, this approach would go one step further and provide incentives to the utility to achieve longer term policy goals and assist in the development of a new utility business model. Under this approach, the utility would profit by achieving concrete distributed resource expansion goals that could be tailored to meet programmatic and/or locational policy objectives helping to minimize system costs.

Historical examples of these innovative approaches can be found in regulatory support for other desirable "behind-the-meter" activities such as energy efficiency or demand response. These provide models for the application of innovative models to the challenges facing regulators in addressing the influx of new technology onto the grid, especially the distribution grid. A useful model for the regulator to follow could be to proceed in a stepwise fashion:

- Identify the specific activities for which nonstandard rate treatment is to be afforded.
- Identify the questions of fact as to the impacts on the distribution system: on the utility, on customer, on other industry participants, and on other stakeholders.

- Identify the value/cost streams and develop a methodology for determining the value.
- Determine the values pertinent to the decision to be made.

By proceeding in this deliberative and logical manner, the regulator can ensure every stakeholder has a fair hearing and full information can be obtained to make a workable policy- and fact-based decision. For example, a recent presentation of value streams for local photovoltaic compared the elements used by Austin Energy and those proposed by the Mid-Atlantic Solar Energy Industries Association. The former included the same categories often used to evaluate energy efficiency (the value from avoided energy, environmental effects, and the avoided capacity value of generation, transmission, and distribution); the latter added value from an energy market price reduction, security enhancement, economic development, and long-term societal benefits.[8] A regulator considering the value of local photovoltaic installations can consider this list, determine which it considers material, determine appropriate methods to calculate value, and perform the study to determine those values.

3 MEDIUM-TERM ISSUES SHAPING THE REGULATORY FRAMEWORK FOR UTILITY 2.0

A vision of the electric power grid that includes dispersed renewable generation and storage coupled with dispersed sensing and control technology is far different from the vertically integrated monopolies of the mid-twentieth century. As the industry moves along this technological path, the business form and model of the industry will be a reoccurring theme. The need for organizations than can assemble large amounts of capital into industrial enterprises to provide an almost essential product (electricity) safely, reliably, and at predictable cost will change. New needs will emerge, and a reoccurring question will be what combination or form of organization best meets society's evolving needs, as utilities, new market entrants, and regulators experiment with new approaches.

While the specific questions we face regarding utility business model and form are driven by today's technologies, the broader questions are not new. The questions policy-makers and RSAs face today mirror those of the 1980s posed by the erosion of monopoly in the generation of electricity through technological advances in generation technology. These advances led to the spread of midsized (30-500 MW) combined-cycle combustion turbine and combined heat and power units. This in turn led to the much-examined changes in statutory framework and state-level industry restructuring and the disaggregation of

---

8. See James Montgomery, "Net Metering Debate Rages On Despite Calls for Calm," in Renewable Energy World.com, February 2013. Accessed 12 November 2013 at http://www.renewableenergyworld.com/rea/news/article/2013/02/net-metering-debate-rages-on-despite-calls-for-calm.

the vertically integrated utility in some regulatory jurisdictions during the 1990s.

But while the nature of the questions is similar, the sheer range and scale of today's technologies present unique challenges to utilities, new market entrants, and RSAs. The range of available disruptive technologies currently available and in the pipeline is unprecedented, and "bundles" of these technologies are in the very early stages of being packaged and marketed to customers.[9] And while the earlier disruptive technologies tended to be appropriate for and available to large industrial and commercial customers, today's technologies are being marketed to—and embraced—across customer categories, including residential customers; the addition of the mass market creates both greater complexity and far greater scale for utilities and RSAs.

In addition, in the past, it was easier to define the rules that represented the public policy component of regulation (the "public good"): universal access, fair rates, low costs, and reliable service. The 1980s and 1990s added conservation, efficiency, and environmental protection—and, in some jurisdictions, customer choice—increasing regulatory complexity, but it was still possible to regulate the utility's obligations and generally manage the transition.

Clear, accurate, well-vetted rate design is a solution to some of the near-term challenges facing the electric industry. Still, technology is advancing and the shift to less polluting generation and a "smarter" grid will continue. With today's range of technologies, together with empowered customers at scale, what is coming is expected to be more challenging to manage. In the near term and at moderate market penetration, the tools of rate design can clearly manage/facilitate the current transition and maintain utility solvency. In the longer term and at high penetration, balancing public policy questions and options will clearly be more challenging: what is fair and equitable, what maintains reliability, and what allows continued support for the medically needy and low-income customers and/or favored resources while funding an ever more complex and critical electric grid?

In the succeeding text are several significant developments that will continue to change the electric industry. First, advances in technology have brought us new, dispersed renewable power generation and is in the process of enabling less expensive storage in the form of much more effective batteries and other storage technologies. Advanced electronics are improving and reducing the cost of metering, while sensors, actuator technology, and more sophisticated control systems will open the way to transactive energy (see Chapters 9 and 10 in this volume).

Transaction-based controls, a key element of transactive energy, are control solutions that allow operational decisions to be based on market signals

---

9. The pace of innovation is illustrated by the recent announcement of bundled PV and battery sales. See http://www.nytimes.com/2013/12/05/business/energy-environment/solarcity-to-use-batteries-from-tesla-for-energy-storage.html.

(e.g., commodity, service, and retrofits) whether it is a direct (e.g., time of day electricity price) or indirect (e.g., price given to the fuel and carbon impact of the existing electricity mix) financially based indicator of the energy system. The market concept in some versions of transactive control reveals an interesting question: will the advance in sensor, control, storage, and other technologies change the characteristics of electricity so that the value it provides can be identified, quantified, and traded in more traditional markets and its social characteristics diminish allowing a reduction in the historic extant of regulation?

A new and growing approach utilizes "smart" or controllable customer-owned assets to help balance energy supply and demand on all dimensions (i.e., the provision of ancillary services) at a lower cost than typical utility system solutions. A solution is to integrate loads in multiple buildings to ensure power demand needs are met, avoid capacity charges, and even keep power consumption below transformer capacity thresholds. These loads (e.g., EVs and HVAC systems) can interact or transact with other loads or generation in a building, between buildings, and with the utility grid, achieving the outcomes the customer desires while providing value to the utility that would have cost significantly more if supplied on the grid itself. The net value of a system of interactive or transactive control is enhanced when it is at a scale (MW), which is most relevant to a utility.

The absence of forward markets, much less futures markets, for ancillary services inhibits investment in technologies that produce those products whether they come from generation sources or commitments from building owners to adjust usage to help support local or wider grid reliability. This new view of ancillary services includes the effects or "products" possible through transactive energy described in the aforementioned text and provides a means to incentivize and finance them if forward markets existed. For example, it could allow customers to invest in on-site energy storage and fast transactive controls using forward transaction to provide some of the financing requirements. Overall this would potentially provide more unique pricing options for utility services. For example a customer uses an electric vehicle at peak times, he/she will pay peak prices for the demand and the energy, or if it is providing energy at peak, he/she could receive payment accordingly. This would have deep implications to allow movement past the split incentive dilemmas that have plagued traditional energy efficiency and demand response.

Transactive systems that are, and are regarded as, *reliable* raise challenges not addressed by only piecemeal technological advances. Currently, there are thousands of equipment providers, utilities, and many data and communication "standards" that can be used. When all grid-reliant electricity users—transmission operators, utilities, and customers—have confidence that transactions can occur nearly instantaneously over large geographies, with outcomes easily quantified, this transactive market can be counted on to more efficiently supply significant grid resources. This includes demand management, frequency and VAR support, and other costly services that are increasingly needed to address system reliability and management. This confidence can likely be built only on a yet undefined common national framework that allows determination of an appropriate value proposition to the building owner, utility, and transmission operator.[10]

4 CONCLUSION

The current system of utility regulation will need to change in order to deliver the services needed by a modernizing grid while supporting the social and other policies of the regulator. In the near-term, the use of more refined cost and revenue analyses and the full range of rate designs should allow the challenges faced by utilities to be met. As technology advances, in the medium term, the regulator will need to consider the future role of the utility and all associated costs, their policy objectives, and lowest-cost solutions. The regulator must be particularly aware of technology changes that impact the economic and social foundations of the current regulatory approach. Some innovative new solutions will be possible, utilizing customer-owned assets, once the systems are in place to scale this across broad geographies and energy users.

---

10. An earlier vision is the "virtual power plant" that is being discussed (and adopted as policy in Korea). A summary of the Korean model is contained in a white paper by Frederick Weston accessible under the pdf icon at the bottom of http://forumees.pl/aktualnosci-ees/art,18,rap-o-wirtualnych-elektrowniach.html.

Chapter 17

Electric Vehicles: New Problem or Distributed Energy Asset?

Glenn Platt, Phillip Paevere, Andrew Higgins and George Grozev

ABSTRACT

In an evolving decentralized future with growing intermittent renewables and distributed generation, any type of storage becomes a valuable resource. While they contain significant energy storage, electric vehicles (EVs) are rarely considered as a potential *resource* to the electricity system—rather, as they represent a very significant additional load, they are usually considered a potential challenge. This need not be the case—with careful management of their battery charging and the ability to discharge in to the grid, EVs may in fact be a very significant distributed energy resource. This chapter explores the impact and potential of EVs—what is clear is that such impacts are quite localized, and careful analysis is needed to determine exactly where and how EVs will impact the grid. With this in mind, this chapter presents a modeling methodology that analyzes the impacts of EVs at an unprecedented level of detail. Such detailed analysis will ultimately be critical to understand the full impact of EVs and whether they will remain a challenge to grid operations or a resource of greater benefit.

Keywords: Electric vehicles (EVs), Vehicle-to-grid (V2G) technology, EV charging, Battery storage, Distributed storage

1 INTRODUCTION

When considering the rise of decentralized energy resources, it is important to consider both supply and demand technologies. Part 1 of this book discusses a range of distributed technologies including solar photovoltaic or gas cogeneration, even battery energy storage systems. On the demand side, typical decentralized technologies are load management of devices such as air conditioners or perhaps new dynamic pricing schemes. One major new technology already causing ripples through the electricity industry, but rarely considered as a distributed energy *resource*, is the electric vehicle (EV).

Most of the world's major vehicle manufacturers are currently developing or have plans to develop mass-market plug-in EVs. The two primary types of EVs on the market are battery electric vehicles (BEVs) that run entirely on a battery

Distributed Generation and its Implications for the Utility Industry. http://dx.doi.org/10.1016/B978-0-12-800240-7.00017-5
335

charged from the grid and plug-in hybrid electric vehicles (PHEVs) that can run on batteries charged from the grid alone, liquid fuel (petrol), or a combination of both. Widespread adoption of these vehicles and replacement of internal combustion engine (ICE) and traditional hybrid electric vehicles (HEVs) are anticipated in Australia and many other developed nations over the next 30 years—most major automakers intend to put plug-in hybrid or fully EVs on the road over the next 2-3 years (Electric Drive Transportation Association, 2013) and one independent estimate predicts an annual sales growth rate of 106% for these cars, between 2010 and 2015, with worldwide sales volumes reaching 1.7 million units annually (Pike Research, 2012).

While projecting the uptake of any technology is always fraught with difficulty, recent sales data give a sense of the trajectory EV uptake is on. Figure 1 shows that plug-in vehicle sales have risen exponentially in recent years, from 345 vehicles in 2010 to 52,835 vehicles in 2012 (Chernova, 2013). This number is expected to double in 2013 (Chernova, 2013). While these numbers are still relatively small (1% of total vehicle sales in the United States), they can have a very localized impact—EV uptake is usually quite clustered, with significant numbers of vehicles occurring in particular discrete locations (Chernova, 2013). Considering these growth trends, EVs may represent one of the most profound technology impacts for our distribution systems and the main cause between many of the market shifts explored in Chapters 2 and 3 of this book. From a technical standpoint, EVs represent a significant concern among the electricity industry regarding the impact that they may have on the electricity network. Drawing their energy from the local electricity distribution grid, EVs

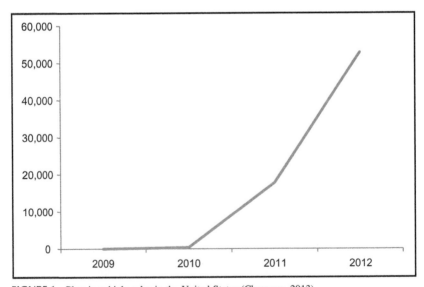

FIGURE 1   Plug-in vehicle sales in the United States (Chernova, 2013).

may consume a significant amount of electricity, increasing energy flow and potentially worsening challenges such as growing peak demand. In short, EVs may have a profound effect on electricity system planning and operation. At the same time, if EVs are charged during periods of low demand, including off-peak, they can increase the network utilization and improve the revenue of electricity utilities. Importantly, EVs may be considered a *resource* to the electricity system, where energy stored in the vehicle's battery can be used to provide system support at critical times.

The impact of EVs in general, and in particular when acting as a distributed energy resource, has received very little detailed attention. Such detailed analysis will ultimately be necessary to understand the actual impact EVs are likely to have on electricity systems, to plan and operate systems with large numbers of EVs, and to take advantage of their potential to act as a distributed energy resource that can benefit the wider network.

This chapter presents an analysis of the challenges and benefits EVs may pose to the electricity system and explores their potential to act as a distributed energy resource. Section 2 explores the "vehicle-to-grid (V2G)" concept, which is key to EVs being a resource rather than an additional load. Section 3 presents a detailed modeling methodology that goes significantly beyond the general analysis commonly available, to providing quantified predictions of the impact of EVs, down to very small geographic areas. Section 4 gives some example results of the V2G modeling that clearly demonstrate the benefits EVs may offer the grid when acting as a distributed energy resource.

2   THE V2G CONCEPT

EVs used for passenger transport may require somewhere in the order of 2 kWh to as much as 20 kWh of electric energy from the grid to meet their daily driving distance requirements. This is a completely new load on the electric grid and potentially represents a significant extra demand that needs to be planned for. This extra demand is of significant concern to electricity network regulators and grid operators—not only is the extra energy consumption of EVs a new load, but also the charging of EVs may exacerbate the peak demand challenges facing many developed grids. For example, while typical household EV chargers may operate at around 10 kW, one EV manufacturers is now rolling out charging stations that operate at 120 kW *per vehicle* (Now Breaking: 120kW charging, 2013).

While EVs can create significant extra demand for the electricity grid, conversely, and of most interest to this book, they can also act as an electricity generator. Very significant electric energy storage sits in the batteries of EVs, and this storage, when not needed for transportation, can potentially be harnessed and used to support the needs of the electricity grid at times of high demand, or constrained generation or distribution, by peak-shaving technologies such as V2G or vehicle-to-house (V2H). The basic premise of such concepts is that the EV, while parked, discharges its battery back into the grid, providing power for adjacent loads. Then, at some other time, the EV will draw power from the grid to recharge its battery, ensuring its availability for transport or later V2G application. Significant energy and power levels are possible with the V2G concept, matching the significant energy storage available in an EV's battery and the power capability of the charging electronics associated with the EV. With perhaps 20 kWh of energy available for "dispatch" into the electricity grid, a V2G-enabled EV represents a significant resource, and a car park full of such vehicles, an incredible quantity of energy, available at no capital cost to the utility.

Acting as generators, EVs may be able to be remotely controlled, dispatching significant quantities of energy at times most worthwhile such as during critical peak periods of demand. The operators of such remote control may be electricity distributors or retailers or energy service companies selling a service (dispatchable capacity) to others.

3 MODELING: WHY "MACRO" UPTAKE FIGURES ARE NOT ENOUGH

The challenges of EVs, and the potential benefits from the V2G concept, have received a reasonable amount of attention from the industry in recent years, including mainstream media coverage such as (Motavalli, 2007). In general though, such attention has been relatively generic, based around rough approximations of the number of EVs the national fleet may be made up of—typical figures are around 10-20% of the vehicle fleet being EVs by 2020 (Plug-in Electric Vehicle Adoption Forecasts, 2012)—or the total quantity of energy across such a fleet that could be available for V2G purposes.

Although the general figures available around EV uptake or V2G benefits are useful to get a sense of potential future scenarios, they do not help with the detailed planning that is necessary to fully prepare for the arrival of EVs—the construction of new infrastructure, the evaluation of where network upgrades may be needed, and so on. Such considerations are quite low-level, affecting the electric network operation in one particular distribution feeder, for example.

When trying to understand the impact of EVs or V2G technologies in detail, it is important to realize that such impacts will vary significantly across different geographies. There are already significant variations in electricity demand across different regions, with local climate, housing design and construction, and population demographics all causing great variations on demand across a particular area. Electricity network infrastructure also varies significantly across regions, based on local factors such as equipment age, amount of distributed generation such as solar, availability of space for capacity upgrades, etc. On top of these variations, EVs will bring their own localized attributes—the uptake of EV technology will vary across regions based on factors such as typical trip distance (EVs may be unsuitable for geographies where people typically drive long distances), population demographics, and individual characteristics such as concern for the environment. Additionally, the daily distance traveled in an EV will significantly affect the energy it requires from the grid on recharging.

Considering the various factors described in the preceding text, getting an accurate and detailed understanding of the impact that an EV, with or without V2G technology, will have on the electricity system is a challenging exercise, requiring comprehensive analysis. Basic steps include the following:

1. Understanding how many people are likely to purchase EVs and where
2. Understanding what trips those EVs will typically undertake, the timing and the distance driven, to estimate how much electric energy the EVs will require and when
3. Predicting the energy consumption of buildings in the area, to understand what the base energy consumption is without the EV
4. Calculating the net energy consumption of the building and EV
5. Performing the analysis at sufficient detail to provide a time series of electric demand, so that the effect of the EV at various times of the day (such as during peak demand periods) can be ascertained
6. Calculating the effect of net electricity consumption by the time of day on existing network capacity in feeders and substations, thus identifying priority demand management or infrastructure components requiring upgrading

Based on these broad steps, the following section presents a novel composite modeling methodology and the results of a case study undertaken for the state of Victoria in Australia (Paevere et al., 2014). The methodology provides spatial and temporal projections of the charging demand from plug-in EVs and the potential benefits from the V2G concept, at exceptional detail (Paevere et al., 2014).

4 DETAILED V2G MODELING

To understand the potential benefits of V2G technology and its application as a form of distributed generation, sophisticated modeling is necessary to understand the impacts of EVs with fine temporal and geographic detail.

The modeling presented in this section links models of future EV uptake, travel by households, household electricity demand, and recharge/discharge of EVs. The analysis is disaggregated to an hourly time step over 365 days of the year and fine-grained spatial mesh blocks of around 250 houses.

The modeling approach is best understood through a case study, in which the residential peak load impacts on the distribution network across the Australian state of Victoria are analyzed under nine different scenarios:

3 × EV uptake scenarios:
1. Base case with no rebates or incentives on EV purchases
2. Rebate case with a $7500 rebate on purchase price
3. Maximum uptake with a 100% penetration of EVs

3 × charging models:
1. Demand charging: EVs recharge solely at their home base on arrival at home.
2. Off-peak charging: EVs recharge solely at their home base with charging delayed until after midnight where possible.
3. Off-peak charging plus V2H: EVs return energy to the grid at the home when possible during peak load times. EVs recharge solely at their home base, daily, with charging delayed until after midnight where possible.

The integrated modeling approach described herein is designed to provide spatial projections of the future impacts of EV usage on the electric grid. The modeling strategy is outlined in diagrammatic form in Figure 2 and incorporates

- *an EV uptake (diffusion) model* for spatially forecasting EV uptake across an urban area at three monthly time intervals, given different financial and policy settings, in spatial mesh blocks of around 250 houses;
- *an EV travel model* for spatially projecting the likely driving distances and times (hourly × 365 days) and the periods of availability for charging and discharging of EVs, spatially across different parts of an urban area;
- *an EV charging and discharging model* for projecting energy requirements (hourly × 365 days) based on the EV travel model and parameters for the various charging algorithms, in spatial mesh blocks of around 250 houses;
- *a residential energy model* for projecting the hourly residential energy usage (×365 days) across an urban area in spatial mesh blocks of around 250 houses.

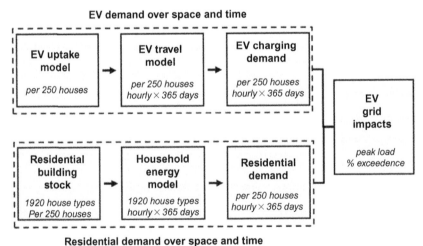

FIGURE 2   Overview of modeling strategy (Paevere et al., 2014).

Each of these modeling components is described in the following sections. While this case study of EVs and V2G impact focuses on the state of Victoria, similar techniques can be used for any other geography, assuming sufficient input data are available.

4.1 EV Uptake

An EV uptake (diffusion) model was used to spatially project EV market penetration across the Australian state of Victoria at three monthly time intervals, given different financial and policy settings. EV penetration levels are projected for spatial mesh blocks of around 250 houses.

The EV uptake model combines features of choice modeling, multicriteria analysis, and diffusion modeling to estimate the uptake rate for different competing passenger vehicle technologies across a landscape of heterogeneous consumers. Vehicle types considered are BEV, PHEV, HEV, and ICE.

The model assumes that the total stock of all vehicles is known over time. The market shares of the respective competing vehicle technologies are then estimated based on a range of criteria relevant to buyers' purchasing decisions. A household's purchasing decision is made when one of its existing vehicles reaches the end of its life span, which is assumed to be 10 years. Seven criteria are used to determine penetration levels: vehicle performance, up-front cost (purchase price), annual cost (maintenance and running costs), household income, demographic suitability, driving distance required, and familiarity. Annual changes in the values of the various criteria and their elasticity with respect to ongoing adoption are incorporated. The model parameters are partly calibrated using data from a large-scale public survey conducted in Victoria (Gardner et al., 2011).

The uptake model is used to estimate the market share of vehicle stock in 2033 for the four different vehicle types (BEV, PHEV, HEV, and ICE), across all of Victoria under the three different scenarios previously mentioned:

1. Base case with no rebates or incentives on EV purchases
2. Rebate case with a $7500 rebate on EV purchase price
3. Maximum uptake with a 100% penetration of EVs

The $7500 rebate to up-front cost applies to every BEV and PHEV purchased between 2013 and 2033. The maximum uptake case represents the situation where traditional hybrids and ICEs are phased out completely, so that the vehicle stock comprises only PHEVs and BEVs. This is intended only to provide an upper bound to the estimates and is not considered plausible for the year 2033. The projections are dependent on the relative costs of the various vehicle technologies and changes in costs of Li-ion battery packs, electricity prices, and oil prices, all of which are subject to significant uncertainty. The assumed up-front and annual costs used are given in Table 1.

TABLE 1 Assumptions for Up-Front and Annual Vehicle Costs

Vehicle Type	Up-Front Cost[a]		Annual Cost[a]	
	Small Vehicle	Large Vehicle	Small Vehicle	Large Vehicle
BEV	$43,000 (−4.5%)	$80,000 (−4.5%)	$2300 (3%)	$2800 (3%)
PHEV	$39,000 (−2.5%)	$70,000 (−2.5%)	$2400 (3%)	$3000 (3%)
PHEV	$35,000 (−2.5%)	$63,000 (−2.5%)	$2700 (3%)	$3300 (3%)
ICE	$18,000 (0%)	$45,000 (0%)	$3500 (3%)	$4500 (3%)

[a]Assumed annual % change in costs is given in brackets.

TABLE 2 Projected EV Penetration for Victoria Under Three Uptake Scenarios

Uptake Scenario	Base Case (2033)	Rebate Case (2033)	Maximum Uptake (2033)
BEVs—total	362,062	661,656	1,439,247
PHEVs—total	619,456	700,198	1,493,019
EVs—total	981,518	1,361,854	2,932,265
BEVs—% of fleet	12%	23%	49%
PHEVs—% of fleet	21%	24%	51%
EVs— % of fleet	33%	46%	100%
BEVs—% of dwellings	18%	33%	74%
PHEVs—% of dwellings	31%	35%	74%
EVs—% of dwellings	49%	67%	148%

Based on Higgins et al. (2012).

These are based on assumptions used in AECOM (2012) and Graham et al. (2008). A summary of the projected Victoria-wide penetration rates for the three scenarios is given in Table 2, and the penetration rates over time to 2033 for the base case and rebate case are shown in Figure 3, which is based on the detailed methodology given in (Higgins et al., 2012). Considering these results, it is

Market share

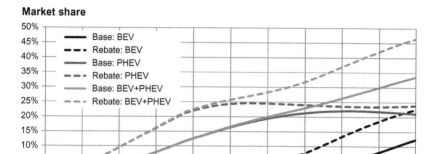

FIGURE 3   Projected EV market share in Victoria for base-case and rebate-case scenarios (Higgins et al., 2012).

worth noting that significant spatial variability could be seen in the data. The primary influences on the spatial variation are driving distance, demographic suitability, existing vehicle ownership, and household income, with urban areas having about three times the proportional uptake of BEVs compared to regional areas. This urban/regional variability is not as strong in the PHEV projections, primarily because the PHEV driving range is not limited by battery size. The examples of the model-projected spatial distribution of EV and PHEV penetration are shown in Figure 4.

4.2   EV Travel

An EV travel model is used to spatially project the likely daily EV driving patterns and time frames (hourly × 365 days) and the periods of availability for charging and discharging across the whole of Victoria. The model is based on the Victorian Integrated Survey of Travel and Activity (Victorian Department of Transport, 2009) and data on traffic volumes provided by the agency responsible for main roads, "VicRoads."

The travel model output provides estimates of EV behavior, including the average distance and frequency of "off-driveway" journeys, the number of journeys, and the numbers of vehicles at home at any given time, disaggregated by location, by time of day, and by date of year. These outputs can be then used to quantify the travel behavior that is relevant for estimating EV charging demand at home and illustrate the proportion of trips for a given home-arrival time and distance traveled. Such inputs have a significant impact on both the timing and duration of home charging of EVs and the shapes of aggregated charging profiles in different geographic regions.

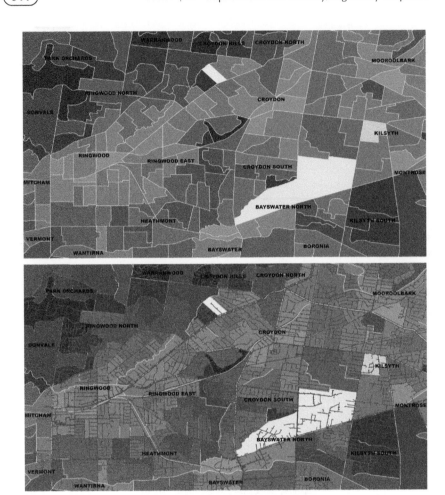

0%   BEV + PHEV Uptake   55 %

FIGURE 4 EV uptake for selected Victorian suburbs (upper) with low-voltage distribution network superimposed (lower) (Paevere et al., 2014).

4.3 Household Energy and Power

A household energy model is used for projecting the hourly residential energy usage by day of year across the whole of the state of Victoria, in spatial mesh blocks of around 250 houses. This approach is necessary to understand the impacts that EV charging and discharging will have on residential demand across different parts of the electricity distribution network at specific times of the day and year—essentially, the EV adds to the total household demand, and so it is necessary to understand what that demand would have been without the EV.

The household energy model is a physics-based bottom-up model that operates by aggregating individual household consumption profiles. The model is driven by simulated hourly weather and uses the heating and cooling load calculation engine from CSIRO's AccuRate software (Ren and Chen, 2010) that calculates the energy required to keep a house of a particular construction and location comfortable for the occupants. AccuRate calculates the energy required by using climate data for a particular geographic region and then calculating the heat flows in and out of a given house construction. Similar software in the United States is EnergyPlus, released by the Department of Energy. AccuRate's estimate of the energy required for heating, cooling, and ventilation is then combined with a recently developed appliance-level residential energy simulation tool that has been used for estimating greenhouse gas emissions in the residential sector (Higgins et al., 2011; Ren et al., 2011). In this way, the total household energy profiles are calculated, including heating, cooling, water heating, lighting, and other appliances. The total Victorian household energy use projected by the model is within 10% of other published estimates.

The household energy modeling process used for this study can be summarized as follows:

1. Develop typologies, which describe the different types of dwellings and their occupancy within each spatial mesh block. The typologies include dwelling type, size, vintage, occupancy type, and family type.
2. Develop a building stock model that quantifies the number of each of the different dwellings and occupancy typologies (defined earlier in the text) within each spatial mesh block. The stock model is derived using national data sources.
3. Simulate hourly energy consumption profiles for each dwelling and occupancy typology. Hourly energy usage is simulated for space heating and cooling, water heating, lighting, and other appliances for the different local climate zones using hourly weather data.
4. Aggregate the energy consumption for each mesh block based on the building stock model and the simulated hourly energy consumption profiles.

In Figure 5, the simulated demand profiles (demand per dwelling) for a variety of Victorian buildings are shown for a hot summer day. The profiles shown are for electricity-dominated (with electricity used for hot water, space heating/cooling, and cooking) and gas-dominated (no cooling, gas used for hot water, space heating, and cooking) dwellings built between 1992 and 2004. The aggregated Victorian average profile for day 33 is also shown for reference. The peak load for the Victorian average load profile is lower than for the individual profiles shown as it averages across a diversity of dwelling types and occupancy scenarios and includes a majority of gas-dominated households, with 70% of dwellings assumed to be gas-dominated across Victoria.

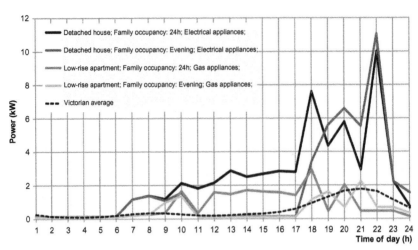

FIGURE 5   Hourly household demand profiles for selected dwelling types on a hot summer day.

4.4   EV Charging and Discharging

An EV charging and discharging model is used for projecting hourly EV charging requirements by day of year at spatial scales down to spatial mesh blocks of around 250 houses.

The EV charging and discharging model uses the outputs from the travel model to determine the magnitude and timing of daily EV charging energy and power requirements at the base household and the magnitude and timing of potential discharging events where relevant. As discussed previously, three charging modes are considered for the purposes of this study:

1. Demand charging (Dem): EVs recharge solely at their home base on arrival at home. Charging energy required is equal to the amount of energy used for the previous off-driveway travel.
2. Off-peak charging (OP): EVs recharge solely at their home base with charging delayed until after midnight where "on-driveway" time window permits.
3. Off-peak charging plus V2H (OPV2G): EVs recharge solely at their home base, daily, with charging delayed until after midnight where on-driveway time window permits. EVs discharge energy to the house between 9 PM and 12 PM when they are at home and battery capacity and charging requirements permit.

Other assumptions used for EV charging and discharging calculations are as follows:

- EV efficiency (including charging efficiency) is assumed 0.18 kWh/km.
- Charging power is 3.6 kW (240 V/15 A).

- Discharging power is 1.8 kW (to allow for longer discharge time and minimal battery damage).
- Charging energy required is based on all previous off-driveway travel.
- Minimum discharge level (depth of discharge) is 40% of battery capacity.
- Battery capacity of BEV is assumed as 25 kWh.
- Battery capacity of PHEV is assumed as 16 kWh.
- 80% of PHEV travel is powered by grid electricity.

Hourly charging profiles are generated using the model for each day of the year for each spatial mesh block. Aggregated, averaged charging profiles for the whole state of Victoria are shown in Figure 6 for the three charging modes.

To check the validity of the projected charging profiles, the aggregated averaged charging profiles (per EV) calculated for Victoria are compared against some measured average charging profiles from EV field trials under way in the United States (US Department of Energy: Vehicle Technologies Program, 2011), Denmark (Energy Agency, 2012), and Melbourne (Victorian Department of Transport, 2012). A summary of these trials is given in Table 3. Figure 7 shows the projected average Victorian off-peak charging profile throughout a day compared with the average demand profiles from the San Diego EV trial, where the majority of the trial participants had an off-peak electricity tariff and typically delay the beginning of charging their Nissan Leaf EV until midnight where possible. For this case, the model shows strong agreement with the measured profiles but slightly underestimates the observed off-peak charging load.

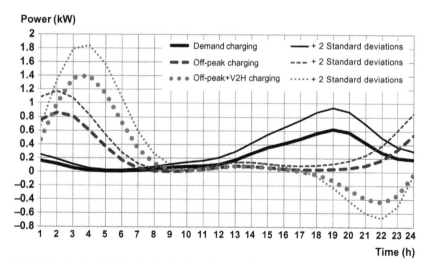

FIGURE 6   Projected daily charging load profiles per vehicle for Victoria (profiles are averaged across all regions in Victoria for all days of year).

TABLE 3 Summary of EV Trial Data Used for Charging Profile Validation

Trial Region	Number of EVs	Type of EVs	Charges per Day	Average Driving Per Day (km)
Oregon (United States)	267	Nissan Leaf	0.7	~30
Washington State (United States)	458	Nissan Leaf	0.71	~30
Tennessee (United States)	229	Nissan Leaf	0.64	~30
San Diego (United States)	458	Nissan Leaf	0.72	~35
Denmark	170	Mitsubishi i-MiEV	Unknown	~35
Melbourne	13	i-MiEV and Leaf	Unknown	~25

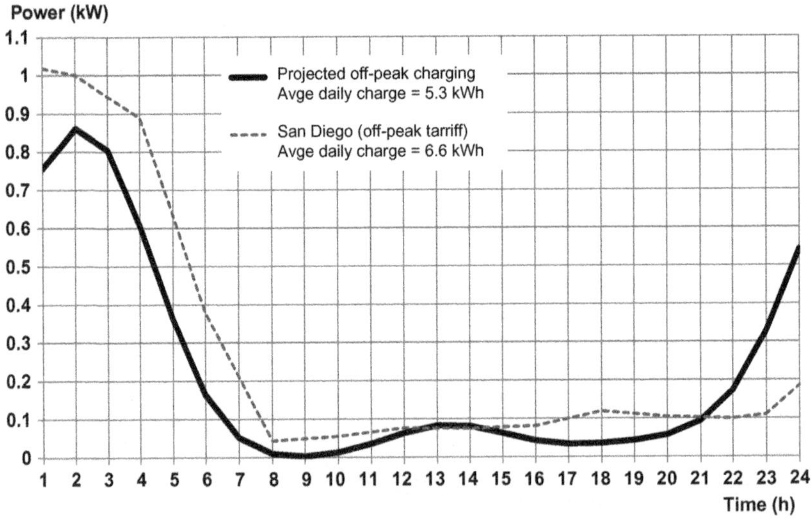

FIGURE 7   Modeled versus measured off-peak charging profiles.

5   MODELING RESULTS: THE REAL IMPACT OF V2G

To determine the impact of EV charging on peak electric loads, the modeled charging load profiles for each charging mode and uptake scenario are combined with the modeled household load profiles across all hours of the year and all spatial mesh blocks. Some examples of the generated load profiles

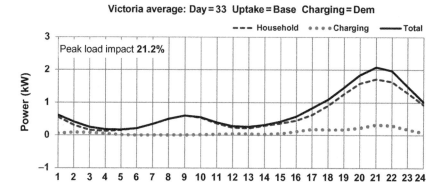

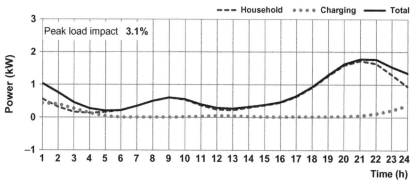

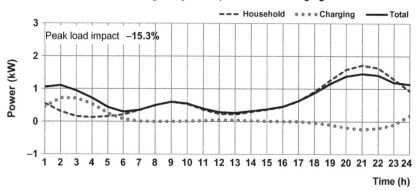

FIGURE 8 Average load profiles per dwelling for three charging modes—area = Victoria, day = 33, uptake = base case—and for demand charging (Dem), off-peak charging (OP), and off-peak charging with vehicle to grid (OPV2G) (Paevere et al., 2014).

are given in Figure 8, which shows the Victorian average load profiles (separated into household (energy just for the house and appliances), charging (energy for the EV), and total) for a hot summer day (day 33) under base-case uptake and for three different charging modes. Statistics for the calculated household energy and power are shown in Table 4.

Considering the average load profiles in Figure 8, the peak load under EV demand charging is increased by around 13% compared to the household-only profile. Peak load is increased by around 2% under EV off-peak charging. Under the off-peak+V2G charging, the peak demand is *reduced* by 15%.

The state average profile in Figure 8 is useful for illustration purposes, but does not provide an indication of outcomes in the areas with highest loads, or most constrained distribution system capacity, which are the most important in the analysis of peak load impacts. In Figure 9, average load profiles for a region

TABLE 4 Statistics of Modeled Daily Household Energy and Power Demand for Victoria

Parameter	Uptake Scenario	Charging Mode	Unit	Maximum	Average	Standard Deviation
He	All	All	kWh	52.51	15.24	3.72
Te	Base	All	kWh	55.11	17.20	3.95
Te	Rebate	All	kWh	55.62	18.06	4.03
Te	Max	All	kWh	60.07	21.49	4.47
Hp	All	All	kW	5.87	1.72	0.44
Tp	Base	Dem	kW	6.38	1.90	0.47
Tp	Base	OP	kW	5.89	1.75	0.44
Tp	Base	OPV2G	kW	5.77	1.59	0.43
Tp	Rebate	Dem	kW	6.44	1.99	0.48
Tp	Rebate	OP	kW	5.90	1.77	0.43
Tp	Rebate	OPV2G	kW	5.71	1.57	0.42
Tp	Max	Dem	kW	6.97	2.41	0.55
Tp	Max	OP	kW	5.93	1.92	0.42
Tp	Max	OPV2G	kW	5.45	2.14	0.56

He, household energy per dwelling; Te, total energy per dwelling=household+charging/discharging energy per dwelling; Hp, household power per dwelling (hourly average); Tp, total power per dwelling=household+charging/discharging power per dwelling (hourly average). Based on Paevere et al. (2014).

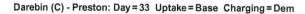

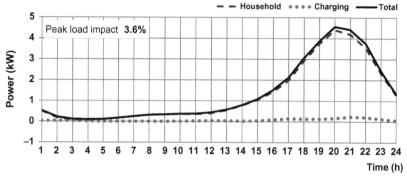

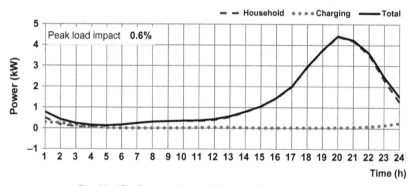

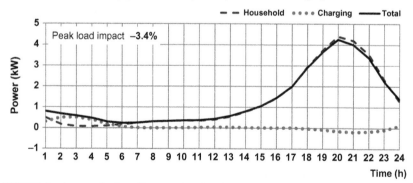

FIGURE 9 Average load profiles per dwelling for three charging modes—area = suburb of Preston, day = 33, uptake = base case—and for demand charging (Dem), off-peak charging (OP), and off-peak charging with vehicle to grid (OPV2G) (Paevere et al., 2014).

with a high per-dwelling household load on a hot summer day (suburb of Darebin/Preston on day 33) are shown. In this geographic area, the peak load is increased by around 5% and 0.4% under EV demand and off-peak charging, respectively. Under off-peak + V2G charging, the peak demand is reduced by 3.4%. Figure 9 illustrates how both the projected peak load increase and also potential for reduction can be quite different in different geographic areas.

Considering the maximum uptake scenario, with off-peak + V2G charging of the EV, the simulation results showed that when the state average profiles are considered, a 38% increase in peak load is projected, even with nighttime charging. Such an increase would have a very significant impact on most distribution systems. This increase occurs because under average load levels, the nighttime EV charging can create a new nighttime peak that is higher than the daytime peak. Conversely, when a high-load geographic area is particularly considered, the model projects a peak load reduction of 11%, as the V2G discharging is able to shave the high daytime peak, and the nighttime charging does not exceed this level. This example reinforces the importance of a spatial modeling approach, by showing quite different peak load impacts for different spatial areas and times of day and year.

To properly quantify and assess the impact of EV charging/discharging on peak electric loads across all days and all regions, a "normalized" ratio of daily peak household loads (with and without EVs) is used. As for previous analyses, three uptake scenarios (base, rebate, and maximum uptake) and three charging scenarios (demand, off-peak, and off-peak + V2G) are considered. The calculation is normalized to the maximum annual household load, in order to limit the impact of low-load days and regions on the analysis. As explained and illustrated in the text earlier, the low-load days and areas are not critical in terms of a peak load analysis, and they are also not as critical in terms of network capacity and investment. Simply stated, the normalized peak load ratio, $R_{peak,i,j}$, indicates the amount that the existing household peak load (on day i in region j) increases under EV charging. The term "normalized" indicates that the days and regions that are not important (i.e., with a low existing peak load) are filtered out of the analysis.

Spatial distributions of the normalized peak load factors for a hot summer day (day 33) are shown in Figure 10, which gives the peak load impacts for the three charging modes under base-case projected uptake in 2033 across the entire state of Victoria on day 33.

These analysis and visualization of the case study results have shown that under expected EV penetration levels for the Australian state of Victoria in 2033, the projected increase in peak electric loads under base-case EV penetration rates and demand charging is mostly <5% for high-demand days but can be up to 15% at the extreme for a handful of days and geographic locations. Under off-peak EV charging, the projected peak load impacts are mostly <2%.

The case study has also explored ways that EVs can act as a distributed energy resource, using the "spare" storage capacity in EV batteries by feeding

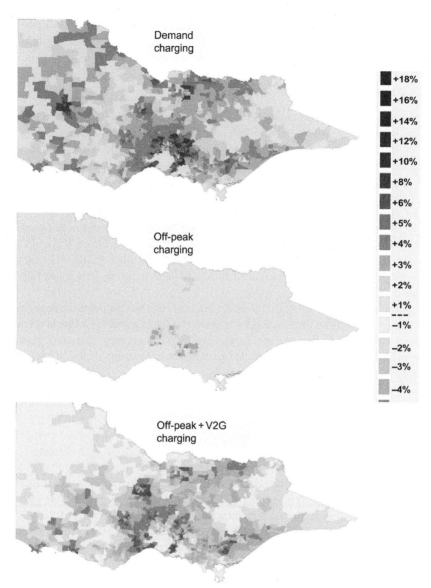

FIGURE 10   Spatially projected normalized peak load increase on a hot summer day (base-case uptake) for the entire state of Victoria.

energy back to the house at times of peak load. It was shown that with judicious use of a V2G discharging strategy, there is widespread potential to reduce peak loads. Under off-peak+V2G charging, a peak load reduction of between 2% and 5% is achievable for the majority of days and the majority of geographic regions. However, it was also shown that under maximum possible EV penetration rates, the benefits of EV discharging are limited to a small geographic and time window, due to the larger charging demands per dwelling.

6   CONCLUSION

Projections from around the world anticipate the number of plug-in vehicles to dramatically increase in the years to come. While it is widely recognized that EVs will have a significant impact on the operation of electricity systems, they are rarely considered as a distributed energy resource that can actually benefit the operation of the electricity network. The V2G concept is fundamentally a distributed energy resource concept, where the battery in the EV is used to support grid operations for brief periods.

Given the absolute number of EVs predicted to exist in the near future, V2G could be a significant grid resource. Alternatively, without V2G, EVs could significantly exacerbate many of the grid operation challenges, such as peak demand, being experienced today, if their charging is not carefully managed.

Most importantly, to understand the impact of EVs and the potential benefits from V2G operations, it is necessary to understand this impact at a fine level of detail—preferably down to street level. Only then can network operation, planning, and upgrade decisions be made with accurate forecasts of the impact of this V2G-distributed energy resource.

Such a detailed understanding of the impact of EVs and their viability as a distributed energy resource requires extensive modeling—from predicting how many EVs will be purchased in a particular area to the trips they will take and the energy consumption of their home base. This chapter presents such a model and a case study on the grid impacts for the state of Victoria in Australia.

Summarizing the results of the Victorian case study, it was found that EVs can significantly increase the peak demand of the distribution network if their charging is not carefully managed. Conversely, if their charging is managed, their impact on peak demand is minimal. Further, if they are enabled for V2G interaction, the EVs can actually reduce peak demand in many areas while still being available for full vehicular use. Perhaps most importantly, these results varied significantly across different geographies—the impact of EVs, and their availability as a distributed energy resource, was highly localized.

Although EVs show great potential as a distributed energy resource and a technology that can actually benefit the grid, careful analysis is necessary when making such claims. The impact of EVs is very localized, and consequently, detailed local modeling is needed to understand their impact and benefits. The modeling techniques presented in this chapter, while demonstrated with one Australian state, rely on commonly available data and so can be used to model the impact of EVs and the potential of V2G as a distributed energy resource, for any other geography around the world.

REFERENCES

AECOM, 2012. Forecast uptake and economic evaluation of electric vehicles in Victoria. Report prepared for Victorian Department of Transport. AECOM Australia Pty Ltd.

Chernova, Y., 2013. Who drives electric cars? Wall Street J.

Electric Drive Transportation Association, 2013. http://electricdrive.org.

Danish Energy Agency, 2012. Midtvejsrapport for Energistyrelsens forsøgsordning for elbiler (Danish EV Test scheme—Interim report).

Gardner, J., Quezada, G., Paevere, P., 2011. Social Study on Attitudes, Drivers and Barriers to the Uptake of Electric Vehicles. CSIRO, Australia.

Graham, P., Reedman, L., Poldy, F., 2008. Modelling of the future of transport fuels in Australia: a report to the Future Fuels Forum. Report No. IR1046, CSIRO, Australia.

Higgins, A., Foliente, G., McNamara, C., 2011. Modelling intervention options to reduce GHG emissions in housing stock – a diffusion approach. J. Technol. Forecast. Soc. Change 78, 621–634.

Higgins, A., Paevere, P., Gardner, J., Quezada, G., 2012. Combining choice modelling and multi-criteria analysis for technology diffusion: an application to the uptake of electric vehicles. J. Technol. Forecast. Soc. Change. 79 (8), 1399–1412, Elsevier.

Motavalli, J., 2007. Power to the people: run your house on a prius. New York Times, September 7.

Now Breaking: 120kW charging, 2013. TESLA release, 30 May, 2013, http://www.teslamotors.com/en_AU/forum/forums/now-breaking-120kw-charging.

Paevere, P., Higgins, A., Ren, Z., Horn, M., Grozev, G., McNamara, C., 2014. Spatio-temporal modelling of electric vehicle charging demand and impacts on peak household electrical load. Sustainability Sci. 9 (1), 61–76.

Pike Research, 2012. Plug in Electric Vehicles: A Practical Plan for Progress. http://www.navigantresearch.com/research/plug-in-electric-vehicles.

Plug-in Electric Vehicle Adoption Forecasts, 2012. EPRI, 31 December, 2012 http://www.epri.com/abstracts/Pages/ProductAbstract.aspx?ProductId=000000000001024103.

Ren, Z., Chen, D., 2010. Enhanced air flow modelling for accurate – a nationwide house energy rating tool in Australia. J. Build. Environ. 45, 1276–1286.

Ren, Z., Wang, X., Chen, D., 2011. Climate change adaptation for Australian residential buildings. J. Build. Environ. 46, 2398–2412.

US Department of Energy: Vehicle Technologies Program, 2011. The EV Project, Q4 2011 Report. Available from: http://www.theevproject.com/documents.php.

Victorian Department of Transport, 2009. Victorian Integrated Survey of Travel and Activity 2007 (VISTA 07). Available from: http://www.transport.vic.gov.au/vista.

Victorian Department of Transport, 2012. The Victorian Electric Vehicle Trial – Information Paper. Available from: http://www.transport.vic.gov.au/projects/ev-trial.

Part III

What Future?

Rethinking the Transmission-Distribution Interface in a Distributed Energy Future

Lorenzo Kristov and Delphine Hou[1]

ABSTRACT

This chapter describes how the proliferation of distributed generation and demand-side resources calls into question the prevailing model of restructured power markets, where an independent system operator (ISO) or regional transmission organization (RTO) delivers energy to transmission grid takeout points and utility distribution companies typically move the energy radially from the grid to the end users. The California Independent System Operator is seeing increasing interest from distribution-connected resources to participate in the wholesale electricity markets, while customers exhibit growing desires to adopt new technologies and self-optimize their energy use. Reliable system operation in a highly distributed electric system requires fundamental rethinking of ISO/RTO and utility distribution company roles around the transmission-distribution interface. This chapter explores two potential models. Under one model, the ISO/RTO extends its operational and market functions to include distribution-level resources; the second model redefines the utility distribution company business as a distribution system operator responsible for real-time balancing at the distribution level.

Keywords: Transmission-distribution interface, Distributed energy resources, Distribution system operator, Local resilience, ISO/RTO operations

1. The models discussed in this chapter for the future transmission-distribution interface and the concept of the open-access distribution system operator reflect the authors' own ideas on these topics and not necessarily the views or policies of the California ISO.

Distributed Generation and its Implications for the Utility Industry. http://dx.doi.org/10.1016/B978-0-12-800240-7.00018-7

1 INTRODUCTION

As part of the great wave of change moving through the electric industry, the growth of diverse types of distributed energy resources is probably having the greatest impact and will continue to do so for years to come. A recent report from Rocky Mountain Institute describes how microgrids could constitute a vital adaptation to extreme climate volatility,[2] while a report from the Edison Electric Institute headlines the "disruptive challenges" of distributed resources and net energy metering and describes how they threaten the traditional electric utility business model.[3] Whether one views distributed resources as a promising adaptation to human-inflicted ecological damage or as a developing threat to existing paradigms, dramatic changes are occurring on the distribution side of the power industry. These changes are largely driven by a confluence of new technology choices and capabilities, a desire to prevent further environmental degradation, greater end-user energy awareness, and state policy initiatives to mitigate climate change.

In this context, the chapter focuses on the interface between the high-voltage transmission grid, which is typically a meshed network that moves bulk power transactions, and the lower voltage distribution system, which is typically structured radially to deliver electricity from the high-voltage grid to end-use customers. The chapter converges several themes that appear throughout the present volume, including the following:

- The likelihood that a major portion of the electricity consumed by end users will be produced by relatively small-scale, variable output-distributed resources and will not rely on the transmission grid;
- Expanding government policies to promote small-scale local renewable supply resources and the declining costs of such facilities;
- The operational challenges facing the grid operator when each transmission-distribution interface point could have energy flowing in either direction in any given interval, and the direction could frequently reverse from one interval to the next;
- The tendency for large amounts of renewable energy to drive down the marginal price of energy while increasing the need and value of flexible and dispatchable capacity to smooth the variability of these resources;
- The desire of customers for local supply resilience when grid service is disrupted due to extreme weather events and the imminent availability of cost-effective technologies that can provide such resilience; and

---

2. Guccione, Leia. "The Micro(grid) Solution to the Macro Challenge of Climate Change." *Rocky Mountain Institute*, October 2, 2013, at http://blog.rmi.org/blog_2013_10_02_microgrid_solution_to_macro_challenge_of_climate_change.

3. Kind, Peter. "Disruptive Challenges: Financial Implications and Strategic Responses to a Changing Retail Electric Business." *Edison Electric Institute*, January 2013, at http://www.eei.org/ourissues/finance/Documents/disruptivechallenges.pdf.

- The adverse impacts of the previously mentioned trends on utility revenue streams that depend on the volume of energy sales to end-use customers.

The authors assume the framework of the restructured electricity market, which features an independent system operator (ISO) or regional transmission organization (RTO) that operates the transmission system and the wholesale spot markets, complemented by separate utility distribution companies and load-serving entities that operate distribution systems and provide retail electric service to end-use customers.[4]

The next section reviews the forces of change, emphasizing California's energy and environmental policies and their anticipated impacts on the power sector. Section 3 focuses on the transmission-distribution interface and sets the stage for the two models explored in more detail in Sections 4 and 5. Section 6 offers some conclusions.

2   THE DISTRIBUTED ENERGY FUTURE

This section surveys the key policy initiatives California has enacted that will drive substantial expansion of distributed energy resources over the coming years and offers some speculative but plausible estimates of the future penetration and impacts of such resources.

2.1   State Goals and Policies Expanding Distributed Energy Resources

California has been aggressive over the last decade in enacting policies to promote distributed energy resources and expand the state's renewable energy penetration. The Global Warming Solutions Act of 2006 (Assembly Bill 32) is often pointed to as a key starting point for the current wave of California's environmental agenda as it established a target for reducing greenhouse gas emissions to 1990 levels by 2020 and designated the California Air Resources Board as the agency responsible to implement specific programs to achieve that target. Although Assembly Bill 32 did not specifically address distributed energy resources, it established a context for other initiatives that did. For example, the state promoted distributed renewable generation with legislation such as the California Solar Initiative (2006), which targeted 1940 MW of newly installed solar capacity by 2016, and Senate Bill X1-2 (2011), which increased the Renewables Portfolio Standard to 33% by 2020.[5]

---

4. Because both authors work within the ISO framework, they do not explore how the transmission-distribution interface may evolve in the context of the vertically integrated utility, though many of the same driving forces of change will likely raise similar issues in that context as well.

5. The Renewables Portfolio Standard specifies the percentage of electric energy delivered to end users on an annual basis that must be obtained from qualified renewable generating resources.

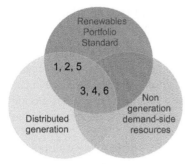

1. State goal of 12,000 MW of renewable distributed generation
2. Feed-in-tariff for small renewables
3. Zero net energy action plan
4. Energy storage procurement targets
5. Zero emissions vehicles
6. Increase renewable procurement, net energy metering, and revise rate structure

FIGURE 1   California actions to promote distributed energy resources. *Source: California Independent System Operator.*

In support of nongeneration demand-side resources,[6] since 2008 the Energy Action Plan has emphasized a "loading order" that prioritizes the state's preferences among different resource types to meet future energy needs. The loading order starts with energy efficiency, demand response and storage as the preferred resource types, followed by renewable energy, and then last, traditional fossil fuel-fired generation. In addition to continued state funding for ongoing energy efficiency and demand response programs, Senate Bill 17 (2009) promoted the use of advanced metering infrastructure by establishing implementation policies and rate recovery for the investor-owned utilities. To further promote distributed resources, the state has enacted specific policies to advance the use of combined heat and power facilities, such as Assembly Bill 1613 (2007) and behind-the-meter self-generation via Senate Bill 412 (2009). To advance the last objective, Assembly Bill 920 (2009) provided compensation to net energy metering customers.

Figure 1 offers a way to visualize the overlaps among the recent state policies to advance renewable energy, distributed generation and nongeneration demand-side resources, and the numerous regulatory proceedings they have engendered. For example, the increase in the state's Renewables Portfolio Standard to 33% also included a goal to install 12,000 MW of renewable distributed generation. Senate Bill 32 (2009) specifically provided a feed-in tariff for small renewables up to a 3 MW size limit. The Zero Net Energy Action Plan (2010) lays out a strategic vision for California to achieve zero net energy consumption for all new residential construction by 2020 and all new commercial construction by 2030.

Under Assembly Bill 2514 (2010), the state will establish targets for and procure energy storage to provide for renewable firming, demand-side management, distributed energy support, and wholesale-level ancillary services.

---

6. In this chapter, "nongeneration demand-side resources" are intended to include energy efficiency, dynamic retail rate structures, and the various types of demand response programs and products.

In fulfilling the directives of this legislation, the California Public Utilities Commission issued a ruling on October 17, 2013 authorizing the investor-owned utilities to procure 1325 MW of storage by 2020, with installations required no later than 2024. There is also a governor's executive order (2012) to have 1.5 million zero-emission vehicles on the road by 2025. The most recently signed into law, Assembly Bill 327 (2013), touches on all three areas by allowing the California Public Utilities Commission to direct procurement of renewables above the current 33% Renewables Portfolio Standard, extend net metering programs beyond the current 2014 deadline and remove caps on participation, and revise the residential energy rate structure to help mitigate utility concerns about the revenue impacts of net energy metering.

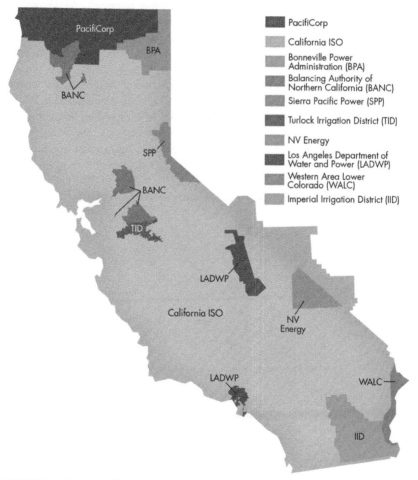

FIGURE 2   Current CAISO footprint and statistics. *Source: California Independent System Operator.*

California's legislation will have far-reaching impacts on the CAISO operations and markets. The CAISO is the only organized electricity market in the western United States, with over 60,000 MW of generating capacity under operational control and serving 30 million people in California and Nevada. In the fall of 2014, the CAISO will begin operating a real-time energy imbalance market that will expand into non-ISO portions of California, Idaho, Oregon, Utah, Washington, and Wyoming. Figure 2 shows a map of the CAISO footprint and provides current statistics.

Currently, 18% of the CAISO capacity comes from renewables, almost half of which is wind generating capacity, as shown in Figure 3.

Looking forward, the installed capacity within the CAISO footprint for wind and solar generation alone is expected to increase about 150% between 2012 and 2020 to meet the 33% Renewables Portfolio Standard. As shown in Figure 4, this is a 37% increase in installed wind capacity—from 5800 MW in 2012 to 7934 MW in 2020—plus a sharp increase in combined solar thermal and solar photovoltaic (PV) generation of over 500%, from 1764 MW in 2012 to 10,789 MW in 2020.

To illustrate the impact of these developments, solar generation in the CAISO peaked at 2071 MW on June 7, 2013. This represented about 5% of the day's demand and more than doubled the 1000 MW record set the prior year. Over time, the CAISO expects the large penetrations of solar and wind generation to change the load shape, leading to significant operational impacts.

Figure 5 provides an illustrative off-peak season example using a typical system load shape peaking at ~44,000 MW. At the bottom of the figure are typical solar and wind generation profiles for the same time period.

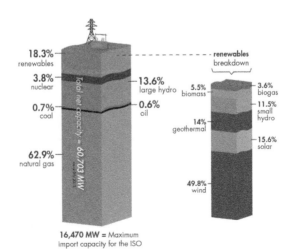

FIGURE 3   CAISO generating resources. *Source: California Independent System Operator.*

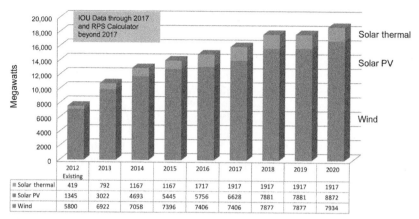

	2012 Existing	2013	2014	2015	2016	2017	2018	2019	2020
Solar thermal	419	792	1167	1167	1717	1917	1917	1917	1917
Solar PV	1345	3022	4693	5445	5756	6628	7881	7881	8872
Wind	5800	6922	7058	7396	7406	7406	7877	7877	7934

FIGURE 4   Existing and projected solar- and wind-installed generation in CAISO. *Source: California Independent System Operator.*

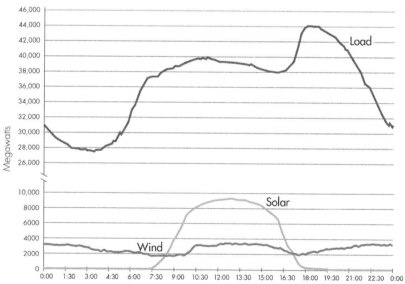

FIGURE 5   Illustrative load, wind, and solar profiles for January 2020. *Source: California Independent System Operator.*

Figure 6 shows the "net load," which is the system load minus the output from solar and wind resources shown separately in Figure 5. The net load increases during the morning ramp, then decreases during midday when solar output is highest, and then increases again for a second time, and quite severely, with the evening peak. Operationally, this illustrates an upward ramp of 8000 MW in the morning, a downward ramp of 6300 MW at midday, and a

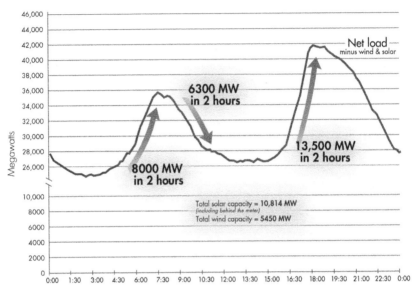

FIGURE 6   Illustrative net load with ramping needs for January 2020. *Source: California Independent System Operator.*

second upward ramp of 13,500 MW in the evening, with each ramp period occurring over only about 2 hours. Additionally, grid operation will be further challenged by the variability and volatility of solar and wind generation output. For example, on a partly cloudy day, rapid changes in solar or wind output can create additional need to dispatch flexible supply-side resources that can respond to net-load changes within 5 minutes or less.

2.2   Distributed Resource Participation in the CAISO Markets

The CAISO expects that certain distributed energy resources will be able to help it manage the anticipated severe ramps and growing supply-side and demand-side variability in the system. In 2010, to enable demand response to participate in the wholesale market and provide some of the needed flexibility, the CAISO introduced the proxy demand resource product for economic load curtailment. This product allows the demand response provider to be a separate entity from the load-serving entity that serves the end users comprising the demand response resource. In addition, the CAISO expects by 2014 to have approximately 1000 MW of emergency-based demand response in the wholesale market. As this chapter is being written, the California Public Utilities Commission is finalizing Electric Rule 24, which will allow retail customers to offer their demand response capability into the wholesale market either directly or through third-party demand response aggregators.

To facilitate distributed generation participation in the wholesale market, in 2013 the CAISO implemented a new annual transmission assessment to identify grid capacity for allocating resource adequacy deliverability status to distribution-connected generation, which enables qualified resources to earn additional revenues by providing resource adequacy capacity to load-serving entities.

Although it is clear that California will continue to encourage the expansion of distributed energy resources of all types, one cannot yet predict the relative amounts and rates of growth of the various distributed energy resource types. The CAISO will continue to develop new vehicles to facilitate the participation of distributed energy resources at the wholesale level. At the same time, while wholesale participation by distributed resources may help meet growing needs for flexible capacity, their integration into the market raises challenging market design and implementation questions such as those around investment, cost recovery, and providing the correct economic signals to end-use consumers as highlighted by Keay et al.[7]

For example, the CAISO relies on wholesale energy prices and ancillary service capacity payments to elicit operationally optimal responses from supply resources to balance the system and meet ramping needs. However, many types of distributed resources are situated behind the end-user meter and are not configured to adjust output based on wholesale prices. Dynamic retail pricing has been identified as a potential mechanism for aligning the responses of distributed resources with grid operating needs, but there is considerable political opposition in California that may preclude enacting dynamic pricing for the foreseeable future. And while participation by distributed resources in the wholesale market may appear to be a desirable approach, there are many technical challenges to incorporating thousands or tens of thousands of small resources, perhaps in the 100 kW size range or smaller, into the CAISO telemetry systems, state estimator, and market optimization. This integration challenge is an important motivating factor for the second transmission-distribution interface model described in Section 5.

2.3 The Rise of Microgrids

Until recently, California state actions and CAISO market design have been focused on single resources or aggregations of similar resources such as solar, storage, or demand response. As customers become more sophisticated in their energy use management, a growing trend is to combine different resources and create a microgrid that is interconnected at the distribution level. The CAISO has been involved in the preliminary phases of microgrid pilot projects with the Los Angeles Air Force Base and the Naval Air Weapons Station at China Lake,

---

7. See Sections 2 and 3 in Keay et al., "Electricity Markets and Pricing for the Distributed Generation Era."

specifically in testing the vehicle-to-grid portion of a larger portfolio of resources that includes on-site solar generation.

On July 8, 2013, the California Energy Commission approved $1.7 million in funding for another military microgrid project at Camp Pendleton in San Diego. The project will integrate "on-site flat-plate and concentrating photovoltaic (PV) technology with energy efficiency, energy storage and other technologies to provide reliable power and support critical base functions."[8] The University of California system is actively pursuing microgrids as well. The San Diego campus can generate up to 92% of its own power[9] while at Irvine, base loading and vehicle fuel cells are combined with a suite of renewable and traditional generation and energy efficiency measures.[10] Another notable microgrid exists at the Alameda County Jail at Santa Rita, which incorporates 1.2 MW of solar capacity, 11 kW of wind capacity, 2 MW of storage, a 1 MW fuel cell, 2.4 MW of backup diesel generation, and an on-site distributed energy resource management system.[11]

These examples demonstrate that customers are seeing value in combining a variety of resources to create microgrids with the potential to participate in the wholesale market and the capability of islanding in situations where external grid service is disrupted. This trend reflects a combination of drivers including cost effectiveness, economic incentives, increasing reliability, local development, and self-reliance. Continued funding from state, federal, and private sources will add momentum to this trend. The challenge for the CAISO is how to optimally interface with these microgrids on the distribution system given the diversity of technologies that comprise them.

2.4   Increasing Demand for Local Supply Resilience

One factor that can be expected to grow in significance is the demand for resilience of electricity supply at the local level or what may be called the "Hurricane Sandy effect." Resilience simply means the ability to maintain at least critical and ultimately all end-use electric service in the event of extreme climate events, major transmission outages, or other disturbances. Traditionally, supply resilience has been viewed as inextricably linked to the reliable

---

8. California Energy Commission. "Energy Commission Awards $1.7 Million for Military Microgrid Project." July 8, 2013, at http://www.energy.ca.gov/releases/2013_releases/2013-07-08_military_microgrid_nr.html.

9. Associated Press. "Microgrid has UC San Diego Generating 92 Percent Of Its Own Power." September 9, 2013, at http://sacramento.cbslocal.com/2013/09/09/microgrid-has-uc-san-diego-generating-92-percent-of-its-own-power/.

10. University of California, Irvine. "UCI Microgrid," at http://www.apep.uci.edu/3/research/pdf/UCIMicrogridWhitePaper_FINAL_071713.pdf.

11. Alegria, Eduardo. "Lessons Learned and Best Practices: Santa Rita Jail Microgrid Project." *Chevron Energy Solutions* presentation, July 30, 2012, at http://e2rg.com/microgrid-2012/Santa_Rita_Jail_Alegria.pdf.

operation of the transmission grid, with local outages either taken in stride by end users due to their relative infrequency and limited duration or mitigated by backup generation for certain vital facilities. Emerging technologies will become game changers in this area, however, by enabling resilience of end-use service to be designed into local distribution systems and, with islanding capability, enabling resilience to be decoupled from reliance on the transmission grid.

Additionally, one can expect that local resilience could follow a fractal design structure, with smaller types of island-capable microgrids nested within larger ones, up to the scale of entire counties and municipalities. At the smallest scale, a single family home may have rooftop solar generation, refrigerator-size batteries, electric vehicles and, with an appropriate control system, be capable of islanding if the local distribution service goes down. At the next level, a neighborhood might share in the development of solar and wind generation and centralized vehicle charging, which could be linked with the residences and small businesses in the area to form an island-capable microgrid. And so on, up to the point where the entire set of distribution facilities below a single distribution interface point with the ISO/RTO grid, with all the end users and resources interconnected to those facilities, could operate as an electric island in the event of a major disruption to the transmission system that would otherwise cause a loss of service in that local area.

The remainder of this chapter examines how the distributed energy future being shaped by these forces requires rethinking of the transmission-distribution interface and offers two conceptual models of how it might be redesigned.

3 TWO MODELS OF THE FUTURE TRANSMISSION-DISTRIBUTION INTERFACE

When the CAISO was formed its roles and responsibilities were explicitly limited to the transmission network and the wholesale markets, in accordance with the traditional distinction between transmission and distribution. The transmission system (or grid) was configured as a networked highway system from which load-serving distribution lines radiated out from a few thousand substations and moved energy in one direction from the grid to end-use customers. Other substations were for the interconnection of generating facilities to the CAISO grid and moved energy in one direction from the generator to the grid. With the advent of the CAISO locational marginal pricing market structure, each substation became a pricing node or "PNode" at which a spot energy price is calculated by the market optimization for each market interval.

Except for a few load-serving PNodes where large end users were directly connected to the CAISO grid, each load-serving substation was an interface point between the transmission and distribution systems where, traditionally, the energy flow was always in the same direction, from grid to customers. Generating facilities and a small number of demand response resources that participated in the CAISO markets were mostly interconnected to the CAISO controlled grid. Any generating facilities that were connected to distribution lines were typically installed and sized to serve the electricity needs of a host commercial or industrial operation and never generated enough electricity to cause a net flow across a load-serving PNode onto the CAISO controlled grid.

Thus, each PNode could safely be classified as either a generation PNode where generating facilities connect to the grid and inject energy or a load PNode where energy is withdrawn to serve end-use customers. Most significantly for this discussion, in the traditional model the CAISO responsibility would in most cases stop at the PNode and thus be limited to the transmission grid, whereas the utility distribution companies would have well-defined responsibilities to deliver energy from the grid to the end users reliably and cost-effectively. Thus, the boundary between transmission and distribution and the corresponding distinction between wholesale and retail were clear. The left side of Table 1 summarizes the traditional features of the transmission-distribution boundary.

As the number and diversity of distributed energy resources continue to grow, the assumptions behind the traditional transmission-distribution interface model will no longer hold. Energy will flow in both directions over distribution facilities, and it is very likely that many PNodes will be bidirectional, that is, will withdraw energy from the CAISO grid at some times and inject energy into the grid at others. Increasing numbers of diversely distributed resources will want to participate in the wholesale markets as supply-side resources, by both

TABLE 1 Traditional and Future Transmission-Distribution Boundary Features

Traditional	Distributed Energy Future
Each PNode is unidirectional, either withdrawing energy to serve load or injecting generation into the grid	Each PNode can be bidirectional, sometimes withdrawing and other times injecting energy
Minimal participation by distribution-connected resources in wholesale markets and ISO/RTO dispatch	Increasing capability of distribution-connected resources to provide resource adequacy capacity, wholesale energy, and operating reserves
ISO/RTO is responsible for the high-voltage transmission grid only up to the PNode; utility distribution company delivers energy from PNode to end-use customers	Need to reconsider roles and responsibilities of ISO/RTO and distribution utilities vis-à-vis the transmission-distribution interface

providing resource adequacy capacity with obligations to be available for commitment and dispatch on a daily basis and providing energy and ancillary services to the grid via the CAISO spot markets. This change is summarized in the right side of Table 1.

Although it is impossible to anticipate with accuracy the rapidity of these changes and the relative amounts of each distributed energy resource type that will be brought online over the next several years, it is not a moment too soon to start considering alternative ways to restructure the transmission-distribution interface and the future roles and responsibilities of the ISO/RTOs and the utility distribution companies to better align with the realities of this new world.

The next two sections describe two conceptualized models for how the transmission-distribution interface could evolve. The two visions are deliberately drawn as extremes in order to highlight their differences, whereas the future may entail an intermediate model to some extent, at least for a transitional period.[12] This chapter does not attempt to examine the process whereby a transition from today's structure to either of these futures might unfold. Rather, the focus is on the two models themselves as potential end states at some future time when the amount and diversity of distributed resources have grown extensively and, as a result, the traditional transmission-distribution interface model may no longer be optimal or even feasible.

4   MODEL 1: TRANSMISSION AND DISTRIBUTION AS A SINGLE INTEGRATED SYSTEM

As the amount and diversity of distributed energy resources increase, perhaps a natural evolutionary path would be for the traditional transmission-distribution boundary to diminish, at least from the perspectives of real-time operation and the spot markets. An ISO/RTO, as the entity responsible for reliable operation of the grid, will naturally tend to view transmission and distribution as an integrated system to be able to predict and monitor how distributed resources affect the net load at each PNode.

To this end, the ISO/RTO would develop modeling capabilities, optimization algorithms, market rules, and related technical standards and requirements to enable distributed resources down to a relatively low minimum size threshold—perhaps as low as 50 or 100 kW installed capacity—to participate in the wholesale markets, provide resource adequacy capacity and reserves, and be scheduled and dispatched via the market and operational systems. It is reasonable to believe that the expansion of distributed energy resources will naturally tend to lead in this direction absent some deliberate policy and design initiatives to head in a different direction and thus to greatly increase the complexity of the ISO/RTO's market and operational systems.

---

12. Felder in this volume also considers two divergent futures and the changing relationship between the transmission and distribution systems.

Under this model of the future, the utility distribution company would continue to function much as it does today, with responsibility for maintaining and operating its monopoly distribution system and for interconnecting new resources to that system. Assuming the distribution system functions remain bundled with the retail load-serving function and absent any major overhaul of retail rate structures, the distribution utility would still be concerned with maintaining kilowatt-hour sales volumes and with conducting the various energy efficiency and demand response programs that come under the prevailing regulatory construct.

Of course, depending on the extent of retail competition in the utility's service territory, the scope and variety of end-use customer services the load-serving distribution utility could offer in the future may expand due simply to the march of technological change, as discussed by other authors in this volume.[13] The main point for this chapter, however, is that the distribution utility would have little or no role with respect to the scheduling and real-time operation of most distributed energy resources beyond interconnecting them to the distribution system.

5   MODEL 2: THE OPEN-ACCESS DISTRIBUTION SYSTEM OPERATOR

In contrast to the vision described in the previous section, this section suggests a major new role for the utility distribution company, which entails relinquishing kilowatt-hour sales as its revenue base and taking on explicit operational and market functions in relation to the forward and real-time management of the distribution system.

5.1   Overview

Under model 1, the transmission-distribution boundary dissolves away so that the resources attached to either system participate in a single wholesale market structure, subject to a single market optimization that schedules and dispatches an integrated transmission and distribution system. In contrast, under model 2, the traditional transmission-distribution boundary at the PNode is maintained— even strengthened. The ISO/RTO continues to manage and optimize the transmission grid and the wholesale market only as far as the PNode, as it does today, and a new entity called the distribution system operator (DSO) manages and optimizes the distribution system below each PNode.[14]

---

13. See, for example, Section 4 in Riesz et al. and various business models proposed by Hanser and Van Horn.

14. This characterization of the DSO is comparable to what the Resnick Institute called an intermediate "distribution system control tier to complement the bulk power system tier and the self-managing DER [distributed energy resource] control tier." See "Grid 2020: Towards a Policy of Renewable And Distributed Energy Resources," Resnick Institute Report, September 2012.

Under this model, the DSO is in some ways analogous to the ISO/RTO itself, but for the distribution system. It provides nondiscriminatory open access to the distribution system for interconnection, operational, and market purposes, reliable distribution system operation via scheduling and dispatch procedures and algorithms, and may even operate local markets for imbalance energy, reserves, and other balancing and reliability services. Importantly, in contrast to today's utility distribution companies, the DSO's business model would not depend on selling kilowatt-hours to end-use customers. A revised business model more suitable to the conceptual DSO design envisioned here is described later in this section.

In the future envisioned here, the distribution system below each individual PNode, which this chapter calls a "local distribution area," will feature a large and diverse set of interconnected resources. The transmission-distribution interface at each PNode then becomes somewhat similar to an intertie between adjacent balancing areas, with the possibility for net energy flows to go in either direction in each market or operating interval and to frequently reverse direction from one interval to the next, depending mainly on the scheduled and anticipated needs of each local distribution area.

The last point reveals a key distinction between a local distribution area and an adjacent balancing area. The balancing area will typically be interconnected to more than one of its neighbors so as not to be entirely dependent on a single neighbor for all its imbalance needs. In contrast, the local distribution area will need to rely on the ISO/RTO when its own interconnected loads and resources are unable to maintain energy balance in an interval or when it has excess supply or excess demand it wishes to transact through the wholesale market.

From the perspective of the ISO/RTO, the transmission system and the wholesale spot markets will operate much as they do today, with two important innovations:

- First, the DSO at each PNode would be a wholesale market participant comparable to today's scheduling coordinator and function as an aggregator of all the loads and resources interconnected in each local distribution area and participating as such in the day-ahead and real-time markets. As a result, most distribution-connected resources would generally not participate directly in the wholesale market and would not be directly scheduled, telemetered, or dispatched by the ISO/RTO.
- Second, the ISO/RTO settlement structure will need to be revised to include charges to reflect the interval-to-interval volatility or variability of each transmission-connected resource and local distribution area.

Regarding the second point, precise definitions for measuring volatility and variability in this context will be an important matter for further development beyond the scope of this chapter. Suffice to say that the concept is intended to capture the extent to which a grid-connected resource or local distribution area relies on the ISO/RTO to provide real-time balancing services, which in turn require the ISO/RTO to procure and dispatch reserves, flexible ramping capacity, regulation, and so forth. In other words, the better the DSO for each local distribution area maintains a predictable schedule that aligns with ISO/RTO operational needs and can follow ISO/RTO dispatch instructions, the less it must rely on and be charged for ISO/RTO balancing services.

The last feature has significant implications for the future business model of the DSO. The settlement between the ISO/RTO and the DSO for each PNode it is responsible for will reflect both the net energy flow in each settlement interval and the variability of the net energy flow from interval to interval. Variability in this context will reflect both the magnitude of interval-to-interval energy changes as scheduled by the DSO and the DSO's real-time deviations from schedules. What this implies, in essence, is that the DSO will need to focus on reliable real-time operation of the local distribution area, including real-time balancing of supply and demand and maintaining voltage and frequency, and, to this end, develop its own balancing capability for the local distribution area rather than rely on the ISO/RTO for such services.

Real-time operation thus would become a basis for the DSO's business model, which it would implement through contractual or tariff-based relationships with distribution-connected resources that either pay the DSO when they add volatility to the local area or are paid (or at least not charged) by the DSO for mitigating volatility. It is likely, of course, that not all local distribution areas will contain a sufficient mix of the resources needed for local balancing, and therefore, the optimal approach for the DSO at each local distribution area will depend on the relative costs of balancing services provided by the ISO/RTO versus the cost and feasibility of developing and utilizing its own local markets and dispatch mechanisms.

One additional factor in favor of the local balancing approach will likely be increasing desire on the part of end-use customers to have resilience of energy supply in the event of system disruptions, as discussed in Section 2. Simply put, a DSO that performs real-time balancing of the distribution system at each local distribution area would likely also have the capability to island an entire area in the event of a transmission system disturbance and thus will be able to provide additional value to both the end users and the distributed resources interconnected in that area.

5.2   A Possible DSO Business Model

The DSO model described here requires a further degree of unbundling of the traditional utility structure than prevails in general today. It entails separating the open-access distribution service function from the retail electric service function. The underlying principle is that the DSO should perform only those distribution functions that warrant a regulated monopoly structure. This will stimulate robust competition in a world where a major portion of consumed electricity is produced locally and end users have the ability, technology, and options to source and manage their own energy needs. This means that some of the activities that comprise today's bundled utility distribution company and load-serving entity would be outside the scope of the DSO and would be offered as competitive services, including the retail sale of energy.

In this configuration, the DSO begins to resemble an ISO/RTO whose balancing area is the local distribution area composed of the distribution facilities below an individual PNode on the ISO/RTO grid and the load and supply resources interconnected to those facilities. For that local area, the DSO provides open-access, reliable wire services to enable distributed supply resources and imports from the ISO/RTO market to serve end-use customers and distributed supply resources to sell and export energy into the ISO/RTO market. There would be a single DSO for each local distribution area, but a single DSO could manage several different local distribution areas. For example, an existing utility distribution company may continue to function as the DSO for all the PNodes that comprise its current service territory.

Also similar to the ISO/RTO, the DSO would administer open-access interconnection procedures for the developers of various types of new resources that enter the market and would support the formation of microgrids and other types of "self-optimizing" customer configurations. The DSO would not, however, in its capacity as the regulated monopoly provider of open-access distribution services, be in the business of offering end-use services and products that self-optimizing customers would be interested in procuring, such as storage equipment and building energy control systems. Finally, the DSO would be responsible for planning and developing enhancements to distribution system infrastructure as needed to support the expansion of distributed resources and would coordinate with the ISO/RTO on joint transmission and distribution planning.[15]

In addition, there is an important and novel function that was mentioned briefly in the text earlier and is needed to make this model work efficiently. Specifically, the DSO would become an ISO/RTO market participant that would aggregate the end users and resources interconnected to its distribution facilities in each local distribution area, similar to the role of the scheduling coordinator in the CAISO market today. This would not preclude end users or load-serving entities from contracting bilaterally for energy to be imported over the ISO/RTO grid, nor distribution-connected supply resources from selling energy into the ISO/RTO market.

---

15. Obviously, planning and upgrading of distribution system infrastructure will be needed under any model of the future distribution utility, given the anticipated growth in distributed energy resources. This raises the further question of whether the DSO should be independent of the entities that own the distribution facilities and earn a regulated rate of return for distribution system investments, just as the ISO/RTO is independent of its participating transmission owners. The authors believe that the DSO probably needs to be independent of the distribution facility owners for the proposed DSO model to function efficiently.

What it would mean is that the interface between the various distribution-connected resources and the ISO/RTO, for purposes of real-time scheduling and operations and any settlements with the ISO/RTO markets, would be through the DSO. The reason this aggregator function is so crucial to the model suggested here is that it enables the ISO/RTO to limit its operational and market optimization roles and responsibilities to the transmission grid. Absent some structure of this kind, the ISO/RTO is back in the framework of model 1, having to schedule, dispatch, and telemeter an integrated transmission and distribution system and potentially thousands of small distributed energy resources.

6   CONCLUSIONS

The chapter offers an initial conceptual vision of a new DSO model for the distribution utility and the rationale for such an entity in the unfolding distributed energy future. The DSO model was developed with five major factors in mind:

- First, unprecedented growth in the amount and diversity of distribution-connected energy resources is happening today and will only accelerate.
- Second, this trend will include microgrids and other configurations capable of islanding, so that electric service reliability or resilience will be provided locally and depend less on the reliability of the high-voltage grid.
- Third, these evolving changes require rethinking the traditional transmission-distribution interface and reformulating the interrelated roles and responsibilities of the ISO/RTO on the one hand and the distribution utility on the other, particularly with regard to operations and markets.
- Fourth, since the traditional kilowatt-hour-based revenue model of utility distribution companies is proving problematic anyway in the face of the declining marginal cost of energy and massive expansion of behind-the-meter generation and net energy metering, it is timely to consider alternative business models for these companies, models more suited to a distributed energy resource world.
- Fifth, the traditional reliability responsibilities of ISOs and RTOs will predispose them to expand their operational and market roles into the distribution systems and to optimize the integrated transmission-plus-distribution system, absent development of attractive and practical alternatives.

Given these factors and the forces of change discussed earlier, it seems obvious that the traditional boundaries between transmission and distribution are rapidly becoming obsolete. Policy-makers, ISOs/RTOs, and other industry leaders need to proactively consider alternative ways to organize and operate electricity systems and markets that may soon be dominated by diverse, relatively small, local renewable resources and demand-side substitutes for kilowatt-hour. Many additional design elements and policy questions need to be addressed to complete the DSO model presented in this chapter. It deserves further development if for no other reason than the current trajectory of change may well move the industry in that direction. The DSO model and the contrasting fully integrated alternative offer a framework for imagining ways to restructure the transmission-distribution interface to design a future power system that embraces the disruptive challenges of distributed energy resources and addresses the economic challenges of transitioning to more sustainable electricity systems.

Chapter 19

Decentralized Generation in Australia's National Electricity Market? No Problem

Rajat Sood and Liam Blanckenberg

ABSTRACT

This contribution examines how participants in Australia's National Electricity Market (NEM) are responding to the challenges of low or falling demand growth plus rising renewable and distributed generation. The authors note how the competitive landscape of the NEM has been altered in recent years by technological developments and policy decisions, including an obligation on retailers to procure renewable energy. These factors have helped increase the pace of vertical integration over the past decade as independent generators and retailers have morphed into "gentailers." The chapter's main finding is that participants have responded in a manner to be expected given the energy-only design of the market in an environment of falling demand, rising renewable generation, and rising retail tariffs. While Australia has not experienced utility-scale renewable plants to date, there is no fundamental reason to believe these developments would undermine the efficient functioning of the market.

Keywords: Distributed energy resources, National Electricity Market (NEM), Renewable energy policy, Energy-only market, Vertical integration

1 INTRODUCTION

Australia's National Electricity Market (NEM) began operating in December 1998, and since that time, it has experienced significant change. The NEM is a gross-pool, energy-only market and is one of the most transparent wholesale electricity markets globally.[1] There are around 48 TW of installed capacity in the NEM. Over $5 billion of electricity, and ~200 TWh, is traded annually to meet the demand of more than 8 million end-use customers who represent the large majority of Australia's population (Figure 1).

---

1. Detailed background on the NEM and its evolution can be found in Moran and Sood (2013).

Distributed Generation and its Implications for the Utility Industry. http://dx.doi.org/10.1016/B978-0-12-800240-7.00019-9

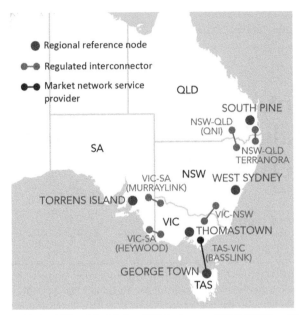

FIGURE 1   The National Electricity Market (NEM). *Source: AEMO (2010), p. 15.*

Due to its transparent market design, the challenges faced by energy utilities in the NEM—particularly in light of recent developments such as large-scale wind deployment and booming household solar photo-voltaic (PV) installations—serve as a valuable case study as to how these significant industry changes can be expected to affect utilities in less transparent or liberalized wholesale electricity markets around the world.

There has been much debate regarding the challenges to the "traditional utility business model" due to the increasing uptake of distributed energy resources (DER) and large-scale grid deployment of renewable technologies such as wind and solar. As the majority of chapters in this volume attest, the focus has tended to be on those segments of the electricity supply chain downstream of wholesale markets: transmission and distribution networks and retailing activities.

This contribution considers how these same forces are affecting upstream supply activities, including how the NEM has responded to the large changes thrown up by the rapid uptake of household rooftop PV units and the installation of utility-scale wind driven by a national renewable energy target (RET). In an energy-only market, the role of wholesale prices in guiding investment and retirement decisions is central to the long-run efficient operation of the market. We focus on the forces that have shaped the structure of the market—particularly the emergence of vertically integrated "gentailers"—and how market prices have so far provided broadly efficient plant entry and exit signals.

The remainder of the chapter is structured as follows: Section 2 briefly recaps NEM market outcomes from market start to the present—over this time, there has been considerable oscillation in the demand-supply balance in the market, with predictable implications for market prices and investment responses. Section 3 considers the emergence of gentailers—the vertically integrated business structure that has been widely adopted across the NEM over the last decade. Section 4 highlights the key challenges posed to the gentailer business model due to the continued deployment of utility-scale renewables such as wind and the rapid uptake of DER such as rooftop photovoltaics. Section 5 considers the implications of these challenges for the traditional gentailer business model and explores whether the existing model is sufficiently robust to adapt to the changes expected over the coming decade. Section 6 concludes the chapter.

2 OSCILLATING DEMAND-SUPPLY BALANCE

The NEM started operating in late 1998 from a position of relative excess capacity, particularly with regard to baseload plant: reserve plant margin at the time was roughly 28% (Figure 2). This excess generation was largely a consequence of uncoordinated and uncommercial investment by state-owned, vertically integrated monopoly suppliers during the preceding decades.

Following the introduction of the wholesale market, spot electricity prices were initially highly divergent across the NEM: relatively low in NSW and Victoria (where generation capacity was most in excess) and relatively high in Queensland and South Australia (where capacity was relatively lacking). However, the development of new plant in South Australia and additional interconnection to Queensland—combined with some investment in peaking plant in Victoria—encouraged wholesale spot prices to fall rapidly (Figure 3).

Throughout the early to mid-2000s, the excess baseload capacity present at the start of the market was absorbed by fairly rapid energy and peak demand growth. From 1999 to around 2006, both energy and peak demand grew on a NEM-wide basis at roughly the same pace, averaging around 2% per annum (pa).

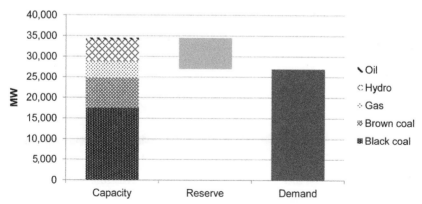

FIGURE 2   Installed capacity, peak demand, and reserve plant margin at NEM start (MW). *Source: AEMO (2013a), ESAA (2013), Frontier Economics analysis.*

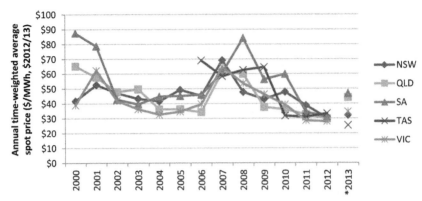

FIGURE 3 Historical time-weighted annual average spot prices ($/MWh, $2012/13). *Source: AEMO (2013a), ABS (2013), Frontier Economics analysis. Note: Nominal prices have been deflated using an Australia-wide CPI. Spot prices for 2013 are presented on a carbon-exclusive basis assuming a $23/tonne carbon price and an average pass-through of 1 tCO_2-e/MWh.*

Consistent with expectation—given that energy and peak demand were growing at broadly comparable rates—the majority of investment over this time was geared toward baseload and intermediate technologies. In particular, this period saw additional investment in baseload coal-fired capacity (predominantly in Queensland), as well as investment in mid-merit gas plant in South Australia and Queensland (Figure 4).

From around 2007, the NEM saw a pronounced acceleration in the rate of peak demand growth: this was driven by increasing penetration of domestic air-conditioning coupled with a string of very hot summers. The hot summers coincided with a severe drought, which limited hydro generation capacity and cooling water for several large baseload coal plant. Prices rose rapidly over this period, as the demand-supply balance tightened further. In response, investment

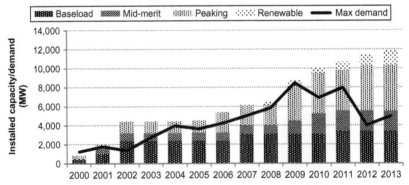

FIGURE 4 NEM demand and new plant investment, relative to 1999 levels (MW). *Source: AEMO (2013a), ESAA (2013), Frontier Economics analysis. Note: Data exclude Tasmanian investment and demand, which entered the NEM in 2006.*

over the period 2007-10 was mainly focused on peaking plant capacity, again consistent with expectation.

From 2008 onward, a confluence of factors led prices on a downward trajectory, resulting in average wholesale prices in 2012 being at their lowest levels in real terms since market start. These factors—discussed in detail in the subsequent sections as well as other chapters in this volume[2]—include the rapid entry of utility-scale wind generation, an explosive uptake in household solar PV, and a falloff in the underlying demand growth.

3   RISE OF THE "GENTAILERS"

In the lead up to the NEM's introduction, most of the former state-owned monopolies were vertically split to separate network activities (which were to be independently regulated) from generation and retail activities (which were to be exposed to new entry and competition). In all cases, generation portfolios were also split horizontally to create a structure amenable to wholesale competition. Subsequently, all the Victorian and South Australian generation and network businesses were privatized. By contrast, the majority of New South Wales and Queensland generation—and all network assets in these states—remained in public hands at the start of the NEM, and most remained so for the first decade in the life of the market.

From about 2002, private energy businesses embarked on a strategy of vertical integration. Unlike the old-style monopolies, the new trend toward vertical integration focused on the combination of generation and retailing activities only—network ownership was only tangentially involved and would have likely been blocked by the competition regulator, the ACCC, had it been pursued more vigorously.

Vertical integration often came about as retailers, concerned about the risk of price spikes, chose to develop their own peaking plant. But in some cases, vertical integration involved the acquisition by retailers of existing generators. The first major transaction to increase vertical integration was AGL Energy's acquisition of an interest in the Loy Yang A power station in Victoria. The ACCC opposed the purchase, on the basis that it would lead to a "thinning" of the wholesale contract market and a resultant increase in the exercise of generator market power and spot prices. However, the federal court disagreed and found the transaction would not substantially lessen competition.[3]

Subsequent transactions took place that reinforced the trend toward vertical integration. In 2002, roughly 8% of NEM capacity was partially or completely owned or controlled by a vertically integrated market participant: by 2013, the

---

2. See, for example, Nelson et al. (2013, in this volume) and Mountain and Szuster (2013, in this volume).

3. *Australian Gas Light Company v Australian Competition and Consumer Commission (no 3)* [2003] FCA 1525.

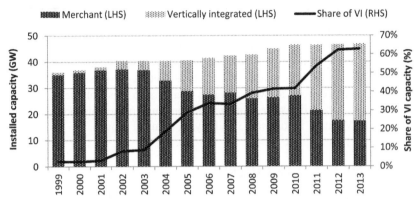

FIGURE 5   Share of capacity owned or controlled by vertically integrated participants. *Source: ESAA (2013), Frontier Economics analysis and research of public reports and records of ownership. Note: Vertically integrated capacity is plant capacity that is either partially or completed owned or controlled by a vertically integrated participant. Adjustments for minority shareholdings have not been made.*

proportion had increased to over 60% (Figure 5). The resulting entities became known as gentailers.

Over the period 2002-10, the majority of vertical integration occurred in South Australia and Victoria. In South Australia, incumbent retailers (AGL and Origin Energy) initially acquired existing or sponsored new mid-merit and peaking gas capacity. Subsequently, International Power acquired the Synergen Peaking Units. In Victoria, two incumbent generation portfolios (Snowy Hydro and International Power) launched retail businesses backed by their generation assets. AGL later acquired a generation portfolio of hydro units, Southern Hydro (Table 1).

Outside of these states, Hydro Tasmania—a large hydroelectric generation portfolio based in Tasmania—entered the retail market in 2008 via its partial (and subsequently complete) acquisition of the retailer Momentum Energy.

Over the period 2011-13, vertical integration further increased primarily via the New South Wales Government's sale of several baseload power stations to incumbent gentailers, TRUenergy (now EnergyAustralia) and Origin Energy. Vertical integration was also progressed by the ongoing entry of renewable generation sponsored predominately by existing retail players.

The rise of vertical integration in the NEM has been attributed to a variety of factors, including the desire to capture operating efficiencies and synergies, reduced transaction and trading costs, and reduced exposure to counterparty default and "holdup" risk when entering hedging contracts (Meade and O'Conner, 2009). However, perhaps the most persuasive explanation lies in the ability of vertically integrated businesses to better tolerate the ups and downs in underlying demand and supply conditions that have prevailed in the NEM at various times. These began with the price spikes of the early 2000s, which were followed by the flat prices across the mid-2000s and later

TABLE 1 Major generation and retail portfolios in the NEM

Generator (capacity share of NEM)	Retailer (small customer market share of NEM)
Origin (12.5%)	Origin (33.5%)
AGL (12%)	AGL (25.5%)
EnergyAustralia (11.7%)	EnergyAustralia (21.5%)
Macquarie Generation (10.2%)	-
Snowy Hydro (9.9%)	Red Energy (2%)
CS Energy (8.6%)	-
Stanwell (8.1%)	-
International Power (7.3%)	Simply Energy (1%)
Hydro Tasmania (5.7%)	Momentum Energy (3.75%)
Delta Electricity (4.3%)	-
Intergen (2.8%)	-
Alinta (2.2%)	Alinta retail (0.5%)
Infratil (0.3%)	Lumo (3.75%)
-	Ergon (7.75%)
Total (95.6%)	*Total (99.25%)*

Source: AEMO, ESAA, AER 2012 State of the Energy Market report. Note: Entities on the same row are vertically integrated between generation and retailing activities.

by the drought-induced high prices in 2007-09. Most recently, the NEM has experienced a slump in wholesale prices due to the slowing demand growth and, as discussed in the succeeding text, the explosion of renewable generation.

4   CHALLENGES TO THE GENTAILER BUSINESS MODEL

The emergence of gentailers came in response to the oscillation in demand-supply conditions to be expected in an industry with relatively inelastic demand and lumpy patterns of investment. However, while gentailers have a greater ability to withstand such swings than non-vertically integrated entities, gentailers are now facing new challenges that are eroding their erstwhile advantages.

These challenges stem from

- the investment flowing from the RET and other pro-renewables policies
- a consumer demand-side response to high-regulated network charges and technological innovation.

4.1   Renewables Investment

Australia has had some form of national RET since 2001, when the mandatory renewable energy target (MRET) was first legislated. In 2007, the scheme was renamed the expanded RET and the scheme's target was significantly increased to 20% of forecast energy demand by 2020. Once accounting for the existing stock of renewables, this equated to 45 TWh of additional renewable generation by 2020.

In 2010, the scheme was further modified to separate tradable certificates created by large-scale and small-scale renewable producers. This change was required in order to prevent the explosion of rooftop PV installations—which, at the time, were eligible to create equivalent renewable certificates and were heavily subsidized through other mechanisms—from collapsing the price of renewable energy certificates and thereby deterring investment in large-scale renewable projects. The result was the creation of large-scale generation certificates under the large-scale renewable energy target (LRET) and the creation of Small-scale Technology Certificates under the Small-scale Renewable Energy Scheme.[4]

From the outset, the Australian RET has been technology agnostic: all renewable technologies receive a single certificate price that is set by the interaction of demand for and supply of the relevant class of certificates. The demand for certificates comes from a requirement for retailers to purchase renewable energy to cover a set percentage of their annual energy sales, while the supply of certificates comes from the construction and subsequent dispatch of renewable generators that are accredited under the scheme.

4.1.1   Wind

Since the RET's introduction, the most cost-effective large-scale technology has been onshore wind, which accounts for the vast majority of new installed capacity that has occurred under the scheme. Wind production currently accounts for roughly 3.5% of total NEM output, with this share expected to rise substantially in order to meet the current RET of 20% renewables by 2020. In earlier years, wind capacity tended to be constructed by merchant renewable generators who signed long-term power purchase agreements with liable retailers to underwrite the investments. More recently, two of the largest gentailers in the NEM (AGL and EnergyAustralia) have moved to develop and fund their own wind assets in-house, thereby capturing the margin previously earned by merchant investors (Figure 6).

This in-house development has coincided with a trend toward much larger projects—typically several hundred megawatts or more—as compared to the smaller projects that were more common during earlier years of the scheme.

---

4. The majority of installed capacity under the SRES has been from household PV units, as further discussed in Mountain and Szuster (2013, in this volume).

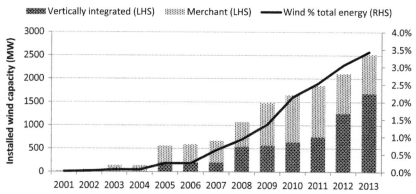

FIGURE 6 Installed wind capacity in the NEM, MW. *Source: ESAA (2013), Frontier Economics analysis of public reports and records of ownership.*

The emergence of utility-scale wind has had several key implications for the NEM. The first arises from the extent to which wind capacity contributes toward peak demand requirements. In the Australian context, the availability of wind during the highest half-hours of demand—on a probabilistic basis—has tended to be well below the average level of wind availability. This has meant that only a fraction of installed wind capacity could be counted toward the capacity requirements needed to meet peak demand. Currently, the market operator AEMO considers only 2-8%[5] of installed wind capacity as "firm" for the purposes of meeting peak demand. In years past when peak demand was rising, the volatility of wind output led policy-makers to the view that 92-94 MW of peaking plant capacity was needed for each 100 MW of wind capacity. In the current environment of flat or falling peak demand, this "capex duplication" has been less of an issue.

The second key implication of the emergence of utility-scale wind has been its impact on the level and volatility of spot prices. At its inception, the current RET scheme was intended to subsidize the entry of a quantum of renewable generation equivalent (in energy terms) to a fraction of the expected growth in energy demand over the life of the scheme. With energy demand growth recently turning negative and forecast to remain subdued for the next decade, the growth in renewable generation being sponsored under the scheme is now likely to either match or exceed the current expected growth in energy demand (Figure 7).

This phenomenon has given rise to the so-called merit-order effect, which is simply the suppression of spot prices in response to a quantity of low marginal cost generation entering the market in quantities that exceed demand growth. This effect has been one of the key drivers of the reduction in spot prices experienced in recent years.

---

5. AEMO (2013c), Table p. 7-1: 85th percentile wind contribution to peak demand.

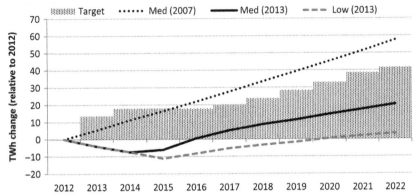

FIGURE 7   LRET target versus previous and current demand forecasts, relative to 2012 values. *Source: AEMO (2013a), WA IMO (2013), CER (2013), Frontier Economics analysis.*

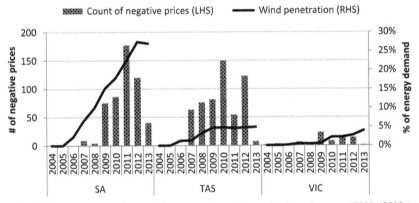

FIGURE 8   Wind penetration and instances of negative pool prices. *Source: AEMO (2013a), Frontier Economics analysis.*

In addition to lowering the average *level* of spot prices, the entry of large quantities of wind appears to have increased the *volatility* of spot prices. For example, there have been an increasing number of negative spot price outcomes in the NEM in recent years over the same time period as wind production has increased (Figure 8).

The reason why the increasing penetration of wind generation has led to more volatile spot prices is the effect of wind output on increasing the volatility of "residual" demand: this being the demand that must be met by thermal plant once renewables such as wind have been dispatched. We note that negative prices have tended to occur when strong wind speeds coincide with periods of low demand. The result is competition for dispatch between

- baseload thermal power stations, which incur large costs in having to reduce dispatch below minimum stable levels, and

- wind generators, which incur zero marginal operating costs and typically receive an output-based subsidy through the RET.

The prevalence of negative prices has been particularly high in South Australia, where wind production currently averages over 25% of underlying annual demand. There is a strong correlation between the coefficient of variation[6] of annual residual demand in South Australia and the penetration of wind in that region (Figure 9). Prior to 2006, the penetration of wind in South Australia was below 5% of annual energy demand, and the coefficient of variation of residual demand had been relatively stable at 2002 levels. From 2007 onward, there has been a strong increase in the coefficient of variation of residual demand in line with the rising penetration of wind in the state (see Figure 8). South Australia has a large amount of flexible mid-merit and peaking gas plant in its generation mix and thus has likely managed this variation to date better than would be expected in other regions going forward (for example Victoria, which is dominated by baseload coal plant).

The volatility of residual demand is also an important influence on the economic and technical suitability of certain forms of thermal technologies (such as mid-merit plant) over others (such as baseload plant). Other things being equal, the more volatile is residual demand, the more likely that flexible mid-merit plant (such as gas-fired CCGT) will be a more efficient choice than baseload plant (such as coal-fired steam turbines), despite mid-merit plant having a higher levelized cost.

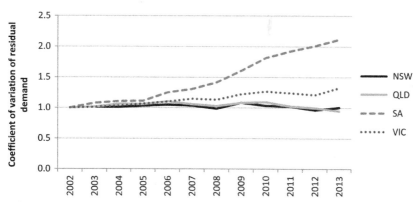

FIGURE 9   Coefficient of variation of "residual" demand, relative to the 2002 levels. *Source: AEMO (2013a), Frontier Economics analysis.*

---

6. Defined as the standard deviation of half-hourly residual demand divided by the mean of half-hourly residual demand, where residual demand is defined as scheduled half-hourly demand having accounted for the half-hourly output of scheduled and large nonscheduled wind plant.

FIGURE 10   Retail prices and NEM demand (LHS) and solar PV installations (RHS). *Source: ABS (2013), AEMO (2013a), ORER (2013). Note: Retail prices are weighted average of 8 capital cities. NEM demand excludes Tasmania. Solar installations are Australia-wide.*

4.1.2   Solar PV

As discussed by Mountain and Szuster (2013) and Nelson et al. (2013) in companion chapters, the past several years have seen a surge in the number of rooftop solar PV installations in the NEM. This has been driven by falling unit costs and state-based subsidies in the form of generous "feed-in tariffs". The growth in PV has also been encouraged by strongly rising retail prices, which can be partially avoided by customers whose PV output enables them to meet all or part of their power needs by consuming "behind the meter" (Figure 10).

While subsidies for solar PV have largely been wound back, continued falls in PV unit costs and rising retail prices are expected to keep installation growth rates strong for the coming decade. This is subject to major changes to the structure of network tariffs toward higher fixed charges, which may slow PV growth. To the extent that solar PV continues to grow, it is expected to temper the growth in demand for grid-supplied energy going forward.

4.2   Demand Response to Rising Prices and Technology

The NEM's experience of rising energy consumption and peak demand ceased abruptly in 2009. What was initially believed to be a temporary response to post-financial crisis economic conditions now appears to be a more prolonged—and possibly secular—trend of falling electricity demand. There are numerous factors that have contributed to the observed drop-off in consumption[7]:

- *Global factors have influenced commercial and industrial load*: a strong Australian dollar, slumps in the price of key resources and metals (such

---

7. Some of these issues in the context of Australia's major load center—Sydney—are discussed in more detail in Smith (2013).

as aluminum), and general cost pressures have led to reductions in Australia's manufacturing competitiveness and the idling or decommissioning of energy-intensive manufacturing capacity. Recent examples include the closure of the Kurri Kurri aluminum smelter and a blast furnace at Port Kembla in New South Wales and the mothballing of capacity at the Portland aluminum smelter in Victoria. These large closures have reduced both energy consumption and peak demand.

- *Ongoing energy efficiency improvements*: state-based energy efficiency schemes, Minimum Energy Performance Standards on appliances, energy efficiency requirements on the stock of new buildings, and other programs targeting energy efficiency in the commercial and industrial sectors have all led to large reductions in energy consumption.
- *Demand declines due to rapid retail price increases*: most NEM jurisdictions have seen large annual increases in retail prices in recent years, driven primarily by large increases in distribution network charges. These price rises have been borne mostly by residential and business customers (as opposed to large industrial) and have reinforced ongoing energy efficiency and conservation efforts.
- *Mild recent summers*: summer temperatures on the east coast of Australia in the last several years have been relatively mild. In particular, the incidence of multiple, sequential, extremely hot days falling during the working week has been unusually low. Yet, it is such heat waves that tend to drive peak demand outcomes due to the strong air-conditioning demand they induce.

4.3   Generation Investment Implications

The recent falls in NEM energy consumption and peak demand have resulted in a significant surplus of baseload generation capacity. This oversupply has been exacerbated, as discussed in Section 4, by the ongoing entry of renewables—predominately wind—driven by the LRET.

The NEM's independent market operator AEMO is now forecasting that no new thermal capacity is required to meet system reserve requirements under either a low or medium demand forecast for the next decade in all regions, except Queensland. AEMO is forecasting a requirement for 159 MW of additional capacity in Queensland in 2019-20 under a medium growth forecast but no additional requirement under a low demand forecast until post 2023 (AEMO 2013b).

As would be expected in an oversupplied market, prices at current levels are well below entrant long-run marginal cost levels (AEMC, 2013a), and many baseload plants are likely to be earning operating profits below levels needed to generate a sufficient return on capital. In response to such low prices—and consistent with a well-functioning energy-only wholesale market—there have been several announcements from market participants indicating either the permanent closure or partial mothballing of plant capacity until trading conditions improve. Several older baseload coal plants (Swanbank B and Collinsville in Queensland, Munmorah in New South Wales, and Playford B in South Australia) have retired capacity citing poor market conditions. Recently announced examples of temporary mothballing of capacity include

- two units (700 MW) at Tarong Power Station in Queensland
- two units (500 MW) at Wallerawang Power Station in New South Wales
- the decision to operate Northern Power Station (520 MW) in South Australia during peak summer months only.

Both the permanent and temporary mothballing of capacity observed in the current environment are consistent with the efficient functioning of a wholesale market such as the NEM. Prices in the NEM are signaling the relative abundance of baseload generation capacity. As a result of these permanent plant retirement and temporary mothballing decisions, the NEM's ratio of demand to installed capacity has narrowed slightly in the most recent year (Figure 11). It is expected that should prices remain at current depressed levels[8] then further mothballing of plant capacity will occur.

4.4   Restoration of Equilibrium

Over time, it is likely that the level of peak demand in the NEM will return to (and exceed) the levels experienced in past years. The level of energy consumption is likely to take longer to recover, because many of the structural and technological changes seen in recent years have had a greater effect on average than

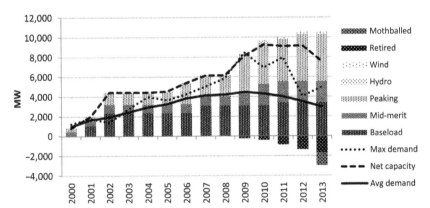

FIGURE 11   NEM demand and net plant capacity, relative to 1999 levels. *Source: ESAA (2013), AEMO (2013a), Frontier Economics analysis of public announcements and reports. Note: Wind has been de-rated to 6% of installed capacity.*

---

8. Accounting for the >50% uplift to pool prices experienced from 1 July 2012 due to the introduction of a $23/tonne carbon price.

on peak demand. If peak demand indeed recovers more quickly than average demand, the level and volatility of wholesale prices is likely to increasingly signal the value of flexible mid-merit and peaking plant capacity—and demand response—over baseload capacity.

In addition, these same signals are also likely to encourage the deployment of large-scale storage, which has the ability to flatten the volatility of spot prices by storing during periods of low prices and discharging at high-price times during periods of supply scarcity.

All of these suggest that gentailers are likely to continue to suffer financially in the short to medium term due to low pool prices and asset write-downs. The long term holds out the prospect of better times as the market gradually returns to a more balanced equilibrium. The key threat to gentailers in the longer term arises from the scope for bypass of centralized power systems altogether or through the emergence of "prosumers," who could potentially supplant and marginalize gentailers over time.

5 CAN EXISTING BUSINESS MODELS SURVIVE?

As noted in the previous section, in the absence of further or ongoing structural or technological changes, wholesale market conditions should eventually stabilize. Under these conditions, the forward-looking viability of gentailers should be restored and the traditional advantages of the gentailer model should again prove worthwhile. In an environment of volatile residual demand and spot prices, the naturally hedged position of gentailers should aid them in riding out cyclical fluctuations in demand-supply imbalances with fewer of the risks of prolonged exposure to high or low spot prices and with less of a need to engage in frequent and costly contracting processes.

The fundamental question is whether the era of major structural and technological changes in the NEM is ending or whether it will continue. As explicit subsidies for renewable generation have been wound back or are scheduled to subside, the likelihood of continued disruption to the traditional utility business model depends on whether organic developments in DER and energy storage can take over from the subsidized developments of the last half-dozen years. Specifically, the question is whether the cost of DER (in particular, solar PV) plus storage will fall sufficiently to

- enable customers to bypass the existing grid altogether, thereby becoming self-sufficient and no longer reliant on traditional utilities. Such an outcome would be catastrophic for the traditional utility model, gentailers, and otherwise, or
- lead to the emergence of "prosumers" who could substitute for or replace traditional utilities.

It would seem that the prospects for complete bypass of the traditional electricity supply chain—including the wholesale market—are remote for the foreseeable future. This is fundamentally because the cost advantages conferred by reserve-sharing across a diversified power system are powerful[9] and are likely to militate against bypass for some time to come. Complete bypass of the grid relies on the benefits of avoided remote generation and network costs outweighing the need for sufficient capacity and storage at each customer's premises to meet the customer's needs under virtually all circumstances. This has profound implications for the prospects for bypass.

For example, consider the situation confronted by a typical modern residential customer consuming roughly 8 MWh per year and with a peak demand of 2.5-3 kW, comprised largely of air-conditioning load on hot summer days. In order to be totally self-sufficient, such a customer would require roughly 3.7 kW of solar capacity to generate sufficient energy across the course of the year. However, due to the typically very "peaky" load shape of residential customers (in the aforementioned example, the ratio of average to peak load—or the load factor—is roughly 37%), considerably more than this would be required to meet peak demand.

Assuming the PV units are outputting at 100% of their utilization, the 3.7 kW of panels would be sufficient to meet this peak demand. However, if the panels are only operating at 50% utilization (for example, if peak demand occurs around 6 pm when solar PV output has begun to drop considerably), then 5-6 kW of panels would be needed to meet peak demand. In such cases, the value and role of storage becomes increasingly clear. To be fully grid independent, a typical residential customer would require PV storage capacity of 2.5-3 kW—plus sufficient PV capacity to generate and feed into storage—to ensure reliable supply during times when peak demand coincides with little or no PV output.

Mountain and Szuster (2013) suggested that storage costs in the range of $160/MWh would be sufficient in the Australian context to justify customers becoming grid-independent. A recent report by the CSIRO forecasts that under certain conditions, distributed generation combined with small-scale storage might become viable—and thus the prospect of "grid independence" achievable—for some customers by 2030-40.[10]

While large declines in the cost of storage may well allow some customers to become fully grid independent in the future, for many customers constraints on the availability of rooftop space will prevent this from happening, at least using PV panels. Multistory housing, houses with steep or poorly oriented rooftops, and commercial premises not suitable for PV panel installations will all continue to require a centralized form of grid generation to meet their demand in the future.

The other key threat to the traditional gentailer business model is the emergence of "prosumers": those customers who both produce and consume

---

9. This is noted and further discussed by Gellings (2013, in this volume).
10. CSIRO (2013).

electricity. However, the rise of prosumers—at least in Australia—appears likely to be checked. This is because policy-making and regulatory authorities are beginning to recognize the implicit subsidies provided to DER through the volumetric structure of most network tariffs.[11] Specifically, by causing lower consumption of grid-supplied energy, DER has been enabling many consumers to avoid contributing as much toward the cost of sunk network assets as consumers without DER. Policy-making institutions appear increasingly willing to promote reforms to address the distortions and inefficiencies being created by existing network tariff-setting methodologies. While policy-makers' interventions appear to be driven by a concern over equity—with DER households tending to be wealthier than other households—rather than efficiency, the likely outcome of any significant reform to tariff structures is that the scope for prosumers to supplant traditional utilities will be set back considerably.

In sum, while the current level of very low market prices is proving financial painful for existing generators, including gentailers, the likely combination of rising load and price volatility going forward could be expected to further solidify the vertically integrated business model. This is due to the inherent ability for gentailers to ride out periods of demand and price oscillations and to efficiently internalize the management of spot market volatility. The seemingly remote prospect for widespread grid bypass in the near future suggests that once wholesale market conditions return to a state of equilibrium, the gentailer business structure—and indeed the market's design itself—will continue to be a viable model into the future.

6 CONCLUSION

As numerous chapters in this volume have explored and discussed, the twin challenges of rising renewable generation and rapid uptake of DER poses massive challenges to the traditional electricity supply chain. However, the Australian NEM's 15 years of experience to date—and this chapter's main contribution—suggests that a well-designed energy-only market should be sufficiently robust to respond to these challenges without significant regulatory or policy intervention.

The changes expected in the coming decade—particularly if the rapid uptake of DER continues—will not be without cost. Indeed, as shareholders of existing baseload thermal capacity in the NEM can readily attest, the process of "creative destruction," whereby a relatively novel technology like solar PV upstages more

---

11. For example, the Australian Energy Market Commission (AEMC) recently highlighted the issue of network tariffs in its *2013 Strategic Priorities for Energy Market Development*, noting the ongoing concern that current network tariff methodologies are not sufficiently cost-reflective (particularly for customers with DER) due to their relatively high reliance on volumetric—rather than fixed—charges (AEMC, 2013b). California's recent Assembly Bill 327 also reflects a move by policy-makers in the direction of more fixed vs. volumetric tariffs, to address the implicit subsidy that DER such as rooftop PV currently enjoy (Greentech Media, 2013).

mature technologies, is never painless for all parties. However, we consider that the Australian NEM should evolve with this new paradigm, providing broadly efficient and timely investment and retirement signals. Within that context, we expect that the gentailer business model will continue to survive due to the underlying transacting and risk-management efficiencies it offers.

For less transparent or liberalized wholesale market designs, the NEM's experience to date points to two key emerging trends. The first is the continued pressure baseload thermal generation is likely to face in a world of rising renewable generation and falling or flat energy demand growth. The second is the increasing value of flexible plant capacity that is able to operate during peak demand periods and/or when intermittent renewable plants are unable to produce. In the more distant future, both distributed and grid-scale storage solutions could also play an increasingly important role.

REFERENCES

ABS, 2013. 6401.0 – Consumer Price Index (Australia, Jun 2013). The Australian Bureau of Statistics. accessed 14 October 2013, http://www.abs.gov.au/ausstats/abs@.nsf/mf/6401.0.

AEMC, 2013a. Final Rule Determination: Potential Generator Market Power in the NEM. The Australian Energy Market Commission. accessed 15 October 2013, http://www.aemc.gov.au/Media/docs/Final-Determination-f1205a60-b0ce-4803-bbfc-118286afec93-0.PDF.

AEMC, 2013b. 2013 Strategic Priorities for Energy Market Development. The Australian Energy Market Commission. accessed 23 October 2013, http://www.aemc.gov.au/media/docs/Strategic-Priorities-for-Energy-Market-Development-2013-9a092f69-5f69-4819-a783-f2dff9c0e2b5-0.PDF.

AEMO, 2010. An Introduction to Australia's National Electricity Market. accessed 9 December 2013, http://www.aemo.com.au/~/media/Files/Other/corporate/0000-0262%20pdf.ashx.

AEMO, 2013a. AEMO Historical NEM Data Files Covering Unit Dispatch and Availability; Regional Spot Prices; and Demand. accessed 14 October 2013, http://www.nemweb.com.au/REPORTS/ARCHIVE/.

AEMO, 2013b. 2013 Electricity Statement of Opportunities. AEMO. accessed 14 October 2013, http://www.aemo.com.au/Electricity/Planning/Electricity-Statement-of-Opportunities.

AEMO, 2013c. 2013 NEM Historical Market Information Report. AEMO. accessed 15 October 2013, http://www.aemo.com.au/Electricity/Planning/Related-Information/~/media/Files/Other/planning/NEM_Historical_Information_Report_2012_13.ashx.

CER, 2013. The Large-Scale Renewable Energy Target (LRET). accessed 14 October 2013, http://ret.cleanenergyregulator.gov.au/About-the-Schemes/Large-scale-Renewable-Energy-Target-LRET-/about-lret.

CSIRO, 2013. Change and Choice: The Future Grid Forum's Analysis of Australia's Potential Electricity Pathways to 2050. CSIRO. accessed 9 December 2013, http://www.csiro.au/Organisation-Structure/Flagships/Energy-Flagship/Future-Grid-Forum-brochure.aspx.

ESAA, 2013. Electricity Gas Australia 2013. The Electricity Supply Association of Australia. http://www.esaa.com.au/policy/EGA_2013.

Gellings, 2013. As the role of the distributor changes, so will the need for new technology. In: Sioshansi, (Ed.), Distributed Generation and Its Implications for the Utility Industry.

Greentech Media, 2013. AB 327: The Dark Side for California Solar. accessed 1 November 2013, http://www.greentechmedia.com/articles/read/ab-327-the-dark-side-for-california-solar.

Meade, O'Conner, 2009. Comparisons of Long-Term Contracts and Vertical Integration in Decentralised Electricity Markets. EUI Working Paper, http://cadmus.eui.eu/bitstream/handle/1814/11029/EUI_RSCAS_2009_16.pdf?sequence=1.

Moran, Sood, 2013. Evolution of Australia's National Electricity Market. Chapter 19, In: Sioshansi, (Ed.), Evolution of Global Electricity Markets: New Paradigms, New Challenges, New Approaches. Academic Press.

Mountain, et al., 2013. Australia's million solar roofs: disruption on the fringes or the beginning of a new order? In: Sioshansi, (Ed.), Distributed Generation and Its Implications for the Utility Industry.

Nelson, et al., 2013. From throughput to access fees: the future of network and retail tariffs. In: Sioshansi, (Ed.), Distributed Generation and Its Implications for the Utility Industry.

ORER, 2013. List of SGU/SWH Installations by Postcode. accessed 14 October 2013, http://ret.cleanenergyregulator.gov.au/REC-Registry/Data-reports?retain=true&PagingModule=2106&pg=1.

Smith, 2013. Crouching demand, hidden peaks: what's driving electricity consumption in Sydney. Chapter 13, In: Sioshansi, (Ed.), Energy Efficiency: Towards the End of Demand Growth. Academic Press.

WA IMO, 2013. 2013 Electricity Statement of Opportunities. WA IMO. accessed 14 October 2013, http://www.imowa.com.au/reserve-capacity/electricity-statement-of-opportunities-%28soo%29.

Chapter 20

What Future for the Grid Operator?

Frank A. Felder

ABSTRACT

This chapter examines how microgrids may affect organized wholesale electricity markets. Two different scenarios are explored. In the first scenario, referred to as "incremental changes," regional transmission organization/independent system operators (RTOs/ISOs) continue and even expand their central role in administering electricity markets, operating the power system, and planning transmission. They develop new wholesale electricity products to address the intermittency of wind and solar resources, extend their dispatch and unit commitment to include retail supply and demand resources, and increase the numbers of their membership. In the contrasting scenario, "decentralization dominates," RTOs/ISOs are relegated to a minimalist grid operator, balancing, when needed, numerous microgrids that infrequently, if at all, rely upon a centralized power system. Implications from these opposing scenarios are discussed.

Keywords: Microgrids, Regional transmission operators, Independent system operators, Wholesale electricity markets, Decentralized generation

1 INTRODUCTION

This chapter assumes that microgrids become a major portion of the electric power system to such an extent that they start to affect, if not impinge, the roles and responsibilities of grid operators, in particular regional transmission operators/independent system operators (RTOs/ISOs). The widespread creation of microgrids, however, is not preordained for several major reasons. It may turn out, although different analysts may assign different subjective probabilities, that microgrids are not the most cost-effective way to meet societies' competing cost, environmental, and reliability objectives. It is also possible that although widespread microgrid adoption may be in the public's interest, existing business and institutional interests thwart it. Or it may be the case that the complexity of the regulatory environment itself and the radical departure that microgrids

Distributed Generation and its Implications for the Utility Industry. http://dx.doi.org/10.1016/B978-0-12-800240-7.00020-5

represent from existing approach effectively hinder their widespread adoption (Marnay et al., 2008).

The purpose of this chapter is to investigate what possible scenarios might occur and what issues and questions do these scenarios raise if microgrids are substantially deployed within a given RTO/ISO system. This chapter does not try to assign likelihoods to different scenarios or discuss every possible scenario or variation, but instead the scenario approach is used as a tool to generate questions and surface issues that need further consideration to assist in analyzing whether widespread microgrids and supporting policies are in the public interests and whether existing organizations may help or hinder their adoption. Kristov and Hou and Riesz et al. (Chapters 18 and 23) also use a scenario approach to investigate possible futures of electricity sector and their implications on utilities.

A useful, although long, definition and description of microgrids is the following (CERTS, 2002, p. 26):

A MicroGrid is a semiautonomous grouping of generating sources and enduse sinks that are placed and operated for the benefit of its member customer(s). The supply sources may be driven by a diverse set of prime movers and/or storage devices. The key distinguishing feature of the MicroGrid is that sources are interconnected by Microsource Controllers. These power electronic devise maintain energy balance and power quality through passive plug and play power electronic inverter features that allow operation without tight central active control or fast (on time scales less than minutes) communication. They also permit connection and disconnection of devices without need for any reconfiguration of equipment, preexisting or new. Overall economic operation within constraints such as air quality permit restrictions, noise limits, etc. as well as maintenance of a legitimate façade to the grid is achieved entirely through slow (on time scales of minutes or longer) communications with a central Energy Manager.

The US Department of Energy has a more succinct definition (Ton and Smith, 2012, p. 84): "a group of interconnected loads and distributed energy resources within clearly defined electrical boundaries that acts as a single controllable entity with respect to the grid. A microgrid can connect and disconnect from the grid to enable it to operate in both grid-connected or island mode." Figure 1 provides an illustration of a microgrid.

The key elements of a microgrid are that it is a semiautonomous grouping that can operate in conjunction with the rest of the power system as well as in islanding mode. The elements of the microgrid—distributed generation such as solar, fuel cells, combined heat and power, energy storage, and controllable loads—act as a unit. Although not stated in the previous definitions, microgrids are connected to the distribution system not the bulk power system, which consists of generation and transmission.

This chapter is organized as follows: Section 2 discusses the history of the transformation of the US electric power industry; Section 3 describes the

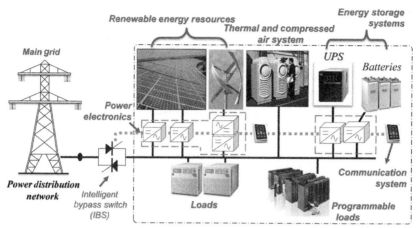

FIGURE 1   Schematic of a microgrid. *Source: Department of Energy Technology, Aalborg University, http://www.et.aau.dk/research-programmes/microgrids/.*

introduction of competition into the industry; Section 4 introduces centralized wholesale power markets and their organizations; Sections 5 and 6 present several scenarios involving the future structure and interaction between centralized electricity markets and microgrids; and the final section offers some conclusions for policymakers to consider.

2   THE TRANSFORMATION OF TODAY'S US ELECTRIC POWER INDUSTRY

At its inception in the 1880s, the fledgling electric power industry did not know the answers to fundamental questions about its technology and industry structure. Edison and Westinghouse, the two biggest names in the industry, debated the merits and demerits of direct current (DC) versus alternating current as described in Chapter 24 by Smith and MacGill. This technological dispute became so acrimonious that it spilled over into the public debate of whether to use the electric chair as a means of capital punishment. The industry began with multiple companies competing to provide electricity, each with its own wires running down streets connecting electric generation to retail consumers, before it transitioned into a regulated industry.

Over the next 90 years, these fundamental questions were thought to be resolved. The industry made a regulatory compact with the government that resulted in the creation of regulated monopolies. Under this compact, utilities were granted an exclusive franchise and were required to serve all customers, referred to as the obligation to serve, within their service territories. To avoid the exercise of market power, governments regulated utilities' rates based on

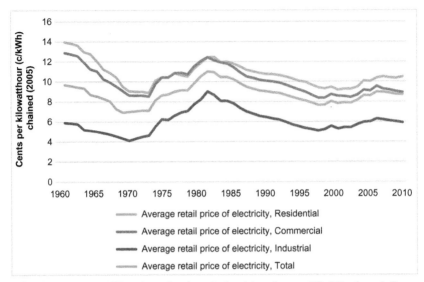

FIGURE 2 Average US real retail price of electricity. *Source: US EIA—Annual Energy Review 2012 (Table 8.10). Real Prices are deflated by EIA's annual Price Deflator estimates (Table D1).*

their costs, including their cost of capital. In return, the utility was granted the opportunity to earn a reasonable return on its investment and protection against entry by competitors.

The industry's organization was a product of its technology: Ever-larger generation units and higher-voltage transmission lines were built to take advantage of economies of scales. Bigger was better because it cost less to build one large power plant than two smaller ones and one high-voltage transmission line rather than two lower-voltage lines. As noted in the introduction and Chapter 1 of this volume, electricity demand was increasing rapidly during this period, and the industry invested in ever-larger facilities, resulting in lower costs, improved reliability, and more customers.

All this changed in the United States and elsewhere in the 1970s. With dramatic increases in energy prices in the 1970s (Figure 2) and the accompanying recession, the industry was also asked to take on an economic development role, which led to the creation of economic development utility rates for large industrial customers. As environmental concerns also gained traction, a new category of generation units, known as "qualifying facilities," was formed; they were not owned by utilities.[1] Instead, nonutilities could develop

---

1. By Public Utility Regulatory Policies Act (PURPA, 92 Stat. 3117). Also see EIA Glossary—http://www.eia.gov/tools/glossary/index.cfm?id=Q.

renewable or cogeneration facilities and sell their output to utilities, which were required to purchase the electricity. Many utilities were required to develop integrated resource plans that included conservation, energy efficiency, and demand-side management, and processes were developed that allowed for extensive public scrutiny of these plans. Later, environmental policy became a tool for economic development by developing renewable resources as a means of increasing employment. Rising rates also triggered concerns about economic equity for low-income customers, the elderly, and other disadvantaged groups.

Over time, the view of electricity has changed from that of being a luxury to being a necessity. Of course, it is not literally a necessity: Humans survived for many thousands of years without electricity. What is really meant is that electricity is not just a commodity but a merit good—a good that takes on social values and has implications of equity and fairness with respect to its acquisition that other goods, such as DVDs, do not. Some would go even further and argue that having access to electricity is a right. Since electricity is a merit good, political analysis cannot be ignored when analyzing electricity markets.

The original narrow objective that the society once had for the industry has become multiple objectives now. Even the attempt to compile a comprehensive list of the industry's objectives is fraught with difficulty. The major categories are

- reliability,
- energy security,
- efficiency,
- economic growth,
- public health,
- environmental quality, and
- public participation,

although there is by no means even consensus that these should be the categories, about their definitions, or their relative priorities. The first six categories are substantive, whereas the last one is procedural. Within each of the substantive categories are numerous subcategories with probably even less agreement regarding their definition and relative importance.

Ambiguous objectives complicate the analysis of electricity markets. It gets even more complicated when one factors in that there are numerous stakeholders with diverse values. Stakeholders behave strategically to define the objectives they think policymakers should pursue. For example, one set of stakeholders may overstate how much they value some objectives and understate others to push decision makers in a certain direction. Of course, other stakeholders with competing values may do the same. So, the seemingly straightforward analytic approach of articulating the objectives, giving weights or values to each objective and then finding an optimal solution, falls apart when there are competing stakeholders with multiple objectives.[2]

For most of the industry's history, its technological approach was to achieve economies of scale with large, centralized power plants that transmitted power over long distances. Over time, utilities built larger power plants and higher-voltage transmission lines as the equipment improved, to capture even greater economies of scale. The demand for electricity was considered an input to the system's operation and planning. Planners and power system operators did not try to reduce demand, assuming instead that it was beyond their control. Electricity was produced exactly when it was needed because the storage of electricity in large amounts was, and remains, cost-prohibitive. It was the ultimate just-in-time supply chain. Communication with retail customers—industrial, commercial, and residential—was limited to monthly bills, except in relatively rare situations when there was a reliability problem.

As reflected in other chapters of this volume, today, the very idea of having large power systems is being questioned. The view that there should only be large centralized power plants is now considered antiquated, and many are advocating microgrids as an alternative or complementary approach. With the development of efficient, reliable, and small gas turbines, larger is not always considered better. Greater emphasis on environmental issues is making distributed generation such as solar, small-scale wind, and combined heat and power more important. Furthermore, developing the ability to store electricity, both in stand-alone devices and for plug-in electric vehicles, is considered critical to promote the generation of intermittent renewable resources such as wind and solar.

Demand is no longer considered a fixed input to power system operations and planning and as described in Sioshansi and Faruqui and Grueneich may be flat or possibly declining in mature economies. Instead, with advances in communication and computer technologies, smart grid technologies promise to enable better monitoring and control of integrated power systems, including the distribution system. Electric loads may be turned on or off depending on system conditions, directly or through price signals to retail consumers, to improve reliability, reduce cost, and match supply and demand using the storage capability of electric vehicles, for example, to counterbalance intermittent generation.

Table 1 contrasts the answers the industry had in the past regarding its objective, technological approach, and industry structure to today's tentative and partial answers.

---

2. See Felder et al. (2011) for a discussion on the difficulty of defining the public interest in the electricity policy context and Vasiljevska et al. (2012) on the use of multicriteria techniques to facilitate the massive deployment of microgrids.

TABLE 1 Today's Electric Power Industry Is Revisiting Earlier Answers to Fundamental Questions

Fundamental Industry Characteristic	Then (1930s-1970s)	Now (1970s to Present)
Objective	Provide electricity to everyone	Achieve multiple objectives, including energy security, reliability, economic growth, environmental quality, and public health, all with substantial public and stakeholder input
Technological approach	Large, centralized power systems that generate, transmit, and distribute electricity when needed to load that is exogenous to system design and operation	Combination of centralized and distributed generation to serve demand that depends in part on system design and operation, leveraging smart grid technologies
Industry structure	Vertically integrated, cost-of-service-regulated utilities	Competition at the wholesale level, although the design of wholesale electricity markets is an open issue, with perhaps competition at the retail level and some aspects of transmission and distribution

3   INTRODUCTION OF COMPETITION INTO THE POWER SECTOR

The introduction of competition has fundamentally altered the industry's vertically integrated regulatory structure. In some countries, competition has been introduced only in the procurement of new generation and perhaps transmission assets. For example, many emerging economies, whose demand for electricity is growing at 5-10% per year, lack the capital to expand their power systems. Through a process run either by the government or utilities, international power plant developers are invited to compete in procurement processes to build, own, and operate new power plants. If their proposals win, they are offered long-term power purchase agreements (PPAs) that stipulate the payment of the facility's fixed and variable costs, which are recovered by the utility through its rates. Competition in this model is limited to the provision of power plants; the rest of the supply chain remains regulated and the government or the utility acts as the single buyer.

Another model of electricity markets comprises a wide range of arrangements from spot markets to long-term sales of power. The centerpiece of this market model is a wholesale spot market that allows the purchase and sale of electricity between multiple buyers and sellers. Investment in new generation is done primarily through contracts and spot market transactions, not via PPAs (although there may be PPAs).

The United States—the context for this chapter—has two major types of wholesale electricity markets: centrally operated and administered markets run by RTOs/ISOs (Figure 3) and decentralized markets for physically based bilateral transactions. In many US states, policymakers permit retail electricity markets in which the final retail customer chooses a provider to supply the electricity commodity. In all these models, the delivery portion—transmission and distribution—remains regulated, although there have been a few instances in which new transmission lines have been built on a merchant or competitive basis.

There are many different electricity market designs (Sioshansi, 2013). In fact, the US models are not necessarily representative of the many other models elsewhere in the world. Not all designs and implementations are equal; some are clearly superior. The point is that electricity markets can be introduced in different parts of the industry's supply chain and, for a given point of

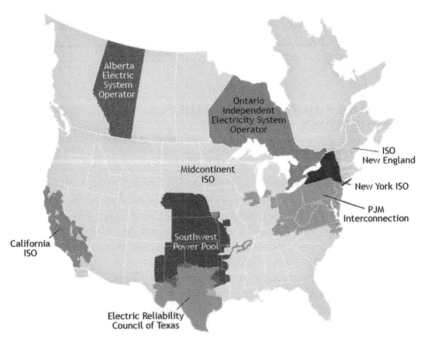

FIGURE 3   Map of RTOs/ISOs in North America. *Source: ISO/RTO Council http://www.isorto. org/site/c.jhKQIZPBImE/b.2604471/k.B14E/Map.htm.*

introduction, can even take on many forms. The extent to which an electricity market model works depends on its ability to integrate coherently power system engineering, economics, public policy, and business strategies.

4   CENTRALIZED OR RTO/ISO MARKETS

RTO/ISO planning, operation, and market administration presume a traditional electric power system, where Figure 4 illustrates a simplified version of that system. Generation units or power plants using water, coal, uranium, natural gas, or other fuels produce electricity that is transmitted long distances through the transmission system at high voltages. The combination of generation and transmission forms the power grid or bulk power system. High voltages are used to minimize the electric losses that occur when transmitting power. These high voltages are then reduced when the electricity is distributed via the electric distribution system to retail customers.

Prior to 1992, the federal policy was that economic regulation, not competition, was the preferred method of structuring the generation, transmission, and distribution of electricity. Economic regulation means that the government, not the market forces, determines the prices, terms, and conditions of the provision of electricity. Federal and state regulators granted electric utilities an exclusive franchise, and in return for that monopoly, the utility had to serve all customers at cost-based rates approved by the regulators.

The federal Energy Policy Act of 1992 created the framework for wholesale electricity markets and fundamentally changed a major portion of the electric utility industry. This act resulted in the development of wholesale electricity markets in which generators, marketers, brokers, and financial participants compete to sell electricity and related services to utilities and other load-serving entities.

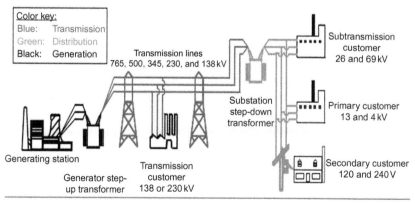

FIGURE 4   Electric power system supply chain (U.S. Canadian Power System Outage Task Force, 2004, final report on the, 2003 Blackout, p. 5.).

In the United States, the federal government regulates and sets policy for the bulk power system and wholesale electricity markets, and individual states regulate and set policy for the distribution system and retail electricity sales. The federal regulatory body that regulates much of the wholesale electric power industry is the Federal Energy Regulatory Commission (FERC).

Federal regulators approve markets for wholesale electricity and its associated products. The delivery portion of the industry, which is the transmission and distribution of electricity, remained under cost-based regulation. The operation of a wholesale power system requires one centralized power system operator to integrate the generation and transmission of electricity in order to ensure reliability. These system operator functions include determining which generation units to start up and shut down, dispatching of operating units, ensuring that the system is being operated reliably, and responding to changing system conditions. The portion of the bulk power grid that the system operator is responsible for is called the control area. There are approximately 130 control areas in the United States.

Prior to the establishment of wholesale electricity markets, utilities were usually the system operators and performed the control area functions. Since utilities were not directly competing against each other, regulators did not have to be concerned that one utility acting as the system operator would discriminate against its competitors as it performed its system operation functions.

With the introduction of wholesale markets, policies are in place to ensure that the system operator performed its functions in a nondiscriminatory manner. This means that the system operator must treat all market participants equally and not favor one market participant or group of market participants over any other. All rules and procedures must be equally and fairly applied. Otherwise, a utility could operate the system so that its generation units operated instead of its competitors' generation units even if the competitors' units were less expensive. System operations must be performed in a nondiscriminatory manner if wholesale electricity markets are to be efficient and reliable.

In many parts of the United States, including the Northeast, the Mid-Atlantic, and Midwest regions, new institutions called Independent System Operators and Regional Transmission Organizations (RTOs/ISOs) were formed to perform these critical functions in an independent and nondiscriminatory manner. RTOs/ISOs must perform their functions in accordance with the RTO/ISO open access transmission tariff and other related agreements, which are approved by the FERC, including any changes to those documents. RTOs/ISOs are the control area operator and are responsible for operating the electric grid reliably.

RTOs/ISOs operate the electric power grid to ensure reliability and they administer the wholesale markets in their respective regions. Based upon the bids and offers submitted by its market participants, RTOs/ISOs operate the system in accordance with their market rules subject to the limitations of the transmission system and generation facilities. RTOs/ISOs calculate and post the prices of various electricity products, settle the accounts of market participants, and ensure that market participants abide by the numerous rules and procedures, including membership and credit requirements.

RTOs/ISOs administer a slate of products including energy, capacity, financial transmission rights (FTRs), and ancillary services such as operating reserves and automatic generation control. Energy markets consist of locational marginal prices (LMPs) that are calculated a day ahead as part of both the centralized unit commitment process and the real-time dispatch process. Capacity markets ensure that there is sufficient generation to meet peak demand accounting for random outages of generation units and the constraints of the transmission system to a stipulated level of reliability, colloquially referred to as the "one time in ten years" requirement (Jaffe and Felder, 1996). FTRs, although they have different names and variations among the different RTOs/ISOs, allow market participants to hedge transmission congestion (the difference in LMPs at the same time but at different locations on the grid). Ancillary services provide additional services that are needed in order to maintain reliability, such as the ability of the power grid to continue to operate within acceptable specifications if a large generation unit or transmission component trips offline.

5 SCENARIO 1: INCREMENTAL CHANGE WITH INCREASED RTO/ISO CENTRALIZATION

Given the current role, importance, and influence of RTOs/ISOs in the United States, these organizations may remain and thrive as more and more microgrids are attempted to be developed. The impetus for this scenario is twofold. First, as a matter of political clout and organizational staying power, RTOs/ISOs are unlikely to allow themselves to fade away into history as more and more microgrids are created. Presumably, these organizations, like most organizations, act to preserve themselves (Felder, 2012) and would respond to microgrid developments, if they cannot thwart them outright, by subsuming microgrids under their purview. Second, RTOs/ISOs also have a very strong public policy rationale that rests on the proposition that in order to optimize the electric power system, more centralization is better than less. Thus, as microgrids develop, so the argument goes, their value can be maximized when their operations and planning are conducted under the auspices of an RTO/ISO. This is not to say that there are not very real practical and organizational limits, some discussed in the succeeding text, that counteract this claim that more centralization is always better than less.

In this scenario, RTOs/ISOs extend their reach through the transmission system to the distribution system and conduct unit commitment and dispatch and real-time operations of almost all generation units (whether central stations or distributed) and load. This scenario is described by Marnay and Lai (2012, p. 8):

"Our power system may evolve from its existing centralized structure to one in which numerous local control centers co-exist at lower-voltage extremities of the network, while the backbone high-voltage meshed grid continues to function as today." Either this could be performed by the RTO/ISO itself or there could be satellite RTOs/ISOs—distribution system operators (DSOs)—located at the distribution level that conduct those functions but under the direction of the RTO/ISO that DSO serves. Such a model may make sense given that many if not most distribution systems are connected radially to the bulk power grid and therefore the interaction among distribution systems is minimal, enabling a DSO to operate a specific system without network interactions with other systems. (See Bauknecht, 2012 for more details.)

Two possible future subscenarios involving the central role of RTOs/ISOs result in fundamentally different outcomes for microgrids. In both subscenarios, the owners of microgrids become direct customers, that is, market participants, of the RTO/ISO. In one subscenario, the RTO/ISO's role is so prominent that there is actually no need for local control of a microgrid. Entities may develop microgrids, but their operation is turned over to the RTO/ISO in a similar way as the operation of transmission and generation is turned over to RTOs/ISOs today: The owner (or its designee) physically operates the asset, but the RTO/ISO dictates what those operations are given the technical and economic parameters provided by the owner. Rahimi and Ipakchi (2012, p. 35) argued that "... a number of microgrids can collectively provide their capability to control demand and supply as virtual power plants that enable them to provide bulk power level tradable products such as ramping, balancing energy, and ancillary services." Under this subscenario, it is not much of a stretch that the RTO/ISO also takes on a major if not primary role in the planning of microgrids, similarly to what it does today with transmission.

In the other subscenario, the owner retains the microgrid's operations and planning functions. The microgrid is operated as a unit by the RTO/ISO as opposed to each generation and load component individually, again in accordance with the technical and economic parameters provided by the owner. Another way of saying this is that the microgrid is a virtual asset or a minipower exchange that would provide a combination of generation and load reduction to the RTO/ISO. In this subscenario, the microgrid owner, not the RTO/ISO, would decide if and when to operate in island mode, perhaps in anticipation of a blackout or even for economic reasons in certain situations. When connected to the grid, however, its operations would be controlled by the RTO/ISO.

Numerous questions and issues need to be addressed for either of these RTO/ISO dominance subscenarios to take hold. First, RTOs/ISOs would have to be able to handle a much larger customer base. For example, PJM is the largest RTO/ISO in the United States and the largest electricity market in the world. Its current number of market participants is about 750, covering an approximate level of installed base of generation of over 180,000 MW (as of 2012). For illustration purposes only, if 25% of PJM was eventually comprised of micro-grids of an average size of 5 MW, then PJM's direct customer base would be 7500 or a tenfold increase.

This 10-fold or even greater increase in the number of market participants raises both practical and fundamental concerns. On the practical side, RTOs/ISOs will have to dramatically increase their business capabilities (e.g., account management, training, and outreach) to handle this increase in their members' numbers. In addition, since market participants also have a major role in RTO/ISO governance, the mechanics of possibly having thousands or tens of thousands of entities potentially participating in stakeholder working groups and committees including voting would have to be thought through. The RTO/ISO unit commitment and dispatch software and systems would have to also be able to accommodate such a large increase in the number of generation and load assets, including storage facilities. This problem would be even greater in the subscenario in which the RTO/ISO is committing and dispatching each generation and load element within a microgrid as opposed to only the micro-grid as a whole.

The possibility of large numbers of microgrids participating in centralized electricity markets also raises many fundamental questions. In the US context, there are important federal-state conflicts due to overlapping jurisdictions and different policy preferences. States have regulatory jurisdiction over distribution systems; RTOs/ISOs are regulated at the federal level. States determine if there is retail electricity competition; the federal government makes the decisions regarding wholesale electricity markets. One example of the type of jurisdictional questions that would come up is whether microgrids and DSOs (if they are implemented) are regulated by the FERC or state utility commissions.

Another fundamental issue that would arise concerns the governance structure of RTOs/ISOs, which was not designed with the possibility of having thousands if not tens of thousands of microgrid market participants. Should this group have their own sector, similar to what generation owners and transmission owners have today, or should they be part of an existing sector such as ones that accommodates other suppliers and marketers? The former option would require a renegotiation and rewriting of the underpinning RTO/ISO governance agreements, which would be a daunting task for the practical reason that no existing market participants would want any part of their voting share and rights diluted. Moreover, RTOs/ISOs may be extremely reluctant to take the lead of such a fundamental governance overhaul that may potentially anger their existing stakeholders.

Another major set of fundamental issues that microgrids introduce is the modification of existing market products and the creation of new ones. With intermittent resources such as wind and solar poised to increase substantially over the next several decades—which is part of the motivation for the introduction of microgrids—RTOs/ISOs are considering a wide range of market product changes such as stochastic unit commitment and dispatch, flexibility market products across different timescales, and capacity market reforms possibility to include ramping requirements, scarcity pricing, and enabling demand response to set wholesale electricity prices. Two general questions arise (also discussed in Chapters 8 and 11): should other products be added to this list such as distribution system LMPs (and losses), reactive power (both at the transmission and distribution levels), finer grain frequency and voltage control to address distribution system power quality concerns, and distribution black start capability and what priority and changes to potential wholesale market designs should occur to be consistent with the types of products and market features that microgrids may need?

Making market or other rule modifications to RTOs/ISOs is not a minor matter. It takes time and effort, and when major rule changes are underway, it foregoes the opportunity to make other changes. The success of microgrids depends in part on whether rule changes that RTOs/ISOs are currently contemplating help or hinder their future development.

6 SCENARIO 2: MICROGRIDS WITH MINIMALIST GRID OPERATOR

Another possible scenario is that microgrids become so prevalent that the need for centralized power plants drops dramatically. The bulk power system remains, perhaps providing power from long-range resources such as wind and a declining coal and nuclear power asset base, but the vast majority of power is produced and distributed locally on the distribution system. The need for centralized unit commitment, management of congestion on the transmission system, and even wholesale electricity markets perhaps does not disappear but fades into the background because with dramatically less centralized power production, the existing transmission infrastructure would now be over built and perhaps even more so with the addition of long DC transmission lines dedicated to the delivery of large-scale wind to load centers.

The center of mass of system operations and planning shifts to the distribution system. The interaction of microgrids—both positive and negative network externalities—on the distribution system becomes the responsibility of the DSO. Under this scenario, RTOs/ISOs monitor and coordinate bulk power operations (perhaps using elements of the residual wholesale electricity markets), but DSOs administer distribution electricity markets, operate the distribution system, and conduct distribution planning. Two possible subscenarios are that the DSO is independent of the distribution company and microgrid owners, similar to how RTOs/ISOs are independent of existing transmission and generation owners, and the electric distribution company acts as the DSO analogously to what utilities did at the bulk power level prior to the introduction of wholesale competition. Table 2 summarizes the two scenarios and their subscenarios.

TABLE 2 Description of Competing Scenarios and Subscenarios for Microgrids in Relationship to Centralized Grid Operations

Scenario 1: RTOs/ISOs dominate through incremental change
Whether through the development of large-scale and long-distance onshore or offshore wind, safe and cost-effective nuclear power, and/or carbon capture and sequestration for coal power plants, the bulk power grid transmits a substantial amount of nondistributed generation to distribution systems along with the widespread deployment of microgrids

Subscenario	Possible Storyline
Scenario 1a: RTO/ISO commits and dispatches microgrids' individual components such as distributed generation and load response	Not all retail electric customers need to be able to operate separate from the grid, and as RTO/ISO demand response programs continue to evolve including the implementation of smart devices in appliances, end users can become direct customers of RTOs/ISOs without being microgrids. Once RTOs/ISOs have these capabilities, end-use customers that want a microgrid can have the RTO/ISO act as the microgrid energy managers as opposed to doing it themselves; perhaps in the future, RTOs/ISOs can dynamically define and implement microgrids
Scenario 1b: Microgrid owner operates microgrid and RTO/ISO dispatches its net output	The deployment of microgrids proceeds faster than the control of distributed generation and loads that are not part of microgrids; microgrid owners and operators learn to participate in RTOs/ISOs, and those organizations respond by providing useful products and services to microgrids and integrating them into their markets

Scenario 2: Decentralization dominates via microgrids
Centralized generation is phased out, leaving the existing grid to transmitting only a small portion of its capability as distributed generation of multiple types provides the majority of energy needed, much of it connected to microgrids

Subscenario	Possible Storyline
Scenario 2a: DSOs take on RTO/ISO-like functions but at the distribution level with RTOs/ISOs managing the less important and residual transmission issues	As the bulk power grid fades in importance, the focus shifts to the distribution system (and therefore to the states) and to figuring out the right combination of market mechanisms and regulation that achieves multiple public policy objectives
Scenario 2b: Electric distribution companies, not separate entities, provide the operating and planning environment that microgrids operate within	Electric distribution companies maintain their authority over their distribution systems and the distribution companies manage microgrids and other distributed generation and load response under a state regulation and regulatory process

7   CONCLUSION

The widespread adoption of microgrids is not a foregone conclusion. Its adoption may be thwarted by other alternatives that surpass its combination of economic, environmental, and reliability benefits or by incumbent institutional forces. Policymakers should recall that microgrids (or any technology for that matter) are not the end but the means to multiple public policy objectives and therefore not lock themselves into a particular solution that may turn out to be a poor choice.

If microgrids start to be deployed in significant amounts, then the responses of RTOs/ISOs will be a major factor, but not the only one, in the extent and speed of their deployment. Policymakers should monitor how RTOs/ISOs respond to microgrids and ensure that these organizations are not using their privileged position as grid operators and market administrators to thwart or limit cost-effective and publicly beneficial microgrid deployment. Other RTO/ISO changes should be considered in light of their potential impact on the future deployment of microgrids.

If the grid appears to be evolving to a situation in which centralized generation units will provide only a small portion of the system's energy needs due to their displacement by microgrids or other distributed generation, policymakers will need to determine whether they want to create the distribution version of RTOs/ISOs (DSOs, i.e., a market-based model) or instead rely upon distribution companies via command and control regulation to provide the interface between microgrids and the distribution system. A possible strategy that might make the deployment of microgrids more palatable to RTOs/ISOs is that if the use of the bulk power system fades into the background, current RTOs/ISOs become the DSOs for a market-based model with microgrids as the centerpiece but on the distribution system.

The public policy discussion of microgrids has started in earnest with much of it centered on definitions, designs, and the relative merits of microgrids compared to other alternatives. To round out the discussion, policymakers should recognize that microgrids are competing with alternatives whose technologies are themselves improving and therefore may be preferable to a widespread deployment of microgrids. They should also keep in mind the importance of considering existing and potentially new institutional arrangements and their influence on microgrid deployment.

ACKNOWLEDGMENTS

I would like to thank Rasika Athawale and Shankar Chandramowli for their research and editorial assistance and Fereidoon Sioshansi for his comments.

REFERENCES

Bauknecht, D., 2012. Transforming the Grid: Electricity System Governance and Network Integration of Distributed Generation.

Consortium for Electric Reliability Technology Solutions (CERTS), 2002. Integration of Distributed Energy Resources: The CERTS MicroGrid Concept, April.

Felder, F.A., 2012. Who watches the ISO watchman. Electricity J. 25 (10), 24–37.

Felder, F.A., Andrews, C., Hulkower, S., 2011. Global energy futures and their economic and environmental implications. In: Sioshansi, F.P. (Ed.), Energy Sustainability and the Environment: Technology, Incentives, Behavior. Elsevier, Amsterdam, pp. 30–61.

Jaffe, A., Felder, F.A., 1996. Should electricity markets have a capacity requirement: if so, how should it be priced? Electricity J. 9 (10), 52–60.

Marnay, C., Lai, J., 2012. Serving electricity and heat requirements efficiently and with appropriate energy quality via microgrids. Electricity J. 25 (8), 7–15.

Marnay, C., Asano, H., Papathanassiou, S., Strbac, G., 2008. Policymaking for microgrids: economic and regulatory issues of microgrid implementation. IEEE Power Energy Mag. 6 (3), 66–77.

Rahimi, F.A., Ipakchi, A., 2012. Transactive energy techniques: closing the gap between wholesale and retail markets. Electricity J. 25 (8), 29–35.

Sioshansi, F.P. (Ed.), 2013. Evolution of Global Electricity Markets: New paradigms, New Challenges, New Approaches. Elsevier, Amsterdam.

Ton, D.T., Smith, M.A., 2012. The U.S. Department of Energy's microgrid initiative. Electricity J. 25 (8), 84–94.

U.S.-Canadian Power System Outage Task Force, 2004. Final Report on the August 14, 2003 Blackout in the United States and Canada: Causes and Recommendations.

Vasiljevska, J., Lopes, J.P., Matos, M.A., 2012. Evaluating the interest in installing microgrid solutions. Electricity J. 25 (8), 61–70.

Chapter 21

Utility 2.0: Maryland's Pilot Design

Dustin Thaler and John Jimison

ABSTRACT

Thanks to the rise of decentralized resources and other important trends, the electric distribution utility of the future will be a different entity from today. A grid of wires for transmission and distribution under central control will still be present and perhaps vital and will still have monopoly characteristics that require regulation, but the energy those wires carry and the services the energy makes possible may be up for grabs in a technology-driven free-for-all. On the other hand, the utilities of today may manage their own evolution in a way that preserves their primacy. This chapter explores the questions of whether utilities of today can adapt and survive to become utilities of the future and, if so, how they should start, chiefly through immediate-term experimentation. The potential is there for the utilities to preserve a role greater than the one analogous to the shriveled status of landline phone companies.

Keywords: Utility, Transition, Adaptability, Evolution, Business models

1 INTRODUCTION

Late in June 2012, a small thunderstorm cell developed in central Iowa. At the time, it was nothing but a standard, run-of-the-mill thunderstorm. Yet, as it traveled east over Chicago, it became worrisome. An exceptionally hot and unstable atmosphere persisted, interacting with the cell and causing it to exhibit instances of severe and eventually extreme weather. By the time it reached Indiana, it was classified as a *derecho*, a cluster of thunderstorms with hurricane-speed winds traveling in the fashion of its Spanish descriptor: *straight*. By the time the derecho reached the Atlantic coast, it had wreaked havoc upon the mid-Atlantic region. 4.2 million Americans lost their power and 22 lost their lives.

Maryland had been battered by a flurry of blizzards, nor'easters, and hurricanes prior to that. The derecho was the knockout blow. In Maryland alone, 1.6 million lost power as a result of downed wires. The cost of residential outages was estimated to be $593 million, and commercial and industrial outages potentially reached billions. Martin O'Malley, the governor of Maryland,

Distributed Generation and its Implications for the Utility Industry. http://dx.doi.org/10.1016/B978-0-12-800240-7.00021-7

understood very well that due to a warming climate, Maryland's system would only continue to deteriorate going forward, not improve. The governor shortly thereafter signed an executive order calling for wide-scale improvements to Maryland's electric distribution infrastructure, part of which was the creation of the Grid Resiliency Task Force.

One accepted recommendation of the task force is the idea that a reliable and resilient electricity system should be designed and constructed as part of a modernized electric utility industry, an industry that is itself "transforming at a pace unseen in its history" due to many of the factors highlighted in other chapters of this book: flattening of electricity sales, rising retail rates, renewable integration, distributed generation, competitive markets, fuel switching, deregulation, behavioral change, higher consumer expectations, and, the theme of this book, the rise of decentralized resources. The recommendation was thus to charge the Energy Future Coalition with the proposal of a Utility 2.0 pilot design that anticipated the contours of the utility of the future as the host entity providing a reliable and resilient electricity system. Six months later, the Energy Future Coalition submitted *Utility 2.0: Piloting the Future for Maryland's Electric Utilities,*[1] a paper that explored ways for which progress toward the utility of the future could occur. The report is premised on a set of generally agreed-upon core principles that the utility of the future should strive for:

- Reliable and resilient service
- A greater range of options for large and residential customers
- Necessary and appropriate utility system upgrades
- Necessary utility business model changes
- Regulatory adjustments that will allow for all of the aforementioned principles to occur in a manner that is affordable to consumers

Many of the other chapters envision a future in which the electric distribution utility as an entity is unrecognizable from what we have today. Indeed, the global rise of decentralized electricity resources creates a starkly different potential alternative to today's standard, centralized system. We do not know how fast or how far the system will evolve in that direction. What we do already know is that the transition to that future has started and will continue as a function of various drivers toward something different, presumably decentralized and presumably better from the customers' and society's perspectives.

In the face of this uncertainty, the safest assumption is that key attributes of the central electric distribution utility will survive and that it will still perform certain natural monopoly functions, even if others are competitive or autonomous. These functions probably include

- owning, erecting, maintaining, and depreciating distribution wires, transformers, and other local infrastructure;
- maintaining power quality, voltage, and frequency;

---

1. Available at www.energyfuturecoalition.org.

- owning, operating, and maintaining the meters, communication infrastructure, and computer equipment that allow customer-by-customer and likely application-by-application data sharing, to allow integration of these millions of devices into an integrated, reliable, and stable grid and to process the enormous data they generate;
- playing an integral role in combating cyber threats to the grid;
- providing service-of-last-resort power supplies to customers not able to make their own arrangements;
- engaging in social energy programs as the entity with customer-by-customer contacts;
- ensuring adequate transmission interties and monitoring bulk power flows; and
- acting as the principal entity involved in restoring service after an outage, whether it be by freak accident, extreme weather event, or cyber-attack.

There may be others, and some of these may be managed by even larger regulated entities at the transmission level. But these functions will continue to be required and to be regulated in the public interest, unless the evolution of technology carries us to the unlikely ultimate extreme of rendering the grid itself superfluous. In other words, while the transition toward more decentralization will occur regardless, one should not assume away the entity providing the remaining central services or ignore that entity's institutional survival instincts. In the electric industry transition, everyone stands to gain if it is more creative than destructive and especially so given the massive amounts of capital already invested in the system.

To explore how to begin creating that future was the goal of *Utility 2.0: Piloting the Future for Maryland's Electric Utilities*. The report attempted to actualize the aforementioned five principles with six key strategies that an electric utility of the future could adopt but might first want to test.

This chapter is organized into seven sections that parallel those in the report for Maryland. Section 2 suggests testing ways in which customers' own notions of the value of their utility services can be embedded in utility compensation schemes. Section 3 proposes that a regulated utility option for implementing smart-grid services be tried. Section 4 encourages utility investments on the customer side of the meter. Section 5 offers an experiment with utility-owned, customer-energized, islandable microgrids. Section 6 suggests the use of electric vehicle charging for system regulation pending developments allowing EV batteries to provide greater value, followed by the chapter's conclusions.

2   LETTING UTILITY CUSTOMERS SAY WHAT THEY WANT, AND PAYING UTILITIES WHEN THEY GET IT

To date, electricity customers have tolerated a passive role in their economic relationship with their utility, blindly paying the price for services that appears on their monthly bill. They have no alternatives to the utility, no elections as to the nature or degree of service, and not much understanding of what was actually provided or at what actual cost.

Complicating the matter is the fact that a majority of customers feel their utility service is poor and overpriced. This despite their immersion in a world that increasingly relies on electricity and interconnectivity, where power outages are becoming costlier and costlier. For the grocery store whose inventory spoils, the medical facility trying to save lives, the manufacturer that must halt production, the investment bank that makes critical transactions by the second, or the professional working from a home office, the cost of electricity when it is there does not come close to the value of electric service when it is not there. Electricity service is much more than just the volume of electrons distributed and received. Attributes such as cleanliness of power sources, cost over time, reliability, on-site customer service, smart-grid adoption, and technology support have real importance to customers and yet are not reflected either in the utility's rates or in the typical customer's awareness of what his monthly bill covers.

A new utility-customer relationship is inevitable as communicative technology replaces dumb system components and as customers begin participating in power markets themselves as sellers and buyers. This makes a new scheme inevitable for rewarding utilities for the services they provide. The challenge is matching the compensation scheme for utilities to the real value of what electricity service provides while keeping it affordable and improving the satisfaction of its consumers.

Further complexity arises because different customers value these aspects of service differently, and so does the utility. Any mechanism by which the utility of the future would adequately integrate such idiosyncratic valuations would have to incorporate both the customers' and the utilities' own perspective. In a world with many new options, what the customer pays and what the utility receives will inevitably be newly sensitive to whether or not the utility meets the performance expected by the customer, with the state public service commissions as the ultimate arbiter of performance.

The Maryland Utility 2.0 pilot proposal was to allow customers to prioritize and weight five attributes of their utility service: cost, reliability, customer service, new technology adaptation, and alternative energy support. Actual utility performance on each would be measured by appropriate metrics, as shown in Table 1.

The priorities could be determined via a survey of ratepayers distributed through an annual utility bill stuffer and sought from any new customers of the utility. Customers would be urged to respond, directly providing their rankings and weightings to the utility of what was most important to them, and would be assured that the utility would be rewarded in accordance with how well it met the priorities stated by those customers who responded. Customers would also be asked to grade the utility's prior year's performance on the same

TABLE 1 Five Desired Attributes of Utility Service, and How They can be Measured

Factors	Description	Metrics
Cost	Utility's responsibilities are to • minimize costs and capital investments; • improve cost effectiveness of operations and maintenance and increase asset utilization; • meet regulatory standards and public policies; • optimize energy use by system itself, including utility vehicles and buildings; • reduce line losses; and • replace inefficient equipment	Cost of service per customer, per unit of electricity delivered, relative to other utilities in Maryland, nation, normalized for weather, per capita income, GDP, inflation, other factors
Reliability	Utility's responsibilities are to • provide continuous service, freedom from outages, anomalies, critical needs, and shed unnecessary load; • restore service promptly after outages; • upgrade system capability to isolate, minimize, and shorten outages; • maintain system voltage, frequency, and other measures of power quality; and • defend system controls, monitors, and data from cyber intrusion and disruption, including customer-operated systems	Number and duration of outages, by region, neighborhood, customer, including momentary blips; Speed of recovery from outages, number and accuracy of outage status reports Voltage and frequency constancy, anomalies, surges; Cyber-related penetrations with damage or delay, detections of malware, viruses, or worms
Customer service	Utility's responsibilities are to • relate productively to all customers; • respond quickly and effectively to customer concerns and opportunities; • provide full, timely, and appropriate information, especially during outages; and • assist customers to achieve energy efficiency	Customer service calls answered, wait time, customer satisfaction surveys, efficiency gains relative to base period by customer by class, by building per square foot, normalized for degree days, economic activity
Smart-grid adoption	Utility's responsibilities are to • select, install, and operate integrated suite of digital equipment on its system to provide information, control,	Extent of system digitally monitored, controlled, automated for contingencies, amount of data managed, number of customers opting

Continued

TABLE 1   Five Desired Attributes of Utility Service, and How They can be Measured—cont'd

Factors	Description	Metrics
	monitoring, billing, and integration with customer applications; • select integrated suite of matching equipment for customers, install, and demonstrate; • train and support customer use and benefits	smart-grid installations, remote appliance operation by utility, customer complaints about smart-grid capability, equipment
Alternative energy support	Utility's responsibilities are to • integrate renewable and other distributed energy generation into system operations, dealing effectively with intermittency; • provide quick and reasonable interconnections; • support development of microgrids; • promote workably competitive markets for energy, capacity, efficiency, and any services other than utility's own monopoly services	Wait time for clean energy interconnection, integration, curtailments from over/under voltage, surveys of clean energy system owners, suppliers, market participants, customers, vendors, costs of competitive offerings relative to other utility areas

attributes. This would provide the utility a detailed, geographically specific database of what its customers valued against which to establish its own priorities and where to work for greater customer approval.

Actual metrics measured over time, not customers' subjective grades, would be collected and analyzed in periodic rate cases. Based on the utility's composite success in achieving the metrics against the customer-stated priorities, perhaps separately in subareas rather than the whole service territory, the utility could be rewarded with or penalized by a 1% upward or downward adjustment, respectively, to its permitted rate of return on equity.

The proposed band gap on return on equity is certainly subject to debate, but a 1% increase or decrease would represent a sizable carrot or stick. This approach allows the utility to *clearly* understand what its customers value, with great geographic detail and with a serious incentive to meet its customers' expectations.

While technical details remain to be hashed out, this approach would address many of the aforementioned broad challenges of the utility-ratepayer relationship: it creates an active role for the customer, places value on their varied needs, and holds the utility accountable for them. It creates a line of communication and feedback between customer and utility in a language that the utility surely understands: shareholder returns.

3   LETTING UTILITIES TRY FIRST TO MAKE THE FUTURE WORK

A growing suite of "smart" technologies is moving the electricity system toward the future that will feature real-time bidirectional information flows and additional controls able to act on those flows. These are being developed and marketed by a host of "smart-grid" companies, web-based service providers, and high-tech equipment manufacturers. This extends far beyond "smart" thermostats like the popular Nest device. It goes to hardware, firmware, software, information standards, and innovative methods of communication and includes offering the hands-on services these technologies enable. That these technologies have arrived is clear and that more are on their way is inevitable. The only questions now are how quickly, directly, and comprehensively electricity will consumers embrace them and how will they "interoperate" with the other devices their customers employ.

In response to the stakeholder solicitation for the Energy Future Coalition's Maryland report, ~80 stakeholders offered technologies or equipment that could provide such capabilities. As the proposals came in, it became increasingly clear that a fully integrated and interoperable programmable power-use environment is probably possible with the *current* technology. However, it was not clear that a typical consumer would have the technical "chops" to select compatible technology and gain the potential benefits or that the consumer's selection of various available third-party vendors would ensure that their equipment would work happily together to deliver the promised benefits.

If the smart grid is going to find eager acceptance by consumers, it must first work and be seen to work, providing the savings and convenience it is touted to offer. Especially in the early days of widespread adoption, success stories have to be heard and seen if smart-grid capability is going to be "the next big thing." At the same time, utilities are struggling to do what they have traditionally done to grow their business and revenues: make investments in new equipment. And utilities are particularly sensitive to the idea that third-party entrepreneurs (read "interlopers") are going to start installing equipment on their grid that will disrupt their system operations and create problems.

Hence, one potential initial step forward is to assign, at least on a pilot basis, the utility itself to take the role of helping its customers select, install, operate, and even provide the capital investment for the attributes of the smart grid. The utility would be motivated to help its customers gain the benefit of the intelligent equipment, controls, and communications that will bring these advantages. This motivation will come not only from the new opportunity to cement a utility-customer relationship in an era when customers will have options to utility service but also from the opportunity to make investments on which a return can be earned for shareholders. Being exposed to regulatory penalties for imprudent or wasteful investments on their customers' behalf would provide a third form of incentive for utilities to get it right.

Customers would be protected from fly-by-night vendors and prematurely released products trying to capture part of the market, along with their traditional protection from regulators, who would have to approve the utilities' offerings and proposed pricing, subject to an opt-in agreement with the customer. Utilities have long had access to an array of smart technologies on the supply side of the meter but have been reluctant or prohibited to manage energy-producing or energy-consuming technologies on the demand side.

With such access and with its customer's agreement, the utility could, for example, manage energy flows to applications that would not interfere with the customer's enjoyment of them: refrigerator defrost cycles, swimming pool pumps, storage water heaters, or air-conditioning compressors. This would allow the utility to better meet peak loads, deal with voltage variations, accommodate variable renewable generation, and generally manage the grid in a cost-effective manner. At least, some of the savings should then be shared with the customer in the form of lower rates, rebates, or other remuneration. To the customer, the benefits of lower electricity consumption would also show up directly on the bill. As well, such utility demand-side controls could allow better overall system management and may avoid the need for additional, costly peaking generation, all of which could result in a lower rate charged to the consumer, lowering all customers' bills.

Leading manufacturers will include or are planning to include smart chips in every major appliance starting after 2014, but most new capabilities provided by a smart chip will go unused unless their use is made easy and beneficial by both utilities and other stakeholders. One way to do this is through standards of interoperability, and indeed, the National Institute of Standards and Technology released the second version of their Framework and Roadmap for Smart Grid Interoperability Standards in February 2012. Documents like this will foster a common language among smart devices and equipment and the parties involved in their use. Additionally, short-range FM band radio has been proposed as a viable, secure, and novel option for sending real-time price and control signals from central to and from free-standing applications within a customer premise. And the increasing success of cloud computing opens up a new realm of possibilities for the transfer of information.

The technology to facilitate a revolution in the way we use our power is here, and such ways of thinking are not disconnected with utility realities. Indeed, whispers of such a setup can already be heard in Europe, as Germany's largest power producer RWE begins to think on such lines (Groot expands on this).[2] The challenge is to ensure that customers not only understand these technologies and their benefits but also *trust* them. Strong communications, therefore,

2. Beckman, Karel. "RWE Sheds Old Business Model, Embraces Transition." *EnergyPost.* 21 Oct. 2013. Web.

would be vital. Other equally important missing ingredients toward a full programmable power-use environment include appropriate regulatory schedules to allow customers to be exposed to variable energy prices, easier and patient sources of up-front investment capital, and the utility's financial incentive to cooperate with all these.

This proposal would not be without serious downsides and controversy, however. Many of the relationship dynamics between customer and utility being discussed fall under the traditional natural monopoly model of the utility and thus already accounted for by regulators. However, many of the new customer-facing services are currently unregulated and may potentially be best offered by competitive third-party entrants in retail electricity service markets and appliance and building management. Everyone understands that it was the blossoming of thousands of "apps" and separate functionalities that led to the utter revolution of the telecommunication sector, so it is controversial to suggest that a similar explosion of competitive entry and innovation might not equally good for the electric utility sector. But the decline and effective demise of landline phone companies cannot be repeated in a sector where the wires remain the only feasible delivery system for the critical commodity, and the utility continues ownership and monopoly control over those wires.

Policy makers face an exceedingly difficult decision between two alternatives: regulators can give the utility exclusive authorization to provide such customer-side investments and services *or* bar the utility from providing services itself but require it to facilitate third-party competition. Allowing the utility to offer competitive services, even as an unregulated affiliate function, while continuing to operate the grid monopoly, seems certain to lead to complaints of undue bias and inside knowledge from competitors for the unregulated business. The regulated and unregulated functions are best fully separated.

The best choice may be best discerned by testing both alternatives in separate pilot projects. The answer may also vary as a function of the consuming sector, as the residential, commercial, and industrial sectors have different dynamics.

For the residential sector, it would be ideal if two sister pilot projects were able to simultaneously test utility exclusivity versus open competition. However, if the choice must be made for only one, the authors would opt for testing utility exclusivity, a decision for which the swing factor is that once the floodgates of competition are opened, they are extremely difficult to shut again. Another important factor in this decision is the creation of a new line of business revenue for the utility—the financial incentive needed for the utility to not only cooperate with a programmable power-use environment but also help it to blossom. It should be emphasized that this would be done on a strictly experimental basis, the results of which would create valuable lessons for how to approach this debate in the future.

The coming transition will present huge challenges to utilities to maintain revenues and services that can avoid shrinking their businesses, shareholder value, and role in their communities. It can only help to allow them regulator-supervised opportunities to work more closely with their customers, make incremental investments in those opportunities on which they earn a return from the customers they persuade to work with them, and give them a true stake in the downstream reliability and efficiency of their service. If utilities can assist their customers with necessary expertise and even necessary investments, paid for by the specific customers that receive them, utilities can create new one-on-one customer relationships of value, can build loyalty that will be important to their own prosperity in a more competitive world, and can improve their own systems' ability to handle contingencies and improve reliability.

Of course, incipient third-party entrepreneurs detest the idea that the utility monoliths they are working to disrupt and displace with their customers will be allowed to extend their monopoly services into potentially competitive domains with reason. More discussion on this can be seen in Brennan. However, it is not realistic to allow utilities to participate through unregulated subsidiaries in a competitive market and expect that competition to be fair, as superior customer access, a litany of historical data, and the ability to restrict interoperability with the existing system all give the utilities a major leg up. In the future, such services will have to be offered *either* by utilities *or* by competitive entrants, not by both. And in that light, a good opening bid is for customers to have their traditional regulated entity try to install and operate a more modernized system under the pressure of a competitive alternative rather than to have these pioneering but dramatic changes be led by a host of undisciplined promoters of new and untested businesses hawking new and untested services and products as add-ons to an old and overtested system—with a painful reassignment of all services to the old monopoly utility as the only fallback in the event of systemic problems.

Again, this is all subject to trial, and should this method clearly fail, full competition can be called upon as the fallback option. The authors themselves reach this judgment reluctantly, fully aware that utilities have never been known for leading innovation or technological advances that displace their own earlier investments. But electric utilities will remain critical players even after the revolution, because the grid itself will remain critical, and to have them among the early casualties of the revolution may lead to its total failure with a chaotic breakdown of service to the common customer.

In addition, utilities are always paying attention to each other, striving to adopt the best practices and improve their balance sheets. So, while utility exclusivity—if it worked and became the standard across multiple service areas—would preclude competition among third parties there, there would still be some sense of "indirect" competition within the electricity distribution sector and comparison between progress and service between utility-served areas and those allowed with free competitive reign.

Another reason for at least testing utility exclusivity is that utility customers—and especially residential utility customers—have traditionally been passive actors and many may not be ready or willing to jump into a competitive market with numerous options, the distinctions between which are explained using technical jargon beyond the normal customer's understanding.

All of that being said, it is far more feasible to go with third-party competition option in the case of commercial and industrial customers. For them, a strong business case for a programmable power-use environment is clear, and many have demonstrated a willingness to dive in to such an environment full bore. Such profit-motivated organizations are far more capable of making wise decisions when choosing between purveyors of smart technologies, provided that the competitors meet the standards to interoperate with the utility's own equipment.

If a cluster of residential customers, perhaps in a multifamily building, have the ability to organize and aggregate and decide to put out a request for proposal to third-party competitors, such an option could be allowed.

Regulators will have to approach the "utility versus third-party" question on a case-by-case basis. The point here is that if they do not at least initially allow the utility an option of offering and operating the customer-side services the new world will bring, they will not have the data or experience to reach an informed judgment.

4   NEW UTILITY REVENUE STREAMS FROM CUSTOMER-SIDE INVESTMENTS

Various structural trends in the electricity distribution industry—as described elsewhere in this book—are offsetting the utility's ability to make upstream investments for which the utility earns a guaranteed rate of return. In the hunt for replacing the loss of these revenue streams, one proposal worth testing would be to allow the utility to finance investments on the customer's side of the meter. Here, there are two sets of options that are not being undertaken at any great scale today but could be with the appropriate regulatory atmosphere.

The first potential set of options are utility-owned investments. Examples include voltage regulators and load controls, which would convey system-wide benefits that not only help the utility maintain its costs during peaks but could also be rate-based with an attached return on equity. This offers benefits from both cost and reliability perspectives thanks to improved operations and revenue streams through an enlarged rate base, even if it might lead to lower sales of electrons. Duke Energy's Save-A-Watt program, though not entirely successful, does set a precedent and offer lessons learned for this kind of action.[3]

Another option is for the utility to finance building-efficiency investments in the customer's premises for those customers who lack the up-front capital required for investments such as smart-grid control systems, smart devices,

---

3. Duke Energy. "How Save-A-Watt Will Work." *2007-2008 Sustainability Report*. N.p., 2007.

and building management systems, of course, but also passive efficiency measures such as insulation, improved fenestration, and efficient lighting systems. The utility would cover the up-front costs of these measures, and the customers would make on-bill repayments. This proposal achieves benefits for the customer, and investment revenues for the utility, and can be conditioned in such a way that the dollar repayment amount is *lower* than the dollar energy savings, thus resulting in an immediate net savings to the customer.

Although utility credit ratings have been deteriorating,[4] utilities continue to have access to capital at relatively low interest rates. Utilities could, in turn, pass that low cost of borrowing through their customer, with considerations for administrative costs, risk, and a modest profit. An added advantage of on-bill financing repayment is that the utility's risk is reduced by reducing its customer's cost of energy, making that customer more able and more likely to pay the bill.

The on-bill repayment option could overcome many of the barriers associated with efficiency and smart-grid measures. The most glaring one, of course, is first cost. Even if customers understand the benefits of such measures, they are often unwilling to pay up-front costs, especially when, in their eyes, the payback is uncertain. Another barrier is that of the split incentive between owner and tenant. A tenant is unwilling to invest in long-term energy-saving options because they will not live in the residence long enough to realize the full benefits. An owner is unwilling to invest because in the majority of situations, they are not responsible for the monthly utility bill. This could be solved by tying the permanent energy-efficiency measures or system controls to the premises, so that the costs and savings are passed down from tenant to tenant. The owner sees his property value increase, the utility gets a new revenue stream, and the tenant sees a lower utility bill even after it includes the repayment of the utility's investment due to lower energy consumption.

Of course, while hurdling some barriers, the introduction of on-bill financing also creates new ones. Regulatory oversight would be necessary to ensure that utilities are not charging exorbitant interest rates and are not abusing their ability to loan capital. On-bill financing also brings utilities into a banking business line that is beyond a typical utility's core competencies and so could require new hiring to be executed at a successful level. The banking industry would not be enthusiastic about competing with utilities to make home improvement loans, even if they are strictly limited to energy-efficiency improvements as they should be. Additionally, new billing software, design, and practices could be needed, which may be troublesome to customers who are both untrustworthy toward utilities and often resistant to change. To quell concerns like these, there is great need for a high-caliber communication and

---

4. Parepoynt, Jon. "Electric Utility Credit Ratings Are Deteriorating." SeekingAlpha.com. N.p., 11 June 2013. Web. 01 Oct. 2013. Look at Economist 12 Oct 2013 to see what is happening in Europe.

messaging effort behind any pilot such as the one being discussed. Such an arrangement would necessarily create a new relationship between the utility and the customers who either took on the debt to the utility or accepted utility-owned equipment on their premises, but such a new relationship is coming anyway; the utilities will need the business and investment opportunities.

5 TESTING AUTOMATED MICROGRIDS FOR ECONOMICS, RELIABILITY, AND INTEGRATION

Following successful demonstrations in cities like Chattanooga and Naperville, two of Maryland's major utilities, Pepco and Baltimore Gas and Electric, are introducing sectionalizing and reclosing into their grids in efforts to improve system-wide reliability and resilience. In the case of distribution-level outages, which are usually weather-related, sectionalizing and reclosing the grid can offer the utility the ability to automatically restore power to portions of the grid not directly affected by the disruption after momentary faults. While it does require significant engineering and installation to put in place, the economic benefit is clear, especially when it comes to avoiding outages in commercial and industrial sectors. Yet, sectionalizing and reclosing subareas of a grid could just be the first step toward improving total system reliability and resilience. The utilities could take their efforts further by implementing automated microgrids in selected areas of their retail service areas.

"Microgrid" has been defined in various ways, including in the chapter by Felder. The definition offered in *Utility 2.0: Piloting the Future for Maryland's Electric Utilities* is "an integrated energy system consisting of distributed generation (ideally both conventional and renewable), multiple electric loads and energy storage, operating either in interconnection with or islanded from the existing utility grid." Microgrids have existed for years in one form or another and are the traditional form of electricity service at sites like university campuses and military installations.

Not until fairly recently did the convergence of factors—greater reliability requirements, distributed generation, smart-grid controls, and more—turn the microgrid discussion from specialized military, medical, and campus applications toward the idea that interconnected microgrids might become the predominant architecture of the electricity sector. When a section of a utility's own service area can function as a microgrid, maintaining critical service to a combination of residential, commercial, or industrial customers, reliability and resiliency can both improve. Given the improved economics of distributed generation, storage, and system controls, however, it is not clear that the costs of service will not also improve compared to the alternative of a further build-out of upstream central generation and transmission with their attendant costs and lead times.

The Maryland pilot design proposes options running from a bare minimum backstop microgrid to a full and complex power system capable of islanding itself for extended periods of time and generating new revenue streams through organized market mechanisms. Given the cost declines observed in recent years for solar-photovoltaic power and for natural gas as a potential energy source for CHP installations, a microgrid could even potentially be competitive or superior in economic terms to the central grid, providing a financial rationale to operate it on an islanded basis, not simply a security and reliability rationale.

At minimum, a utility could establish a section of their service territory capable of surviving off the grid with backup power generation in the case of a disruptive weather or cyber-security event. Ideally, the section would feature some mixture of residential, commercial, and industrial loads, so that the backup power generation could be provided by a combined heat and power plant already on or to be installed on the premises of a major customer. Another option would be to install a standby generator at the area's designated utility substation. The ability of a section of utility service territory to self-generate, and thus avoid outages, is a valuable economic proposition, an expression of which can be seen in Figure 1, which estimates the direct costs of outages to customers in PG&E's service territory.

A "bare" microgrid of this caliber would require the ability to self-generate long enough to maintain power until outages are corrected. To that end, it could use remote load controls to ensure that only critical needs are powered. Relatively low-voltage loads such as lighting, telecommunication, and electronics could be kept on continuously; higher-voltage loads such as furnace fans, air-conditioning compressors, water-heating coils, clothes drying, electric cooking, clothes washing, and refrigeration could be remotely controlled, permitted at announced intervals, or cycled during contingencies. For example, heating in winter and cooling in summer could be cycled so that the system does not take on too much load at any one time. Introducing smart-grid technologies such as distribution and substation automation could increase operational and reliability efficiencies, and using smart management systems to control them can result in a fully functional "local utility" capable of optimizing its service to customers. If conditions warrant, such a setup could even operate off the grid in times that it is economically advantageous to do so, such as during peak

Customer Class	$/kWh Unserved
Industrial	$12.70–$424.80
Commercial	$40.60–$68.20
Agricultural	$11.50–$11.70
Residential	$5.10–$8.50

FIGURE 1 Estimated direct costs of outages in PG&E territory. *Source: EPA.*

hours. Renewable generation could be used to heat water or buildings when it was available.

Storage could be charged when power was available, used when it was not, all to maintain critical needs. In organized electricity markets like the PJM Interconnection (which includes Maryland), the arrival of the Federal Energy Regulatory Committee's (FERC) Orders 755 and 784 has made fast-responding energy storage more valuable than ever in frequency regulation markets and has allowed storage to participate as a resource in ancillary services markets. Potentially feasible technologies range from flywheel to batteries. Electric vehicle battery storage, discussed in Section 6, is another option.

To operate such a microgrid, again, a new relationship between the utility and the customers within the microgrid area would be required. It would still be the utility's wires, and presumably, the utility would retain the responsibility for voltage control, frequency, service-of-last-resort, outage restoration within the microgrid, cost recovery from customers, payment for power generated to generators, and other bookkeeping and control functions. However, the utility would have arrangements with customers that would also allow for curtailment of noncritical loads at necessary times to maintain critical services and arrangements to accept and compensate distributed generation from customers or others with generating capacity, perhaps having its own standby generation or the ability to bring portable truck-mounted generation into the area, for which it would be compensated. The ability to serve as generation for a microgrid area could encourage customers with major thermal energy requirements, such as hospitals, industrial parks, or district heating or cooling areas, to invest in combined heat and power equipment, selling surplus electric energy to the grid under normal circumstances, while serving it as baseload power during broader grid outages.

Microgrid customers would be obliged to accept a suite of controls and monitoring of major uses.

Such a "full-suite" microgrid could however offer offsetting economic benefits:

- Avoided costs of outages through added reliability, which typically mount to many multiples of the cost of power itself.
- On-site generation, which, besides having a backup generation function, could be used to avoid peak energy costs.
- Enrollment in demand response programs could both reduce energy costs and earn additional revenue streams through organized market mechanisms.
- Participation in capacity, ancillary, and frequency regulation markets adds additional revenue streams.

All of these benefits are further economically justified within the context of a microgrid, purely due to the ability to operate on *or* off the grid. Lawrence Berkeley National Laboratory attempted to quantify these economic benefits by first creating a framework for doing so and then plugging in data from a "typical, large, Canadian, semi-rural feeder with 10 MW peak load and 6.2 MW average load." In their full microgrid scenario, customers within the microgrid saved $327,142 per year. Further positive economic externalities due to society amounted to $94,469 per year.[5]

Any utility of the future should study and employ microgrid options to the feasible extent. Testing them now is already feasible and will accelerate their benefits.

6   ELECTRIC VEHICLE CHARGING

While most storage technologies have not yet reached the mass commercial viability that they one day will, utilities may be able to find battery storage peppered throughout the residential garages of their own service territories. Penetrations of electric vehicles are rising with technology improvements, sustained high costs of gasoline, greater awareness of individual environmental footprint, and the intent of all major automobile companies to market all-electric vehicles. The world's electric vehicle fleet is still quite a humble one, with a global EV stock of about 180,000 (0.02% of all total passenger cars),[6] but policy, market, and environmental drivers continue to push in the direction of a massive energy conversion for person transportation.

Here, there is certainly mutual opportunity for utility, customer, and automaker. It is well documented that the greatest barrier to greater electric vehicle penetration is the high battery cost, which, even when factoring in avoided costs of gasoline, government incentives, and industry rebates, leaves an electric vehicle several thousand dollars more expensive than gasoline-powered counterparts. If, however, a utility offered a sizable contribution of $500–$1500 toward an electric vehicle purchase for customers within its pilot project territory, it would certainly entice at least some customers to buy electric vehicles. In exchange, the pilot customer who utilizes such a contribution would have to agree to allow the utility to use the car battery—60 kWh in the standard Tesla Model S, for example—as a grid asset. To give an idea of what a battery of that size could accomplish, average residential electricity consumption in the United States is about 30 kWh per day.[7]

The utility could easily justify such a contribution through the functions of an electric vehicle battery as a grid asset, which range the entire spectrum of storage applications: backup power, peak shaving, voltage regulation, load

5. Morris, Greg Young, Chad Abbey, Geza Joos, and Chris Marnay. *A Framework for the Evaluation of the Cost and Benefits of Microgrids*. Tech. Ernest Orlando Lawrence Berkeley National Laboratory, 15 Sept. 2011. Web.

6. International Energy Agency, Clean Energy Ministerial, and Electric Vehicles Initiative. *Global EV Outlook: Understanding the Electric Vehicle Landscape to 2020*. April 2013.

7. EIA. "How Much Electricity Does an American Home Use?" *U.S. Energy Information Administration*. N.p., 19 Mar. 2013. Web.

control, and frequency regulation. Indeed, electric vehicles in a test system at the University of Delaware are currently earning $150 per month per vehicle by offering voltage regulation services through dischargeable batteries. This sort of practice certainly qualifies as one of those coveted revenue streams utilities are so diligently hunting for. Platt et al. go into electric vehicles as a grid asset in much more detail.

For any undertaking of this sort, a key partnership between the utility and the automaker must be forged. Because electric vehicle batteries were not designed for the express purpose of acting as a grid asset, there are warranty issues protecting against battery wear that would need to be overcome. In that same test system at University of Delaware, Professor Willett Kempton has discovered ways to minimize wear by keeping the battery charge "in the middle." In other words, battery wear occurs most when a battery is fully charged or fully discharged, so if a battery is kept in a 40% through 60% charge range, it is possible that it wears even *less* than it would with normal use.[8] In short, this is an issue that can be overcome with proper agreements between automaker and utility, ensuring that battery wear is being adequately managed. In addition, the automaker would be asked to offer special retail prices to people living or working in the pilot region in order to create further competition between plug-in electric vehicles and conventional gas-powered vehicles. In return, the automaker receives a strong marketing and publicity push, as electric vehicles would have to be guaranteed to play a significant role in any such pilot's communication efforts.

At a minimum, there is no damage to a vehicle's battery from charging it whenever the power is available and there is room to accept the charge, so utilities can use electric vehicles as a variable load to match against variable renewable power or variable peak and valley swings in their other loads, again with a benefit to the owners of the vehicle, who could also be assured of at least a minimum charge sufficient for an assured minimum number of miles.

A parallel challenge is installing the necessary infrastructure for charging when vehicles are parked outside of home garages. Utilities could take on responsibility for connecting their grids to electric charging stations or even supply those charging stations. In most cases, the property owner would be willing to pay the costs of powering such a station, as the value the customer brings to the office building, shopping center, or other facility is greater than the cost of supplying power to the station. However, if this were not the case, a deal between the utility and the property owner could be brokered. With longer-term parking spots (e.g., full workday) or even overnight parkers, utilities can save capital costs by installing "slow" 120 V level 1 chargers as opposed to "fast" 240 V level 2 chargers.

---

8. Curwood, Steve. "Electric Cars to Buffer the Grid." *Living on Earth*. N.p., 3 May 2013. Web.

7   CONCLUSION

Many of the recommendations made in this pilot entail utility business model changes that fall into regulatory gray areas, and therefore, much of the pilot hinges on the cooperation of state public service commissions, which must strike a delicate balance in their obligation to protect public interest. In addition to adopting the five performance factors mentioned in Section 2, the public service commission would have to adopt new metrics that appropriately measure those factors. The public service commissions would also have to oversee utility financing of customer-side investments, an idea that falls not only outside the core competencies of the utility but also for the commission. And in order to take the best advantage of smart-grid controls, new and different time-sensitive rate structures would be appropriate. Clearly, there are challenges.

Hypothetically, if a utility were to pilot every new function mentioned in this chapter, the operational synergies between them would provide additional new learning and potentially new opportunities. One possible (and superlative) outcome would be that the utility would achieve a robust level of savings and several additional revenue streams, additional money that would be passed to the customer indirectly as the utility's profitability allowed it a lower cost of capital and its cash balance allowed it to make further investments in reliability and efficiency. Another potential benefit is that the system achieves a level of reliability previously unseen.

The reason utilities should at least experiment with one or more of these options is that the alternative to doing so is to head into a certain future of declining load, invasive competition from innovators and entrepreneurs, necessary major investments in status quo infrastructure while customers are being drawn to alternatives that potentially lead them to disconnect from the grid, and economic circumstances the utility industry itself is referring to as a "death spiral."

Successful pilots of this sort could serve as a model for the future; unsuccessful ones would provide lessons learned without paying the death penalty. As it always does, the process of transitioning the utility to a vastly different electricity landscape will happen through creative destruction. Again, the goal of these proposed pilots is to help advance toward that landscape more creatively than destructively, to add functionality and survivability for all key players, utilities, and customers, in an increasingly decentralized world.

Chapter 22

Turning a Vision to Reality: Boulder's Utility of the Future

Kelly Crandall, Heather Bailey, Yael Gichon and Jonathan Koehn

ABSTRACT

This chapter describes the unique attributes of the city of Boulder, Colorado, and how these traits have contributed to a new vision for localizing energy supply. The authors describe the community's goal to ensure that Boulder residents, businesses, and institutions have access to reliable energy that is increasingly clean and competitively priced. This underpins the need for an electric utility that delivers energy as a service rather than as a commodity. Boulder is currently engaged in a process to determine the path toward this vision through the formation of a municipal utility or the development of a new partnership with its current electricity provider. Ultimately, this chapter asks: what would a utility look like when developed in collaboration with a community to meet its diverse needs?

Keywords: Localization, Utility of the future, Energy as a service, Boulder, Colorado, Municipalization

1 INTRODUCTION

Most observers now recognize that the global climate will experience radical changes in the coming decades, without immediate and far-reaching action. The response to this threat will have enormous implications for the power sector that requires nothing short of a fundamental rethinking of how electricity is produced, transmitted, and used. While communities remain hopeful for a federal response, it has become increasingly clear that local jurisdictions will need to lead. Cities have the flexibility, capacity, creativity, and motivation to become resilient in the face of climate change; moreover, facing its impacts, they have the urgency and the imperative. Boulder now has the opportunity and resources to lead by example.

This chapter discusses the perspective of Boulder, Colorado, on the critical shift to prepare for a post-carbon era, the driving factors behind Boulder's

Distributed Generation and its Implications for the Utility Industry. http://dx.doi.org/10.1016/B978-0-12-800240-7.00022-9
435

process to municipalize its electric utility service from its incumbent investor-owned utility, and its vision of a twenty-first-century electric utility business model. It tells the story of one community that is forging a path to realize the energy future it wants—a future that minimizes cost, risk, and environmental impact and maximizes opportunity, options, and societal benefit. This vision is being developed through an ongoing dialogue with the community. Rather than being the final word, this chapter describes the start of a process by which communities, utilities, policymakers, regulators, investors, analysts, and advocates can consider how utility decisions and behaviors support local goals, including decarbonization.

The chapter consists of seven sections in addition to Section 1. Section 2 discusses Boulder's long-time commitment to sustainability. Section 3 touches on climate action in Boulder, including its extensive energy efficiency offerings. Section 4 describes Boulder's changing relationship with its energy supply, focusing particularly on the development of goals to guide its Energy Future. Section 5 highlights some of the flaws in the existing utility business model to contrast them with Section 6, which lays out several elements that should underpin the future of energy services in Boulder. Section 7 provides the next steps for the project followed by conclusions.

2   BOULDER'S GREEN HERITAGE

Boulder, Colorado, is 25 miles northwest of Denver, located in the Boulder Valley at the foothills of the Rocky Mountains (Figure 1). Boulder has ~98,000 residents; over 70% have a bachelor's degree or higher. The median household income is about $54,000, slightly higher than the national figure of $52,000, according to 2007-2011 US Census Bureau data.

Boulder is a mix of mountain town (Figure 2) and high-tech powerhouse, the site of world-class federal laboratories, large manufacturing companies, a vibrant startup community, and Colorado's flagship university. Frequently top-ranked in US surveys of happiest, healthiest, and best-educated cities, it remains "25 miles^2 surrounded by reality" to some because of its liberal, environmentally conscious bend in a somewhat conservative state.

Boulder's sustainable heritage largely began with an acquisition of mountain open space in 1898. In the years since, this acquisition has built to over 45,000 miles^2, with a large part of it open for public recreation. As a Colorado "home rule" municipality, Boulder has significant powers to manage its land and growth. In 1959, the "Blue Line" limited city water service to below 5750 ft., and building height limits and solar easements maintain access to unimpaired mountain views. In 1976, the Danish Plan limited the city's population growth. Although controversial, contributing to higher property values and housing constraints, many of these policies have led the Boulder community to focus on smart growth, green building, and alternative transportation

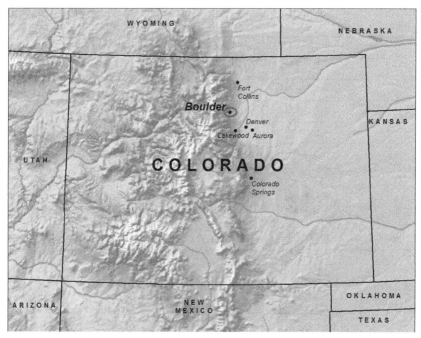

FIGURE 1   Map of Colorado identifying Boulder. *Source: GIS data.*

FIGURE 2   The distinctive flatirons of Boulder, Colorado. *Source: City of Boulder.*

TABLE 1 Selected Sustainability Initiatives in Boulder

1967	Boulder begins taxing itself for open space preservation
1977	Boulder Valley Comprehensive Plan adopted
1982	Solar access ordinance adopted
1993	Boulder adopts integrated pest management (IPM) policy
1996	Adopts first mandatory green building standard, Green Points
1996	Adopts transportation goal of "no long-term growth in vehicle traffic"
2006	Adopts Master Plan for Waste Reduction and Climate Action Plan
2008	Beginning of Boulder's curbside composting program
2011	Launch of Boulder B-cycle bicycle rental service
2013	Boulder begins implementing disposable bag fee

options that have increased Boulder's desirability from a quality of life perspective.[1] A selection of the actions Boulder has taken related to environmental sustainability is summarized in Table 1.

The environmental consciousness that pervades Boulder turned to climate action in the early 2000s. It is notable that among Boulder's elected officials, few—if any—have doubted the legitimacy of human-induced climate change. This may be in part because Boulder seems to have a higher concentration of climate scientists than other cities. According to Brennan (2013), more than a dozen scientists from local research institutions, including the National Center for Atmospheric Research, the National Oceanic and Atmospheric Administration, and the University of Colorado, contributed to the 2013 Intergovernmental Panel on Climate Change Fifth Assessment Report. That knowledgeable group, combined with a progressive activist community and outdoors enthusiasts concerned about negative impacts to the natural environment, set Boulder on the path to taking aggressive actions related to climate change. Boulder City Council (2002) adopted a resolution by which Boulder committed to try to meet the Kyoto Protocol goal of a 7% reduction in greenhouse gas (GHG) emissions from 1990 levels by 2012.

---

1. A short history of Boulder is available on the city's website: https://bouldercolorado.gov/visitors/history. More information on Boulder's history, environmental heritage, and economic characteristics can be found through the Boulder Convention and Visitors Bureau (http://www.bouldercoloradousa.com/about-boulder/green-lifestyle/) and the Boulder Economic Council (http://www.bouldereconomiccouncil.org/).

3   IMPLEMENTING BOULDER'S CLIMATE ACTION PLAN

By 2006, the city and concerned residents had drafted a comprehensive Climate Action Plan that touched on energy, buildings, transportation, waste, and natural environmental impacts. Carbon mitigation efforts would be funded by the nation's first voter-approved carbon tax, the "CAP Tax," a per-kilowatt tax placed on the bills of electricity customers in Boulder except for those who purchased wind-powered renewable energy credits. Brouillard and Van Pelt (2007) described the CAP Tax in greater detail, as does the city's Web site.[2]

The CAP Tax has funded an extensive suite of services and initiatives since 2006. Because Boulder's GHG emissions derive primarily from electricity—generally 55-60%—followed by natural gas (Figure 3), CAP Tax funding was primarily directed to energy-related programs, specifically those efforts geared at reducing consumption through efficiency and conservation. Over the years, Boulder piloted and implemented energy efficiency rebates, compact fluorescent light bulb handouts, financing programs, and advising services. The city frequently leveraged state and federal funding and the regulated demand-side management programs offered by the energy utility that currently serves Boulder, Xcel Energy. CAP Tax-funded programs faced significant barriers, including access to energy consumption data from the utility, the split incentive between landlords who pay for improvements and tenants who pay the resulting cheaper bills, and the challenges associated with encouraging participants to move from energy audits to energy efficiency upgrades.

The city's policies addressed these barriers head-on. For example, a 2010 city ordinance called "SmartRegs" requires landlords to ensure rental units meet baseline energy efficiency standards by 2019 as a condition of rental licensing. This is significant, as about one-half of Boulder's 46,000 residential units are rentals. Also in 2010, Boulder piloted a program called "Two Techs and a Truck" to offer

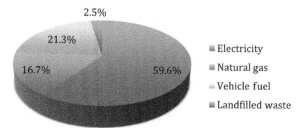

FIGURE 3   Breakdown of greenhouse gas emissions in Boulder community (2010).[3] *Source: Boulder (2011b).*

---

2. For more information, visit https://bouldercolorado.gov/climate.

3. Boulder has been unable to provide a GHG inventory since 2010 due to data quality issues with Xcel Energy.

energy advising services. Without advisors to walk residents and businesses through the process of identifying efficiency opportunities and selecting reliable contractors, energy audits can sit on the shelf, forgotten for other priorities. In 2010, Boulder collaborated with Boulder County, Denver, and Garfield County to apply for and receive $25 million in federal Better Buildings funding to roll out energy advising programs for residents and businesses. These programs are implemented as Boulder County EnergySmart, the Denver Energy Challenge, and Garfield Clean Energy. Boulder (2013c) showed that this type of advisor service has dramatically increased the "audit to action" ratio in the city, leading to upgrades for over 70% of owner-occupied residential units and 60% of businesses after receiving EnergySmart advising services.

The success of these programs reinvigorated enthusiasm and built relationships throughout the community—they also correlate with a reduced upward trend in Boulder's GHGs—leading Boulder voters to approve the extension of the CAP Tax in 2012 by a significant margin. Meanwhile, Boulder was embarking on a larger, community-driven process to reexamine its overall energy future, which is particularly relevant to the present volume.

4   BOULDER'S ENERGY: THE FUTURE IS NOT WHAT IT USED TO BE

Adding natural gas, energy-related GHGs make up about 75% of Boulder's annual emissions, with the balance coming from transportation and landfilled waste. The energy efficiency programs Boulder has implemented correlate with a slightly reduced trend in overall emissions over the last few years. Yet they pale with the possibilities from reducing emissions by increasing renewable energy as a proportion of Boulder's electricity. This difficult lesson came out of Boulder's experience with climate action programs and led to an increased focus on renewable, and local, energy supply. This process over the last few years has increased the energy literacy of both city staff and the Boulder community. Indeed, much of it has come out of the significant efforts of community volunteers who have offered their time and expertise to inform Boulder's decision-making.

4.1   The Franchise: An Opportunity to Forge a New Partnership

Xcel Energy (Xcel) was formed in 2000, by way of an extensive series of mergers that created one of the nation's largest investor-owned utilities. Colorado has a traditional, vertically integrated utility structure, with a public utilities commission (PUC) that regulates two primary investor-owned utilities: Public Service Company of Colorado, an Xcel subsidiary, and Black Hills Energy.

Like many cities in Colorado, Boulder has had a series of 20-year franchise agreements with Xcel and its precursors in interest. Boulder has considered forming a municipally owned and operated electric utility since 1897 for cost and reliability reasons, but generally rejected the idea. The Colorado Association of Municipal Utilities (2013) noted that there are 29 municipal utilities in Colorado, most formed in the early twentieth century, and the Colorado Constitution grants municipalities significant authority to that effect.

In 2005, with the 1990-2010 franchise agreement drawing to a close, Boulder once again began exploring forming a municipal utility more specifically by analyzing a series of financial and operational issues. This time, however, the key driver related to having greater input in energy decisions and emissions from Xcel's resource portfolio. In 2008, to dissuade Boulder from pursuing municipalization, Xcel offered Boulder the opportunity to become the nation's first SmartGridCity™—an idea that initially excited the community by offering possibilities to innovate through increased local solar, electric vehicles, smart customer-facing technologies, and grid intelligence to ensure greater reliability. The outcome of SmartGridCity™ has not been wholly negative, as Xcel has gained increased fluency with grid-facing modernizations like distribution voltage optimization. However, the conceptual promise never materialized for customers.[4]

In 2010, the city was faced with a choice: to renew its franchise or to develop an alternative. Extensive negotiations over the years had attempted to find common ground between the parties while providing Boulder with more green power and a path to shared decision-making about issues like reliability, carbon, data, and programs. While Xcel rejected city requests to incorporate new provisions into the franchise agreement, the company offered Boulder the option of signing a new franchise by which Xcel would allocate to the city the output of a new wind farm it was developing, Limon II. Tying the wind deal to the franchise was unacceptable to much of Boulder's elected officials and vocal community activists—as evidenced by experience with smart grid infrastructure in Boulder, the utility industry was on the verge of significant changes. In lieu of a franchise, Boulder voters approved a 5-year "utility occupation tax" that would take the place of a franchise fee and potentially fund municipalization.[5] This tax may

---

4. In 2013, the Colorado Public Utilities Commission denied Xcel the remaining one-third of their cost contribution to SmartGridCity™, citing in large part the lack of a strategic vision for fulfilling the customer-facing benefits that were initially promised. Decision No. C13-0436 can be found in Proceeding No. 11A-1001E, at https://www.dora.state.co.us/pls/efi/EFI.Show_Docket?p_session_id=&p_docket_id=11A-1001E.

5. Utilities often pay a fee to local governments as a condition of having a franchise to provide exclusive electric service. By the end of its last 20-year franchise agreement with Xcel, Boulder received about $4 million annually that funded core city services; this arrangement provides a strong incentive for local governments to enter into franchises. Because Xcel retains the obligation to serve, the "utility occupation tax" was developed to provide similar funding for city services in the absence of a franchise agreement. Both the tax and the fee are treated as direct pass-throughs to Boulder utility customers. The utility occupation tax was approved in 2010 and subsequently increased in 2011 to fund the municipalization process.

PART | III   What Future?

be an important tool for communities looking to avoid long-term agreements with their power providers.

Thus, by 2011, Boulder was embarking on a renewed effort to change its electricity supply. This process has been guided by the community's commitment to energy localization.

4.2   Localization and the Energy Future Goals

It has often been asked what the Boulder community is trying to achieve by postponing a franchise agreement with Xcel. Xcel has developed significant expertise in managing large amounts of wind power, it maintains lower electricity rates than the national average, and it has implemented a quality of service plan in response to reliability concerns in the early 2000s.

Yet the changing global climate may demand more aggressive, and immediate, actions than Xcel and other regulated utilities are willing or able to take at this time. Indeed, the traditional utility business model must transform to address this seminal issue in addition to bedrocks like safety and reliability. Ultimately, Boulder's Energy Future Project is about looking at the options the community has for transitioning its electricity supply while maintaining Boulder as a desirable place to live and work. Thus, both the climate imperative and the ability for local decision-making to translate into local economic vitality motivate this process. Additionally, while Boulder has the means to investigate these opportunities, a core aspect of the process is creating something replicable for other communities.

In 2010 and 2011, city staff worked with community volunteers and stakeholders, reached out to local businesses, and talked with the city's Environmental Advisory Board and the City Council to understand what people want from their energy supply. The following key themes arose: Electricity should be renewable; it should be affordable; and it should be highly reliable. Boulder has an economically diverse community, with manufacturers, researchers, and affordable housing constraints, and an environmental ethic, so none of these themes were a surprise.

What did emerge as a new theme, however, was the idea of more localized energy as a means of local empowerment. Boulder's history of buying local, from community gardens to boutique shops, began to inspire the community's discussions about energy as well. Clean, locally sourced energy—with services developed in collaboration with community stakeholders—emerged as a key way to differentiate Boulder and enhance economic vitality. Burger et al. similarly examines how local energy generation in German villages was a source of self-sufficiency and empowerment for residents that was just as critical an issue as affordability. Boulder began to look at how a local utility could provide services that maintain core values like affordability and reliability while enhancing the community's access to clean, local energy.

In 2011, Boulder hired Local Power Inc., to examine the community's ability to source energy locally. Local Power Inc. (2011) presented a "localization portfolio standard," akin to Colorado's Renewable Energy Standard,[6] which concluded that Boulder could meet 30% of its electricity needs locally. Two-thirds would be met through efficiency, primarily an innovative program that would retrofit all buildings in Boulder with smart technologies, and one-third would be met through renewable energy, including hydroelectricity, solar gardens, small-scale wind, and waste to energy developed in Boulder and Boulder County.

The localization report offered a first glance at how much could be done within Boulder to manage the community's energy demand and supply. However, the report also articulated the limitations that Boulder faces in implementing such an effort because changes to utility operations, including system upgrades, and even some customer-facing changes, such as microgrid development, would require ownership of the system or collaboration with Xcel.

As the localization report was being developed, the City Council approved six "energy future goals," with an overall goal of ensuring that Boulder residents, businesses, and institutions have access to reliable power that is increasingly clean and competitively priced. Boulder (2011a) presented these goals:

1. Ensure a stable, safe, and reliable energy supply.
2. Ensure competitive rates, balancing short-term and long-term interests.
3. Significantly reduce carbon emissions to improve environmental quality.
4. Provide Boulder customers with a greater say about their energy supply.
5. Promote local economic vitality.
6. Promote social and environmental justice.

Other chapters, such as Chapter 21, articulate similar considerations. As these goals suggest, access to energy is key to human health, safety, security, and well-being. This access can impact businesses' bottom lines and provide opportunities for them to differentiate their products. And energy that is clean and renewable is a crucial step in the transition to a low-carbon economy.

Importantly, the goals themselves are agnostic in that the community has recognized that there may be multiple pathways to accomplish them. The city's role is in identifying opportunities to collaborate with the Boulder community to achieve those goals. This could occur through multiple actions: legislative changes (such as drafting the state's first solar gardens bill in 2010), an innovative new franchise agreement (a partnership for utility performance rather than a contract for rights-of-way), or a locally owned and operated city utility.

---

6. Colorado's Renewable Energy Standard requires investor-owned utilities, and certain municipal and cooperative utilities, to meet percentages of retail sales with renewable resources. Investor-owned utilities like Xcel must have 30% renewable energy by 2020 (however, there is a multiplier for in-state renewable energy). The Database of State Incentives for Renewables and Efficiency (2013) provided a summary.

Boulder is pursuing all of these options at this time. However, without significant changes to the energy industry, thinking about how a utility could or should operate in pursuit of these goals is the logical next step.

5   WHAT'S "BROKEN" ABOUT THE CURRENT MODEL

There are well-documented challenges inherent in the traditional, regulated utility model, as articulated in Chapters 11 and 14, and others that cite the recent study for the Edison Electric Institute on "disruptive challenges" to the utility industry (Kind, 2013). Energy continues to be treated as a commodity with incentives to sell more and own large, central station generation assets; utilities frequently lack transparency; they are slow-moving entities; they require a return on investment as part of the regulatory compact for providing a public good; they face difficulties with embracing renewables and new technologies because of their large infrastructure investments. Despite Xcel's remarkable advances in integrating wind energy into its portfolio, its Colorado service territory produces one of the dirtiest energy supplies in the country, with over 50% coal generation according to Xcel Energy (2013). Its baseload includes the Comanche 3 plant, which began operation in 2010 and has a life span into the 2060s.

Simultaneously, the utility industry is changing. Several chapters—Chapters 5, 9, and 16, in particular—note the increasing need for grid intelligence and advanced data analytics to understand and manage distributed solar, microgrids, energy storage, and electric vehicles. Futurist thinkers might point to even more dramatic changes that could be on the horizon: space-based solar power, thorium reactors, and 3D printers replacing equipment inventories. The prices of renewable resources are quickly dropping, according to Wiser and Bolinger (2013), as Xcel's recent electric resource planning process showed; the Colorado PUC approved a preferred, low-cost resource plan that included 450 MW of wind and 170 MW of utility-scale solar.[7]

Customers are developing widely varying expectations and providing a high quality of service and a unique customer experience will be key to managing diverse changes. Yet regulated utilities are typically required to practice a "most favored nations" policy[8] by which they must treat all customers alike, regardless of local conditions that may support an alternative approach.

Unquestionably, to weather these coming changes, utilities will need to demonstrate foresight, flexibility, and responsiveness. The Boulder community

---

7. Xcel's 2011 Electric Resource Plan is Colorado PUC Proceeding No. 11A-869E; documentation can be found at https://www.dora.state.co.us/pls/efi/EFI.Show_Filing?p_fil=G_185369&p_session_id=.

8. A reference to the international political term that describes bilateral trade relationships between nations that treat each other nondiscriminatorily.

began to ask: What would a utility that exhibits those traits look like? And if not Xcel, could a city-owned entity be the catalyst for this change?

6   ENVISIONING THE ELECTRIC UTILITY OF THE FUTURE

With that history, Boulder has begun developing a more detailed and concrete vision of how a future "Boulder utility"[9] could operate. This vision is not fully formed; ratemaking issues have not been analyzed, nor has a full suite of energy services been developed. However, it incorporates many concepts that have long been implemented piecemeal by different power-supplying entities nationally and internationally; Attachment E of Boulder (2013b) references numerous progressive practices. These ideas have come about through research into those best practices, by looking at pilots and new ideas coming from local electric utilities, cooperatives, and investor-owned utilities and from leading-edge minds in the Boulder community and elsewhere. As described by Thaler and Jimison and Hanser and van Horn, changing the utility business model is key.

The underpinning of Boulder's vision of this future utility is a city known for energy innovation

- where individuals and businesses can easily select from an array of energy services that meet their needs, even from a smartphone;
- where customers have access to detailed information about how they use electricity and how much it costs, helping inform their choice of services;
- where clean, local energy helps businesses thrive, with time of use being more important than consumption;
- where a network of electric vehicle-charging stations can be used as backup storage;
- where community solar and distributed generation abound and are affordable to all customers;
- where microgrids help provide areas of high reliability and resiliency to natural disasters; and
- where the utility facilitates these services as a partner in the transition to a low-carbon economy.

These changes are taking root nationally and internationally, as this volume emphasizes—Boulder desires to bring them together in one place.

Ultimately, moving toward an electric utility of the future is about developing a mechanism to help the Boulder community meet its overall goals related to a sustainable and economically beneficial energy supply. This section describes two approaches Boulder is taking to developing this vision: looking at how electric consumers want to engage with their energy supply and determining how to meet those needs in ways that sync with Boulder's environmental ethic.

---

9. For the following discussion, "Boulder utility" is used to mean the electric utility of the future implemented in Boulder, not specifically a municipally run utility.

6.1   Engaging the Customer and the Community

Utilities and customers have traditionally had a buyer-seller relationship. Blansfield and Jones describe how the "prosumer" is challenging the old belief that customers rarely think about where their electricity comes from provided that the lights come on and the beer stays cold. Rate increases, significant outages, and heat waves make it difficult for taxpayers to ignore the consequences of fossil fuel consumption and deferred grid maintenance. Local governments may be called upon to address the resulting harms to health, public safety, and local economies. These diverse consequences suggest interacting with customers based on their unique needs and, increasingly, based on local conditions. In their current form, large multistate utilities simply may not have the ability or the motivation to address specific local conditions and to consider the externalities that negatively impact local communities. Interstate Renewable Energy Council and Rabago (2013) echoed this sentiment when they recommended incorporating environmental and societal benefits in valuing distributed solar generation because of "the reality that society is made up of utility consumers."

Electricity customers' changing expectations are also based on changes to other types of markets and services. Anderson (2004) articulated a concept that has increasingly caught on to characterize the evolution of markets: the long tail. Anderson foresaw that instead of meeting all of customers' needs with a few broad-appeal products, fields like entertainment would begin appealing to niche markets instead. The same concept may be applied to energy as renewable energy becomes more diverse and technology offers customers greater opportunities for energy management.

A Boulder utility, with its local focus, could offer customization to meet its customers' diverse needs. Regulated utilities are broadly limited to treating customers alike, outside of special (generally industrial) situations. The traditional service model relied on engineers to design a limited number of universal services without gathering customer input. Yet increasingly, most energy customers are in unique situations: They rent, they are beneath the poverty line, they have home medical equipment, and they run tech start-ups. Some customers want to manage their energy use actively and others do not. At the same time, in Boulder, some companies have found that renewable energy is part of branding a socially responsible business.

Customization is part of the vision of a services-oriented utility, as distinct from the current model of a commodity-oriented utility. Anderson (1999) described how Interface Flooring was at the forefront of a similar industry transformation when it began selling the service of high-quality carpet, complete with closed-loop manufacturing and recycling, rather than the carpet itself. Instead of kilowatt-hour, a Boulder utility could sell highly reliable, clean, and affordable electricity sufficient to meet customer's needs that are large or small. Increasingly, industry commentators analogize this approach to that of smartphone applications, where the utility provides the framework that facilitates customers' access to a wide variety of diverse services.

A goal of a Boulder utility should be to help customers transition to arrangements that meet their energy and reliability needs efficiently and effectively and help them weather a changing industry. In some cases, this may mean transitioning customers into independent, self-sustaining systems. Currently, large industrial users must develop their own backup generation or pay hefty premiums for the utility to provide the service. The consequences of minor fluctuations in power supply or quality could be the loss of millions of dollars due to interruptions in a pharmaceutical process or microchip manufacturing. Looking to industrial ecology examples like Kalundborg,[10] a public-private partnership and "Smart City," a Boulder utility might facilitate microgrids that enable industrial customers to share backup generation and energy storage. For residential customers, it might offer proactive energy management programs, such as enabling them to opt-in to receive a notification when a water heater or home medical device malfunctions. Hauser and Crandall (2012) noted that differentiated reliability is likely the first foray into offering "long tail" customer services.

A key factor differentiating a Boulder utility from the traditional mold is its basis in an engaged and educated community. A Boulder utility will need to value the diverse opinions and expertise in the community and collaborate with residents and businesses to meet their needs—and address their environmental concerns—rather than simply deliver services. The utility of the future will be expected to cultivate relationships and partnerships as much as to provide safe and high-quality service.

6.2   Delivering Local, Renewable Energy

Local, renewable energy will be key to a Boulder utility, as demonstrated by the community's long commitment to environmental sustainability and climate action. Boulder desires to maximize the amount of renewable energy in its resource supply, as quickly as is affordable. At the same time, the City Council and the community have expressed a preference that renewable energy be generated and used locally rather than deemed "green" based on renewable energy credit purchases. The implementation of this vision may be challenging but is the opportunity for learning.

Boulder can look for inspiration to European examples like the Danish island of Samsø, which is now 100% powered by renewable energy, including a large amount of local solar and biomass. Malmö, Sweden, integrates diverse renewable resources including combined heat and power, biogas, waste energy, solar, and ground-source heat pumps. However, there are examples closer to home: Aspen, Colorado, has a significant hydroelectric supply and a goal of

---

10. For more information, visit http://www.symbiosis.dk/en and http://www.smartcitykalundborg.dk/.

being 100% renewable by 2015; Palo Alto, California, has similar hydroelectric resources and has chosen to source all the town's electricity needs from renewable sources; and Oak Park, Illinois, chose an all-green power portfolio through community choice aggregation.[11]

Transitioning to a localized, decarbonized energy mix will be challenging. As other chapters note, renewable resources currently need to be balanced with backup generation, generally natural gas, or storage.[12] Additionally, there are challenges to increasing renewable energy based on regulatory and relationship constraints, such as centralized dispatch that leads to wind curtailments rather than reducing generation from coal baseload.[13] The technological and capacity limitations of the grid are still being evaluated.

However, technological changes and more granular data are increasingly improving the effectiveness of decentralized resource management. Diesendorf (2007) pointed to a future where renewable resources with diverse generation profiles can be carefully paired to meet demand. Miller et al. suggest that utilities might incentivize customers to install energy management technologies—for example, smart inverters—that allow them to provide ancillary grid services. Undoubtedly, the need for highly reliable power will increase as society becomes more technologically advanced, and new technologies will be needed regardless of whether cleaner, but more variable, energy resources are incorporated.

Given the fact that Boulder has over 300 days of sunshine annually, distributed local solar will play a vital role in a localized energy portfolio. Finding ways to properly value and encourage distributed solar will be critical to Boulder's efforts. According to the Rocky Mountain Institute (2013), there are significant disparities in how utilities attribute the costs and benefits of distributed solar. Numerous chapters acknowledge the disruption local solar creates for traditional utilities, although Mountain and Szuster challenge the prevailing view that solar erodes utilities' ability to recover the costs of managing the distribution grid. The Interstate Renewable Energy Council and Rabago (2013) offered recommendations on how to create a standardized and transparent methodology for valuing distributed solar generation—one example Boulder is looking to in developing a utility centered around local energy.

Thaler and Jimison credit the resiliency benefits of distributed solar in emergency situations, such as the recovery from Hurricane Sandy, but they also note that microgrids are becoming increasingly commercially feasible (see also

---

11. Although these US examples allow RECs to be used to meet renewable energy goals, these communities help signify the increasing value placed on renewable energy.
12. At this time, Boulder anticipates natural gas playing a role in its transition to decarbonization under either a municipal utility or a partnership with Xcel Energy; this continues to be a concern as Boulder voters recently approved a temporary moratorium on hydraulic fracturing within city limits.
13. Xcel Energy adds the costs of coal cycling onto wind contracts when planning for resources. This approach continues to enshrine fossil-fueled baseload as the long-term vision for electricity supply.

LaMonica, 2012). Microgrids may provide greater opportunities for effective coordination of local renewable resources. Boulder residents and businesses have indicated interest in producing, and sharing, renewable electricity across microgrids to create zero-carbon districts. This concept pairs with the broader sustainability planning initiative of EcoDistricts, of which Boulder is a founding member.[14] EcoDistricts allow for a more holistic approach to community planning by looking at performance areas beyond energy to include water, waste, access and mobility, health and well-being, human behavior, and ecosystem function.

To continue fleshing out these concepts, Boulder (2013e) laid out two key steps. First, a city staff team has formed working groups of community experts to examine the role of local solar in the utility of the future and to identify and encourage best practices related to natural gas extraction and use. Second, Boulder is identifying a series of energy services pilots—such as a new partnership with Pecan Street Inc.—that will examine ways to increase local energy efficiency and renewable energy. These efforts will provide valuable information to help develop the appropriate business model for a Boulder utility, whether a municipal entity or a new partnership. Information on these efforts is available at www.BoulderEnergyFuture.com.

7   WHERE BOULDER IS NOW

At the time of publication, Boulder is pursuing two paths. Either has the potential to be historically unique, and hopefully influential, because of the overarching focus on environmental and climate concerns and the recognition of the importance of local decision-making.

First, the city is engaged in discussions with Xcel regarding opportunities to develop new products and services that would be provided not only to Boulder but also to the company's entire Colorado—and perhaps even its eight-state—service territory. The city released a policy paper, Boulder (2012), proposing a series of innovative ways that Boulder and Xcel could collaborate to implement energy localization concepts: for example, forming a renewable energy-focused Xcel-Boulder subsidiary, developing a cleantech incubator, and enhancing local investments in distributed energy and energy efficiency. These discussions are occurring with a group of 12 community leaders who have contributed numerous innovative ideas and recommendations. This process began in April 2013, and although it has not yet borne fruit in terms of a realized proposal from Xcel, it may yet provide ways to achieve the Boulder community's goals.

Second, Boulder is continuing to pursue municipalization of the local electric system. Feasibility modeling laid out in Boulder (2013a) and (2013b) demonstrated that there are opportunities to meet the conditions that voters approved in the City Charter as prerequisites for forming a municipal utility. Among these are the following: the utility must maintain a certain level of financial margin,

---

14. For more information, visit http://ecodistricts.org/.

increase renewable energy and decrease GHGs compared to Xcel, ensure equal or greater reliability, and be able to charge rates that do not exceed Xcel's at the time of acquisition.[15] Indeed, Boulder's modeling showed that it could form a local electric utility that met the conditions of the City Charter and could double, to 50% or more, the renewable energy projected to be available on Xcel's system. This effort was conducted with the participation of over 75 community volunteers, many with deep subject-matter expertise.[16]

In August 2013, the City Council approved Boulder moving forward to acquire Xcel's property within the service territory described in Boulder (2013d). In November 2013, Boulder voters once again affirmed their commitment to moving forward while approving a limit on the amount the city can pay to acquire the system that supplements the preexisting City Charter requirements. Boulder is thus preparing for legal processes and developing a work plan to transition smoothly to utility operation.

8   CONCLUSION

This chapter described how Boulder's unique legacy of sustainability and conservation has led to a process for reenvisioning the relationship between an energy utility and a community. The utility of the future will expand the traditional obligation to provide safe, reliable, and affordable energy to include other interests, such as environmental stewardship and economic vitality. It can do this best by recognizing customers' diverse attributes and providing a forum for collaboration with those customers through the development of energy services—a very different approach from the traditional buyer-seller relationship. Moreover, for Boulder, the utility of the future will increasingly decarbonize its energy supply, relying as much as possible on local energy resources, to address the climate change imperative.

Boulder's utility of the future vision is still being developed, and significant research must be performed and vetted with the public collaboratively to develop a transformative business model. It is Boulder's hope that the lessons being learned here can be used by other communities on the path to a prosperous, sustainable, and secure energy future.

ACKNOWLEDGMENTS

The authors would like to acknowledge the city of Boulder's Energy Future Project team for their contributions to this process, in particular recognizing Sarah Huntley and Debra Kalish for their assistance with this chapter.

---

15. The relevant section of the City Charter is available at http://www.colocode.com/boulder2/charter_articleXIII.htm.
16. The findings were also verified by an independent third-party evaluator, at Boulder (2013d).

REFERENCES

Anderson, R.C., 1999. Mid-Course Correction: Toward a Sustainable Enterprise: The Interface Model. Peregrinzilla Press, Atlanta, GA.

Anderson, C., 2004. The long tail. WIRED Magazine 12.10. Available at, http://www.wired.com/wired/archive/12.10/tail.html (Oct. 2004).

Boulder City Council, May 2002. Resolution 906: A Resolution of the City Council of the City of Boulder Establishing a Policy to Take Cost-Effective Actions that Benefit the Community by Reducing Local Greenhouse Gas Emissions. Available at, https://www-static.bouldercolorado.gov/docs/City_Bldr_Resolution_906-1-201307081512.pdf (May 2002).

Brennan, C., 2013. Boulder Scientists Have Large Voice in Climate Document. Boulder Daily Camera. Available at, http://www.dailycamera.com/news/boulder/ci_24192727/boulder-scientists-have-large-voice-climate-document (Sept. 27, 2013).

Brouillard, C., Van Pelt, S., 2007. A Community Takes Charge: Boulder's Carbon Tax. Available at, https://www-static.bouldercolorado.gov/docs/community-takes-charge-boulders-carbon-tax-1-201305081136.pdf.

City of Boulder, 2011a. Boulder's Energy Future: DRAFT Purpose, Framework, Goals and Objectives. Available at, https://www-static.bouldercolorado.gov/docs/DRAFT_Purpose_Framework_Goals_Objs-1-201306061222.pdf.

City of Boulder, 2011b. 2010/2011 Community Guide to Boulder's Climate Action Plan. Available at, https://www-static.bouldercolorado.gov/docs/2010-2011-community-guide-to-boulders-climate-action-plan-1-201305081156.pdf (Sept. 2011).

City of Boulder, 2012. Roundtable Discussion on Exploring Alternative Opportunities for Reaching Boulder's Energy Future Goals. Available at, https://www-static.bouldercolorado.gov/docs/Dec_2012_Options_to_work_with_Xcel-1-201306061248.pdf (Dec. 6, 2012).

City of Boulder, 2013a. Study Session: Boulder's Energy Future Municipalization Exploration. Available at, https://www-static.bouldercolorado.gov/docs/BEF_SS_Feb26_2013_Final_Packet-1-201306201201.pdf (Feb. 26, 2013).

City of Boulder, 2013b. Study Session Update on Boulder's Energy Future Municipalization Exploration Project. Available at, https://www-static.bouldercolorado.gov/docs/Energy_Future_SS_Memo_07232013-1-201307241011.pdf (Jul. 23, 2013).

City of Boulder, 2013c. Boulder City Council Study Session: Boulder's Climate Commitment, Energy Efficiency Programs and Market Innovation Updates. Available at, https://www-static.bouldercolorado.gov/docs/July_30_FINAL_Study_Session_Packet-1-201307240903.pdf, (Jul. 30, 2013), Att. B & D.

City of Boulder, 2013d. City Council Agenda Items 3A & 3B. Available at, https://documents.bouldercolorado.gov/WebLink8/0/doc/121433/Electronic.aspx (Jul. 24, 2013).

City of Boulder, 2013e. Boulder City Council Study Session: Update on Boulder's Energy Future Municipalization Exploration Project. Available at, https://www-static.bouldercolorado.gov/docs/Dec17CCMemo-1-201312121106.pdf (Dec. 17, 2013).

Colorado Association of Municipal Utilities (CAMU), 2013. Public Power in Colorado. Available at, http://coloradopublicpower.org/Public-Power-in-Colorado/public-power-in-colorado.html, last visited Dec. 19.

Diesendorf, M., 2007. The Base-Load Fallacy. Available at, http://www.ceem.unsw.edu.au/sites/default/files/uploads/publications/MarkBaseloadFallacyANZSEE.pdf.

Database of State Incentives for Renewables and Efficiency (DSIRE), 2013. Colorado Renewable Energy Standard. Available at, http://www.dsireusa.org/incentives/incentive.cfm?Incentive_Code=CO24R, last review Jun. 25.

Fenn, P., et al., 2011. Boulder's Energy Future: Localization Portfolio Standard: Electricity and Natural Gas. Local Power, Inc., Oakland, CA. Available at, https://www-static.bouldercolorado. gov/docs/LPI_BoulderLPS_elecandgas_13July2011-1-201306171446.pdf.

Hauser, S.G., Crandall, K., 2012. Smart grid is a lot more than just 'Technology'. In: Sioshansi, F.P. (Ed.), Smart Grid: Integrating Renewable, Distributed, & Efficient Energy. Academic Press, New York.

Interstate Renewable Energy Council, Rabago, K., 2013. A Regulator's Guidebook: Calculating the Benefits and Costs of Distributed Solar Generation. Interstate Renewable Energy Council, New York. Available at, http://www.irecusa.org/wp-content/uploads/2013/10/IREC_Rabago_ Regulators-Guidebook-to-Assessing-Benefits-and-Costs-of-DSG.pdf.

Kind, P., 2013. Disruptive Challenges: Financial Implications and Strategic Responses to a Changing Retail Electric Business. Edison Electric Institute (EEI), Washington, DC. Available at, http://www.eei.org/ourissues/finance/Documents/disruptivechallenges.pdf.

LaMonica, M., 2012. Microgrids keep power flowing through sandy outages. MIT Technology Review. Available at, http://www.technologyreview.com/view/507106/microgrids-keep-power-flowing-through-sandy-outages/ (Nov. 7, 2012).

Rocky Mountain Institute (RMI), 2013. A Review of Solar PV Benefit & Cost Studies, second ed. Rocky Mountain Institute, Boulder, CO. Available at, http://www.rmi.org/elab_empower (Sept. 2013).

Wiser, R., Bolinger, M., 2013. 2012 Wind Technologies Market Report. Lawrence Berkeley National Laboratory (LBNL), Berkeley, CA. Available at, http://www1.eere.energy.gov/ wind/pdfs/2012_wind_technologies_market_report.pdf (Aug. 2013).

Xcel Energy, 2013. Corporate Responsibility Report for 2012. Xcel Energy, Minneapolis, MN. https://www.xcelenergy.com/staticfiles/xe/Corporate/CRR2012/index.html.

Chapter 23

Perfect Storm or Perfect Opportunity? Future Scenarios for the Electricity Sector

Jenny Riesz, Magnus Hindsberger, Joel Gilmore and Chris Riedy

ABSTRACT

This chapter applies futures thinking to explore possible scenarios that electric utilities may face in the coming decades. The chapter applies a top-down approach to identify the key drivers that could influence business models. It describes three possible futures in detail. Firstly, the "centralized" future moves toward decarbonization but retains the centralized model present in most power systems today. In contrast, the "decentralized" future moves toward greater decentralization while retaining a significant role for the grid. The "disconnected" future moves to complete decentralization, with most customers disconnecting from the grid entirely. The chapter concludes that all three scenarios are possible and will have important implications for electric utilities. Wise businesses will adopt a risk management approach.

Keywords: Future scenarios, Futures thinking, Decentralized, Disconnected, Electricity supply industry

1 INTRODUCTION

Thinking about the future rigorously and creatively is hard. When we do think about the future, it is common to imagine futures that are fundamentally similar to the present. Alternatively, it can be easy to underestimate the inertia of the present and imagine rapid revolution when sedate evolution is more likely.

It appears clear, however, that change is coming in the electric supply industry. Change itself is not new, but the rate of change has increased, as has the number of concurrent changes, as discussed throughout this book.

Around the developed world, growth in utility-delivered electricity has declined and even appears to have reversed in some places (Sioshansi, 2013). This has been caused by a combination of slowing economic growth, customers responding to price increases, investments in energy efficiency, and—the key topic of this book—increased levels of distributed energy resources (DER). In addition

Distributed Generation and its Implications for the Utility Industry. http://dx.doi.org/10.1016/B978-0-12-800240-7.00023-0

to this, we observe a growing focus on reducing greenhouse emissions, affecting supply-side investments. Furthermore, developments in information technology, "smart grids," and new ways for consumers to more directly engage in their electricity supply present new opportunities.

These changes present new challenges for electricity utilities. For example, traditional generation companies providing centralized power from emission intensive assets are likely to see revenues increasingly threatened by declining energy sales and growing competition from renewable generation. For some regions, these challenges remain in the future, while for others, they are a present reality, as highlighted in the chapter by Burger and Weinmann on Germany.

Some have termed this "the perfect storm" for the electricity supply industry (ESI), although the severity will vary across countries and business areas. Some businesses are already well into the hurricane and must adapt to the changing environment under difficult circumstances.

With spending on energy services remaining relatively constant in many developed nations, this is a zero-sum game with a first-mover advantage. As is often the case, those who are likely to be most successful in navigating the stormy waters will have the vision to turn the threat into the "perfect opportunity."

Throughout this chapter, the various components of the ESI are referred to by functional area, as illustrated in Figure 1. While each area is discussed independently as in a fully liberalized market, the same observations will also generally hold true for the corresponding parts of a vertically integrated utility.

While other chapters have explored in depth the individual issues, this chapter aims to put them in context by applying "futures thinking" tools holistically to explore possible scenarios that might eventuate for electricity industries around the world.

The chapter consists of four sections in addition to "Introduction." Section 2 uses a "futures triangle" to provide an overview of the drivers that contribute to plausible global futures for the electricity sector. Section 3 describes three possible future scenarios, distinguished by the degree of centralization or decentralization of the power supply. Section 4 explores the possible roles of various

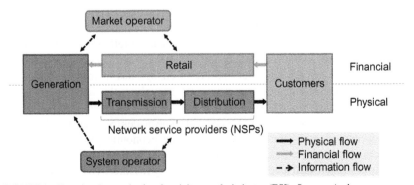

FIGURE 1 Functional areas in the electricity supply industry (ESI). *Source: Authors.*

stakeholders in creating these future scenarios and describes how they might be impacted followed by the chapter's conclusions.

2   PLAUSIBLE GLOBAL FUTURES

This section uses a mapping tool—the futures triangle (Anthony, 2007; Inayatullah, 2008)—to summarize the perspectives, drivers, tensions, and barriers that contribute toward plausible global futures. As shown in Figure 2, the futures triangle draws attention to three forces that shape plausible futures. Table 1 outlines these three forces and the types of questions associated with each. By considering these questions, and how the three forces interact, it is possible to build up a map of plausible futures. The sections later provide a brief futures triangle analysis for the global electricity sector, drawing on insights discussed in earlier chapters.

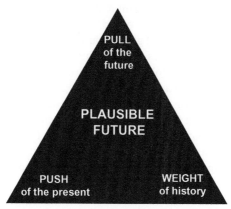

FIGURE 2   The futures triangle. *Source: Based upon (Inayatullah, 2008).*

TABLE 1   Questions in the Futures Triangle

Force	Questions
Push of the present	• What *trends* are pushing us toward particular futures? • What *quantitative drivers* are changing the future? • What disruptive innovations are emerging? (for example, new technologies, globalization, and demographics)
Pull of the future	• What is pulling us toward particular futures? • What are the compelling or aspirational *images* of the future that we would prefer? • Are there competing images of the future?
Weight of history	• What is holding us back or getting in our way? • What are the *barriers* to change? • What are the *deep structures* that resist change?

Source: Authors.

2.1   Push of the Present

There are many drivers "pushing" power system development in particular directions at present, as outlined later.

2.1.1   Electrification

1.2 billion people still do not have access to electricity and 2.8 billion have to rely on wood or other biomass to cook and heat their homes (Banerjee et al., 2013). Electrification, particularly in developing Asia and sub-Saharan Africa, could expand centralized electricity networks or support greater reliance on distributed energy solutions, depending on how it is delivered.

2.1.2   Decarbonization

It is now likely that future development in the electricity sector will be carbon-constrained (Sioshansi, 2009). Most of the world's existing fossil fuel reserves cannot be burned if we are to limit global warming to $<2°C$, increasing the risk of stranded assets for companies with substantial investments in fossil fuels (CTI and Grantham, 2013). Policy responses to climate change are diverse and remain uncertain (e.g., carbon pricing vs. regulatory limits on emissions) but tend to increase the cost of electricity from traditional carbon-intensive sources or smooth the path for DER. Mechanisms such as feed-in tariffs for distributed photovoltaics have been a key driving factor in the uptake of DER in many nations, as discussed in the chapter by Mountain et al. Whether or not future climate change policies favor DER remains an important area of uncertainty.

2.1.3   Energy Security

International fossil fuel markets are uncertain and potentially volatile (Sanyal, 2011), creating a strong driver for nations to seek local energy sources to maintain energy security.

2.1.4   Reliability

Electricity consumers continue to seek reliable electricity supplies, although the level of reliability they require varies greatly. In some countries, such as Australia, higher reliability standards established by regulators have driven substantial investment in electricity networks. This has increased electricity network charges.

2.1.5   Affordability

Policies for decarbonization and reliability, along with other factors, have increased electricity prices in many countries in recent years. Consumer concern about electricity affordability is growing, leading to increased political interest in ways to limit price rises. Maintaining affordability of energy services in a time of substantial transformation for the electricity sector will be an ongoing challenge, and incumbents that fail to maintain affordability may be outcompeted.

2.1.6   Disruptive Technologies and Innovation

Disruptive innovations and technologies redefine the status quo in a market and typically lead to the failure of incumbents and the rise to prominence of new entrant firms (Christensen, 1997). The ESI is in a period when rapid development of technologies for power generation, transmission, distribution, and storage increases the likelihood of disruptive innovation. For example, the cost of photovoltaics has fallen significantly in recent years, allowing more and more consumers to become producers of electricity and threatening business models based on constant growth in centralized supply. Anticipated developments in energy efficiency, electric vehicles, energy storage technologies, and technologies to actively manage demand are likely to make it increasingly feasible to go "off-grid". Any of these technologies could act as disruptive innovations that undermine the current business model of ESI incumbents, as discussed in the chapter by Sioshansi.

2.1.7   Social and Cultural Change

Substantial social and cultural changes are also under way that will shape plausible futures in the electricity sector. In developed countries, demographic trends such as an aging population and reduced household size will impact on electricity consumption patterns. Some developed nations have seen reductions in electricity demand in recent years, undermining existing electricity forecasting, planning, and supply models. In contrast, in the developing world, the electrification and the growth of a more affluent middle class are driving rapid growth in electricity demand.

The rise of the Internet, social networking, social media, and smartphones has created an information society in which consumers are empowered through access to data and peers. An emerging sharing economy (Botsman and Rogers, 2010) is also challenging existing business models by promoting peer-to-peer economic transactions instead of business-consumer transactions. Empowered consumers are demanding more from the technologies, services, and organizations they engage with. In the energy sector, this is driving the emergence of a plethora of devices and apps to allow consumers to take more control of their energy use. It creates a more volatile market, where consumers are more likely to switch away from incumbent suppliers toward alternatives.

As Burger and Weinmann discussed in their chapter on Germany, empowered consumers are attracted by self-sufficiency. Many are concerned about rising electricity prices and would prefer to avoid dependence upon corporate entities and market regulators. Self-generation of electricity, now possible and affordable through technologies such as photovoltaics, offers an attractive way to meet these desires. Scenarios along these lines are also discussed in the chapter by Sioshansi and Weinberg.

2.2   Pull of the Future

The pull of the future refers to compelling, aspirational images of the future that pull us in particular directions. These images are often in competition, as different people value different futures. Inayatullah (2008) identified five generic images of the future, summarized in Table 2. The table explores the implications

TABLE 2 Images of the Future for the Global Electricity Sector

Image of the Future	Description (Inayatullah, 2008)	Potential Electricity Sector Implications
Evolution and progress	• "More technology, man as the centre of the world, and a belief in rationality" (p. 7) Utopian, science fiction future in which humans expand into world-spanning cities and eventually into space	• High-tech energy system using most cost-effective technologies to supply ongoing growth in energy demand • Agnostic about precise technologies as long as they facilitate continued economic progress—possibly favors centralized systems as more likely to meet continuing growth in electricity demand
Collapse	• Humans have overshot the Earth's carrying capacity, eventually leading to the collapse of civilization • Dystopian future, which is a staple of popular culture	• The possibility of collapse in response to climate change and other global challenges has been thoroughly explored and is certainly plausible (e.g., Slaughter, 2010)
Gaia	• "The world is a garden, cultures are its flowers, we need social technologies to repair the damage we have caused to ourselves, to nature and to others, becoming more and more inclusive is what is important. Partnership between women and men, humans and nature and humans and technology is needed" (p. 7) • Potentially high tech, but technologies blend into the background and material consumption plateaus as humans pursue well-being through relationships with nature and each other	• Vision of restorative and regenerative development, in which technologies are more consistent with natural cycles and scales and create space to improve ecological health and natural capital (Birkeland, 2007) • Buildings generate more energy than they consume and put the excess back into the grid—more likely to be distributed and to reject risky technologies like carbon capture and storage and nuclear power in favor of renewable energy • Current proponents of this image of the future include Sustainia and the Living Building Institute

Continued

TABLE 2 Images of the Future for the Global Electricity Sector—cont'd

Image of the Future	Description (Inayatullah, 2008)	Potential Electricity Sector Implications
Globalism	• The free flow of technology and capital brings riches to all and old borders begin to break down • Assumes that emerging global system will be modeled on the Western ideal, allowing little space for alternatives	• Similar to "evolution and progress" mentioned earlier • Technology transfer allows developed and developing countries to share in new energy technologies • Free trade encourages the development of energy resources where there is a competitive advantage (e.g., solar in Australia and northern Africa) • Continent-scale electricity networks that transfer electricity from large-scale solar plants in desert areas
Back to the future	• Humans return to simpler times, abandoning material consumption and disruptive technologies • Romantic vision of the past that conveniently ignores the fact that for much of human history, life was nasty, brutish, and short	• Energy descent, perhaps dominated by small-scale cooperatives operating isolated, distributed energy systems • Centralized technologies would be abandoned in favor of low-tech renewable energy, human labor, and animal labor • Arguably the least likely to eventuate of the five images presented here

Source: Authors.

of each image of the future for the global electricity sector. These five images of the future constitute possibilities or expectations that are held by individuals, who may work actively or passively to bring about or prevent them. As such, contestation between these images of the future shapes plausible futures for the global electricity sector.

2.3 Weight of History

Finally, plausible futures are shaped by the weight of history—the inertia in the system that acts as a barrier to the emergence of alternative futures. The "inertia" in present power systems should not be underestimated. Key sources of inertia are outlined later.

2.3.1   Centralized Electricity Networks

The vast majority of power systems, infrastructure, and associated institutions are currently structured around a centralized system. In particular, most developed nations have invested in large-scale power stations and extensive transmission and distribution grids. These assets have long lifetimes of up to 40 years or more. A move to decentralized power would cause this substantial investment in physical power stations and grids to become stranded. The owners of these assets will work hard to extract as much benefit from this infrastructure as they can, well into the future.

2.3.2   Conservative Electricity Industry

Electricity businesses largely grew out of government electricity authorities that were initially charged with electrification and later with "keeping the lights on" as electricity demand continued to grow. These organizations traditionally had systems and procedures that changed slowly and low tolerance for risk. While this is changing, many electricity businesses have a cultural legacy of conservatism that makes them slow to adapt.

Associated institutions were designed to support a centralized system and can create unnecessary barriers to emerging distributed systems. For example, many electricity markets do not have appropriate mechanisms and incentives in place to allow consumers to participate actively. This may inhibit uptake of demand-side participation and DER or, in some cases, provide perverse incentives that encourage detrimental behavior. These mechanisms take time to adapt, as explored in more depth in the chapters by Kristov and Hou and Sood and Blanckenberg. Further, in many systems, the tools and methodologies used for system planning and operation may not be able to readily accommodate new technologies. This can tend to prefer future scenarios that look similar to the present, since planning models cannot predict futures that are outside of their capability.

2.4   Summary: Plausible Futures

When the push of the present, pull of the future, and weight of history are considered together, some key tensions emerge that will shape plausible futures. Primary among these are the pace and type of decarbonization and the balance between centralized and distributed electricity systems. If decarbonization does not proceed rapidly, then dystopian images of the future become a real possibility. However, like Paul Gilding (2011), this chapter proposes that humanity will eventually get its act together and move down a decarbonization pathway. The rest of this chapter assumes decarbonization will take place.

The type of decarbonization remains uncertain, but renewable energy appears to have some important advantages at present. Given the opposition to nuclear power in many countries and the slow development of carbon capture and storage technologies, renewable generation is likely to play a major role in many nations. A range of significant studies have recently demonstrated that systems with very high renewable proportions are technically feasible and economically competitive (AEMO, 2013; NREL, 2012; Riesz et al., 2013). The "push" of climate change, energy security, and new lower cost renewable technologies and the consistency of renewable energy with several of the key images of the future are likely to make a renewable energy future a reality. Demand management and storage technologies could be additional key enablers.

Less clear, however, are the ultimate outcomes of the tension between centralized and decentralized power systems. These alternatives are associated with quite different images of the future. The weight of the existing centralized electricity network provides a great deal of inertia, and incumbents will try to resist stranding of existing centralized assets. However, there is a lot of push behind distributed energy, which can facilitate decarbonization and empower consumers to take control of their energy use. This tension forms the basis for the three scenarios outlined in the next section and explored in more detail in the rest of this chapter.

3 FUTURE SCENARIOS

Since introduced by Shell in the 1970s for strategic business planning (Schwartz, 1991), scenarios have been widely used in the energy sector. Scenario planning provides a method for analyzing, discussing, and communicating challenges the industry might face. It can serve to make the target audience aware of critical factors that, if ignored, could significantly disrupt the energy supply (whether that audience is the company executive, national policy makers, or the global community).

This section introduces three scenarios, with the purpose of discussing the implications for the ESI in the developed world. The scenarios under consideration are illustrated in Figure 3. The scenarios have been selected with a particular focus on understanding the possible impacts of the growth in DER. As discussed in the previous section, a move to a low-carbon power supply is assumed to be inevitable, with scenarios that remain in the two left quadrants seen as infeasible in the long run. Thus, the key differentiator between scenarios is the share of DER in each, as represented by their locations along the vertical axis.

The three scenarios centralized, decentralized, and disconnected are presented in that order since they introduce increasing degrees of difference to the ESI of most markets today.

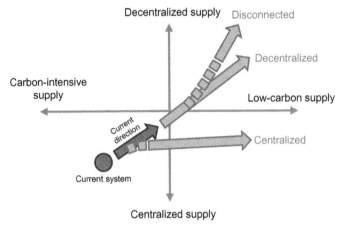

FIGURE 3   Possible electricity system futures. *Source: Authors.*

All scenarios have the predetermined elements of lower-carbon-intensity, new technologies such as smart grids, empowered consumers, demand-side participation, energy service companies, and greater uptake of energy efficiency. Thus, all three scenarios could be considered "smarter and greener" than most power systems at present.

3.1   Centralized Scenario

In the centralized scenario, large-scale centralized generation (whether renewable, nuclear or carbon capture, and storage-based) maintains a large proportion of market share. The current trend toward DER saturates, and generation in the longer term continues to be supplied predominantly from centralized power sources. A large proportion of this is assumed to be from utility-scale installations of renewable technologies such as wind, photovoltaics, and solar thermal. Nuclear power may retain or grow market share in nations where politically accepted, and carbon capture and storage technologies may play a role in the event that the technology matures sufficiently to be commercially available and cost-competitive with renewables.

Overall, this scenario will be the closest to most power systems in operation today, although the ESI will need to ensure the lower-carbon supply system including variable renewable sources that can be operated securely and reliably. DER would remain a relatively minor contributor, and the majority of consumers would continue to source their electricity from the grid via electricity retailers. Network service providers (NSPs), generation companies, retailers, and market operators would retain roles similar to present.

3.2   Decentralized Scenario

The decentralized scenario is a continuation of the current trend toward DER, to a point where the majority of energy is supplied locally. Consumers become increasingly engaged in their electricity supply, generating meaningful quantities of energy at the site of consumption and using energy efficiency and demand-side management to tailor their costs and electricity services to their preferences.

In this scenario, the vast majority of consumers remain grid-connected, but they purchase much smaller quantities of energy from the grid than at present. The grid is used primarily for "balancing and backup," rather than as the primary source of electricity.

Microgrids may evolve to support local balancing of supply and demand. Small local networks may assist with balancing both energy and ancillary services via the traditional grid, as envisioned by van Overbeeke and Roberts (van Overbeeke and Roberts, 2002).

In the decentralized scenario, in addition to managing a lower-carbon supply, the industry will also face the challenges of operating a dispersed system with many actors. This will be combined with a significant reduction in the volumes of centrally supplied electricity. This will have important consequences for the ESI as discussed further in Section 4 and in the chapter by Kristov and Hou.

3.3   Disconnected Scenario

The disconnected scenario is a more extreme DER scenario. In this scenario, a significant proportion of consumers elect to entirely remove their load from the grid and become largely self-sufficient. Completely self-contained home generation through technologies such as rooftop photovoltaics combined with home energy storage systems becomes commonplace. Other customers may be supplied through independent microgrids supplying the local community through shared, localized generation sources and storage.

The key driver for this scenario would be a reduction in enabling costs, such as the cost of storage solutions, making a compelling business case for individuals or local communities. Furthermore, consumers in this scenario would need to be prepared to take a much more active role in organizing and optimizing their energy needs.

This scenario would have dramatic consequences for the ESI. Electricity retailers and centralized generation companies would lose the majority of their market share, and other utilities may have a significantly reduced role. Companies that survive this transition will be those that innovate and discover new business models, as discussed further in Section 4.

3.4   Driving Factors

The driving factors that could lead to each of these three scenarios are outlined in Table 3.

The centralized scenario could eventuate if grid-connected electricity prices remain relatively low compared with DER alternatives and will be further facilitated if policy makers implement mechanisms that favor investment in utility-scale generation. This could include market settings that prevent DER and demand response from providing ancillary services essential for operating a system with a large penetration of variable generation. It would also be facilitated by a low degree of consumer engagement in electricity issues or possibly by a relatively higher degree of consumer trust that the government and corporate entities involved will continue to provide cost-effective, reliable, safe, sustainable, and secure electricity for the foreseeable future.

If DER becomes a cost-effective alternative to grid-connected electricity, a move to the decentralized scenario may be observed. This could be assisted by government policy that supports the development of DER.

The development of low-cost storage, such that sole reliance upon DER becomes an economically competitive option for consumers, is likely to be a critical factor for the development of the disconnected scenario. NSPs could also play a role in bringing about this outcome. As pointed out in the chapter

TABLE 3 Driving Factors that Could Lead to the Three Scenarios

Driving factor	Centralized	Decentralized	Disconnected
Price comparison—DER versus grid electricity	Grid-connected electricity is cheaper than DER	Grid-connected electricity is more expensive than DER	
Distributed storage costs	Remains expensive		Becomes cost-competitive with grid-connected electricity
Policy mechanisms	Favor utility-scale generation (little support for DER)	Support DER	Support DER and possibly also distributed storage
Consumer engagement in electricity	Low	Moderate	High
Consumer trust in government and utilities	High	Moderate	Low

Source: Authors

by Nelson, if NSPs provide competitive connection fees, the majority of DER customers would rather remain connected to take advantage of the cost-effective reliability offered by the grid. Consumers in this scenario are likely to be highly engaged on electricity issues and may have a low trust of government and utility services. If not, even though off-grid installations may be economically attractive, the majority of consumers may remain grid-connected to avoid the hassle of ensuring their systems perform reliably on an ongoing basis.

Finally, a hybrid scenario could eventuate where different groups of customers respond differently. For example, residential customers may move to off-grid operation driven by a preference for self-sufficiency, while industrial customers may be more economically driven and remain grid-connected for added reliability.

It is important to bear in mind that the ESI and policy makers are not helpless bystanders to this process. The decisions and approaches adopted by these organizations are also key determining factors. Innovative pricing arrangements and a willingness for flexibility and adaptability in the ESI could facilitate a slowing of the transition to a disconnected scenario, allowing at least partial recovery of extensive sunk costs and a more gradual transition to new business models. This is discussed further in the following section.

4   INSIGHTS FOR STAKEHOLDERS

This section explores the key insights for relevant stakeholders, including the various functional areas of utility businesses (as defined in Figure 1). Insights include the potential consequences of the three scenarios and also the ways in which stakeholders may have influence over which scenarios eventuate and how their business fares in that new future.

4.1   Network Service Providers

NSPs are identified as being one of the key stakeholders. NSPs are highly exposed to the consequences of a shift to greater DER and also typically have a relatively higher degree of control over a key influencing factor: the pricing of network services.

In a centralized scenario, NSPs could expect to retain a similar role to present. The transmission NSPs may need to supply additional transmission network to connect remote renewable generation to the load centers, but otherwise, their role would remain relatively unchanged.

The distribution NSPs may see increased competition from retailers and aggregators in signing up customer load for demand-side participation. This may lead to a loss of load control capability currently used to manage local network constraints, therefore potentially bringing forward network investments. One solution could be to use localized and dynamic network tariffs to improve price signals for aggregators and retailers to use the DSP where optimal from a societal point of view. This may also require more coordination between distribution NSPs and the system operator, as discussed in the chapter by Kristov and Hou.

In the decentralized scenario, NSPs are placed in an interesting conundrum. In many nations, the majority of network costs are recovered from consumers via regulated cents per kilowatt hour tariffs. To maintain cost recovery with declining energy sales, these tariffs will need to increase, assuming that costs remain relatively static. This then drives more customers to utilize more DER and energy efficiency, thus consuming less centralized electricity, resulting in what some have termed a "death spiral" (Simshauser and Nelson, 2012). This could be further exacerbated if peak demand growth continues while energy usage declines. In this case, NSPs would need to continue to expand the network to meet rising peak demand, increasing costs, while that network is being utilized for a declining proportion of time, reducing revenues.

Through careful structuring of tariffs, possibly using some combination of time-of-use components, fixed charges, and capacity charges as discussed in the chapter by Nelson, it may be possible to limit this effect. The approach adopted would need to be carefully balanced to provide accurate price signals to consumers to encourage the use of existing assets with sunk costs while limiting increases in peak demand and thus avoiding costly network augmentation where DER or demand-side participation may be a cost-effective alternative.

However, cost recovery is no longer the only relevant consideration in network pricing. With growing availability of DER and storage options, customers now have an increasingly realistic alternative to network services. This disturbs the "natural monopoly" long held by NSPs. Thus, the way in which networks are priced and regulated may need to change dramatically.

Two possible situations can be envisioned as follows:

1. *Network remains lowest cost*—In this situation, the lowest-cost way of supplying reliable electricity to consumers, in aggregate, remains the grid. This suggests that a decentralized or centralized solution remains lower total cost than a disconnected solution.
2. *Distributed solution becomes lowest cost*—In this situation, it is genuinely lower cost in aggregate for consumers to source reliable electricity from local DER and storage options, assuming all existing network costs are sunk. This suggests that a disconnected solution becomes lower total cost than a decentralized or centralized solution.

In either case, NSPs, with consent of regulators and policy makers, may have a choice on how to set network tariffs and thus could bring about either the decentralized or the disconnected scenario. Thus, four possible combination scenarios eventuate, as illustrated in Table 4 and discussed further later.

If a disconnected scenario is ultimately lower cost than a decentralized scenario, as in the right column in Table 4, it is likely that the system will inevitably

TABLE 4 Alternative Combination Scenarios

	Decentralized is Lower Cost	Disconnected is Lower Cost
Decentralized scenario eventuates	NSPs adapt tariffs, providing innovative pricing structures that reflect the lower cost of network solutions and are attractive to consumers	Likely to be a temporary transition to disconnected scenario, perhaps perpetuated by network asset write-downs or government subsidies
Disconnected scenario eventuates	Could be driven by NSP failure to provide attractive offering to consumers ("death spiral"), which reflects the lower cost of this solution	Could occur rapidly and cause stranding of existing network assets, if NSPs do not provide an attractive offering. Transition could be slowed if NSPs respond with "shadow pricing" approach

Source: Authors

trend toward disconnected. However, at least in the short run, continued utilization of the existing network is likely to be beneficial to both consumers, deferring capital expenditure, and NSPs, continuing revenue and avoiding stranded assets.

If the network-connected solution remains lowest cost in aggregate, it is in the long-term interests of consumers to remain grid-connected, leading to the decentralized or centralized scenarios. However, this will only occur if each individual consumer is provided with an attractive offering from NSPs. If NSPs fail to adjust their tariffs to reflect the lower cost of the grid-connected alternative, customers progressively may elect to leave the grid, leading to the disconnected scenario—an unfavorable outcome for both NSPs and customers, with significant stranded assets and higher aggregate consumer costs.

In this situation, NSPs may be better advised to adopt a "shadow pricing" approach. This would adjust network tariffs associated with the use of the existing network to be just below the cost to customers of moving to DER and storage alternatives.

This approach acknowledges that full cost recovery of the sunk costs in the existing network may no longer be possible in this scenario, but seeks to utilize the existing infrastructure to the maximum benefit of consumers and recover as much of the sunk cost as possible. For government-owned assets, this would represent a significant reduction to government revenues, while for private NSPs or for equity investors, it would require a major write-down of asset value.

For networks owned by governments, a reduction in government revenue, while continuing to supply network services at a price below cost reflectivity, represents a government subsidy. Governments could subsidize tariffs for all consumers or, alternatively, just the most marginal customers. Although funded from government revenues, the total cost to consumers—taxes plus electricity—would still be lower than the disconnected scenario and each consumer should, at least monetarily, "prefer" these subsidies.

Even if the disconnected scenario is lower cost, such that the system will inevitably trend in that direction, the shadow pricing approach will slow the transition, ensuring that the maximum value is extracted from existing network assets until they are fully retired.

In any case, since network tariffs are heavily regulated at present, policy makers and regulators will need to be actively involved in these decisions that may require the freedom to implement innovative solutions that do not exist under present regulatory frameworks.

The shadow pricing methodology requires distinction between existing network assets with sunk costs and investment in new network assets. For example, regulators and policy makers should be careful to avoid implementing network subsidies that encourage new infrastructure to be installed where allowing a decentralized approach to evolve would be more cost-effective in the long term.

The specific structure of the tariffs is also likely to be extremely important and nontrivial to optimize. Differing combinations of cents per kilowatt hour tariffs, capacity charges, time-of-use charges, and other innovative pricing methodologies may be appropriate for different customer groups, depending upon the local alternatives for DER and storage, the local costs of network augmentation, and the amount of "headroom" available in the existing network capacity. This could be highly locationally specific, perhaps extending as deeply into the network as the individual feeder level. This creates new challenges for regions that have previously smoothed prices over large areas, ensuring that remote customers are not disadvantaged. For example, in the Australian state of Queensland, the Australian Community Service Obligation subsidizes rural networks out of government revenue (QCA, 2012). Equity between customers and protection of vulnerable consumers are likely to be key issues for consideration. The structuring of network tariffs is discussed in more depth in the chapter by Nelson and others, while retailer strategies are discussed in the chapter by Faruqui and Grueneich.

The greater degree of demand-side participation could also mean that in the future, individual customers may have the ability to select their desired level of reliability and corresponding cost. Rather than the market operator making a judgment on the customer value of reliability in aggregate, customers would be able to tailor their energy services to meet their individual needs. Analysis in Australia suggests that while the average value of customer reliability may be around \$95,000/MW h, residential customers may value reliability at a much lower level of around \$20,000/MW h, with small businesses valuing reliability at a much higher level (Oakley Greenwood, 2012). Thus, network businesses may have opportunities to offer tailored reliability options to customers, specific to their individual needs. The chapter by Thaler and Jimison considers the possibility of customer-driven incentives for retailers.

4.2 Policy Makers and Regulators

As indicated in Section 4.1, policy makers and regulators have an important role to play in working with NSPs to ensure sufficient regulatory flexibility for innovation and market responsiveness. Regulatory policies that could allow the necessary transition to new business paradigms are discussed in more detail in the chapter by Miller et al. Ultimately, if DER and storage alternatives provide a realistic alternative to grid connection, it may be possible to reduce regulatory controls on NSPs and allow market forces to motivate their actions.

More broadly, policy makers have influence over the types of technologies that may become cost-effective. For example, subsidies for renewable energy can deliberately or inadvertently support uptake of DER or instead preference investment in utility-scale generation. Policies that accelerate the uptake of electric vehicles may also have the secondary effect of promoting research and development in storage technologies, making off-grid options more affordable.

Another key question for policy makers is around customer equity and the protection of vulnerable customers. Low-income families, apartment dwellers, or renters who do not have access to capital, rooftops, or low-cost distributed generation could face higher costs, particularly if they are required to fully fund the remaining network, while other customers move to off-grid operation; the chapter by Burger and Weinmann quantifies costs in the German system. Cross subsidies that incentivize remaining grid-connected, or support vulnerable customers, may be justified in some cases. These subsidies could be funded via general taxation. Alternatively, some governments (such as Spain) tax rooftop photovoltaic installations, slowing uptake and providing additional revenue.

4.3 System Operator/Market Operator

In the centralized scenario, the system operator and market operator roles are similar to today, though with the added complexity of managing an increasing share of large-scale renewable generation. An increasing share of variable generation will necessitate evolution in system operation practices and market design considerations (Riesz et al., 2013).

Under the decentralized and disconnected scenarios, the system and market operator roles will need to evolve, managing the remaining centralized generation, facilitating the operation of loosely connected microgrids, responding to increased customer participation, and potentially managing a system where a declining proportion of generation is under direct control. The chapter by Kristov and Hou further discusses the interface between the system operator and the distribution system.

An increase in DER may facilitate the retirement of large thermal plants. In most systems, these plants currently provide important ancillary services, including some that are not explicitly priced at present, such as inertia. The establishment of new markets or ancillary service types may be required to pay eligible generators, including DER, for services previously supplied for free. Furthermore, increasing levels of demand-side participation may become important. As energy supplied via the grid declines, there will be a strong push from grid users for system and market operator efficiency gains, so that the cost per unit of energy served does not grow significantly faster than other consumer prices.

The future role of the grid operator is discussed in more detail in the chapters by Kristov and Hou, Felder, and others.

4.4 Electricity Retailers and Generation Companies

In many markets, electricity retailers and centralized generation companies are vertically integrated or bundled, as a strategy for managing volume and price risks.

In the decentralized and disconnected scenarios, both retailers and generation companies will see declining energy sales and thus declining revenues. DER, storage, and energy efficiency will act as competing suppliers of electricity services. In many markets, increasing penetration of renewable technologies with low operating costs will "undercut" incumbents, further decreasing the market share of conventional generators. Where there is significant penetration of variable renewables, this could exacerbate uncertainty and price risk across all scenarios.

Generation companies, in particular, have large sunk costs in existing generation assets that could become stranded in a decentralized or disconnected scenario. Generation companies could attempt a "shadow pricing" approach similar to that described for NSPs in Section 4.1 but in competitive markets may already be offering generation at close to operating costs. Generating at market prices below operating costs simply exacerbates financial losses. Thus, generation companies that are already offering conventional generation at the lowest price possible and with the maximum degree of flexibility may have relatively little ability to respond to changing market conditions. Diversification into new business areas, as described earlier and in Section 4.5, may be the best option to ensure continued profitability.

Large utilities may need to respond through diversification, in either markets or technologies. Retailers may have a larger ability to take advantage of new business opportunities involving further empowerment and engagement of consumers, as described in Section 4.5.

Lower demand growth will also affect investments in lower carbon emitting plants. Investments in nuclear power and carbon capture and storage may be considered too risky in a lower growth environment given the high capital costs. Investment in peaking plant may also be considered high risk compared with investing in demand-side participation. Demand-side participation is typically based on shorter term contracts with customers and carries minimal up-front capital costs, and neglecting to pursue demand management opportunities means missing out on one of the few parts of the electricity industry that is likely to grow in coming years (Doom, 2013).

4.5   Emerging Business Opportunities

With the decline of certain business models, others emerge. This section highlights several business opportunities identified as having growing potential in the three scenarios considered.

In many regions, utility businesses have not been able or encouraged to engage in entrepreneurial activities. Some are already recognizing the limitations of their present skill set and have sought to partner with investment firms to help establish start-ups, pilot programs, demonstrations, and test beds (McCue, 2013).

4.5.1   Investment in Low-Carbon Technologies

Investment in low-carbon technologies may offer diversification and profitability in markets where those technologies are adequately supported. However, these technologies will be threatened by a shift to DER and storage in a similar manner to conventional centralized generation.

In many European countries, the initial development of renewables was typically done by smaller market players, but the larger utilities are expected to provide about half of all new large renewables projects (Lorubio et al., 2013). However, there are large differences between utilities, with some investing little and some heavily in renewables. An example of the latter is DONG Energy, where investing in off-shore wind capacity is one of the three strategic pillars of the company. At the beginning of 2013, DONG had led the construction of 38% of all installed off-shore wind capacity in northern Europe and owned 1700 MW (DONG Energy, 2013).

4.5.2   Supplying DER and Storage Technologies

The supply of DER and home storage technologies is a clear emerging business opportunity. With a move to a decentralized or disconnected scenario, there would be large market take-up of these technologies, suggesting that companies with competitive offerings in this area could operate successfully. Many electricity retailers have already moved into this space, supplying DER to their customers in the form of rooftop photovoltaics as an alternative to grid-connected electricity.

Some companies are moving into innovative financing models, such as offering solar leasing arrangements. These "pay as you go" alternatives can make DER options feasible for customers with constraints around access to capital (Martin, 2012).

4.5.3   Energy Management Technology

Technologies that assist customers in managing their electricity consumption and generation are likely to play a growing role in future electricity systems. This could include more energy-efficient appliances, as well as software and hardware that allow automated or preplanned response of customer load to signals from the system operator or direct response to system prices.

Innovative financing alternatives are also being explored for energy efficiency investments. Models that allow customers to "pay back" energy efficiency investments over a period of time through their power bills could become increasingly popular (Pentland, 2013).

4.5.4   Demand Aggregators

There is likely to be a growing place for demand response in future power systems, particularly where there is an increase in variable renewable generation. Demand response allows greater flexibility and potentially lower system costs, while supplying customers with the level of reliability they desire.

Demand aggregators could play an important facilitating role in bringing demand response to the market. Aggregation of many small participants could reduce transaction costs and provide greater certainty to the system operator that a certain degree of response is available with a known degree of confidence. Frequency control services could potentially be provided, in addition to energy services. For example, many frequency control services are already provided by demand response in New Zealand (Zammit, 2012).

4.5.5   Minigrids

While some customers may prefer to be entirely self-sufficient, loosely connected minigrids may evolve to take advantage of the benefits of sharing demand and generation variability over a larger customer base and geographic area. Thus, new business opportunities could arise in the provision and management of minigrids and in the technologies necessary for their efficient operation. Customers are likely to continue to want high reliability without significant investment of personal time, creating opportunities for in-home energy management services. The chapter by Felder provides further discussion on microgrid scenarios.

4.5.6   Telecommunications

With the evolution of "smart grids," NSPs could have an opportunity to expand into the provision of fiber-optic networks, often rolled out alongside smart meters. This could provide additional services to customers such as video on demand. In Australia, SA Power Networks, the local distribution NSP, has secured a 3-year contract to deliver fiber-optic broadband to around 300,000 south Australian households as a part of the National Broadband Network rollout (Swallow, 2013).

5 CONCLUSIONS

An analysis of the key drivers of change in power systems at present suggests that while a shift to lower carbon intensity supply appears inevitable, there is far less certainty around the amount of DER that could operate in future grids. In some ways, a move to extensive DER could be far more transformational for the electricity sector than a move to lower emissions energy. It would dramatically change the structure of the industry, eliminating the need for some utility roles and opening up opportunities for new ones.

This creates a challenging environment for utility businesses, particularly those with the potential for costly stranded assets, and a business model that, at worst, could become defunct, such as NSPs. These businesses will need to adapt and seek innovative approaches. In some circumstances, it may be necessary to accept an approach of slowing an inevitable transition toward DER to allow managed diversification into alternative business models.

REFERENCES

AEMO, 2013. 100 Per Cent Renewables Study—Modelling Outcomes. Australian Energy Market Operator, Melbourne.

Anthony, M., 2007. The new China: big brother, brave new world or harmonious society? J. Futures Stud. 11, 15–40.

Banerjee, S.G., Bhatia, M., Azuela, G.E., Jaques, I., Sarkar, A., Portale, E., ... Inon, J.G., 2013. Executive summary, Vol. 1 of Global Tracking Framework. Sustainable Energy for All, World Bank, Washington D.C. Retrieved from http://documents.worldbank.org/curated/en/2013/01/17747194/global-tracking-framework-vol-1-3-global-tracking-framework-executive-summary.

Birkeland, J., 2007. Positive development: design for net positive impacts. BEDP Environment Design Guide. 1 (4), 1–10.

Botsman, R., Rogers, R., 2010. What's Mine Is Yours: The Rise of Collaborative Consumption. HarperCollins, New York.

Carbon Tracker Initiative, Grantham Research Institute on Climate Change and the Environment, 2013. Unburnable Carbon 2013: Wasted Capital and Stranded Assets. Carbon Tracker & The Grantham Research Institute, LSE. Retrieved from http://carbontracker.live.kiln.it/Unburnable-Carbon-2-Web-Version.pdf.

Christensen, C., 1997. The Innovator's Dilemma. Harvard Business Review Press.

DONG Energy, 2013. DONG Energy Enters into Agreements on 6 Megawatt Offshore Wind Turbines with Siemens AG for German Projects. DONG Energy, Denmark.

Doom, J., 2013. NRG May Follow Demand-Response Purchase with More Renewable Buys. Bloomberg, New York.

Gilding, P., 2011. The Great Disruption. Bloomsbury Publishing, London.

Inayatullah, S., 2008. Six Pillars: futures thinking for transforming. Foresight 10 (1), 4–21.

Lorubio, G., Schlosser, P., Nies, S., 2013. Utilities: powerhouses of innovation—full report. Eurelectric—Electricity for Europe.

Martin, J., 2012. Solar Leasing "Free Solar Panels"/Pay-as-you-go solar agreements now available in Australia. Retrieved from Solar Choicehttp://www.solarchoice.net.au/blog/free-solar-panels-pay-as-you-go-solar-systems-australia/ April 23.

McCue, J., 2013. 2013 Outlook on Power & Utilities—My take. Deloitte Center for Energy Solutions, Washington, D.C.

NREL, 2012. Renewable Electricity Futures Study. National Renewable Energy Laboratory, Golden, CO.

Oakley Greenwood, 2012. NSW Value of Customer Reliability. Australian Energy Market Commission, Australia.

Overbeeke, F.V., Roberts, V., 2002. Active Networks as facilitators for embedded generation. In: IQPC Conference on Embedded Generation within Distribution Networks, London.

Pentland, W., 2013. The Battle for your Energy Bill. Retrieved from Forbes, http://www.forbes.com/sites/williampentland/2013/06/14/the-battle-for-your-energy-bill/, June 6.

QCA, 2012. Final Determination—Regulated Retail Electricity Prices 2012-13. Queensland Competition Authority, Australia.

Riesz, J., Gilmore, J., Hindsberger, M., 2013. Market design for the integration of variable generation. In: Sioshansi, F. (Ed.), Evolution of Global Electricity Markets: New Paradigms, New Challenges, New Approaches. Elsevier, Amsterdam/Boston.

Sanyal, D., 2011. How crisis can shape the treasury for tomorrow's risks and opportunities. In: Speech to International Treasury Management Conference, Rome, Italy.

Schwartz, P., 1991. The Art of the Long View. Doubleday, New York, NY.

Simshauser, P., Nelson, T., 2012. The Energy Market Death Spiral—Rethinking Customer Hardship. AGL Energy, Australia.

Sioshansi, F.P., 2009. Introduction—carbon constrained: the future of electricity generation. In: Sioshansi, F. (Ed.), Generating Electricity in a Carbon Constrained World. Academic Press, Burlington, p. xxxiii.

Sioshansi, F.P., 2013. Energy Efficiency: Towards the End of Demand Growth. Academic Press, Oxford, UK/Waltham, MA.

Swallow, J., 2013. SA Power in major NBN rollout deal. Adelaide Advertiser. 37.

Zammit, M., 2012. Submission to RIT-T: Project Specification Consultation Report—Heywood Interconnector Upgrade. EnerNOC, Melbourne, Australia. Retrieved from: http://www.aemo.com.au/Electricity/Planning/Regulatory-Investment-Tests-for-Transmission/~/media/Files/Other/planning/0179-0305.pdf.ashx.

Chapter 24

Revolution, Evolution, or Back to the Future? Lessons from the Electricity Supply Industry's Formative Days

Robert Smith and Iain MacGill

ABSTRACT

Is the electricity industry facing unprecedented revolutionary change? Do emerging drivers including climate change, distributed energy technologies, energy efficiency, and growing customer engagement undermine the industry's existing centralized generation, large grids, and the business models they supported? This chapter explores these questions by looking back to the early history of the electricity industry, starting with Edison's light bulb. It describes the uncertainties and difficulties that initially faced the grid we now take for granted. It highlights that the revolution some see as imminent might better be viewed as a counterrevolution, back to before the development of nation-spanning grids and utilities to run them. Yet, history also cautions that current arrangements and business models have been remarkably adaptive for over a century and may well prove more resilient than imagined. It concludes that, whether evolutionary or revolutionary, the industry needs to prepare for change and find opportunities to be part of the transition, should a largely decentralized grid prove to be better placed to meet society's future energy needs.

Keywords: Electricity industry history, Utility business models, Evolution of electric grid, History of Edison's light bulb, History of electric vehicles, Thomas Edison and Samuel Insull

1 INTRODUCTION

It is a wonderful story. In 1879, Thomas Edison, the boy genius, the Wizard of Menlo Park, already famous as the inventor of the phonograph, applies "1% inspiration and 99% perspiration" and invents the light bulb.[1] Only three years

---

1. Edison did much to feed the invention myth; see Jonnes, Klein, and DeGraaf.

Distributed Generation and its Implications for the Utility Industry. http://dx.doi.org/10.1016/B978-0-12-800240-7.00024-2
475

later, the Edison Illuminating Company is distributing power to Lower Manhattan from the coal-fired Pearl Street central station. The electricity grid is born and sweeps the globe.

As the grid matures, it continues to grow with ever-expanding end uses for electricity and the scale economies that arise from ever-larger power stations. Then comes middle age, a long period of stable technology and institutional arrangements as a natural monopoly service provider, as either publicly owned or regulated vertically integrated private utilities. Amazing new end-use technologies emerged in air-conditioning, radio, television, computers, and smartphones yet the network utility soldiered on largely unchanged until. Over 130 years later as the incandescent bulb becomes obsolete, has the time come for the grid to retire as well? Do the emerging industry trends and drivers described in earlier chapters including climate change, distributed energy technologies, and growing end-user engagement signal a revolutionary transformation away from centralized generation and nation-spanning grids and the business models they supported? Or, instead, can the grid stand in the face of such change and see these developments smoothly integrated into current arrangements? Or is this indeed actually a possible counterrevolution, back towards an earlier time when the grid had not achieved its present dominance?

This chapter discusses that the future of the electricity industry, like its history, is likely to be more rich, complex, and dynamic than many imagine. In particular, it revisits the history of the electricity industry to put these revolutionary "new" drivers into context. The intent is not to dismiss the importance and potential scale of change ahead for current arrangements but, instead, highlight the many uncertainties facing the industry. Is the change facing the industry best seen as a revolution, evolution,[2] or revisiting the past? And how should the industry respond?

A common feature of future thinking is a tendency towards expecting extremes, most explicitly when seeing choices as between utopias and dystopias.[3] Common themes that tend to accompany simplified versions of our energy future include the following:

- Change is more rapid, radical, and unrelenting now than it has been in the past.
- Today's new ideas and inventions represent a revolution in thinking that requires a complete reorganization of the electricity industry, the economy, our institutions, and society.

---

2. Both terms appear in this book; see Burger (Chapter 3) and Hanser and van Horn (Chapter 11).
3. See, for example, Huber and Mills, Hawken et al., and Sabin for Paul Ehrlich and Julian Simon's bet and Strathern and Smith (2007) for a list of "new economic" futures as a mishmash of computers, the Internet, energy, services, the environment and associated changes in the laws of nature, human nature, and the rules of economics.

- New technologies have a momentum of their own. New is better. The economics underpinning current technologies either no longer matters or will be swept aside in the change.
- Local decentralized solutions are better, small is beautiful, and centralized systems are inefficient, cumbersome, and ugly.
- Energy services, not commodity provision of electricity, are what matters now.[4]

However, the birth of the electricity industry suggests that a more interesting and less straightforward future lies ahead. Countering claims that "it's different this time" and "the old rules no longer apply" is the evidence of history that:

- Change is continuous but its impacts are neither instantaneously nor necessarily bigger now than in the past. Rearview mirror myopia affects how we view change. The electric light, the electric grid, and the telegraph[5] before them had impacts more profound than what photovoltaics (PV), electric vehicles (EVs), fuel cells, and smart grid technologies may well have in the future.
- Evolution rather than revolution is a more common path for change with competition among solutions but also long periods of complementary and coexisting outcomes.
- Economics and broader social drivers, rather than technology, ultimately decide which outcomes rule—costs need to be outweighed by benefits as judged by consumers, not inventors, engineers, entrepreneurs, bureaucrats, or economists.
- Centralized or decentralized approaches are neither inherently good nor bad. Instead, they, like the electricity grid, are only more or less appropriate for the uses they serve and the markets and institutions they operate within.
- Energy services, what customers want, are certainly the reason for having an electricity industry; however, there are still questions of how best these services can be supplied, priced, and delivered and on what role consumers wish to play in the process.

This chapter supports this view of a more complex future by looking back to the birth and early days of the grid. Section 2 looks back to Edison's time when light bulbs and grids were new, highly promising, yet still unproven. Section 3 examines "new" disruptive technologies, particularly EVs and PV, and how current promise has often been seen before. Section 4 looks at an often overlooked part of the grid's birth—Samuel Insull and the development of the centralized utility

---

4. Moody and Nogrady's *The Sixth Wave* is a good example.
5. See Standage on the telegraph as "the Victorian Internet."

business model that underpinned the grid's growth and maturity. Section 5 provides observations of the tussle between centralized and decentralized solutions—a recurring theme throughout the book. This chapter concludes with some thoughts on what the future may hold for the industry.

2   ILL-LIT BY MOONLIGHT: REMEMBERING THE IMPACT OF ELECTRIC LIGHT

Electrification was voted as the greatest engineering achievement of the twentieth century by the National Academy of Engineering,[6] ahead of the automobile (2nd) and the airplane (3rd) and well in front of the computer (8th) and the Internet (14th). Yet, back at the turn of the nineteenth century, the same grid, despite its obvious benefits and promise, was undergoing a difficult birth.

An incandescent light bulb is now the symbol for a "bright idea,"[7] but in its earliest days, the long-term benefits of electricity were not plain for all to see. There is an apocryphal story of the UK prime minister visiting Faraday's lab to admire his electromechanical apparatus asking "of what use is it?" to which Faraday is reported to have replied "Why sir, there is every possibility that you will soon be able to tax it."[8]

Before the grid, lighting was a problem for even the richest and most technologically savvy. Boulton and Watt, the inventors of the practical steam engine, were members of the "Lunar Society," a Birmingham dinner club of scientists, businessmen, and philosophers. The group was so named because, in a world without street lighting, their monthly meetings were organized around the light of the full moon so they could travel home safely.[9] Boulton and Watt's efficient steam engines freed the industrial revolution from its reliance on wind and water mills, which could only deliver localized and intermittent energy, and started the path to the centralized electricity grid. Yet, a world-changing invention does not guarantee instant riches or success. Making

---

6. See http://www.greatachievements.org/.

7. The light bulb symbol remains even as it is being phased out by regulation and falling costs of CFL and LED alternatives—technologies that have had their own development challenges—see Johnstone and Waide.

8. Electricity and light were key scientific themes of Faraday's day with his most famous public lectures *The Chemical History of a Candle* and his mentor Sir Humphry Davy, famous for discovering the electric arc lamp in 1808 and inventing the Davy safety lamp for miners.

9. In Blainey, organizing social events around the full moon was still common around the turn of the century; see Uglow. Boulton's quote, "I sell here, Sir, what all the world desires to have – Power!" and portraits of Boulton and Watt can now be found on the English fifty pound note. But, in a preview of Edison's travails with the light bulb, Boulton initially struggled with financing, conducted a trial for performance contracting, coined the term "horsepower" as a marketing tool, fought bitter patent battles and frequently teetered on the edge of bankruptcy.

something work is one thing; having it adopted and making it pay are another.[10]

In retrospect, Edison's light bulb, first patented in 1879, was a "killer app," a sure thing. It delivered 762 lumen-hours per 100Btu, a step change in lighting efficiency compared to its contemporaries: the Welsbach mantle gas lamp at 255; the kerosene lamp at 15; and its predecessors whale oil at 39 and the tallow candle at 22.[11] But Edison's bulb also delivered a major increase in the quality of light, convenience, and safety from fire and a significant improvement in indoor air quality, albeit a light less soft and romantic than the flame of gas and candles. The convenience of electric light was not lost on early adaptors, particularly compared to kerosene lamps, which were finicky, smoky, and sooty, and the cleaning of which was "one of the least pleasant of the house-keeper's duties"[12] (Figure 1).

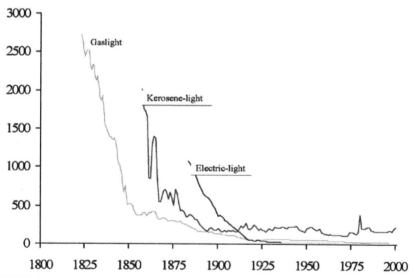

FIGURE 1   Price of lighting from gas, kerosene, and electricity in the United Kingdom (per million lumen-hours) 1800-2000.[13] *Source: authors' own estimates – see Section 2.1.3-5 and 2.3.*

---

10. This can be seen in China's failure to turn their world's first inventions into an industrial revolution in the Middle Ages; see Moykr, Winchester, and, for a more light-hearted view, Wilson.
11. Nordhaus notes that to write a play, Shakespeare would have spent the equivalent of UK £15,000 on tallow candles (updating estimates from Waide and Tanishima). Whale oil and whale bone were among Australia's first major export industries, and debates about peak whale oil soon followed, predated Hubbert's peak.
12. Brox.
13. Fouquet and Pearson.

Edison's incandescent bulb was not the first electric light.[14] It prevailed due to its durability and quality and his fame but ultimately because of its supporting infrastructure, the grid. Yet, the invention of the practical light bulb and the grid did not assure Edison's success.

Others in the field were not as certain as Edison when it came to the future of electric power. For one thing would every customer even want electrical power? Some imagined each home with its own mini-generating plant that ran on coal. For many the storage battery was very much seen as the future of domestic electricity.[15]

One scheme proposed to sell "milk bottle" plans to deliver batteries just as milk was, while another proposed two batteries in the home—one discharging while the other is recharging from a central station. These schemes seem less fanciful when it is remembered that at the time, most houses in the America had their own wells and outhouses and generated their own power from coal or wood for heating and cooking—all independent of centralized public utilities. Even today, a "milk bottle" model continues for LPG bottled gas, and coal, oil, and wood are still delivered to houses for self-power "generation." The current call for decentralization and an "energy services" model to replace kilowatt-hour sales is part of a long history of competition between solutions. The possible but not preordained spread of microgrids discussed by Felder and the role of bioenergy villages in Gemany's "Energiewende" by Burger and Weinmann are recent bouts in a long-running brawl.[16]

Edison's light and the grid had to fight for its eventual success. Just as whale oil fought to replace tallow candles, to be then replaced by kerosene and gas, electric light from central stations had competition. The gas lighting industry responded to Edison's light bulb with the invention of the Welsbach mantle gas light that kept gas lighting cheaper than electricity into the twentieth century. Price was a major consideration as Edison's light bulbs cost around $1 each in the late 1880s, equivalent to over $200 in present dollars adjusted on an average income basis.[17] While the first news of Edison's incandescent light had sent the price for US gas lighting company stocks tumbling, they recovered soon after and many continued to be successful until the new business model for electricity distribution emerged and prices fell. However, even as central electric stations took off in America in Britain, where gas was cheaper and conditions were different, the industry started, faulted, and had effectively died away by 1890. Meanwhile, in faraway Tamworth, in rural NSW, the technology

---

14. Arc lights were already an established but struggling technology, and Swan and others had already made claims to the incandescent globe; see Fara, Brox and Jonnes.
15. Schlesinger.
16. Felder in Chapter 20 puts microgrids in the context of competition among solutions to achieve policy aims not as an end in themselves.
17. A range of measures can be calculated at http://www.measuringworth.com/uscompare/rela tivevalue.php.

spread with remarkable speed, turning on Australia's first electric town lighting in 1888.[18] Nevertheless, gas streetlighting persisted elsewhere, only beginning to be replaced in Sydney in 1904. Indeed, Sydney Council actively legislated against electricity's introduction into central Sydney, delaying it to the point where it was the last major municipality in the country to be electrified.[19]

It is difficult to imagine now but at the time of its inception, "electric lighting is best thought of as a niche market, for people and cities looking for a higher quality product, and willing to pay a higher price."[20]

Edison's biggest competition however came not from gas, alternative energy sources, or stand-alone systems. "The seven years' incandescent light bulb war"[21] was fought over Edison's patents involving massive legal bills and hampering the industry's investment and development. The acrimony of the patent dispute, now largely forgotten, then extended into the better remembered "War of the Currents," pitting Thomas Edison's direct current (DC) for electric power distribution against George Westinghouse and Nicola Tesla's alternating current (AC). The war was a bare knuckle affair with Edison sponsoring a large-scale lobbying, disinformation, PR, and propaganda campaign of false claims about the safety of AC. This included electrocuting "Topsy," a rogue circus elephant, with AC and sponsoring the invention and adoption of the electric chair to have death row prisoners "Westinghoused" with AC.[22] The War of Currents also proved to be a battle of business models.

Despite his fame and efforts to push DC, Edison eventually fell victim to the better economics of large-scale centralized systems over more decentralized solutions. The advantage shifted to Westinghouse and Tesla's AC model as the industry moved to ever-bigger generators supplying power via high-voltage transmissions from remote locations. Edison's DC systems however lingered on in specialized applications, most notably not only for railways but also for elevators, with DC systems only being removed from New York, Chicago, and Melbourne in the last 20 years and living on still for some applications in San Francisco.[23]

Ultimately, Edison, Westinghouse, and Tesla, the technological fathers of grid, all fell victim to the recurring financial panics of the late 1900s losing control of their patents, their businesses, and the industry they helped invent. As the following chapters outline, forecasting technology change is fraught with difficulty, and a sustainable business model can be as hard to discover as a new technology.

---

18. A readable history is provided by Tamworth Regional Council.
19. A history of the Australian electricity is provided in CIGRE. See also Wilkenfeld and Spearitt for insights into how institutional and political factors can hamper a bright idea, and Anderson.
20. Shiman.
21. Jonnes.
22. See McNichol particularly for Edison's promotion of an AC electric chair.
23. Fairley.

3   ELECTRICITY AND THE SHOCK OF THE OLD

As humans, we seem to be wired for the new, the next, the bigger, the better. And since the scientific and industrial revolutions of the eighteenth century, our neophobia[24] has fed upon a continuous stream of invention and product innovation. But as historian David Edgerton points out, our emphasis on invention and novelty means that the most important technologies in use today are often overlooked. "A history of how things were done in the past, and the way past futurology has worked, will undermine most contemporary claims to novelty."[25] With an incomplete, technology-centered view of what works in the present, it can be easy to overestimate the scope for change in the future:

Use-centred history is not simply a matter of moving technology time forward. . .We work (ed) with old and new things, with hammers and electric drills. In use-centred history technologies do not only appear, they also disappear and reappear, and mix and match across the centuries.

Alexander Graham Bell would not recognize his original telephone in today's smartphones, but Edison would arguably find today's electricity supply industry much as he left it: incandescent light bulbs remains in use; Tesla's induction motor of 1888 still drives most domestic appliances; household refrigerators emerged in the 1920s and individual window-mounted air-conditioners appeared in 1932 (and the first power-generating plant built specifically for air-conditioner loads in New York only 10 years later); and electric resistance water heating, first explored in 1841, is still a major domestic load. What has changed more are the economics and public policy of use. Today's appliances are much cheaper both to buy (through mass production and technology learning) and to run, but lifecycle costs (now often including carbon costs) are shifting more towards operation rather than purchase cost.[26] Government energy efficiency policies to drive down kilowatt-hour usage are now being supported by both technology and economics, but impacts continue to be mixed.

Some "new" lighting technologies including CFLs and now LED globes can offer fourfold efficiency improvements over incandescent globes together with far longer lifetimes. Others like reverse cycle air-conditioners create worrisome summer peaking load but with winter heating efficiency four times that of the traditional resistive heater and probably better greenhouse impacts than gas heaters in some contexts. As such, old and new technologies are often found side by side, with effects positive and negative for peak demand, energy use, and the environment. The household refrigerator (see Chapter 1; Figure 9) is a classic case study with increased size and numbers being offset by greater

---

24. Gallagher.
25. Edgerton.
26. Molotch and Boxshall. An electric heater that can be bought today for less than the minimum hourly wage could have cost two weeks wages even as late as the 1960s.

FIGURE 2   Uses of electricity—Rural Electrification Administration in 1938.

energy efficiency and falling appliance prices. The fridge's success however came as part of a bigger innovation "the cold chain," which involves business, transport, and institutions as well as people's tastes, behavior, and preferences combining as part of a total value chain[27] (Figure 2).

By contrast, many grid technologies have changed remarkably little. In many countries, wires on wooden poles still feed end users from predominantly coal-fired generation only modestly better in terms of costs and efficiency than those being built forty years ago. The "modern" closed-core transformer was first patented in the mid 1880s, and units more than 50 years old can still be found in service around the world. Some cables in the Australian electricity

---

27. Rees.

industry can date back to the 1920s and 1930s. The standard induction watt-hour accumulation meter came into general use in the 1890s and is still in wide use for residential and small business customers in some electricity industries.

The most recent "rocket science" innovation for centralized generation has perhaps been open-cycle gas turbines (and associated combined cycle plants) for peaking plant that came from the aerospace industry. In this context, renewable generation is sometimes portrayed as a revolutionary modern development. In reality, there were an estimated 200 hydro power plants in the United States in 1889, and the Adams hydro station at Niagara Falls was the proving ground for large-scale AC. Even with wind and solar energy, small stand-alone wind generators with battery storage for electricity provision in rural areas without grid connection were commonplace in the 1930s and 1940s, particularly in the United States and Australia, while solar steam engines were developed in the 1800s.[28] The first megawatt-scale wind generator was installed in 1941 in the United States.[29]

There are, however, two "new" disruptive technologies—with implications for the broader question of whether the electricity industry is likely to remain physically centralized, transform itself towards a decentralized industry, or likely end up somewhere in between—that deserve particular attention: EVs and PV.

3.1 Everything Old Is New Again: An EV Road Trip Through Time

It seems that the smart money is betting on EVs. Major automobile companies are producing or developing EV models, governments are subsidizing research and sales, and Warren Buffett has invested in Chinese EV manufacturer BYD. Tesla Motors' CEO is Fortune's 2013 Businessman of the Year and its "Model S" the car of the year. EVs, storage batteries, and V2G technologies look set to take off and take over. Platt et al. discuss the implications in Chapter 17. However, a user-centered history of electric transport suggests a different story.

The long dominance of internal combustion engine (ICE) vehicles was not obvious at the start of the twentieth century. The bicycling craze of the 1870s had lead to better roads that paved the way for the automobile; firstly only as dangerous rich kid's toys, until around 1908 when the electric starter and the Model T Ford brought safety, convenience, and affordability to motoring. In America, around 1908, electric and steam vehicles were as common as internal combustion and included taxis, motor bikes, trucks, and buses. Many saw quiet

---

28. See Madrigal for one view. Similarly, Augustin Mouchot's concentrated solar power parabolic trough system of the 1860s was funded by the French government facing concerns about the future availability of coal. While his work lost favor as coal prices fell, the underlying technology reappeared after the oil crises of the 1970s in the United States and has now returned again there and in Southern Europe.

29. Note, however, that it failed shortly after entering service; see the Wikipedia.org page on the Smith-Putnam wind turbine for more information and Gipe for a useful historical overview.

FIGURE 3   Edison and the electric car.

clean EV as ideal "town cars" and noisy, sooty, smelly ICE better suited for
longer country trips and "Edison truly believed the ICE was nothing more than
a bridge technology that would eventually lead to the electric car."[30] (Figure 3).

Edison, America's most successful inventor, and Henry Ford, America's
most successful businessman, worked together to develop an electric car. Eventually the lack of an effective battery defeated Edison's inspiration and perspiration and his belief that "I don't think nature would be so unkind as to withhold
the secret of a good battery if a real earnest hunt for it is made."[31]

So, while the Model T Ford went on to sell over 16.5 million, the electric car
disappeared from view for a century.[32] But even as electric cars became rarities
on roads, they never completely disappeared holding on in niches like milk
floats, golf carts, wheelchairs, forklifts (with batteries as counterweights),
and moon buggies. In a larger sense, EVs on roads are again an exciting novelty,
but electric transport has been ubiquitous the whole time as trains, trams, elevators, travelators, cable cars, and lifts continue to move billions daily.

---

30. Schlesinger and DeGraaf; also, see Carsson and Vaitheeswaran.
31. Meanwhile, ICE cars continue to use the same basic lead acid battery technology that Edison
hunted to replace to run starters motors and ever more complex electronics.
32. The most successful motor vehicle is the Honda Super Cub motorbike with sales over 60 million. The most successful vehicle is the "Flying Pigeon," the basic bicycle of China's communist
years boasting its over 500 million sales.

And as a final "back to the future" twist, bicycle sales outstrip car sales world-wide[33] and electric bicycle sales are projected to reach 26 million in 2013, out-stripping electric cars in users, kilometer and kilowatt-hour.

What of new V2G technologies, as discussed by Platt et al.? In fact, the concept behind V2G, "power takeoff" (PTO) from cars, is not new. For rural owners a Model T Ford was as much a portable engine as a means of transport. With a belt drive attached, it became a water pump, ran farm tools, or acted as an electric generator. PTO is still a common feature in tractors and trucks, but stand-alone generators are generally preferred these days, as stand-alone batteries may be in the future.

As Faruqui and Grueneich point out, electrification of transport via EV could let utilities flatten load and grow their way out of current problems—a solution straight out of Samuel Insull's playbook for trams and trains. Yet, EV's impact is still uncertain and mass take-up still seems many years away even on optimistic forecasts. Moreover, when batteries are cheap enough to make EV economic they are cheap enough for stand-alone use as well—a different challenge and an impetus for capacity tariffs.

The idea of distributed electric energy storage is also not new either. A paper in the Journal of the Society of Telegraph Engineers and Electricians in 1888 proposed the use of DC power distribution with customer-located accumulators (batteries) to better manage the peakiness of London electric lighting loads. And again, in faraway Tamworth in 1907, electric lighting services for households and shops were based on charging batteries overnight when the town's power station was running for electric street lighting[34] (Figure 4).

Will distributed energy storage, as V2G or household batteries, be a game changer for the grid? Perhaps, but it will compete and coexist with the myriad old unexciting energy storage solutions currently in households as:

- rechargeable batteries in laptops, phones, and mobile devices;
- traditional batteries in toys, tools, torches, and minor appliances;
- thermal storage in electric, gas, and solar hot water and in building mass and in fridges and in some cases geothermal;
- bottled gas where main gas is unavailable but also for most people's BBQs;
- wood, oil, and coal for heating and petrol-powered mowers and tools; and
- even propane, kerosene, and tallow/wax that are still present in candles, lanterns, and camp stoves.

Most importantly, all these storage solutions will compete with the 100 years invested in developing and expanding a ubiquitous grid.

---

33. http://www.npr.org/blogs/parallels/2013/10/24/240493422/in-most-every-european-country-bikes-are-outselling-cars.
34. http://www.tamworthregionalgallery.com.au/museum.php.

JOURNAL

OF THE

SOCIETY OF

Telegraph-Engineers and Electricians.

Founded 1871. *Incorporated* 1888.

VOL. XVII.   1888.   No. 73.

CENTRAL STATION LIGHTING:

TRANSFORMERS V. ACCUMULATORS.

By R. E. CROMPTON, Member.

The present paper is the outcome of the discussion which took place on Messrs. Kapp's and Mackenzie's papers on transformers, recently read before this Society. I was asked to give facts and figures in support of the statement I then made, that I believed the distribution of electricity by transformers offered no special advantages over other methods, particularly over distribution by means of accumulators used as transformers.

Table No. 3.

COST OF 10,000 LIGHT, OR 600-KILOWATT, PLANT.

A.T.—ALTERNATING TRANSFORMER DISTRIBUTION.	£	B.T.—ACCUMULATOR TRANSFORMER DISTRIBUTION.	£
Generating Station, Buildings, Chimney Shaft, Water Tanks, and General Fittings	11,000	Generating Station, Buildings, Chimney Shaft, Water Tanks, and General Fittings	8,000
Dynamos — 600 Kilowatts, in 6 sets of 100 Kilowatts each, including spare sets, divided as convenient ...	5,540	Dynamos — 600 Kilowatts, in 6 sets of 100 Kilowatts each...	4,800
Motive Power, *i.e.*, Engines, Boiler, Steam and Feed Connections, Belts, &c., at £3 12s. per I.H.P.	12,470	Motive Power, *i.e.*, Engines, Boilers, Steam and Feed Connections, &c., at £3 12s. per I.H.P.	8,600
500 Transformers, *i.e.*, one to every pair of houses, at £15 each	7,500	4 Groups of Accumulators, in all 240 cells, in series, at £40 per cell, including Stands ...	9,600
2,000 yards Primary or Charging Main, exterior to area of supply, at £306 per 100 yards	6,160	2,000 yards Charging Main, at £306 17s. 6d. per 100 yards (see Table 2)	6,137
20,000 yards Distributing Main, 50 m/m. sectional area, at £91 7s. (*see* Table 1) ...	14,270	20,000 yards Distributing Main, 161·25 m/m. sectional area, at £100 12s. 6d. (*see* Table 2) ...	20,125
Regulating Gear	500	Regulating Gear	2,600
	£57,440		£59,762

3.2   Here Comes the Sun: The Rise of PV

Could anything be more revolutionary than PV? PV is an extraordinary technology—a solid-state device, with no moving parts, highly scalable and able to generate electricity powered by nothing but sunshine. Edison did not foresee PV but predicted "I'd put my money on the sun and solar energy. What a source of power! I hope we don't have to wait until oil and coal run out before we tackle that."[35] The transformational potential of PV was recognized immediately upon the announcement by Bell Laboratories of the first silicon solar cell in 1954. The New York Times had a front-page story on "the beginning of a new era, leading eventually to the realization of one of mankind's most cherished dreams -- the harnessing of the almost limitless energy of the sun for the uses of civilization." [36] The first commercial PV cell followed only a year later, albeit priced at over US$1800/W. Many governments began to support PV research and development, particularly following the oil crises of the 1970s, which for a period, as noted earlier, focused attention on renewable energy alternatives to fossil fuels.

Major grid deployment of PV commenced in key markets only in the 1990s under significant policy support. The technology continued to improve and costs lowered modestly. And then, it took off. Households and businesses can now all join the supply side of the industry. The story of PV is widely discussed through this book: German PV's impact is well described by Burger and Weinmann and Australia's experience by Mountain and Szuster. While a remarkable story, PV is still just weaning itself from the generous feed-in-tariffs, net metering, and "infant industry" support that were largely responsible for its initial success. Nelson et al. look at PV as the new "partial substitute" product for the grid, and Faruqui and Grueneich place emphasis on the importance of energy efficiency alongside PV as draining away load growth, previously the "rising tide that lifts all boats."

Due to the "availability heuristic," PV's role as the new, exciting fast-growing technology undermining the scale economies of Insull's utility business may have been overemphasized compared to energy efficiency. It is easy to see panels, count rooftops, and measure in kilowatt-hour solar generation compared to the unseen negawatts of energy efficiency saved. For example, for Australia, despite the enormous growth of PV, energy efficiency impacts are forecast to be the major factor slowing future demand.[37]

PV's future seems assured in the many countries where it is now understood to be at or close to the cost competitive against retail electricity tariffs without

---

35. Attributed to Edison in 1931 shortly before his death and quoted in Chendo but possibly apocryphal.
36. Perlin.
37. AEMO.

policy support.[38] However, it is PV acting together with energy efficiency that is exposing weakness in current arrangements, particularly tariffs, which remain largely rooted in the early utility business model for the electricity industry.

4   ELECTRICITY AND THE SHOCK OF THE NEW: HOW THE BUSINESS MODEL EVOLVED

The other chapters of this book outline the current threats to the existing grid business model, especially in "mature" jurisdictions. A century of almost uninterrupted electricity demand growth has halted (see Faruqui and Grueneich and Nelson et al. in this book and Sioshansi, 2013). The predictable one-way flow from large centralized power stations is becoming a more complex intermittent two-way flow from multiple locations hosting a range of distributed generation sources, particularly trigeneration and rooftop PV. EVs and battery storage are looming on the horizon. Smart technologies are bringing to the electricity industry the digital information that revolutionized the communications industry. Old carbon-intensive fuels are scheduled to be on their way out and renewables on their way in. Energy efficiency has reduced kilowatt-hours, while demand peaks and the costs of supply remain. It could well feel like unprecedented change is here, "it's different this time," and the "old rules no longer apply." The grid faces uncertainty and change, but is this all so different from the past?

While residential electric lighting was the "killer app" that spurred on the growth of the electricity grid, it was far from an ideal product around which to build a business model. In its early days, electric lighting was expensive for customers and a peaky, undiversified load for utilities leaving costly grid infrastructure idle for much of the day.

And financing the capital-hungry electricity industry was also contentious. Edison's Pearl Street Station was finished in 1882 only after lengthy delays, much angst from his investors and costing $600,000, double the original estimate. Edison initially struggled to find backers for his larger central stations. Bankers found financing distributed generation, with its quicker paybacks on smaller investments in individual buildings and businesses, more attractive. Even after Pearl Street opened, it took years to turn a profit, and meanwhile, Edison continued to install hundreds of stand-alone generators in high-profile locations like the New York Stock Exchange while waiting for his central station concept to be widely adopted. Making an invention is one thing; making it pay is another. Edison's friend and collaborator of Henry Ford called him "the

---

38. Different measures of PV "socket parity" (Groot) or "grid parity" and its timing are discussed throughout the book. The issues of PV net metering as free storage (Keay et al.) and appropriate pricing of intermittent distributes generation are not addressed here.

world's greatest inventor but the worst businessman," and Peter Drucker noted, "Edison so totally mismanaged the businesses he started that he had to be removed from every one of them to save it."[39]

4.1 Samuel Insull and the Grid Business Model

The credit for creating the business model for the modern grid goes largely to Edison's one time clerk/business secretary/general factotum Samuel Insull. Largely unremembered today, Insull eventually became among the most famous businessmen of this time, a Steve Jobs of his day, an innovator who shaped the future of the most exciting technology of his time.[40] He learned the business from the ground up, was part of the Pearl Street Station start-up, and then managed Edison's business and financial affairs until Edison lost control in 1892 whereupon Insull moved to Chicago to run the struggling Commonwealth Edison (Figure 5).

FIGURE 5 Samuel Insull.

39. DeGraaf.
40. Insull appeared on the cover of Time magazine 3 times compared with Steve Jobs 8 times, but then, Time only started in 1923 toward the end of Insull's career. Insull's remaining fame is as one of the models for Orson Welles' "Citizen Kane."

Electricity at the time was full of promise but neither a stable nor a secure industry. ComEd, like the rest of the early grids, was struggling with finding capital for expansion, finding customers, coping with technology change, setting standards, competing with other electricity companies, competing with gas and distributed generation, fighting hostile public and political opinion, and creating effective regulation. Insull's business model for ComEd was to:

- get big to lower costs and prices through economies of scale and "mass production;"[41]
- diversify and build load to flatten load curves and use capacity; and
- create regulation that allowed private utilities to operate as monopolies.

The magic of Insull's business model for the modern grid was in marrying the technology and its economics with social, policy, and institutional change. Not just an invention or a technology seen in isolation, the business model removed uncertainty and allowed for the large long-term investment needed to build the modern grid.

Insull pursued these strategies vigorously, soliciting transport loads in rail and trams, offering aggressive pricing for daytime industrial loads, pushing manufacturers for ever-larger generators, giving away 10,000 irons to attract customers and grow load, and copying the two-tier pricing he saw in Brighton, England, to flatten and manage demand. Recognizing the value of diversity that farm loads brought, Insull was even aggressive in electrification of rural areas, which most companies saw as unprofitable, ahead of government initiatives. Diversity, continued growth, and scale economies allowed Insull to cut prices ahead of regulators' demands and still grow profits. In the footsteps of his old boss, Insull backed these strategies with innovation in some of the earliest, largest, and most professional marketing, advertising, lobbying, and PR campaigns in America to shape public opinion and policy.

Eventually, Insull's continuous financial innovation led to his undoing. His success with door-to-door selling of "safe" utility company stock to electricity customers combined with the complex holding company model he developed to reduce the control of New York bankers rebounded on him during the Wall Street Crash of 1929. A poster boy of the pre-crash business boom, Insull lost his businesses, his fortune, and the savings of the small "widow and orphans" investors he had enticed to buy his stock. While not convicted, Insull faced public trial and was disgraced.[42]

---

41. According to McDonald, the concept of "mass production" entered the language from Insull well before being attributed to Henry Ford. Utilities' efforts to woo large constant profile flattening loads like aluminum smelters, subsidize off peak hot water, sell electric appliances to customers, and introduce time-based pricing have all been seen before.
42. For a view of Insull as innovator and business guru, see Wasik, Jonnes, and McDonald; for an opposite view, see Beder or Sakis. Insull's extensive writings and speeches are accessible on the web.

Ultimately, Insull's model of scaling up, diversifying load and generation, and establishing regulated monopolies—private or government-owned—became the dominant grid model for electricity industries worldwide. Yet, it still took almost 60 years of steady progress—with an actual falloff in electrification rates during the Great Depression—to electrify around 80% of US households.

4.2   The Old Business Model and the Twenty-First Century

By the end of the twentieth century, the traditional grid business model started to change. A number of jurisdictions restructured their electricity industries, amid more general microeconomic reform agendas, by breaking up vertically integrated utilities into generation, network, and retail (supplier) functions and introducing competition where possible. In practice, generation and retail proved far more amenable to the introduction of competitive electricity markets, and the networks continued to be operated as monopoly infrastructure.

Over the same period as well, a number of jurisdictions have looked to transform their electricity industries towards less reliance on fuel imports and lower greenhouse emissions through a range of policies including renewable energy policy support, energy efficiency efforts, and carbon pricing. And in response to these and other drivers, demand growth has fallen or even reversed, and a range of new technologies have emerged that challenge existing market and regulated arrangements.

Again, many of these challenges are not entirely new to the electricity industry. The industry was forced to respond to its reliance on oil in the 1970s (a key driver for the nuclear industry in countries including France and Japan), regional air pollution impacts in the 1970s and 1980s (a key driver for increased reliance on gas generation in Europe and the United States), and a range of technologies such as residential and commercial air-conditioning.

Is falling demand an entirely new phenomenon? It is certainly a challenge for existing business models as highlighted in growing discussion about the "death spiral" (see the introduction, Hanser and Van Horn, and Nelson et al.), including this example:

There is a new buzz word surfacing in Pacific Northwest electric utilities these days. It is the 'death spiral'... A death spiral occurs during periods of rising electric rates. The theory is that as electricity demand increase, electric utilities are forced to build expensive new power plants. This causes electric rates to rise and consumers to use less power. Electric utilities have large fixed costs, so as demand – thus revenue – is reduced, rates must be increase again, causing further reductions in consumption and the cycle is repeated: a death spiral[43]

---

43. Extracted from the Boca Raton News, 4 August 1983. Available at http://news.google.com/newspapers?id=gDVUAAAAIBAJ&sjid=rIwDAAAAIBAJ&pg=6870%2C700394.

The death spiral may be real but it has been slow, given the quote in the preceding text was made in 1983. If conceived as predominantly a price response, it is difficult to see the death spiral acting alone when electricity has traditionally had the lowest price elasticity of consumer products (see Sioshansi, 2013 for a discussion).[44]

In the opening chapter, Sioshansi discusses challenges to volumetric kilowatt-hour energy pricing for distribution and transmission and the flexibility of Cazalet's transitive retail tariffs is described as a potential solution. Yet, some flexibility already exists in the old business model to build upon. Keay et al. identify a precedent in the United Kingdom's "Economy 7" for separation between "as-available" and "on-demand" tariffs, and the traditional business model already encompasses capacity charges for big business, interruptibility tariffs, and time-based pricing. As discussed by Nelson et al., the necessary tariff reform for a new business model may be less of a technical but more of a policy and customer acceptance issue.

Distributed generation brings new stakeholders—millions in the case of PV—into the supply side of the industry with potentially revolutionary intent. But customers with PV still need the grid more than the grid needs them. Business models will have to be found that appropriately reflect not only the many values—particularly environmental—but also costs that disruptive technology brings. Similarly, "smart" IT technologies allow energy users to play a much more active role in their energy service provision. Some businesses have responded with smartphone apps and online monitoring tools to assist households and businesses better manage their energy consumption and, of course, save money. However, there are still questions about the level of engagement that end users are really seeking—by 2012, both Google PowerMeter and Microsoft Hohm had entered and exited the space. In Australia, interval and "smart" meters connect all large business customers and soon all residential customers in Victoria, yet real changes from a smarter grid have been slow to emerge. Business models need to be invented as much as technologies do. A spontaneous metamorphosis of customers from apathy and inertia into engaged prosumers needs more than assumptions as drivers.

Engineers, economists, entrepreneurs, and environmentalists may be sold on a new grid and a new energy future but customers need to buy in and come along with them. This means not just the creation of new technologies but the hard work of understanding customers' wants and creating business models and supporting structures to deliver them successfully. Edison's light bulb needed Westinghouse and Tesla's AC standard, Insull's business model, and a myriad of social, policy, economic, and institutional changes to cement its place as the "bright idea" icon. Technologies typically come of age when people want them and when they can afford them. Even though energy services are what

---

44. Jääskelä.

customers ultimately consume, many still seem content to have their energy packaged and sold independently in incomprehensible kilowatt-hour units. Economics and broader social drivers, rather than technology drivers alone, ultimately decide which outcomes rule.

Central to Insull's business model is an understanding of the end of the value chain, satisfying customers as well as managing and setting the chain's component parts. Faruqui and Grueneich mentioned the role of "soft" psychographic factors affecting customers' tastes, behaviors, and perceptions. Burger and Weinmann discussed the emotionalization of energy to compensate for higher tariffs and replace the policies and subsidies that have initially funded most prosumer behavior. Nillsesen et al. asked, "where the value is and who captures it?" The answer might be a major shift in the traditional value chain, not from technologies or industry introspection but from adding a new emotional value for customers at the end. This, rather than at the end of a technology, is where the prosumer may emerge.

The consensus across this book's authors suggests that it seems too early to call an end to the traditional network business model while supply still needs to balance demand at all times and locations across the grid. The old grid may become less valuable with growing distributed energy options and storage, or it might become more valuable as a means of stitching these growing distributed options together. Conventional centralized generation has the advantage of sunk investments with no alternative use and low scrap value, which may keep them surprisingly competitive against distributed options. The role of large retailers, also known as suppliers or load-serving entities, in the electricity industry is less clear; are they simple energy sellers, aggregators, and brokers of risk management across energy users?

Finally, the grid's introspection is a risk. Revolutionary change in the wider economic, institutional, and societal context, such as a concerted global effort to tackle climate change, may drive the technical and business changes that create the grid of the future. At present, the industry appears far from being ready, willing or able to embrace such change and the new business models it may require.

5   HERE, THERE, AND EVERYWHERE: A CENTRALIZED OR DECENTRALIZED FUTURE?

As noted in the preceding text, the early phase of the electricity sector was as a distributed industry with local generators supplying local loads just as windmills, watermills, and steam engines had delivered energy. The creation of the national, or even international, grid was an outcome of economies of scale with key generation technologies and the value of diversity (in both generation and load) for an industry where supply must equal demand at all times. However, even this industry was inherently decentralized to some extent as "all loads are local."

As many of the book's chapters describe, we are seeing a threat to the old centralized grid in a number of "disruptive" distributed technologies including more efficient end-use equipment, new-generation technologies including PV, EVs, distributed energy storage with promising progress including batteries, microgrids and controllable loads, and new "smart" technologies to coordinate it all. All of these hold the promise of another approach. Which will prevail?

For many people centralized or decentralized grid it is still an unasked question. Today, an estimated 1.3 billion people are without access to electricity, roughly the same number as when Edison was born. The IEA's "Energy Access for All" suggests that both centralized and decentralized (household PV systems and mini-grids) grid extensions have a key role to play. In July 2012, over 620 million Indians, about 9% of the world's population, experienced the largest blackout the world has ever seen. From most reports, no one died and the impact was minimal because Indians, used to intermittent electric supply, were prepared. Distributed generation helped India manage, and the mixed approaches adopted in new grids may a good model for viewing the problems of the old.

Where the grid does exist, can one envisage circumstances where distributed energy resources markedly change the role that the grid plays? Might some end users start to leave the grid entirely? Of course, both outcomes are possible, and the first seems reasonably probable. However, again, the existing grid represents around 100 years of investment and effort, and its value should not be underestimated. It is possible to envisage PV's future as largely centralized utility-scale plants in locations with excellent solar resources versus a highly distributed future located at points of end use. Likely, we will see a mix of both as evident in a PV pioneer, Germany. Many battery technologies are also highly scalable from household to utility scale. They may make considerable progress from their current emerging status to play a key role in supporting energy users leaving the grid behind. Or perhaps, another technology will emerge as a game changer—fuel cells are often put forward as one such possibility although progress has been a long way from what some were promising in the 1990s.

Paradoxically, as centralization is being challenged in the grid, it is being championed elsewhere. For the first time in history, more than half of the world's population live in cities, driven by the advantages of exchange, close contact, interconnection, agglomeration, and economies of scale and scope. People are moving together; population is becoming more not less centralized. While "atoms are local, bits are global" and while food miles are fashionable, international trade is booming around the lower transport cost from container shipping.[45] Decentralization is in many ways going back to the future and in some ways against the grain.

---

45. See Levinson, and while other established industries look to Silicon Valley to find new business models, technology writer Nicholas Carr sees Insull and the electricity utility model as the future of computing in the centralized "cloud." Also, see Bradfield Moody for a forecast of ex Silicon Valley software business executive Shai Agassi's plans for the EV company Better Place (now in receivership) to have the impact of the Model T Ford based on mobile-phoneplan-like contracts and battery swapping.

Finally, if centralized electricity infrastructure is to fall away, it may not be alone. Or be the first. Natural gas because of its lower greenhouse impact and the possibility of fuel cells is a potential driver of the grids' future and a big winner compared to centralized electricity generation. One can envisage fuel cells and PV allowing customers to leave the grid while maintaining reasonably reliable electricity services. However, with high-efficiency reverse cycle air-conditioning making winter heating with electricity cheaper and greener and customers' preference for electric ovens, PV, efficient batteries, electric-boosted solar hot water, and bottled gas for the BBQ, a household's gas use may be limited to stove-top cooking. Without hot water and heating loads, the economics of gas supply may start to look marginal and, in jurisdictions such as Australia with high gas prices and high standing charges, prohibitive. So for some customers, going "off-gas" may prove a more likely scenario than going "off-grid" if gas fuel cells cannot pay their way.

As the book's Introduction makes it clear, pressure for a new grid is, to date anyway, mostly emerging as an issue in a few nations with high retail electricity tariffs and low growth and incentives for renewables. But while strong in some places, these trends are not universal and the future path of the grid is not yet set.

6 CONCLUSIONS

The 21st century still has another 86 years left, so it is likely to see renewable energy, distributed generation, battery storage, PV, EVs, and even V2G become bigger parts of the grid mix. All this together with some technologies not yet imagined, some waiting in the wings and many invented back in Edison's day.

Science fiction writer and foreseer of the Internet William Gibson observed that "The future is already here, it's just not very evenly distributed."[46] This is likely to be true but needs the corollary, seen from the birth of grid, that this is because the "already here" future still has bugs, people do not want it or cannot afford it just yet, and even when they can, it may still be a long time coming. How new disruptive technologies are distributed in time, quantity, and location, centralized or decentralized, remains an open question.

Even disruptive new entrant technologies need to fit into an existing ecosystem of competing and complementary niche application rather than just replacing one big old dominant centralized energy monopoly. Centralized or decentralized approaches are neither inherently good nor bad. The book's proceeding chapters have wisely avoided thundering forecasts of utopian or dystopian futures, but all predict change. Yet, just as the electricity industry's marketing push for the "all electric house" never eventuated and the computer did not produce the "paperless office," the grid of the future is less likely to involve a revolution than an evolution and repositioning.

---

46. Strathern.

The most valuable lesson from the birth of the grid and a user-centric view of its history is how for any new technology the economic/societal challenges and a sustainable business model must be solved alongside the technical challenges and how they can be just as hard. Evolution rather than revolution is a more common path for change with competition among solutions and long periods of complementary and coexisting outcomes.

While proponents of particular solutions may see dystopia or utopias and look for the industry to take a proactive approach to forecasting grid changes, history cautions against picking winners or extrapolating too far out based on current trends. Eurelectric's call for utilities to be "powerhouses of innovation" or even a "fast second" response would require a major change.[47] Being proactive, visionary, fearless, and bold were the paths of the grid's inventors, but, at least for now, being watchful, heedful, and mostly reactive may still prove the best strategy for electricity utilities to transition to the grid's next incarnation. Yet, even this cautious approach will be a big shift for a mature industry, comfortable with its recent history as a stable regulated monopoly and which has forgotten the turbulent times of its birth. Evolutionary or revolutionary, the industry needs to prepare for change and find opportunities to be part of the transition, should a largely decentralized grid prove to be better placed to meet society's future energy needs.

REFERENCES

AEMO, 2013. National Electricity Forecasting Report (NEFR). http://www.aemo.com.au/en/Electricity/Forecasting/2012-National-Electricity-Forecasting-Report.

Beder, S., 2003. Power Play: The Fight to Control the World's Electricity. The New Press, New York.

Blainey, G., 2004. Black Kettle & Full Moon: Daily Life in a Vanished Australia. Penguin, Melbourne.

Boxshall, J., 1997. Every Home Should Have One: Seventy-Five Year of Change in the Home. Ebury Press, London.

Moody, J.B., Nogrady, B., 2010. The Sixth Wave How to Succeed in a Resource-Limited World. Random House Australia, Sydney.

Brox, J., 2010. Brilliant: The Evolution of Artificial Light. Houghton Mifflin Harcourt, Boston.

Carr, N.G., 2008. The Big Switch: Rewiring the World, from Edison to Google. W. W Norton & Company, New York.

Carsson, I., Vaitheeswaran, V., 2008. Zoom: The Global Race to Fuel the Car of the Future. Penguin Books, London.

Chendo, J., 1989. Uncommon Friends: Life with Thomas Edison, Henry Ford, Harvey Firestone, Alexis Carrel, and Charles Lindbergh. Harcourt Brace Jovanovich.

CIGRE/AHEF – Australian contribution to A dictionary on electricity.

---

47. Attractive as the "fast second" strategy sounds it requires the wisdom to recognize when first has happened and to react before "late last" is the only option left.

DeGraaf, L., 2013. Edison and the Rise of Innovation. Sterling Signature, New York.

Edgerton, D., 2008. The Shock of the Old: Technology and Global History Since 1900. Profile Books.

Fairley, P., 2012. San Francisco's Secret DC Grid: the last direct-current power lines are being dismantled just as DC distribution seems headed for a comeback. Posted 15 Nov 2012, http://spectrum.ieee.org/energy/the-smarter-grid/san-franciscos-secret-dc-grid.

Fara, P., 2002. An Entertainment for Angels: In the Electricity Enlightenment. Icon Books.

Fouquet, R., Pearson, P.J.G., 2006. Seven Centuries of Energy Services: The Price and Use of Light in the United Kingdom (1300–2000). The Quarterly Journal of the IAEE's Energy Economics Education Foundation, Energy J. 27 (1), 139–177.

Gallagher, W., 2012. New: Understanding Our Need for Novelty and Change. The Penguin Press, New York.

Gipe, P., 1995. Wind Energy Comes of Age. Wiley, New York, USA.

Hawken, P., Lovins Amory, B., Hunter, L., 2003. Natural Capitalism: The Next Industrial Revolution. Routledge, New York.

Huber, P., Mills, M., 2005. The Bottomless Well: The Twilight of Fuel, the Virtue of Waste and Why We Will Never Run Out of Energy. Basic Book.

Jääskelä, J., Windsor, C., 2011. Insights from the Household Expenditure Survey, Reserve Bank of Australia Bulletin – December Quarter 2011.

Johnstone, B., 2007. Brilliant!: Shuji Nakamura and the Revolution in Lighting Technology. Prometheus Books, New York.

Jonnes, J., 2003. Empires of Light: Edison, Tesla, Westinghouse, and the Race to Electrify the World. Random House, New York.

Klein, M., 2008. The Power Makers, Steam, Electricity and the Men Who Invented Modern America. Bloomsbury Press, New York.

Levinson, M., 2008. The Box: How the Shipping Container Made the World Smaller and the World Economy Bigger. Princeton University Press, Princeton, NJ.

Madrigal, A., 2011. Powering the Dream, the History and Promise of Green Technology. Da Capo Press.

McDonald, F., 1962. Insull: The Rise and Fall of a Billionaire Utility Tycoon. University of Chicago Press, Chicago, IL.

McNichol, T., 2006. AC/DC: The Savage Tale of the First Standards War. Jossey-Bass, San Francisco, CA.

Molotch, H., 2003. Where Stuff Comes From: How Toasters, Cars, Computers, and Many Other Things Came to Be as They Are. Routledge.

Nordhaus, W., 1998. Do Real-Output and Real-Wage Measures Capture Reality? The History of Lighting Suggests Not. Crowles Foundation paper No. 975, Yale University, New Haven.

Perlin, J., 2002. From Space to Earth the Story of Solar Electricity. Harvard University Press, Cambridge, MA.

Rees, J., 2013. Refrigeration Nation: A History of Ice, Appliances, and Enterprise in America. Johns Hopkins University Press, accessed at http://muse.jhu.edu/books/9781421411071.

Sabin, P., 2013. The Bet, Paul Ehrlich, Julian Simon, and Our Gamble over the Earth's Future. Yale University Press.

Sakis, C., 2001. Frauds, Deceptions and Swindles. Checkmark Books.

Sioshansi, F., 2013. Energy Efficiency, Towards the End of Demand Growth, Academic Press.

Schlesinger, H., 2010. The Battery: How Portable Power Sparked a Technological Revolution. Smithsonian Books.

Standage, T., 1998. The Victorian Internet: The Remarkable Story of the Telegraph and the Nineteenth Century's On-line Pioneers. Walker & Company.

Strathern, O., 2007. A Brief History of the Future: How Visionary Thinkers Changed the World and Tomorrow's Trends are 'Made' and Marketed. Running Press.

Shiman Daniel, R., 1993. Explaining the Collapse of the British Electrical Supply Industry in the 1880s: Gas versus Electric Lighting Prices. Economic and Business History 23, no.1 fall 1993.

Smith, R., 2007. The economy. In: Paper Presented at the Australian Conference of Economist, Hobart 2007. http://www.ecosoc.org.au/files/File/TAS/ACE07/presentations%20(pdf)/Smith.pdf.

Tamworth Regional Council, 2006: Tamworth and Districts Early History.

Uglow, J., 2002. The Lunar Men: Five Friends Whose Curiosity Changed the World. Faber & Faber, London.

Waide, P., Tanishima, S., 2006. Light's Labour's Lost, Policies for Energy-Efficient Lighting, in Support of the G8 Plan of Action. International Energy Agency, Paris.

Wasik John, F., 2006. The Merchant of Power: Sam Insull, Thomas Edison, and the Creation of the Modern Metropolis. Palgrave Macmillan, New York.

Wilkenfeld, G., Spearitt, P., 2004. Electrifying Sydney: 10 Years of Energy Australia, Sydney.

Wilson, D.H., 2007. Where's My Jetpack? A Guide to the Amazing Science Fiction Future that Never Arrived. Bloomsbury, USA.

Winchester, S., 2009. Bomb, Book and Compass: Joseph Needham and the Great Secrets of China. Penguin.

As other chapters of this volume explain, the rapid rise of distributed generation, propelled by technological advancements on the one hand and continuously falling costs on the other, has caught many in the electricity supply industry by surprise. For the first time in the industry's history, consumers will soon be able to generate some or all of their power needs through self-generation at prices that rival those offered through retail utility electric tariffs. The prevailing reaction from the industry is to view distributed generation as a threat to the traditional utility business model and to defend that existing model as an entitlement. A minority of utilities, however, are seeking to develop new models to provide for an orderly transition from their traditional business structure to a new one that is profitable for them and workable for their consumers.

Utility regulators—I was one until recently—also surprised by the rapid developments beyond their direct control, are doing their best to digest what is at stake and what may be the best way forward. They must decide what is ultimately good for the customers, the environment, and the society at large. The billions invested over the decades on an extensive and expensive infrastructure to serve the customers—central station generation plants plus the transmission and distribution network—while currently critical to serve customers' energy needs, is expected to diminish in importance and value as distributed energy resources grow. In a decentralized, bidirectional future where growing numbers of customers will decide to operate in ways that will reduce rather than increase consumption of kilowatt hours from the grid, we must all consider what will be the impact and effect on traditional nondistributed resources and how will the stranded investments for those resources be recovered if at all.

Keussen, in the book's Foreword, talks about the rapid pace of technological advancement and its impact on the industry, specifically the imperative for the incumbent players to adapt to changing business environment rather than attempting to hold it back. Cavanagh, in the book's Preface, proposes that investor-owned utilities' earnings opportunities should include performance-based incentives tied to benefits delivered to their customers by cost-effective initiatives to improve energy efficiency, integrate renewable and distributed generation, and improve grid's reliability. The book's other contributors have taken different perspectives based on their expertise and insights—providing a rich and varied context for anyone interested in the subject matter.

It is fair to say that the rapid pace of technological advancements in the energy services delivery space coupled with falling costs due to mass production and penetration is *inevitable and unstoppable*. Just as the cost of rooftop PV systems has fallen dramatically over time, the same can be expected for storage technologies, micro-grids, natural gas fueled micro-gen, and other distributed resource supply-side technologies. And similarly, we can also be certain that technologies to improve demand-side efficiency of energy use, both behind the customer meter and on the grid side, will continue to expand at an ever-accelerating pace. Witness the rapid rise of the adoption of LED lighting as one example. These developments all reflect the power of functional and effective markets that provide goods and services that are desired by consumers and are beneficial for the environment and for the society. Traditional utility models that rely on monopolistic policies and practices cannot hold back the forces unleashed by these competitive markets.

In this context, how should we address or do we need to address the plight of the incumbent monopolist utility who claims they are being harmed by revenue erosion when *consumers* seek and adopt more attractive market choices and lessen their dependence on traditional grid resources? As an extreme example of this trend, there are now instances of groups of consumers going off-grid, operating in virtual isolation from the grid, as evidenced in a number of so-called bio-energy villages in Germany described in chapter by Burger and Weinmann. Yet, is this any different from small island republics that have been "independent" from a larger grid for decades? No doubt as technology advances in sophistication and comes down in cost, these market forces driving grid independence will be more prevalent. Couple the technology and the economics with the desire for security and reliability that a grid-tangled system can no longer provide in the wake of natural and human sourced disasters and you have a future prediction for the potential obsolescence of large portions of the traditional electric utility infrastructure. This will leave incumbent utilities with potential for massive levels of stranded assets, investments once made in good faith to serve customers who may no longer need it, need it less, or unable and unwilling to pay for it.

So, what if everybody stopped drinking milk tomorrow? Did not milk producers make investments in dairy herds, milking machines, trucks, processing and packaging facilities that would instantly all be worth a fraction of their original value? Often when posing such analogies, the electric industry will shoot back—electricity is an essential commodity and what about the "regulatory compact"? Yes, electricity is essential, but if markets provide mechanisms for consumers to obtain the commodity in a cheaper and more reliable manner, should the consumer be penalized? And as for the regulatory compact, state and federal governments did agree to allow private entities to provide monopoly services and receive a rate-regulated return on their investment in return for an agreement to provide service to all consumers in their monopoly service territory. Will new cheaper distributed technologies eliminate the need for that regulatory compact? Perhaps.

In any case, there is no easy answer and there will not be a simple uniform solution. Regulated utilities in the United States are governed by fifty plus state commissions and the Federal Energy Regulatory Commission at the federal level. And there are state and federal legislative bodies as well that also have jurisdiction in these matters. Similar arrangements exist in other parts of the world. All will play a part in the process of transitioning from our historical monopolistic, central station, rate-regulated world for delivery of electric services to a market-driven, distributed world of low-cost advanced energy technology providing consumers reliable energy services in the specific quantity and quality that they desire.

Various contributors have noted their own suggestions and solutions on ways that we may move forward, from different forms of retail tariffs to more elegant—and complicated—ways of addressing the emerging future through concepts such as transactive tariffs. I am not convinced that we need to agree initially on "a solution." Rather, I think what is needed is a fair and open process or a series of processes in the various states and countries to discuss the issues and provide a forum to make rational, substantive decisions that minimize costs for consumers and maximize efficiency for the society. If we can do this, I have confidence we can arrive at the solutions we seek.

Jon Wellinghoff
Partner, Stoel Rives LLP
Former Chairman, Federal Energy Regulatory Commission

Index

Note: Page numbers followed by *b* indicate boxes, *f* indicate figures and *t* indicate tables.

Printed and bound by CPI Group (UK) Ltd, Croydon, CR0 4YY

03/10/2024

01040427-0009